THE COMPLETE OFFICIAL TRIUMPH SPITFIRE MK III, MK IV & 1500

Model Years 1968–1974

Selected Titles From Bentley Publishers

Engineering

Robert Bentley's Repair Manual for British Cars *John Organ*
ISBN 978-0-8376-0041-3

Vintage Racing British Sports Cars *Terry Jackson*
ISBN 978-0-8376-0153-3

Scientific Design of Exhaust & Intake Systems *Philip H. Smith and John C. Morrison*
ISBN 978-0-8376-0309-4

The Design and Tuning of Competition Engines *Philip H. Smith* ISBN 978-0-8376-0140-3

New Directions in Suspension Design: Making the Car Faster
Colin Campbell
ISBN 978-0-8376-0150-2

Racing Car Design and Development *Len Terry with Alan Baker* ISBN 978-0-8376-0080-2

Austin-Healey

Austin-Healey 100/6 NS 3000 Workshop Manual: 1956-1968
British Leyland Motors
ISBN 978-0-8376-1743-5
Hardcover, 416 pages

The Complete Official 1275 cc Sprite/Midget: 1967-1974
British Leyland Motors
ISBN 978-0-8376-1748-0
Hardcover, 400 pages

The Complete Official 948 cc & 1098 cc Sprite/Midget
British Leyland Motors
ISBN 978-0-8376-1747-3
Hardcover, 256 pages

MG

The Complete Official MGB: 1962-1974 *British Leyland Motors*
ISBN 978-0-8376-1761-9
Hardcover, 480 pages
Includes: Driver's Handbook, Workshop Manual, Special Tuning Manual

The Complete Official MGB: 1975-1980 *British Leyland Motors*
ISBN 978-0-8376-1760-2
Hardcover, 304 pages

The Complete Official MG Midget 1500: 1975-1979
British Leyland Motors
ISBN 978-0-8376-1744-2
Hardcover, 384 pages

The MG Workshop Manual: 1929-1955 *W.E. Blower*
ISBN 978-0-8376-1746-6
Hardcover, 602 pages

Triumph

The Complete Official Triumph GT6, GT6+ & GT6 MK III: 1967-1973 *British Leyland Motors*
ISBN 978-0-8376-1749-7
Hardcover, 480 pages
Includes: Driver's Handbook, Workshop Manual

The Complete Official Triumph TR6 & TR250: 1967-1976
British Leyland Motors
ISBN 978-0-8376-1762-6
Hardcover, 608 pages
Includes: Driver's Handbook, Workshop Manual

The Complete Official Triumph Spitfire MK III, MK IV & 1500: 1968-1974 *British Leyland Motors*
ISBN 978-0-8376-1763-3
Hardcover, 480 pages
Includes: Driver's Handbook, Workshop Manual

The Complete Official Triumph Spitfire 1500: 1975-1980
British Leyland Motors
ISBN 978-0-8376-1745-9
Hardcover, 536 pages
Includes: Driver's Handbook, Workshop Manual

The Complete Official Triumph TR7: 1975-1981
British Leyland Motors
ISBN 978-0-8376-1742-8
Hardcover, 544 pages
Includes: Driving Manual, Workshop Manual

The Complete Official Triumph TR4 & TR4A: 1961-1968
British Leyland Motors
ISBN 978-0-8376-1741-1
Hardcover, 404 pages
Includes: Driving Manual, Workshop Manual, Competition Preparation Manual by R.W. "Kas" Kastner

The Complete Official Triumph TR2 & TR3: 1953-1961
British Leyland Motors
ISBN 978-0-8376-1750-3
Hardcover, 456 pages
Includes: Driver's Instruction Book, Service Instruction Manual

Jaguar

The Complete Official Jaguar "E" Series 1 & Series 2: 1961-1971
British Leyland Motors
ISBN 978-0-8376-1764-0
Hardcover, 790 pages
Includes: Driver's Handbook, Workshop Manual, Special Tuning Manual

BentleyPublishers®
Automotive Reference

Bentley Publishers has published service manuals and automobile books since 1950. For more information, please contact Bentley Publishers at 1734 Massachusetts Avenue, Cambridge, MA 02138 USA, or visit our web site at `BentleyPublishers.com`

THE COMPLETE OFFICIAL TRIUMPH SPITFIRE MK III, MK IV & 1500 1968–1974

Comprising the official

Driver's Handbook

Workshop Manual

BentleyPublishers®
.com

BENTLEY PUBLISHERS® | Automotive Reference™

Bentley Publishers, a division of Robert Bentley, Inc.
1734 Massachusetts Avenue
Cambridge, MA 02138 USA
800-423-4595 / 617-547-4170

Information that makes the difference®

BentleyPublishers®.com

Technical contact information

We welcome your feedback. Please submit corrections and additions to our technical discussion forum at:

http://www.BentleyPublishers.com

Updates and corrections

We will evaluate submissions and post appropriate editorial changes online as updates or tech discussion. Appropriate updates and corrections will be added to the book in future printings. Check for updates and corrections for this book before beginning work on your vehicle. See the following web address for additional information:

http://www.BentleyPublishers.com/updates/

WARNING—Important Safety Notice

In this book we have attempted to describe repairs, modifications and accessories which may be used with Triumph vehicles, using examples and instructions which we believe to be accurate. However, the examples, instructions, and other information are intended solely as illustrations and should be used in any particular application only by personnel who are experienced in the repair and modification of Triumph vehicles and who have independently evaluated the repair, modification or accessory. Implementation of a modification or attachment of an accessory described in this book may render the vehicle, attachment, or accessory unsafe for use in certain circumstances. **REPAIR AND MODIFICATION OF MOTOR VEHICLES CAN BE DANGEROUS UNLESS UNDERTAKEN WITH FULL KNOWLEDGE OF THE CONSEQUENCES.**

Do not use this book unless you are familiar with basic automotive repair procedures and safe workshop practices. This book illustrates procedures required for some service and modification work; it is not a substitute for full and up-to-date information from the vehicle manufacturer or aftermarket supplier, or for proper training as an automotive technician. Note that it is not possible for us to anticipate all of the ways or conditions under which vehicles may be serviced or modified or to provide cautions as to all of the possible hazards that may result.

The vehicle manufacturer and aftermarket suppliers will continue to issue service information updates and parts retrofits after the editorial closing of this book. Some of these updates and retrofits will apply to procedures and specifications in this book. We regret that we cannot supply updates to purchasers of this book.

We have endeavored to ensure the accuracy of the information in this book. Please note, however, that considering the vast quantity and the complexity of the information involved, we cannot warrant the accuracy or completeness of the information contained in this book.

FOR THESE REASONS, NEITHER THE PUBLISHER NOR THE AUTHOR MAKES ANY WARRANTIES, EXPRESS OR IMPLIED, THAT THE EXAMPLES, INSTRUCTIONS OR OTHER INFORMATION IN THIS BOOK ARE FREE OF ERRORS OR OMISSIONS, ARE CONSISTENT WITH INDUSTRY STANDARDS, OR THAT THEY WILL MEET THE REQUIREMENTS FOR A PARTICULAR APPLICATION, AND WE EXPRESSLY DISCLAIM THE IMPLIED WARRANTIES OF MERCHANTABILITY AND OF FITNESS FOR A PARTICULAR PURPOSE, EVEN IF THE PUBLISHER OR AUTHOR HAVE BEEN ADVISED OF A PARTICULAR PURPOSE, AND EVEN IF A PARTICULAR PURPOSE IS INDICATED IN THE BOOK. THE PUBLISHER AND AUTHOR ALSO DISCLAIM ALL LIABILITY FOR DIRECT, INDIRECT, INCIDENTAL OR CONSEQUENTIAL DAMAGES THAT RESULT FROM ANY USE OF THE EXAMPLES, INSTRUCTIONS OR OTHER INFORMATION IN THIS BOOK. IN NO EVENT SHALL OUR LIABILITY WHETHER IN TORT, CONTRACT OR OTHERWISE EXCEED THE COST OF THIS BOOK.

Your common sense and good judgment are crucial to safe and successful automotive work. Read procedures through before starting them. Think about how alert you are feeling, and whether the condition of your vehicle, your level of mechanical skill or your level of reading comprehension might result in or contribute in some way to an occurrence which might cause you injury, damage your vehicle,or result in an unsafe repair or modification. If you have doubts for these or other reasons about your ability to perform safe work on yourvehicle, have the work done at an authorized Triumph dealer or other qualified shop.

Before attempting any work on your Triumph, read the Warnings and Cautions on page vi and any warning or caution that accompanies a procedure or description in the book. Review the Warnings and Cautions each time you prepare to work on your Triumph.

ACKNOWLEDGEMENT

Special thanks are due David G. Head, New England Parts Representative, Leyland Motor Sales, Leonia, N.J.

This book is prepared, published and distributed by Bentley Publishers, 1734 Massachusetts Avenue, Cambridge, Massachusetts 02138 USA.

The Complete Official Triumph Spitfire MK III, MK IV & 1500: 1968–1974

ISBN 978-0-8376-1763-3

Job Code: X123-13

Library of Congress Catalog Card No. 74-20004

The paper used in this publication is acid free and meets the requirements of the National Standard for Information Sciences-Permanence of Paper for Printed Library Materials. ∞

Manufactured in the United States of America.

Foreword

This Manual has been compiled in order to provide complete, accurate, and comprehensive maintenance and repair data to both car owners and professional mechanics. The *Driver's Handbook,* which comprises the first part of this Manual, is similar to the handbook provided with every new Triumph sports car. The *Workshop Manual,* which comprises the second — and largest — portion of this Manual, is the official factory manual and was originally intended for use by dealer service departments. The Publisher has added new material to the factory data so that the present Manual is suitable both for persons of limited mechanical experience and for those trained mechanics who may not be altogether familiar with Triumph cars.

Home Maintenance and Repair

Though professional mechanics will find this Manual indispensable as a source of specifications, testing procedures, and precision fitting data, such mechanics usually carry out a basic assembly and disassembly work guided mainly by their years of training and experience. For example, professional mechanics know without being told that engine, transmission, and rear axle parts must be properly pre-lubricated when they are reassembled following an overhaul. Therefore, such information is not commonly included in factory workshop manuals. On the other hand, non-professionals often need to be reminded of such things since mechanical work may not be a part of their daily routine. Wherever possible, the Publisher has included *Notes* and *Cautions* to alert the non-professional mechanic to those additional considerations that a professional mechanic would take into account as a matter of course.

Though simple maintenance procedures — such as those described in the *Driver's Handbook* portion of this Manual — require only a minimum of tools, the car owner will sooner or later find it desirable to obtain a fairly complete set. The basic tools include an assortment of open- and box-end wrenches and a selection of screwdrivers, pliers, one or two adjustable wrenches and other miscellaneous items such as feeler gauges, a spark plug socket, etc. A socket wrench set — preferably ⅜-in. drive — will speed up many common jobs and may be indispensable to others. A quality set of tools can be a good investment; they will last a lifetime if not abused. Cheaply made tools will not last long without wear and breakage and often damage bolts and fittings which cost more to replace than the money saved in buying inferior tools.

Liberal reference is made in this Manual to the special service tools which are recommended by the factory. Such special tools are not supplied by British Leyland Motors but can be ordered from Messrs. V. L. Churchill & Company Limited, P. O. Box No. 3, London Road, Daventry, Northants, England — information that may prove of little value to a car owner in the U.S. who must repair his car by the next weekend! Fortunately, generic equivalents for most of these special tools are available from American tool manufacturers that supply professional mechanics. Do not, however, overlook the possibility of obtaining the needed tool from a tool rental agency. In many U.S. cities, there are also do-it-yourself garages that rent or loan special tools to those who patronize the shop.

This Manual describes certain precision reconditioning jobs which even Authorized Dealers commonly farm out — either to an automotive machine shop or to a specialty shop. These jobs include cylinder reboring, crankshaft reconditioning, valve grinding, starter rebuilding, sheet metal repair, etc. Instead of delivering the entire car to a professional mechanic, who may take the parts to an outside machine shop, you can save considerable money by removing the worn or damaged components and taking them to the machine shop yourself.

Advice for Professionals and Car Owners Alike

Assembly work on small, highly-tuned machines such as the Triumph sports cars must be carried out with greater precision than is commonly practiced on large American cars. Particular emphasis must be given to the proper use of torque wrenches and to the strict adherance to tightening torque specifications which are given in this Manual. Though all of the torque specifications may fall within the range of a 0 - 100 ft. lb. (0 - 14 mkg) torque wrench, torque wrenches operate most efficiently when used in the middle two quarters of their scales. Thus, a small torque wrench with a range of 0 - 50 ft. lb. (0 - 7 mkg) is recommended in addition to a larger torque wrench with a 150 ft. lb. (21 mkg) capacity. A fastener that is too tight can be worse than one which is too loose — especially on a lightweight sports car. Stretched or broken bolts and distorted parts, which result from overtightening by musclebound mechanics, become a serious concern where the precision fitting of light alloy and thin-wall iron castings is involved.

Lastly, the importance of cleanliness cannot be overemphasized. Under no circumstances should an engine or gearbox be repaired on the ground or on a garage floor. Thoroughly clean the exteriors of major components prior to disassembly in order to keep road dirt and other grime out of the working parts. No more than a pinch of abrasive dust in a gearbox can cause rapid failure of the synchronizers and bearings.

During the final assembly of an engine or gearbox, the cleaned parts should be laid out on a clean workbench which has been covered with clean sheets of new cardboard or wrapping paper. The engine or gearbox itself, if not mounted on a special stand, should likewise be placed on a clean workbench. Sandpapering, valve grinding, or the use of bench grinders should not be permitted near the area where final assembly is taking place. If assembly cannot be completed in a day, enclose the partially-assembled engine or gearbox in a large plastic bag — such as a new trash bag or a dry cleaner's garment return bag — so that dust and dirt will be excluded until assembly work resumes.

Please read these warnings and cautions before proceeding with maintenance and repair work.

WARNING—

• Do not re-use any fasteners that are worn or deformed in normal use. Many fasteners are designed to be used only once and become unreliable and may fail when used a second time. This includes, but is not limited to, nuts, bolts, washers, self-locking nuts or bolts, circlips, cotter pins. For replacements always use new parts.

• Never work under a car unless it is solidly supported on stands designed for the purpose. Do not support a car on cinder blocks, hollow tiles, or other props that may crumble under continuous load. Do not work under a car that is supported solely by a jack.

• If you are going to work under a car on the ground, make sure that the ground is level. Block the wheels to keep the car from rolling. Never work under the car while the engine is running. Disconnect the battery ground strap to prevent others from starting the car while you are under it.

• Never run the engine unless the work area is well ventilated. Carbon monoxide kills.

• Friction materials such as brake or clutch discs may contain asbestos fibers. Do not create dust by grinding, sanding, or by cleaning with compressed air. Avoid breathing asbestos fibers and asbestos dust. Breathing asbestos can cause serious diseases such as asbestosis or cancer, and may result in death.

• Tie long hair behind your head. Do not wear a necktie, a scarf, loose clothing, or a necklace when you work near machine tools or running engines. If your hair, clothing or jewelry were to get caught in the machinery, severe injury could result.

• Disconnect the battery ground strap whenever you work on the fuel system or the electrical system. When you work around fuel, do not smoke or work near heaters or other fire hazards. Keep an approved fire extinguisher handy.

• Illuminate your work area adequately but safely. Use a portable safety light for working inside or under the car. Make sure the bulb is enclosed by a wire cage. The hot filament of an accidentally broken bulb can ignite spilled fuel or oil.

• Catch draining fuel, oil or brake fluid in suitable containers. Do not use food or beverage containers that might mislead someone into drinking from them. Store flammable fluids away from fire hazards. Wipe up spills at once, but do not store the oily rags, which can ignite and burn spontaneously.

• Finger rings should be removed so that they cannot cause electrical shorts, get caught in running machinery or be crushed by heavy parts.

• Keep sparks, lighted matches and open flame away from the top of the battery. If hydrogen gas escaping from the cap vents is ignited, it will ignite gas trapped in the cells and cause the battery to explode.

• Always observe good workshop practices. Wear goggles when you operate machine tools or work with battery acid. Gloves or other protective clothing should be worn whenever the job requires it.

• Do not attempt to work on your car if you do not feel well. You increase the danger of injury to yourself and others if you are tired, upset or have taken medicine or any other substance that may impair you from being fully alert.

• Some aerosol tire inflators are highly flammable. Use extreme caution when repairing a tire that may have been inflated using an aerosol tire inflator. Keep sparks, open flame or other sources of ignition away from the tire repair area. Inflate and deflate the tire at least four times before breaking the bead from the rim. Completely remove the tire from the rim before attempting any repair.

• Greases, lubricants and other automotive chemicals contain toxic substances, many of which are absorbed directly through the skin. Read manufacturer's instructions and warnings carefully. Use hand and eye protection. Avoid direct skin contact.

CAUTION—

• If you lack the skills, tools and equipment, or a suitable workshop for any procedure described in this Manual, we suggest you leave such repairs to an authorized dealer or other qualified shop. We especially urge you to consult an Authorized Dealer before beginning repairs on any car that may still be covered wholly or in part by the new-car warranty.

• British Leyland Motors was constantly improving its cars and sometimes these changes, both in parts and specifications, were made applicable to earlier models. Therefore, part numbers listed in this manual are for reference only. Always check with your Authorized Parts Department for the latest information.

• Before starting a job, make certain that you have all the necessary tools and parts on hand. Read all the instructions thoroughly, do not attempt shortcuts. Use tools appropriate to the work and use only replacement parts meeting specifications. Makeshift tools, parts and procedures will not make good repairs.

• Use pneumatic and electric tools only to loosen threaded parts and fasteners. Never use these tools to tighten fasteners, especially on light alloy parts.

• Be mindful of the environment and ecology. Before you drain the crankcase, find out the proper way to dispose of the oil. Do not pour oil onto the ground, down a drain, or into a stream, pond or lake. Consult local ordinances that govern the disposal of wastes.

Contents

1 OFFICIAL TRIUMPH SPITFIRE MK. III DRIVER'S HANDBOOK

2 OFFICIAL TRIUMPH SPITFIRE MK. IV & 1500 DRIVER'S HANDBOOK

3 OFFICIAL TRIUMPH SPITFIRE MK. III, MK. IV & 1500 WORKSHOP MANUAL

TRIUMPH SPITFIRE
MK III
DRIVER'S HANDBOOK

PART 1

Important

In all communications relating to Service or Spares, please quote the Commission Number (Chassis Number) Paint and Trim Numbers

LOCATION OF COMMISSION AND UNIT NUMBERS

Note. L.H. and R.H. refer to Left-hand and Right-hand side of the vehicle viewed from the driving position.

Commission, Paint and Trim Numbers—On L.H. Scuttle Side Panel (May be seen by lifting the bonnet)
The Commission Number is also stamped on a plate attached to the L.H. windscreen pillar, and visible through the screen
Engine Number—On L.H. side of Cylinder Block
Gearbox Number—On R.H. side of Clutch Housing Flange
Rear Axle Number—On Hypoid Housing Flange

Issued by British Leyland Motor Corporation

Introduction

The "SPITFIRE" is designed and engineered to provide outstanding performance and reliability and conforms with U.S.A. Federal Regulations, relating to safety, exhaust emission, and fuel evaporation.

Read carefully the contents of this book which gives, in the simplest possible terms, information vital to the proper operation, care and regular maintenance of the car.

Because of the stringent Federal Regulations relating to emissions, owners are particularly urged to make use of the Maintenance Voucher scheme operated by Triumph Dealers and described in a separate booklet supplied with the car. This will ensure that the emission control system and other vital parts are correctly serviced by trained personnel.

Important

In the interests of safety, the importance of maintaining correct tire pressures cannot be overemphasized. Pressures should be checked at least every two weeks or 1,000 miles, and maintained in accordance with recommendations given on page 29.

TRIUMPH SPITFIRE—U.S.A. MODEL

STANPART

Spare Parts Service

Replacement parts are not supplied from the factory direct to the general public, but are directed through Distributors who in turn, supply their Dealers.

Genuine spare parts are marketed under the trade mark "Stanpart" and carry the me guarantee as the original part. The same high quality material is used and the strictest accuracy maintained during manufacture. You are advised, therefore, to insist on the use of these parts should replacements be necessary. Remember, parts which do not carry the trade mark "Stanpart" will invalidate the guarantee if fitted to your vehicle.

The description and illustrations appearing in this book are not binding. The MANUFACTURER, therefore, reserves the right—whilst retaining the basic features of the Models herein described and illustrated—to make at any time, without necessarily bringing this book up-to-date, any alteration to units, parts or accessories deemed convenient for improvement or for any manufacturing or commercial reason.

CONTROLS, INSTRUMENTS AND INDICATORS

The bracketed figures in the text cross-refer with the key to Fig. 2 below.

Windscreen Wiper and Washer Control (1)

A dual purpose control operates the two-speed wiper motor and screen washer jets. From the "off" position (full anti-clockwise), turn the control clockwise to the next position for slow wiping, or fully clockwise for fast wiping. The wipers park automatically at the base of the windscreen when the control is returned to the "off" position. The wipers operate only when the ignition is switched on.

Clear mud from the screen by repeatedly pushing the control to spray clean fluid as the wiper blades operate. Refer to windscreen washer (page 16).

Turn Signal Monitor (2)

A green monitor at the left-hand side of the facia glows intermittently when the ignition is switched on and the turn signal control operated. Should the monitor fail to operate, this indicates a broken filament in one of the turn signal lamp bulbs or a faulty monitor bulb.

Turn Signal Control (3)

The turn signal lamps are controlled by moving the lever (3) to either of the arrowed positions. Operation of the turn signal lamps is indicated by the intermittent flashing of a green indicator monitor light (2). The "turn signal" control operates only when the ignition is switched on.

Speedometer (4, Fig. 3)

Calibrated in miles and kilometres per hour, the speedometer also combines the following indicators and warning devices:

Trip Indicator (*a*)

When initially set at zero, figures within the aperture above the centre of the dial will record the distance of a particular journey. To zero the trip indicator, turn the knob (19) attached to the end of a flexible cable which extends downwards from behind the instrument, as directed by a "transparent" label attached to the face of the instrument.

Fig. 3

Mileage Indicator (*b*)

Figures within the aperture below the centre of the dial record the total mileage of the vehicle.

Main Beam Warning (*c*)

An indicator at the bottom left-hand side of the dial glows

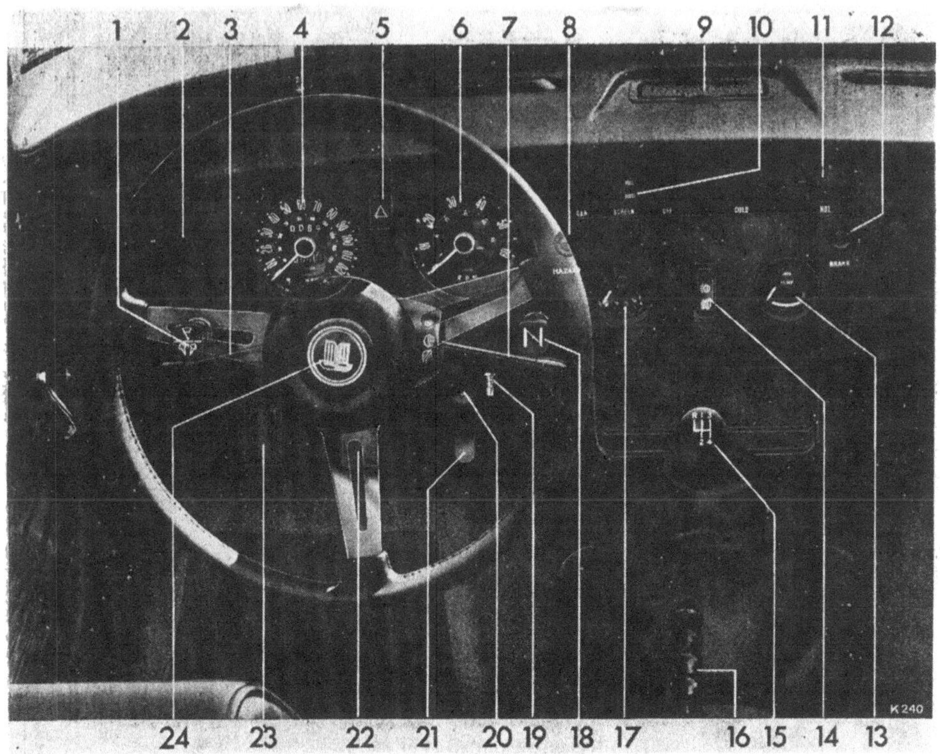

Fig. 2. Cockpit controls

Key to Fig. 2 (Facia)

1 Windscreen wiper and washer control
2 Turn signal monitor
3 Turn signal control
4 Speedometer
5 Hazard warning switch
6 Tachometer
7 Headlamp dip and flasher switch
8 Hazard warning monitor
9 Ash tray
10 Air distribution control
11 Heater control
12 Brake-line failure indicator
13 Fuel gauge
14 Lighting switch
15 Gear shift lever
16 Handbrake lever
17 Temperature gauge
18 Choke control
19 Trip release control
20 Ignition and steering lock switch
21 Accelerator pedal
22 Brake pedal
23 Clutch pedal
24 Horn push button

blue when the headlamp main beams are selected and is extinguished when the headlamps are "dipped".

Oil Pressure Warning (d)

An indicator at the bottom centre of the dial glows green when the ignition is switched on and is extinguished as the engine speed increases. Should the indicator continue to glow when driving away, stop the engine and check the sump oil level. If this is satisfactory, have the lubrication system checked immediately.

Ignition Warning (e)

An indicator at the bottom right-hand side glows red when the ignition is switched on and is extinguished as the engine speed increases. Should the indicator continue to glow when driving, this indicates a fault in the charging system which should be rectified without delay.

Hazard Warning Switch (5)

If the vehicle is immobilised and constitutes a hazard to other vehicles, warning may be given by using the "hazard warning system". To operate, depress the upper portion of the switch (5) when all turn-signal lamps and the hazard warning monitor (8) will flash intermittently.

Tachometer (6)

The tachometer indicates engine speed in revolutions per minute and is calibrated in increments of 10×100, extending to 6,000 r.p.m. The speed range within the coloured segment is subject to special precautions. These are given on page 9.

Headlamp Dip and Flasher Switch (7)

With the "lighting switch" (14) at position 3, move the switch lever "down" for "dipped beams" or "up" for "main beams". See main beam warning (4c). "Pulling" the lever towards the steering wheel rim will "flash" the main beams, regardless of the "lighting switch" position.

Hazard Warning Monitor (8)

Refer to "hazard warning switch" (5).

Ash Tray (9)

Raise the flap for access. Lift out the complete unit to clean.

Air-Distribution (10)

The air-distribution control operates independently of the heater control. Hot or cold air is directed to the screen, or to the car interior, by moving the control to the desired position.

Air Booster Motor (10a)

A two-speed motor, for boosting the supply of hot or cold air, is controlled by a 3-position pull-switch mounted on the end of the air-distribution lever (10). From the "off" position (fully forward), pull the switch to the mid-position for slow speed, or pull it fully out for maximum speed. The motor can be operated only when the ignition is switched on.

Heater Control (11)

The lever controls a valve which disrupts the supply of hot water to the heater when the lever is at "cold" or opens to progressively increase the water supply as the lever is moved from "cold" to "hot".

Brake-line Failure Indicator (12)

When the ignition is switched on, the "brake-line failure" and "low oil pressure" indicators (4d) glow faintly; these are extinguished when the engine is running. Should failure of the front or rear brake lines occur, the indicator (12) will glow brightly.

A broken bulb filament is indicated by the warning light failing to glow when the ignition is switched on, before starting the engine.

Fuel Gauge (13)

The fuel gauge indicates the contents of the fuel tank. When the ignition is switched "on" the needle moves slowly across the scale, taking up to one minute to reach a steady reading which is maintained regardless of vehicle movement, until the ignition is switched "off".

Lighting Switch (14)

To operate the three-position lighting switch:

1. Fully press the top of the switch bar to extinguish all external driving lamps.
2. Move the switch bar to its level position for illuminating the front and rear parking lamps and licence plate.
3. Fully press the bottom of the switch bar to illuminate all external driving lamps.

Gear Shift Lever (15)

Moving the gear shift lever from "neutral", the gear positions are selected as follows:

1st	Move lever left and forward
2nd synchromesh ..	Move lever left and rearward
3rd synchromesh	Move lever right and forward
4th (top) synchromesh ..	Move lever right and rearward
Reverse	Press the lever downwards, move it to the extreme left and forward. Engage reverse gear only when the vehicle is stationary.

Handbrake Lever (16)

To apply the rear wheel brakes, pull the handbrake lever upwards; it is retained by a ratchet and pawl. Release the handbrake by pulling it slightly upwards before depressing the button to free the pawl; then allow the lever to move downwards to the "off" position.

Temperature Gauge (17)

When the ignition is switched "on", the pointer moves slowly across the dial, taking up to one minute to reach a steady reading.

Normal operating temperature is reached when the pointer registers in the central sector of the dial. Should the pointer reach the highest mark, stop the engine immediately, allow it to cool and check the level of coolant in the radiator. Refer to page 11.

Choke Control (18)

The choke control is used to enrich the fuel mixture for easy starting from cold. Pull the control fully out for maximum enrichment. The control is automatically restrained in the selected position. Refer to "starting the engine" on page 8.

Trip Release Control (19)

Refer to "trip indicator" (4a).

Ignition, Starter and Steering Lock Switch (20)

The combined ignition/starter/steering lock switch is operated by a special key.

Incorporated in the switch is an "Audible Alarm System". The alarm system is fitted to encourage the driver to remove the ignition key from the lock before leaving the vehicle.

Separate keys are supplied for locking the driver's door. The switch has four positions (Fig. 4) as follows:

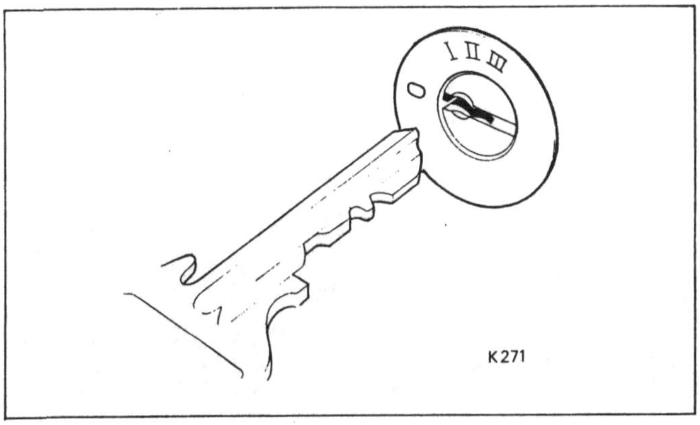

Fig. 4

Symbol 0	"OFF" in which position the key may be inserted or withdrawn (see Audible alarm system).
Symbol I	"Auxiliary", in this position the ignition circuit is isolated to allow the use of a radio when the vehicle is stationary and the ignition is switched off.
Symbol II	"Ignition" (see ignition warning).
Symbol III	"Start" (Refer to "Starting the engine" on page 8).

Audible Alarm System

The Audible Alarm System only functions when the ignition key is positioned in the switch and the driver's door is open. The "warning" denoted by a continuous buzzing sound will terminate when the driver's door is closed or the ignition key is completely removed. (See Audible Alarm Switch page 15).

1. **To disengage Steering Lock and Start Engine**

Insert the key and turn in clockwise direction. If difficulty is experienced in turning the key, this can be rectified by simultaneous movement of the steering wheel.

Turn the key clockwise to symbol II (Ignition) the ignition will be switched on.

To start the engine, the key should be turned a little more against spring pressure to symbol III ("start"), as soon as the engine fires release the key which will return automatically under spring pressure to the ignition position (symbol II).

2. **To stop engine and engage steering lock**

Turn the key in an anti-clockwise direction from the "ignition" position (symbol II) to the "lock" position (symbol 0). This action stops the engine.

Removal of the key in this position automatically actuates the steering lock mechanism. (See Audible Alarm System).

Accelerator, Brake and Clutch Pedals (21, 22 and 23)

These are conventional items which should require no further explanation.

Horns (24)

The horns are operated by pressing the button in the centre of the steering wheel.

Reversing Lamps

Reversing lamps are operated when the ignition is switched on and reverse gear is selected.

SEATS AND SAFETY HARNESS

Seat Adjustment (Fig. 5)

Each seat pivots on its forward attachment to the seat-runner.

To alter the angle of the squab raise and hold the lever (1), and, using controlled body pressure, move the back rest to the desired position. Release the lever to lock the back rest at the chosen angle. The seat back, if unrestrained, will return to the upright position when the locking lever is raised.

Moving a spring-loaded lever (2) inwards permits forward or rearward movement of the seat as required. Releasing the lever causes its locking pawl to re-engage in the seat-runner but slight additional movement of the seat may be necessary to achieve this.

Safety Harness

Two quick-release hooks attached to the end of the harness, clip into the eye-bolts provided (3) Fig. 5, the third attachment is bolted to the wheel arch shown on Figs. 6 and 7.

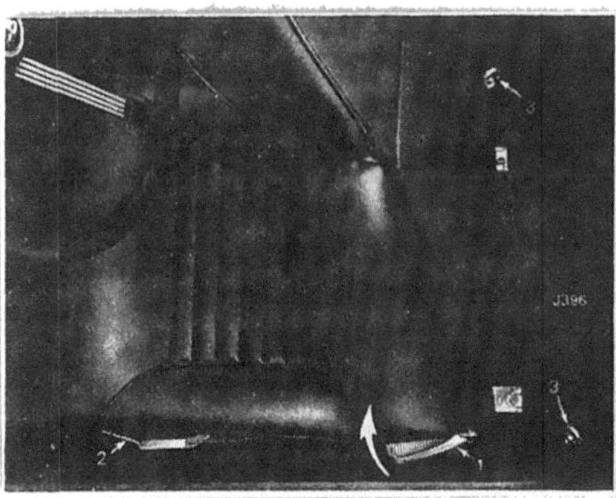

Fig. 5

Fig. 6

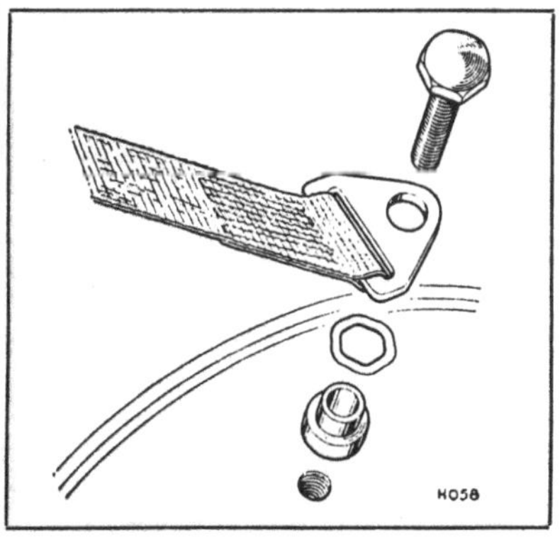

Fig. 7

To use the harness pass the shoulder strap over the shoulder and the buckle ends of the harness around the hips; fasten by pushing the locking plate into positive engagement with the buckle. This is denoted by a "click". Adjust the length of the harness by moving the plastic loop and passing the belt around the self-locking knurled rollers.

NOTE: When adjusting the harness, ensure that the buckle is positioned on the side of the body nearest to the centre of the vehicle.

To release the harness, lift the buckle through 90°.

LOCKS AND KEYS

The vehicle is provided with three sets of keys, the spare set being obtained from the selling dealer. The large key is used to operate the ignition and steering lock, the other two sets are for the luggage compartment and door locks.

CAUTION: The large key is delivered complete with a key identification disc. This must not be mislaid for it provides the only record of the three large keys. In the event of loss of keys the disc will have to be submitted to your selling dealer to obtain a duplicate set.

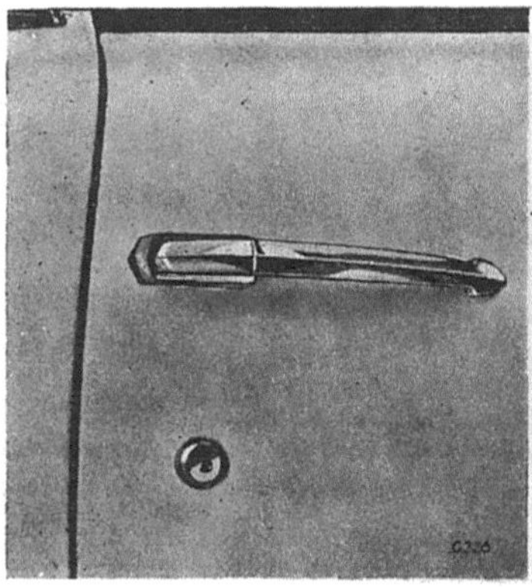

Fig. 8

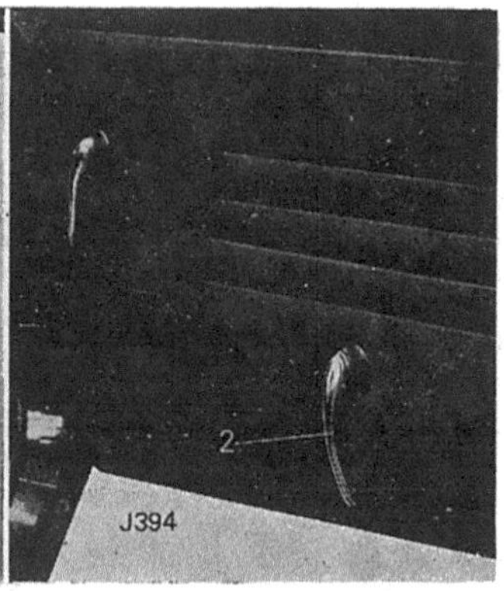

Fig. 9

Door Locks (Figs. 8 and 9)

"Anti-burst" locks fitted to both side doors are opened by depressing a pushbutton on the outside or by pulling the remote-control lever (2) on the inside. To lock the door from the inside, push the lever forward; to lock the door from the outside, insert the key and turn it a quarter turn away from the shut face. The key will return under the influence of spring loading to a vertical position when it may be withdrawn.

Lubrication

Once a month, particularly during freezing weather, apply a few drops of thin machine oil into the latch and key slots.

IMPORTANT: Do not apply grease to the lock cylinders.

Fig. 10

Fig. 11

Bonnet Lock (Fig. 10)

The bonnet is opened by raising, as far as possible, a lever on each side to release the catches and lifting the bonnet at its rear. This permits the bonnet to pivot on its anchorage to a vertical position where it is held by a folding strut.

To close the bonnet, pull the centre of the strut (arrowed Fig. 12) simultaneously supporting and lowering the bonnet. Press each lever flush with the side of the bonnet to lock.

Luggage Compartment Lid (Fig. 13)

To open the luggage compartment lid, turn the unlocked handle (Fig. 11) counter-clockwise to a vertical position and raise the lid to its limit before lowering it on to the telescopic support.

Close the lid by raising it slightly to release the catch in the telescopic support, lower, and turn the handle which may be locked by turning the key a half turn counter-clockwise.

Fuel Filler Cap (Fig. 14)

The fuel filler cap, located forward of the luggage locker lid, is opened by compressing a small lever at the side of the cap. Press the cap to close.

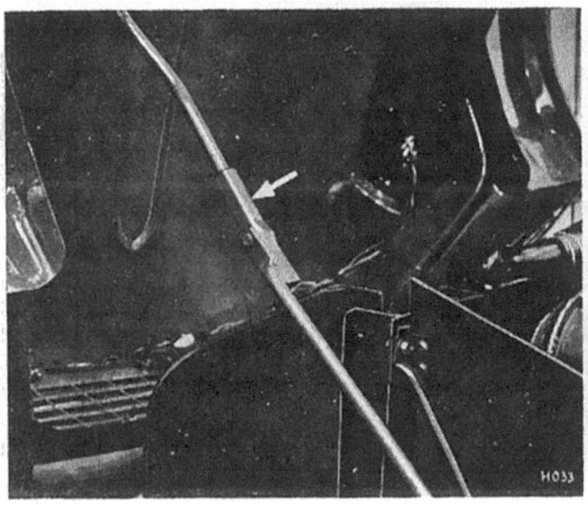

Fig. 12

Fig. 13

Fig. 14

SOFT TOP

The soft top is made from P.V.C. material, and is supported by a hinged frame. The assembly folds down into the rear of the car and is protected by a soft top cover.

Lowering the Soft Top

Unclip the fasteners securing the fabric to the second hoodstick (Fig. 15) and tilt the seats forward. Release the soft top header rail from the windscreen frame by turning the catch levers in the direction shown in Fig. 16 and the fasteners (four each side Fig. 17) securing the edges of the hood to the body.

Push the soft top header rail rearwards and slightly upwards while knocking the second hoodstick (arrowed Fig. 15) forwards, until the assembly begins to fold. **DO NOT PULL the second hoodstick downwards.** Continue lowering the frame and pull the fabric flat over the luggage compartment lid (Fig. 18).

Fold the fabric forward over the hoodsticks and turn the sides inward (Fig. 19). Ensure that the Vybak windows are free from distortion and the hood fabric is not trapped by the hoodsticks.

Soft Top in the Down Position

Retain the soft top in position by fitting the cover (Fig. 21) as follows:

Attach the cover to the fasteners and continue working towards the centre and locate the three hooks (Fig. 20) on the cover under the bottom hoodstick.

Raising the Soft Top

Unfasten and remove the hood cover. Fold the sides of the hood fabric outwards and pull the fabric rearwards over the luggage compartment lid. Lifting the soft top header rail, raise the assembly sufficiently to allow the fabric to lie evenly over the frame.

Secure the fasteners (four each side Fig. 17) to the body. Locate the soft top header rail on the windscreen header rail and turn the levers (Fig. 16) inwards. Knock the second hoodstick (arrowed Fig. 15) rearwards as far as possible, and secure the fastener (Fig. 15).

Sun Visors (Fig. 16)

Two adjustable sun visors, padded to reduce the risk of impact injury, may be unclipped from their inner attachment blocks and swung to eliminate side glare. The passenger's sun visor combines a vanity mirror.

The driving mirror is mounted on a ball socket for easy adjustment.

Opening Rear Window (Fig. 22)

To open the rear window, release the zip-fastener and roll the Vybak panel downwards. Retain the rolled panel in position by

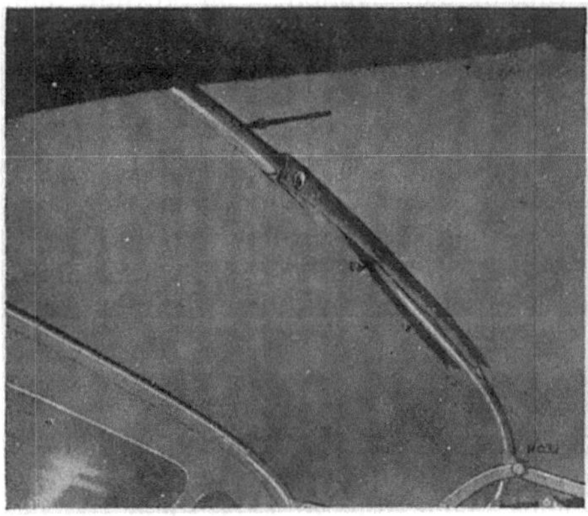

Fig. 15

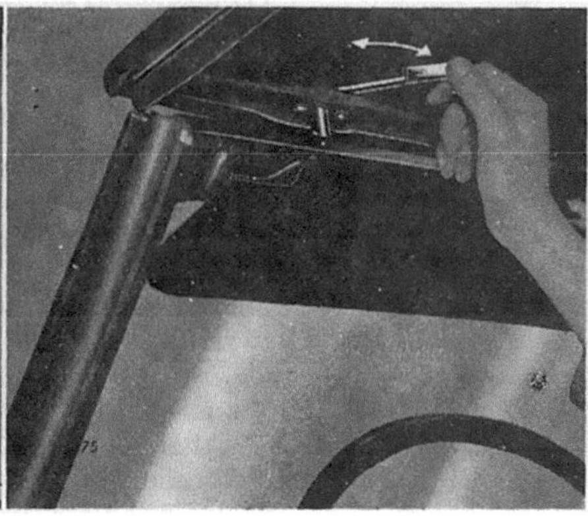

Fig. 16

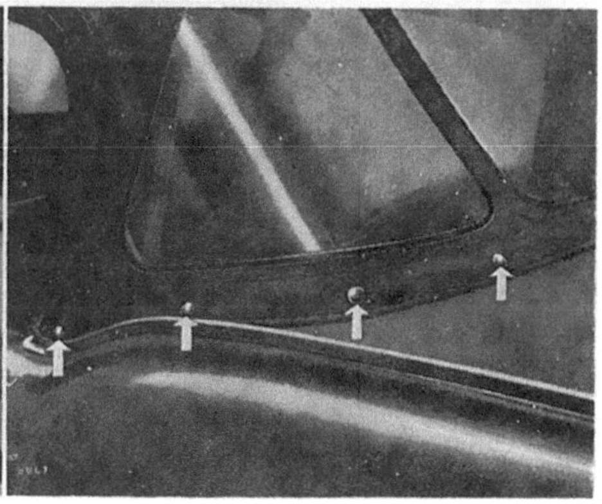

Fig. 17

Fig. 18

Fig. 19

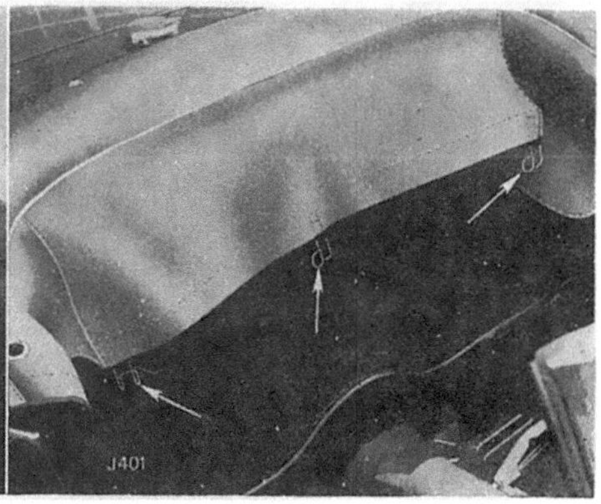

Fig. 20

Fig. 21

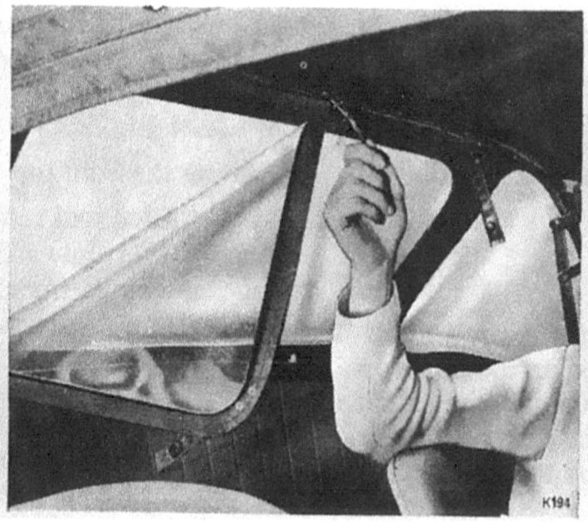

Fig. 22

Fig. 23

attaching the straps to the press fasteners located on the back wall of the rear compartment.

Tonneau Cover (Optional) (Fig. 23)

The tonneau cover provides weather protection for the vehicle interior when the soft top is lowered. It incorporates press-studs for securing to the car and a zip fastener which permits access to either or both of the front seats.

HARD TOP

HARD TOP (Optional)

The vehicle may be used as an open sports car by removing the hard top assembly as follows:

Unscrew the domed-head bolts securing the hard top side brackets to the door pillar brackets (Fig. 24).

Remove the domed-head bolts and washers securing the hard-top to the windscreen header rail (Fig. 25) and the rear deck panel (Fig. 26).

With the aid of a second operator carefully lift off the hard top assembly.

To refit the hard top, reverse the foregoing procedure.

STARTING THE ENGINE

Preparation

Check and if necessary top-up the engine sump oil level and replenish the radiator overflow bottle to half-full (cold). Apply the handbrake and ensure that the gear lever is positioned at "neutral".

Choke Control

In cold weather pull the choke control fully out; in warm climates pull to the mid-position and in hot climates do not use the control.

Ignition Switch

Insert the ignition key and turn it to the "ignition" (symbol 11) position (Fig. 4), causing the "ignition", "oil pressure" and "brake-line failure" indicators to glow, and all electrical circuits controlled by the ignition switch to become operative.

Starting

From the (symbol 11) position, turn the key clockwise against spring pressure to operate the starter motor. Immediately the engine fires, release the key and allow it to return to the ignition position. Should the engine fail to start at the first attempt, do not re-operate the starter switch until the starter motor has come to rest.

Warming-up

After starting the engine, push the choke control "half-in" (cold climates), or "fully-in" (warm climates) and warm the engine at an approximate idling speed of 1,100 r.p.m. At this speed the "ignition", "oil pressure" and "brake-line failure"

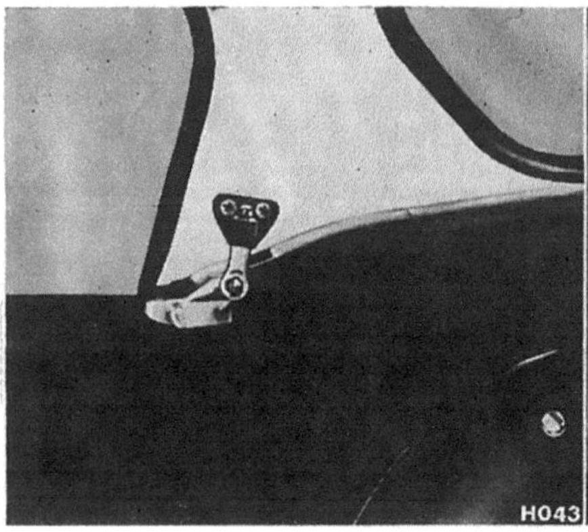

Fig. 24

Fig. 25

Fig. 26

indicators should cease to glow. Should a warning indicator continue to glow, stop the engine and establish the cause, otherwise serious damage to the engine may result.

Driving

When driving a cold engine from rest, maintain engine speed of 1,100 r.p.m. until the choke control can be pushed fully-in and the engine becomes fully response to the accelerator pedal. Avoid prolonged use of the choke and do not use full-throttle during the warming-up period. A thermostat enables the engine to be quickly warmed from cold.

Hot Re-starting

To re-start a hot engine, depress the throttle pedal to about one-third of its travel before operating the starter switch. The choke control should not be used.

Recommended Fuel

The "Spitfire" engine is designed to operate on Premium grade fuel (100 octane).

When travelling outside the United States and such fuels are not available, and it is necessary to use fuels of lower or unknown rating, the ignition timing must be retarded from the specified setting, to prevent audible detonation (pinking), under all operating conditions, otherwise serious damage to the engine may occur. The use of lower octane fuels will result in overheating, loss of engine power and efficiency.

Conversely, when the car enters the "United States" the ignition timing must be advanced to suit the use of Premium fuel and to comply with the Federal Regulations on emissions from the crankcase and exhaust.

DRIVING FROM NEW

Running-in

The importance of correct running-in cannot be overemphasised for during the first few thousand miles of motoring, particularly the first thousand, all working surfaces of the vehicle, including clutch and brakes are "bedding-in" and the eventual performance of the vehicle is affected by the care it receives during its early life.

Most important, avoid the use of full throttle, particularly at low road speeds and when the engine is cold. Running-in should be progressive and no harm will result from allowing the engine to "rev" fairly fast for short periods provided that it is thoroughly warm and not pulling hard.

Use up to about half-throttle initially and select a lower gear as necessary to avoid overloading the engine. As the engine becomes increasingly responsive, wider throttle openings can be used, but the engine should not be subjected to full throttle in any gear, until at least 1,000 miles (1,600 km.), have been covered and even then, full throttle should be used only for short periods which may be gradually extended as running-in progresses.

Maximum Engine Speeds

After "running-in" has been completed, owners are advised not to drive the car continuously at engine speeds in excess of 6,000 r.p.m. This speed is indicated by the beginning of an orange coloured segment on the tachometer dial. Speeds within this segment are permissible for short periods, such as when overtaking another vehicle, but under no circumstances must the engine be allowed to rev beyond 6,000 r.p.m., which coincides with the beginning of a red segment.

Overdrive Unit (Optional fitment)

An overdrive unit serves as a convenient method of providing alternative gear ratios to reduce engine speed which with proper use can effect fuel economy.

Greatest benefit will occur from judicious use of the overdrive, the governing factor being that, when changing into overdrive, the vehicle continues to run easily without sign of engine labouring, combined with the minimum throttle opening necessary to maintain the same road speed. Maximum economies are effected under "motorway" conditions.

The overdrive unit is remotely controlled by an electrica switch, mounted on the steering column, which is moved downwards or upwards to engage or disengage the unit. Do not change from overdrive at engine speeds in excess of 4,500 r.p.m. This corresponds with engine speeds within the orange segment in normal gears. Disengagement of overdrive at higher engine speeds will cause the engine to over-rev and increase the possibility of damage to the engine.

WHEELS AND TYRES

Spare Wheel and Lifting Jack

The tools and spare wheel are housed in the luggage compartment as shown on Figs. 27 and 28.

NOTE: A variant of the jack illustrated may be provided.

To remove the spare wheel, lift off the cover and unscrew the retaining nut (Fig. 28).

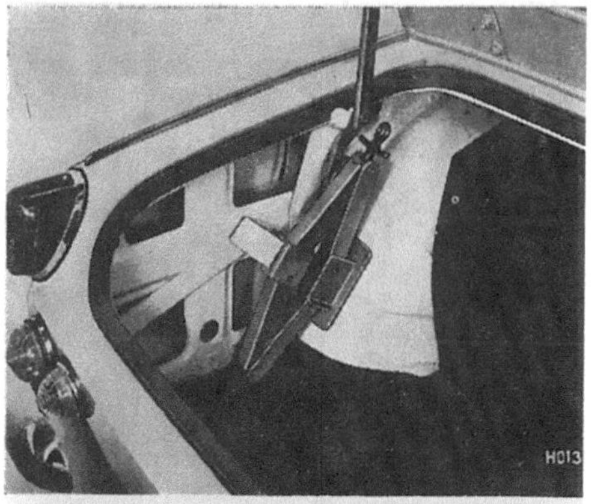

Fig. 27

Fig. 28

The Jack (Fig. 29)

Locate the nut of the fixing bolts (rearwards of the front wheel and forwards of the rear wheel) in the head of the jack for safety when lifting a wheel.

Assemble the handle into the jack and turn it to lift the ıequired wheel from the ground.

Fig. 29

Wheel Changing Procedure (Pressed Steel Wheels)

1. Firmly apply the handbrake and chock the wheel diagonally opposite the one being lifted.
2. Remove the spare wheel from the luggage compartment and make sure that its tyre pressure is correct (Fig. 28).
3. Using the lever provided in the tool kit, lever off the wheel trim as shown and slightly loosen the wheel nuts (Figs. 30 and 31).

Fig. 30

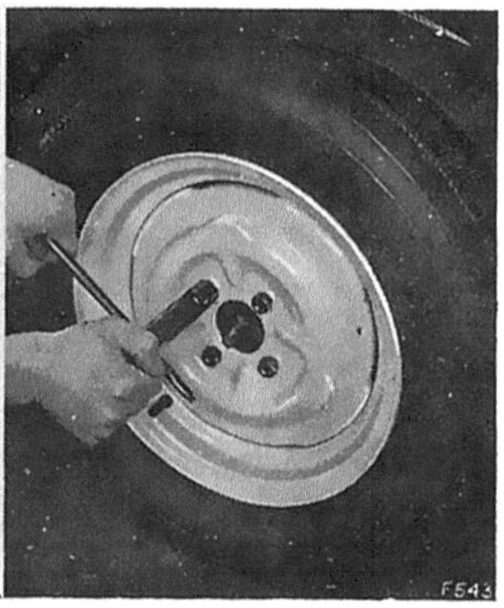

Fig. 31

4. Place the jack in position and lift the wheel clear of the ground. Should it be necessary to lift the vehicle whilst it is on sloping ground, exercise the greatest care.
5. Completely remove the wheel nuts, exchange the road wheels and replace the nuts.
6. Lower the jack, give the wheel nuts a final tighten and refit the wheel trim by placing its edge over the wheel projections and giving the plate a sharp tap with the hand to spring it into position.

Wheel Alignment

The correct front wheel alignment is $\frac{1}{16}$" (1·6 mm.) to $\frac{1}{8}$" (3·2 mm.) toe-in (kerb condition). Excessive misalignment caused by kerb impact or other accidental damage will result in severe tyre wear and faulty steering.

Wheel Run-out and Ovality

The maximum tolerances for both run-out and ovality are as follows:

Press steel wheels .. .. 0·070" (0·18 mm.)

Excessive run-out and ovality will result in severe tyre wear and faulty steering.

Tyre Pressures

Refer to page 29.

General

Wheels and tyres, of correct types and pressures, are an integral part of a vehicle's design. Thus the regular maintenance of the tyres contributes not only to the safety but to the designed functioning of the vehicle, as road holding steering and braking are especially vulnerable to the use of incorrectly pressurised, badly fitted or worn tyres.

Radial-ply and Cross-ply Tyres

It is dangerous to use, on the public highways, a vehicle fitted with unsuitable combinations of tyres. The following recommendations should therefore be observed.

1. Do not mix radial-ply and cross-ply tyres on the same axle.
2. Do not fit radial-ply tyres to the front wheels and cross-ply tyres to the rear wheels.
3. With suitable tyre pressure adjustments it may be possible to obtain acceptable handling with cross-ply tyres on the front wheels and radial-ply tyres on the rear wheels, but this combination is not recommended.

Winter Tyres

Winter tyres are designed primarily to give improved traction and braking in mud and snow. Their performance on hard surfaces may, however, be inferior to normal road tyres and extra care is required when using them under normal conditions.

Pressures

Adjust tyre pressures in accordance with the recommendations given on page 29. These pressures are satisfactory for sustained speeds up to 110 m.p.h. (175 km.).

NOTE: Should the vehicle be tuned to increase its maximum speed, or be used for racing, consult the respective tyre company regarding the need fo r tyres of full racing construction.

Tyre Wear

To prolong tyre life, avoid severe braking, sudden changes of direction at speed, and driving over or against high kerbstones, as this can result in severe damage to the tyre walls. Clean off any soil or grease which may be on the tyres by usingcloth moistened in petrol.

Fins on the inside or outside of the pattern ribs indicate excessive toe-in or toe-out respectively. Because sharp pattern edges may also be caused by road camber, it is better to make sure by having the track checked.

"Spotty" tread wear or flats can result from grabbing brakes or unbalanced wheel assemblies. The original degree of balance is not necessarily maintained, but may be affected byuneven tread wear, repairs, tyre removal and refitting, or by wheel damage and eccentricities. The vehicle may also become more sensitive to unbalance due to normal wear of moving parts.

Excessive wear in the centre of the tread results from over-inflation. Excessive wear at the outer edges of the tread results from under-inflation, a condition which causes excessive heating and premature tyre failure.

COOLING SYSTEM

The pressurised "no loss" cooling system incorporates a translucent plastic overflow reservoir (Fig. 32) which collects excess coolant from the radiator as the coolant in the system expands with heat. Depression created as the system cools, causes the coolant to flow back from the reservoir into the radiator. The fluid level which is visible through the translucent reservoir should be maintained at least half full when cold.

CAUTION: If the engine is hot avoid danger from scalding by exercising extreme care when removing the radiator filler cap. Turn it a half-turn and allow pressure to be fully released before completely removing the cap.

Draining

To drain the system, move the heater control (11) Fig. 2, to the hot position, remove the radiator filler cap Fig. 32, open the tap in the bottom of the radiator (Fig. 33) and the tap at the rear right-hand side of the cylinder block (Fig. 34).

Flushing

Efficient cooling is maintained by thoroughly flushing the system once each year before adding anti-freeze. When carrying this out, it is advantageous to remove the drain tap completely and to use plenty of clean running water.

Allowing anti-freeze solution to remain in the system throughout the summer period affords anti-corrosion protection. The solution, however, should be changed at the beginning of each winter period as the inhibitor becomes exhausted.

Filling

Close both drain taps, open the heater control fully and remove the radiator filler cap. Fill the cooling system with clean (soft) water and run the engine at approximately 1,500 r.p.m. for 1 to 2 minutes. Top up the radiator and replace the filler cap. Completely fill the plastic overflow reservoir with clean water.

Frost Precautions

The car heater cannot be completely drained by normal methods. Therefore frost damage will not be prevented by merely draining the radiator.

For your protection during freezing weather, an approved anti-freeze solution should be added to the coolant in the radiator. Because of the searching effect of these solutions, advise your Dealer to check the system for leaks before adding the anti-freeze.

At certain temperatures glycol water solutions adopt a "mushy" state with a viscosity which impairs circulation and can immobilise or damage the water pump. Therefore, consult the following chart before adding anti-freeze, for the degree of frost protection required.

ANTI-FREEZE CONCENTRATION	25%	30%	35%
Complete Protection Vehicle may be driven away immediately from cold	10°F. −12·2°C.	3°F. −16°C.	−4°F. −20°C.
Safe Limit Coolant in mushy state. Engine may be started and vehicle driven away after short warm-up period	1°F. −17°C.	−8°F. −22°C.	−18°F. −28°C.
Lower Protection Limit Prevents frost damage to cylinder head, block and radiator. Engine should **NOT** be started until thawed out.	−14°F. −26°C.	−22°F. −30°C.	−28°F. −33°C.

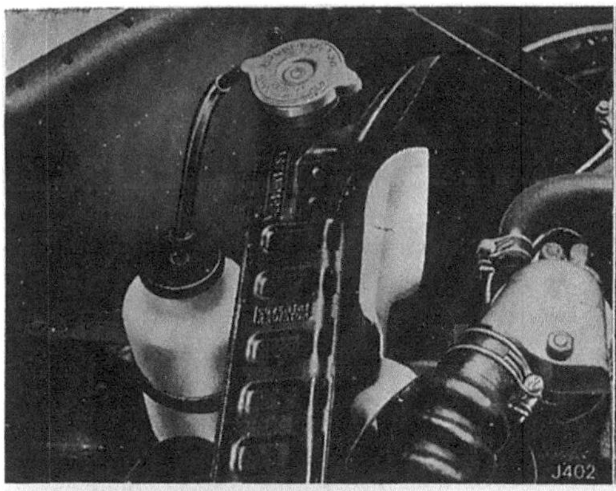

Fig. 32

Fig. 33

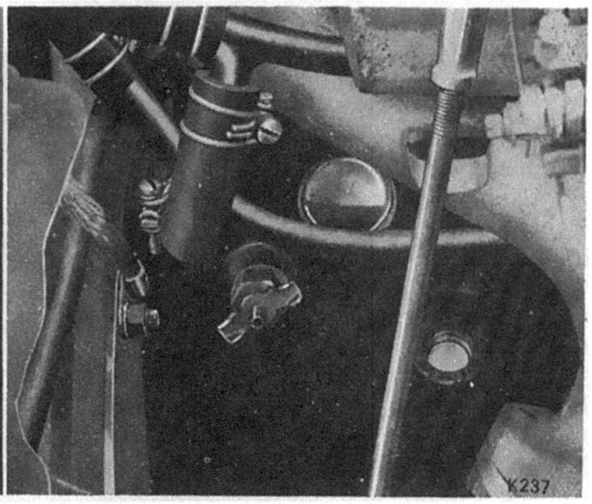

Fig. 34

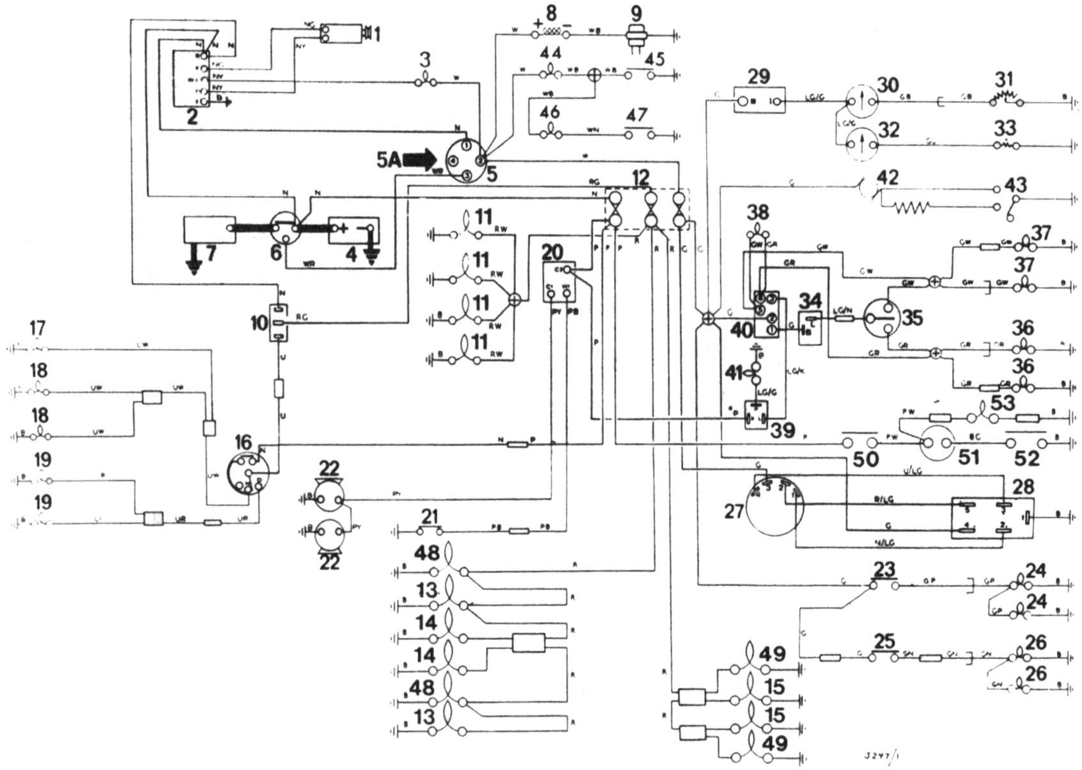

Fig. 35. Wiring diagram

KEY TO WIRING DIAGRAM

CAUTION: THIS VEHICLE IS FITTED WITH A NEGATIVE EARTH ELECTRICAL SYSTEM. ENSURE THAT THE BATTERY EARTH LEAD IS ALWAYS CONNECTED TO THE BATTERY NEGATIVE TERMINAL.

EXERCISE CARE WHEN CONNECTING INTO CIRCUIT ANY ACCESSORY THAT MAY CONTAIN SILICON DIODES OR TRANSISTORS. IRREPERABLE DAMAGE MAY RESULT TO SUCH POLARITY SENSITIVE COMPONENTS IF INCORRECTLY FITTED.

NOTE: No polarity sensitive components are fitted to the vehicle during production. However any of the following accessories—approved or unapproved by Leyland Triumph—may contain such components: Alternator systems, Automatic anti-dazzle mirrors, Automatic dipping systems, Automatic parking lamp systems, Electronic ignition systems, Electronic tachometers and Radios.

1 Generator
2 Control box
3 Ignition warning light
4 Battery
5 Ignition starter switch
5A Ignition starter switch—radio supply position
6 Starter solenoid
7 Starter motor
8 Ignition coil
9 Ignition distributor
10 Master light switch
11 Instrument illumination
12 Fuse assembly
13 Tail lamp
14 Plate illumination lamp
15 Front parking lamp
16 Column light switch
17 Main beam warning light
18 Main beam
19 Dip beam
20 Horn relay
21 Horn push
22 Horn
23 Stop lamp switch
24 Stop lamp
25 Reverse lamp switch
26 Reverse lamp
27 Windscreen wipeı switch
28 Windscreen wiper motor
29 Voltage stabilizer
30 Fuel indicator
31 Fuel tank unit
32 Temperature indicator
33 Temperature transmitter
34 Turn signal flasher unit
35 Turn signal flasher switch
36 L.H. flasher lamp
37 R.H. flasher lamp
38 Flasher warning light
39 Hazard flasher unit
40 Hazard switch
41 Hazard warning light
42 Heater motor
43 Heater switch
44 Brake line failure warning light
45 Brake line failure switch
46 Oil pressure warning light
47 Oil pressure switch
48 Rear marker lamps
49 Front marker lamps
50 Door switch
51 Key warning buzzer
52 Key switch
53 Courtesy light

COLOUR CODE

N.	Brown	LG.	Light Green
U.	Blue	W.	White
R.	Red	Y.	Yellow
P.	Purple	S.	Slate
G.	Green	B.	Black

FUSE SYSTEM

The fuse assembly is a component of the main harness. It is secured to an aperture provided on the bulkhead by integral plastic clips. The unit contains three operational fuses and has provision to house two spares. The fuses are protected by a pull-off transparent plastic cover.

Failure of a particular fuse is indicated when all the circuits protected by it become inoperature. If a new fuse fails establish the cause and rectify the fault before fitting a second replacement.

Fuse

Manufacturer	Lucas
Rating	35 amp.
Lucas Part No.	188218
Stanpart No.	58465

Fig. 36

Circuits

The top fuse fed by a white cable from the ignition/starter switch protects the following circuits:

- Stop lamp circuit
- Reverse lamp circuit
- Windscreen wiper circuit
- Fuel indication circuit
- Temperature indication circuit
- Turn signal circuit
- Heater circuit

The centre fuse fed by a red/green cable from the master light switch protects the following circuits:

- Instrument illumination circuit
- Rear marker lamp circuit
- Tail lamp circuit
- Plate illumination lamp circuit
- Front marker lamp circuit
- Front parking lamp circuit

The bottom fuse fed by a brown cable from the battery protects the following circuits:

- Headlamp flasher circuit
- Key warning circuit
- Courtesy light circuit
- Hazard warning circuit
- Horn circuit

CHARGING SYSTEM

Generator

The Lucas C40-1 unit is driven by a vee belt which should be adjusted as detailed on page 19. The generator rear bearing should be lubricated as detailed on page 17.

Control Unit

The Lucas RB340 current voltage regulator adjusts the current flow through the generator field winding to control the generator output to suit the electrical requirements of the vehicle and the state of charge of the battery. Maximum output is 22 amps.

Fig. 37

BATTERY

CAUTION: REFER TO "CAUTION" ON PAGE 12.

A conventional battery is located on the bulkhead. Battery data is given on page 29.

Ensure that the battery top and terminals remain clean and dry. Coat terminals with petroleum jelly (Vaseline) to prevent corrosion.

Check electrolyte level monthly and if required replenish with pure water as detailed on page 17. If electrolyte has been spilled clean the affected area with a cloth moistened with ammonia to neutralize the acid and prevent acid corrosion.

Ensure that the battery is always firmly clamped in position by the retaining assembly. When fitting battery leads do not hammer terminals to terminal posts. Such action may damage battery.

The battery will deteriorate rapidly if left in a discharged condition. If the unit is reduced to a low state of charge it should be recharged at the first opportunity.

STARTER SOLENOID AND STARTER MOTOR

The starter solenoid is normally solenoid operated by remote control from the ignition/starter switch. It may also be actuated manually from the engine compartment by depressing the rubber cap shown arrowed on Fig. 38.

WARNING: BEFORE ACTUATING THE STARTER SOLENOID ENSURE THAT THE GEAR LEVER IS IN NEUTRAL AND THE HANDBRAKE IS FIRMLY APPLIED.

Fig. 38

Fig. 39

If the starter motor fails to operate when selected the starter drive may be jammed in mesh with the engine flywheel ring gear. This condition can normally be freed by turning the squared extension of the shaft shown arrowed on Fig. 39 using a spanner.

If when selected the starter motor runs at high speed but fails to crank the engine the starter drive pinion may not be meshing with the engine flywheel ring gear. This condition can normally be cured by removing the starter motor and washing the starter drive in petrol or paraffin and lubricating lightly with thin engine oil.

HEADLAMPS

NOTE: TO AVOID BEAM AIMING DO NOT DISTURB BEAM AIMING SCREWS A AND B.

Beam Aiming

Insert a large screwdriver behind rim adjacent to clip as shown on Fig. 40. Twist screwdriver to release rim from clip. Lift rim from upper retainers. Screw A positions the beam in the horizontal plane. Screw B controls beam height.

Beam aiming can best be accomplished using equipment such as Lucas "Beamsetter" or "Lev-L-Lite". This service is available at Triumph distributors or dealers and will ensure maximum road illumination with minimum discomfort to other road users.

Filament Failure

In the event of a filament failure the sealed beam light unit must be replaced. Insert a large screwdriver behind rim adjacent to clip as shown on Fig. 40. Twist screwdriver to release rim from clip. Lift rim from upper retainers. Remove three screws 1, 2 and 3 to release retaining rim and sealed beam light unit. Pull connector from light unit. Renew light unit and reassemble.

LAMPS—BULB RENEWAL

Bulb renewal for the majority of lamps is conventional. Remove lens by unscrewing required screw/screws. Renew bulb and reassemble.

Front and Rear Marker Lamps

Use a screwdriver to turn back rubber lip to remove chrome rim. Remove lens in a similar manner. Renew bulb. Locate lens

BULB CHART

Lamp	Watts	Lucas Part No.	Stanpart No.	
Headlamps—R.H. Dip—U.S.A.	50/40	54522231	——	*
France	45/40	411	510219	
Other R.H. Dip markets	45/40	410	510218	
L.H. Dip	60/45	54521872	512231	*
Front parking lamps	6	207	57591	
Front flasher lamps	21	382	502379	
Front marker lamps	4	222	501436	
Rear marker lamps	4	222	501436	
Tail/stop lamps	5/21	380	502287	
Rear flasher lamps	21	382	502379	
Reverse lamp	21	382	502379	
Plate illumination lamps	5	501	514797	
Instrument illumination	2·2	987	59492	
Courtesy light	2·2	987	59492	
Warning lights	2·2	987	59492	

* Sealed beam light unit

Fig. 40

Fig. 41

Fig. 42

to lip and "run round" with screwdriver to secure lens flange under lip. Fit chrome rim in a similar manner.

Plate Illumination Lamps

Remove cowled rim by unscrewing two screws. Use screwdriver to turn back lip to remove lens. Pull out bulb. Renew bulb and reassemble. Note that the cowled rim must be positioned so light is directed towards plate.

Instrument Illumination and Warning Lights (Figs. 43 and 44)

To obtain access to bulb holders on or adjacent to the centre panel withdraw the panel as follows. Remove one battery lead to isolate electrical power. Slacken small Allen screws in heater control knob sides and withdraw knobs. Remove four screws. Carefully withdraw the panel.

Pull bulb holder—which is a component of the main harness—from instrument or housing. Unscrew bulb from holder. Renew bulb and re-assemble.

Fig. 43

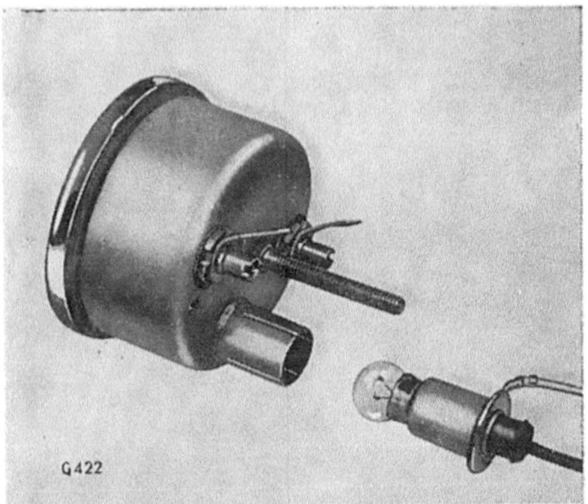

Fig. 44

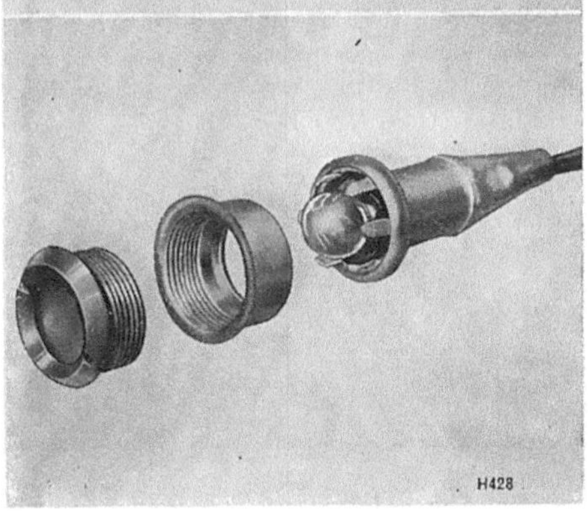

TURN SIGNAL FLASHER UNIT—RENEWAL

Locate unit behind facia on right hand side of bulkhead. Pull unit from socket. Fit new unit to socket.

Fig. 45

KEY WARNING SYSTEM

This system is designed to encourage the driver to remove the ignition key from the lock before leaving the vehicle.

If the driver's door is opened while the ignition key is in the lock an audible buzzer will sound. Removing the key or closing the door will cause the buzzing to cease.

The buzzer is housed in a cylindrical container located behind the extreme left hand side of the facia.

The system should prevent encouragement of theft but is not intended as a comprehensive anti-theft device.

Associated with the key warning system is the courtesy light which illuminates the lock to facilitate key entry. The door switch controls the courtesy light and also features in the key warning circuit.

Lubricants

When carrying out the following maintenance work, use high grade lubricants only. The lubricants listed on page 28 have maintained a high standard of quality for many years and are approved only after extensive tests in collaboration with the oil companies concerned. In countries where these oils are unobtainable, use similar oils having the same characteristics.

Engine

When a new car is delivered the engine sump contains a quantity of special running-in oil, sufficient for the running-in period. Provided that the level does not fall below the low mark on the dipstick, topping-up is unnecessary. If replenishment is required, top-up with one of the engine oils listed on page 28.

At the "Free Service", or after completing the first 1,000 miles (1,600 km.), drain the running-in oil from the sump and replenish with new oil, of the appropriate grade, to the high mark on the dipstick.

Transmission, Overdrive and Rear Axle

Rear axles, transmission and overdrive units fitted to new cars are filled with a special oil, formulated to give all necessary protection to new gears. This oil should not be drained but may be topped up with any of the approved oils.

Braking System

In addition to adjustment and examination/renewal of shoes and pads at the intervals recommended in the following pages, it is strongly recommended that the brake system is overhauled and the brake fluid renewed every 36,000 miles (60,000 km.) or 3 years (whichever is sooner). This involves dismantling, examining and renewing all seals and defective items.

Owners are urged to seek the assistance of a Triumph Dealer who will be pleased to estimate for the work which, for reasons of safety, should be entrusted only to skilled workshop personnel.

Preventive Maintenance

To ensure continued efficiency and prolonged vehicle life, the maintenance voucher scheme, produced by Standard-Triumph engineers, offers a carefully designed plan of lubrication requirements and adjustment checks at pre-determined periods.

Operated by all Triumph dealers, and specifically recommended to owners wishing to obtain the greatest pleasure from their motoring, the scheme involves the use of a series of Maintenance Vouchers contained in a booklet supplied with the car. Service operations appropriate to mileage or periods of time are listed on pages preceding the vouchers.

The space provided on the counterfoil of each voucher should be filled in by the dealer to constitute proof of regular servicing, should this be required when making a claim under the warranty, or when selling the vehicle.

PERIODIC CHECKS

Engine

Prior to starting out on a long run, or every 250 miles (400 km.), check the engine oil level and, if necessary, add oil until the level reaches the high mark on a fully inserted dipstick while the vehicle is standing on level ground. Should the level be at the lower mark on the dipstick add 2·4 pints via the cap (Fig. 46), to bring the level to the top mark.

Ensure that the filler cap (1) and dipstick (2) are replaced.

Radiator Water Level—Weekly (Fig. 32)

The level of water, visible through the translucent plastic reservoir, should be maintained at least "half full" by adding soft water, when required, via the screwed cap. Refer to "filling" on page 11.

Screen Washer

Examine the water level in the plastic windscreen washer container. To refill the reservoir, hold the plastic tubing while unscrewing the cap. Fill with clean soft water to the bottom of the neck and replace the cap. If the wiper blades tend to smear, a little household detergent added to the water in the reservoir will help to prevent this.

To avoid frost damage during cold weather, either empty the reservoir or fill with a mixture of 1 part de-natured alcohol (methylated spirits) to 2 parts water. Do not use glycol anti-freeze.

Should a screen washer jet become obstructed, a strand of Bowden cable, or a fine pin not greater than 0·030 in. diameter can be used to clear the jet orifice.

The use of a Lucas Battery Filler will aid topping-up. Ensure that the battery filler is filled with distilled water and insert it into a filler plug orifice until it rests gently on top of the separators. Sufficient water will pour into the cell to bring the electrolyte to its correct level. Check each cell in turn.

CAUTION: Never use a naked light when examining the battery. The mixture of oxygen and hydrogen given off by the battery is dangerously explosive.

Brake Master Cylinder (1, Fig. 47)

Every week check the level of fluid in the brake master cylinder reservoir. The fluid level is visible through the translucent

Fig. 46 *(upper)* **Fig. 47** *(lower)*

casing of the reservoir, **do not remove the cap.** A gradual lowering of the level over a long period is caused by brake pad wear and does not require topping-up. A sudden appreciable drop in the level must be investigated, the cause ascertained and rectified immediately.

Do not allow the level to drop below the danger line on the side of the casing (see Fig. 47).

To avoid dirt entering the system ensure that the reservoir is clean externally before removing the cap. Use only new fluid taken from a sealed container and re-seal the container after use. Replace the reservoir cap immediately after filling.

Clutch Master Cylinder (2, Fig. 47)

Every month, check the level of fluid in the clutch master cylinder. To prevent dirt entering the system, clean the cap and surrounding area prior to removing the cap. Top-up the fluid until is is level with the line on the side of the reservoir.

Tyres

The maintenance of correct tyre pressures, tread and thickness and freedom from cuts and other tyre damage, are important safety factors which affect the behaviour of steering and suspension. It is, therefore, expedient to examine the tyres at periods not exceeding two weeks and to ensure that pressures are maintained in accordance with the figures given on page 59.

Battery—Monthly (Fig. 48)

Examine the level of the electrolyte in the cells and, if necessary, add distilled water via the filler orifices to bring the level up to the top of the separators.

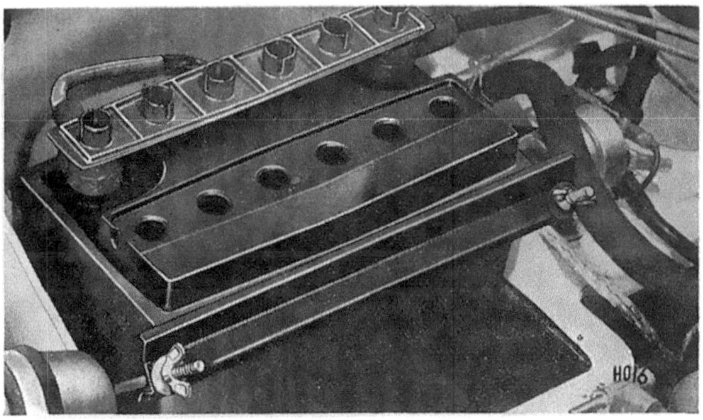

Fig. 48

Engine Oil Sump (Fig. 49)

Every 6,000 miles (10,000 km.), remove the plug (1) to drain the oil. Refit the plug and refill to the correct level, via the filler cap (page 36, Fig. 46). Reduce this period for short journeys involving frequent stop/start driving, particularly during cold weather when greater use is made of the choke control.

If the vehicle is used for competitions or sustained high speed work, the use of higher viscosity oil is recommended because of increased oil temperature.

Oil Filter (Fig. 50)

Every 12,000 miles (20,000 km.), unscrew the old filter from the cylinder block and replace it with a new one. Ensure an oil-tight seal by cleaning and smearing the joint faces with oil before screwing the new filter unit tightly home.

Generate Rear Bearing

Inject a few drops of medium viscosity oil through the hole marked "oil".

Fig. 49 (*upper*) **Fig. 50** (*lower*)

Fuel Line Filter (Fig. 51)

Every 12,000 miles (20,000 km.) renew the fuel line filter (shown arrowed Fig. 51) situated underneath the fuel pump in the tank/fuel pump line.

Crankcase Ventilation

Every 12,000 miles (20,000 km.) remove the piping connecting the rocker cover/evaporation canister and carburettor.

Remove the rocker cover and wash it in clean fuel, allow it to dry and re-assemble.

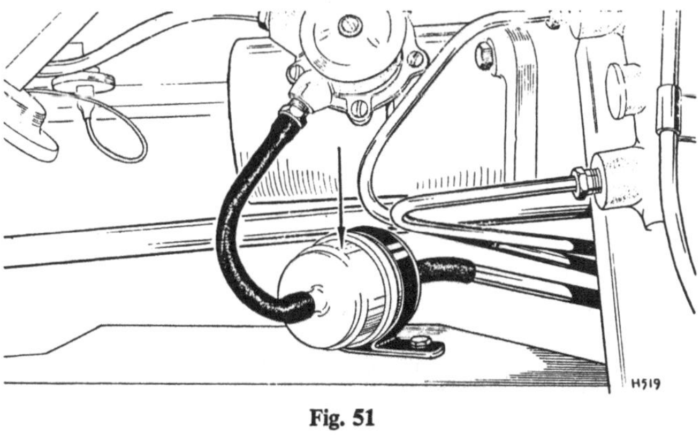

Fig. 51

1,000 MILES—FREE SERVICE

The engine oil sump is initially filled at the factory with a special running-in oil which should be drained after completing the first 1,000 miles (1,600 km.) and refilled with one of the high grade oils recommended. During this period many of the components, including the brakes, fan belt, gaskets, studs and nuts, settle down, thus necessitating slight adjustment and an overall check.

The owner is, therefore, urged at the completion of this period to return the vehicle to the selling dealer who will perform the following operations free-of-charge, except for oil and grease.

ENGINE
Coolant—Check level
Oil sump—Drain and refill
Throttle control linkage and pedal fulcrum—Oil
Mounting bolts—Check tightness
Cylinder head—Check tightness
Manifold—Check tightness
Valves—Adjust rocker clearances
Fan belt—Adjust tension
Oil filter—Check for oil leaks
Fuel pump—Clean filter
Distributor—Lubricate. Adjust points and check ignition timing at idling speed
Carburettor—Top-up damper and adjust engine idling speed using air/fuel ratio or CO meter

TRANSMISSION
Transmission, Overdrive—Check level and top-up
Rear axle—Check level and top-up
Universal joint coupling bolts—Check tightness

STEERING AND SUSPENSION
Front wheel alignment—Check with aid of tracking equipment
Rear wheel alignment—Check by conditions of tyre tread
Steering unit attachments and "U" bolts—Check for tightness
Tie rods and levers—Check for tightness
Lower steering swivels—Oil

BRAKES AND CONTROLS
Parking brake cable and linkage—Lubricate
Hydraulic pipes—Check for leaks, chafing and for hose clearance
Master cylinder—Check level and top-up
Brake shoes and handbrake cable—Adjust as necessary

WHEELS
Wheel nuts—Check tightness
Tyres—Check and adjust pressures

ELECTRICAL EQUIPMENT
Battery—Check and adjust electrolite level. Check charging rate
Generator and starter motor—Check fixing bolts for tightness
Headlights—Check alignment and adjust if required
Lights, heater, windscreen washer, wipers and warning equipment—Check operation

BODY
Door strikers, locks and hinges—Oil and check operation
Body mounting bolts—Check tightness
Door handles, controls and windscreen—Wipe clean
Car—Road test and report any defects

Air Cleaner — Single Carburettor (Fig. 52)

Every 6,000 miles (10,000 km.) remove and clean the paper element (4), more frequently where conditions of extreme dust prevail.

To gain access remove the bolts securing the unit to the carburettor intake flange. Remove two bolts (1), detach the closing plate (5), lift out the element (4) and, using a low pressure air line or a soft brush, clean between the folds of the paper element.

Adopt the reverse procedure to refit, ensuring that the gasket (6) is in good condition and that the slots in the closing plate (5) and the gasket (6) align with those in the carburettor flange.

Every 12,000 miles (20,000 km.) renew the paper element.

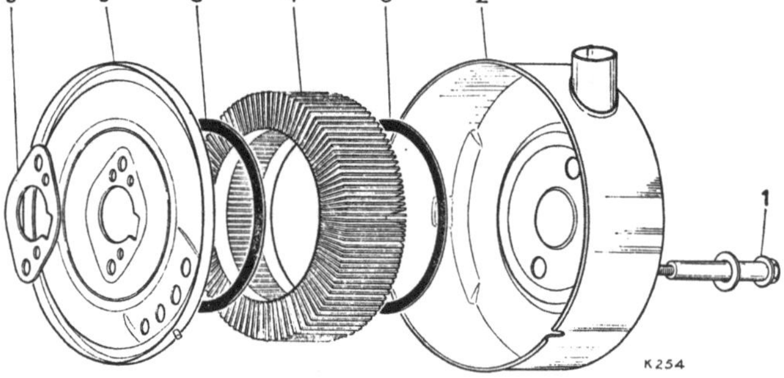

Fig. 52

Air Cleaners — Twin Carburettor (Fig. 52a)
Remove and de-dust the paper elements.

Remove the bolts attaching the assembly to the carburettor flanges, and detach the assembly.

Remove the centre bolt from the body, take off the cover plate and lift out the element. Clean out the body and use a foot pump, low pressure air line, or soft brush to remove foreign matter from between the folds of the paper element.

Reverse the foregoing procedure to assemble the unit, renewing the gaskets as necessary.

If the engine is operating in dusty conditions, clean the filters more frequently.

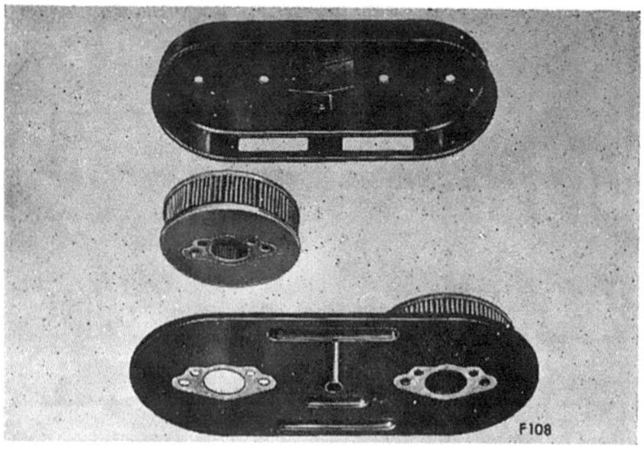

Fig. 52a

Fuel Pump (Fig. 53)

Every 12,000 miles (20,000 km.) clean out the fuel pump.

Access to the fuel pump bowl and filter is gained by unscrewing the bolt (1) and removing the domed cover (2). Lift the filter gauze (4) from its seating and wash it in petrol.

Using a small screwdriver, loosen the sediment in the bowl and blow it clear by using a jet of compressed air. A foot pump used for tyre inflation is ideal for this purpose.

Renew the cork gasket (3) if it has hardened or is broken. Assemble the filter gauze (4) into its seating, taking care to place the gauze face downwards so that it can be removed easily when required.

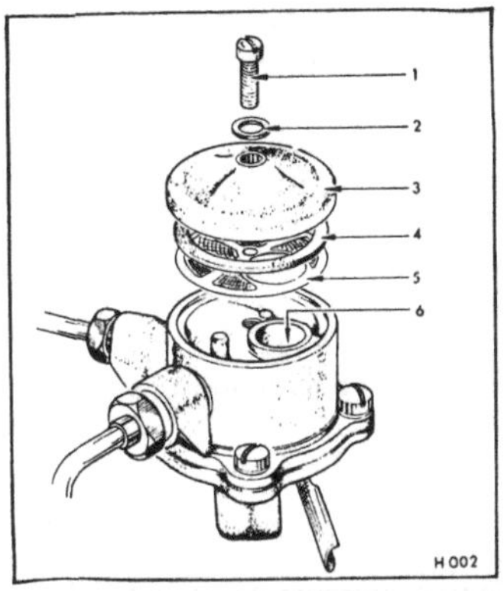

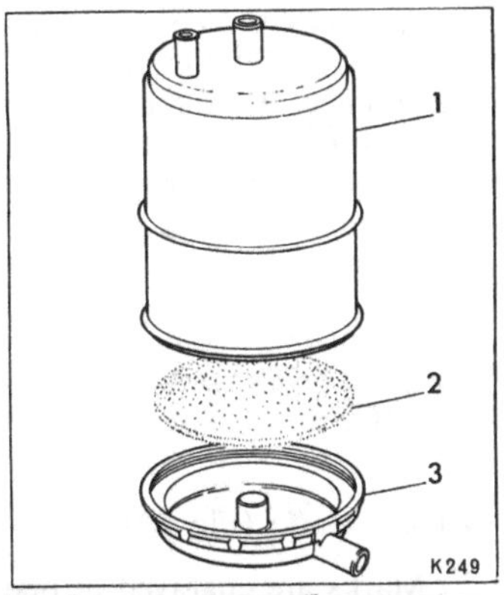

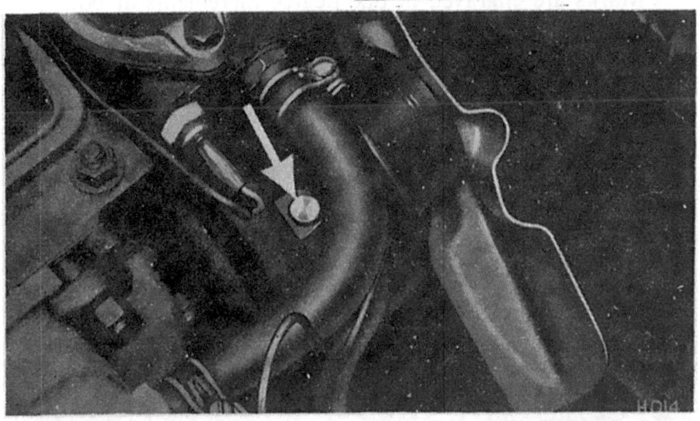

Fig. 53 (*left*) Fig. 54 (*lower*) Fig. 55 (*right*)

Water Pump (Fig. 54)

Every 12,000 miles (20,000 km.) remove the sealing plug from the water pump and replace it by a grease nipple ($\frac{1}{8}$ in. Briggs taper). Apply a grease gun until grease exudes from a pressure release hole in the side of the water pump. Replace the sealing plug.

Evaporation System—Carbon Canister (Fig. 55)

Every 12,000 miles (20,000 km.) replace the filter in the carbon canister as follows:

Remove the inlet and purge tubes from the top of the canister (1). Slacken the clip securing the canister to its mounting bracket and remove the canister.

Unscrew the base cover (3) and remove the filter gauze (2). Clean the base cover, fit a new gauze, replace cover and refit the canister.

Ensure that all piping is not chaffing and free from kinks.

Every 48,000 miles replace the canister adopting the method described above.

Steering Unit (Fig. 56)

Every 12,000 miles (20,000 km.) remove a sealing plug from the top of the steering unit and replace it by a grease nipple ($\frac{1}{8}$ in. B.S.P. Parallel). Apply the grease gun and to avoid damaging the bellows give 5 strokes only. Remove the nipple and refit the plug.

Lower Steering Swivel (Fig. 58)

Every 6,000 miles (10,000 km.) remove the plug (arrowed). Fit a grease nipple and apply a grease gun until oil exudes from the swivel. Remove the nipple and refit the plug.

Fan Belt Adjustment (Fig. 57)

Every 12,000 miles (20,000 km.) slacken the pivot bolt nut (1) and the adjustment bracket bolts (2). Pivot the generator away

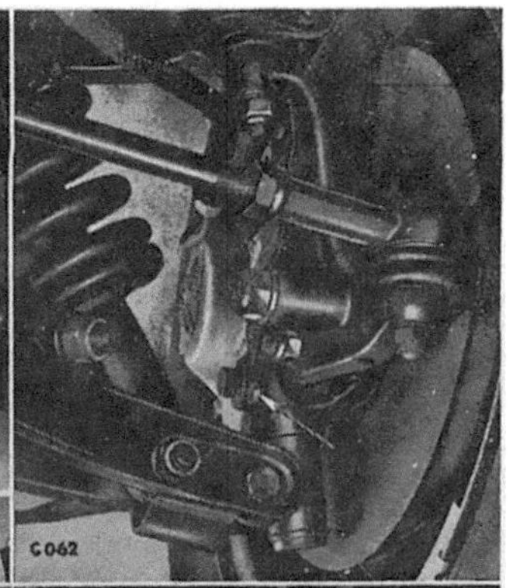

Fig. 56 (*left*) Fig. 57 (*lower*) Fig. 58 (*right*)

from the engine until the belt can be moved $\frac{3}{4}$ in. to 1 in. (19 to 25 mm.) at the mid-point of its longest run. Maintaining the generator in this position, tighten the bolts (2) and nut (1).

Valve Seat Attention

Every 6,000 miles (10,000 km.) have the compression pressures checked by your Triumph dealer. Providing that the engine is functioning satisfactorily, and the compression pressures of all the cylinders are equal, you are advised not to disturb the engine.

The need for decarbonising arises when the build-up of carbon, a product of combustion, becomes excessive. If premium grade fuels and high quality lubricants are used, carbon deposit is so minimised that frequent decarbonising is unnecessary. Carbon removal may, therefore, be restricted to occasions when the cylinder head is removed for attention to the valves and seats.

Valve Clearances—Adjustment (Fig. 59)

Every 12,000 miles (20,000 km.) remove the rocker cover and turning the engine clockwise, check and adjust the valve clearances

Fig. 59

to 0·010 in. (0·25 mm.) if required, in the following sequence while the engine is cold:

Adjust	Nos.	1	and	3	valves	with	Nos.	8	and	6	valves	open.
,,	,,	5	,,	2	,,	,,	,,	4	,,	7	,,	,,
,,	,,	8	,,	6	,,	,,	,,	1	,,	3	,,	,,
,,	,,	4	,,	7	,,	,,	,,	5	,,	2	,,	,,

Cylinder Head Nuts (Fig. 60)

Tighten the cylinder head nuts in the order shown. Slacken them by reversing the sequence.

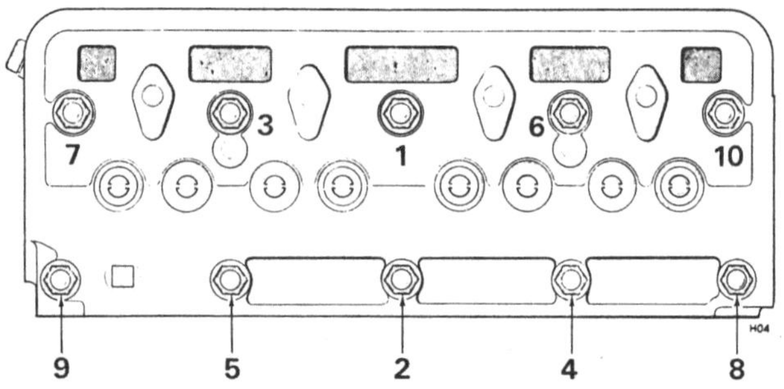

Fig. 60

Ignition Distributor (Figs. 61 and 63)

Every 6,000 miles (10,000 km.) release the clips and remove the distributor cap and rotor arm. Apply a few drops of thin oil to points (4), (5) and (6). Lightly grease the cam surface (7) and inject approximately 5 c.c. (one teaspoonful) of engine oil through the hole (8).

Adjustment

Turn the engine until the contact breaker lever is operating on the highest point of the cam lobe, *i.e.*, gap at its widest.

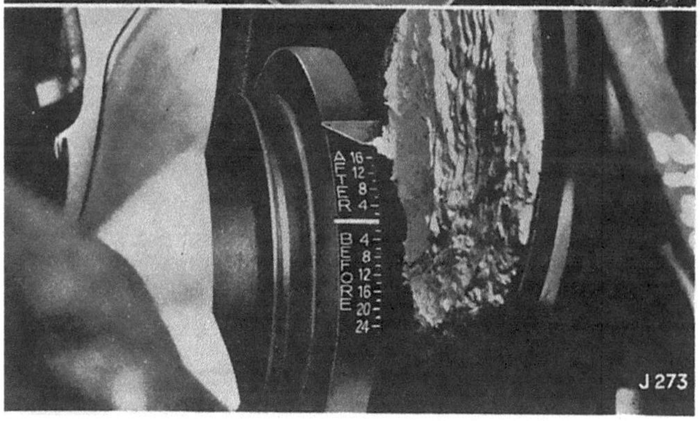

Fig. 61 (*upper*) **Fig. 62** (*lower*)

Slacken the fixed contact screw (2) and turn the eccentric screw (3) to obtain 0·015 in. (0·4 mm.) gap using a feeler gauge between the contacts (1), and retighten screw (2).

Renew worn or damaged points when required.

If the distributor has been removed from the engine, use a static ignition timing of 6° ATDC for starting the engine only. As this setting cannot achieve the extreme accuracy required for the proper functioning of the emission control system, it is vitally important that the final ignition setting is made dynamically to achieve a precise setting of 2° ATDC at an engine idling speed of 800/850 r.p.m.

Timing Marks (Fig. 62)

Marks are engraved in one degree increments at both sides of zero on the periphery of the crankshaft pulley. Numbers one and four pistons are at top dead centre when the zero mark on the pulley is turned to coincide with the pointer on the timing cover.

Plug Lead Position (Fig. 64)

Ensure that the plug leads are attached to the sparking plugs as shown. Firing order is 1, 3, 4, 2, taken in anti-clockwise order, as viewed from the top of the rotor arm.

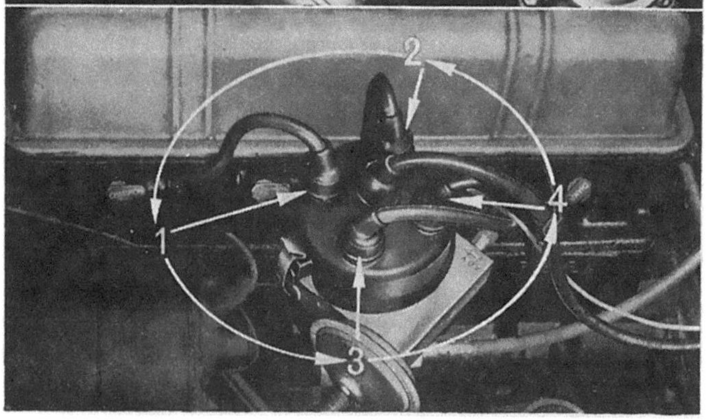

Fig. 63 (*upper*) **Fig. 64** (*lower*)

High Tension Cables

All high tension cables fitted to the ignition system are made from carbon impregnated nylon or cotton cords encased in rubber or neoprene to form a high resistive conductor. Replacement cables should be obtained from a Triumph Dealer and must be the same length and type as the original.

Sparking Plugs

Every 6,000 miles (10,000 km.) remove the sparking plugs for cleaning and reset the gaps to 0·25 in. (0·63 mm.). Clean the ceramic insulators and examine them for cracks or other damage likely to cause "H.T." tracking. Test the plugs and renew those which are suspect.

Every 12,000 miles (20,000 km.) renew all the sparking plugs. Ensure that they are of the correct type (page 59) and that the gaps are set to 0·25 in. (0·63 mm.).

Replace the H.T. leads in the order shown in Fig. 64.

Rear Hubs (Fig. 80)

Every 12,000 miles (20,000 km.) remove the plug and fit a screwed grease nipple. Apply a grease gun until grease exudes from the bearing. Remove the nipple and refit the plug. Repeat, with the opposite rear hub.

Transmission (Fig. 49)

Every 6,000 miles (10,000 km.) with the vehicle standing on level ground, remove the oil filler (2) and top up the gearbox until the oil is level with the bottom of the filler plug threads. Allow surplus oil to drain away before refitting the plug and wiping clean. An oil transfer hole between the gearbox and overdrive unit provides a common oil level. Maintenance of the overdrive unit is thus limited to ensuring that the correct oil level is maintained in the gearbox.

Rear Axle (Fig. 65)

Every 6,000 miles (10,000 km.) remove the oil level plug (shown arrowed) and top up the rear axle until the oil is level with the bottom of the filler plug threads. Allow surplus oil to drain before refitting the plug and wiping clean.

Carburettor Dampers — Single or Twin Carburettors (Fig. 66)

Every 6,000 miles (10,000 km.) unscrew the plug from the top of the carburettor and withdraw the plug and damper assembly. Top-up the damper chambers with engine oil. The oil level is correct when, utilising the damper as a dipstick, its threaded

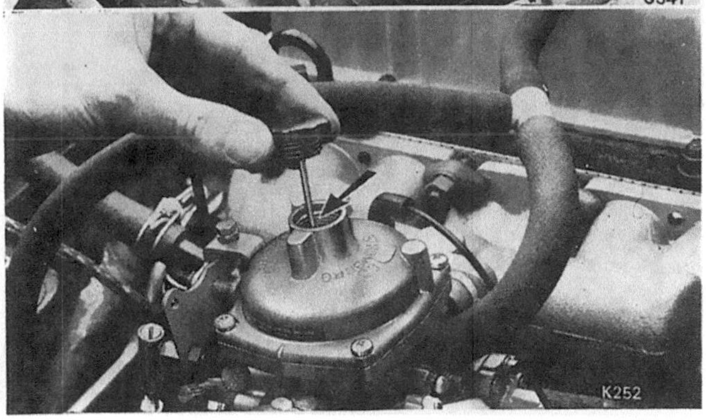

Fig. 65 (*upper*) **Fig. 66** (*lower*)

plug is $\frac{1}{4}$ in. above the dash-pots when resistance is felt. Refit the damper and hexagonal plug. Using an oil can, apply oil to the throttle and choke control linkages.

Front Hubs

Every 12,000 miles (20,000 km.) check and if necessary adjust the front hubs.

At major overhaul periods or every 12,000 miles if the car is being used for competition work, re-pack the front hubs with grease as follows:

1. Jack up the front of the car and remove one front road wheel.
2. Without disturbing the hydraulic pipe unions, unscrew the caliper securing bolts (arrowed Fig. 67) and lift the caliper from the disc, tying it to a convenient point to prevent it hanging by the attached hydraulic pipe. Note the number of shims fitted between the caliper and the vertical link.
3. Referring to Fig. 68, prise off grease cap (11), withdraw the split pin (12) and unscrew the slotted nut (10). Remove the "D" washer and pull the hub assembly from the stub axle.

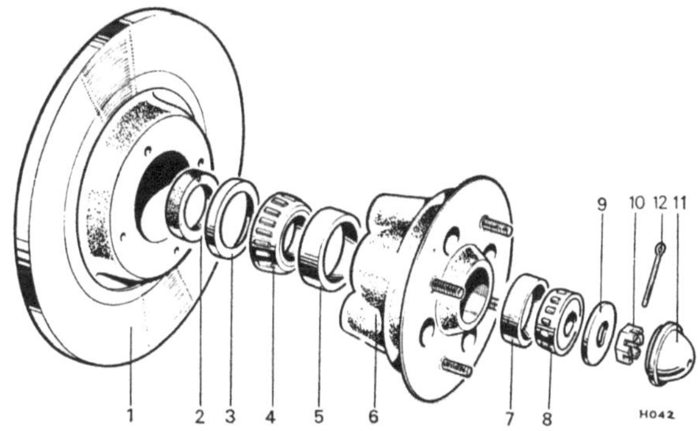

Fig. 67 (*left*) **Fig. 68** (*lower*) **Fig. 69** (*right*)

4. Wash all grease from the hub and bearings. Pack the hub and bearings with grease, working it well into the rollers and re-assemble.
5. Spin the hub and tighten the nut until resistance is felt; then slacken off the nut one half flat and fit a new split pin.
6. Re-assemble the brake caliper unit, refitting any shims removed during dismantling.
7. Refit the road wheel, lower the jack, and repeat the above operations with the opposite wheel hub.

Brakes

The front self-adjusting disc brakes, and rear drum brakes having leading and trailing shoes, are operated hydraulically by the foot pedal. The handbrake is connected by cables and operates the rear wheel brakes only.

Every 6,000 miles (10,000 km.) check and adjust the rear brake shoes as necessary.

Every 12,000 miles (20,000 km.) check the front wheels, jack up the rear of the car and remove both road wheels and brake drums. Examine the brake linings for wear and renew worn or grease contaminated linings (Fig. 72).

Using compressed air, blow all loose dust from the brake mechanism and, with a dry clean cloth, wipe clean the inside of the drums. Avoid touching the braking surface with greasy hands.

Front Brake Pads (Figs. 70 and 71)

Periodically inspect the friction linings of the front disc brake pads for wear and when they are reduced to a thickness of ⅛ in. (3 mm.) renew them as follows:

1. Apply the handbrake, jack up the front of the car and remove the front road wheels.

Fig. 70

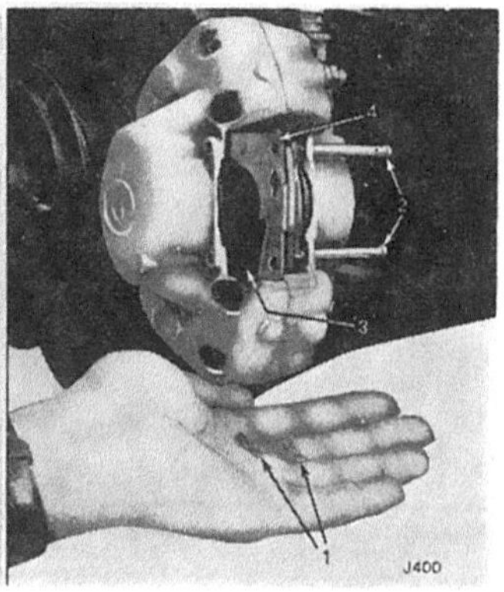

Fig. 71

2. Release the retaining clips (1) and remove the pad retaining pins (2).
3. Lift the friction pads (3) and the anti-squeal plates (4) if fitted, from the caliper.

   IMPORTANT: Do not depress the brake pedal while the pads are removed.

4. Clean the exposed faces of the pistons and the recesses into which the pads fit, then carefully push the pistons back into the calipers.

NOTE: This action will displace fluid back into the master cylinder reservoir. To prevent overflowing, syphon off surplus fluid.

5. Fit the brake pads and anti-squeal plates, ensuring that the arrows on the plates are pointing in the direction of wheel rotation.
6. Insert the pad retaining pins and secure them with the spring clips.
7. Pump the brake pedal several times to adjust the brakes and check the level of fluid in the reservoir.
8. Replace the front wheels and remove the jack.

Adjustment (Fig. 76)

Excessive foot pedal and handbrake travel indicates the need for rear brake adjustment.

Each rear brake is provided with an adjuster on the rear of the backing plate. To adjust the shoes, turn the adjuster (3) clockwise until the shoes are hard against the drum; then slacken the adjuster by one notch increments until the drum is free to rotate.

NOTE: There is a constant drag on the rear wheels caused by the action of the differential and axle oil. Do not confuse this with brake drag.

Handbrake Adjustment

Handbrake "free-travel" is reduced to normal when the rear brake shoes are correctly adjusted.

Handbrake Cable Guides (Figs. 73 and 74)

Apply grease around the cable guides and the compensator sector, arrowed.

HYDRAULIC SYSTEM

A split hydraulic system employs a tandem master cylinder to pressurise independent front and rear brake lines.

A brake line failure in either line will be indicated to the driver by illumination of the brake line failure warning light mounted on the facia panel (12, Fig. 2). The dangerous possibility of continual motoring with a brake line failure, therefore, is prevented.

The warning system is actuated by a pressure differential warning actuator (3, Fig. 47).

The system circuit is amalgamated with the oil pressure indication circuit—see Wiring diagram—so that when the ignition circuits are energised both warning lights will illuminate faintly

Fig. 72

Fig. 73

Fig. 74

to indicate no bulb filament failure.

When the engine is started the oil pressure will raise causing the oil pressure switch to break the earth contact. Both warning lights will extinguish.

Should pressure loss occur in either brake line the pressure differential warning actuator switch will actuate. The "BRAKE" warning light will illuminate brightly.

Should the oil pressure fall below the safe operating pressure while the engine is running the oil pressure switch will actuate. Both the "OIL" and the "BRAKE" warning lights will illuminate faintly.

Bleeding the hydraulic braking system

General

If air has entered either of the hydraulic braking systems then only the system affected need be bled. During bleeding, exercise care, as described in the following procedure, to avoid moving the shuttle from its mid-position. However, if the shuttle has moved during bleeding or subsequent to a fault condition, centralise the shuttle by performing operations 5 and 9 page 23.

Preparation for bleeding

Before commencing to bleed the brakes ensure that all the bleed nipples (Figs. 75 and 76) are clean and, taking care to avoid dirt entering the fluid reservoir, remove its filter cap and top-up with new hydraulic fluid. During the bleeding operation keep the level of fluid above the dividing partition in the reservoir. Do not use fluid bled from the system for topping-up.

Use new fluid from a sealed container, resealing the container after use.

Fig. 75

Fig. 76

Procedure

Commence with the brake, of the pair being bled, farthest from the master cylinder. If both systems are to be bled, bleed the rear brakes first. When bleeding the rear brakes, release the handbrake and turn the brake adjusters to lock the shoes against the drums. When the bleeding is completed adjust the brakes as detailed on page 22.

1. Attach a rubber tube of approx. $\frac{1}{4}''$ (6 mm.) bore to the brake bleed nipple allowing the other end of the tube to hang submerged in a jar containing a quantity of clean brake fluid.

2. Unscrew the bleed-screw enough to allow the fluid to be pumped out (half a turn is normally sufficient).

3. Depress the brake pedal and allow it to return slowly noting that only a LIGHT pedal effort is required and the pedal must NOT be pushed through at the end of the stroke. (In addition, never "try" the pedal until all air has been dispelled and the action is fully bled, as either action will cause the shuttle to move and actuate the clutch.) Passing between each depression of the pedal, continue pumping until all air has been dispelled from the bleed-screw (denoted by the absence of bubbles in the fluid being pumped into the jar).

4. With the pedal depressed, close the bleed-screw nipple and repeat the operation on the other brake.

Procedure for re-centralising the P.D.W.A. Shuttle*

If, for reasons previously described, the P.D.W.A. shuttle requires to be re-centralised, adopt the following procedure.

5. Fit a rubber tube, as described in 1 above, to a brake bleed-screw at the opposite end of the car to that which has just been bled.

6. Open the bleed-screw.

7. Switch the ignition on but DO NOT START THE ENGINE. (The brake warning light will glow but the oil warning light will remain extinguished).

8. Exert a steady pressure on the brake pedal until the brake light dims and the oil light glows. (A click should be felt on the pedal as the shuttle returns to its mid-position.)

9. Tighten the bleed-screw.

NOTE. If the pedal has been pushed too hard the shuttle will move to the other side of the valve, thus requiring the procedure to be repeated on a brake at the opposite end of the car.

*Pressure Differential Warning Actuator

Clutch

The clutch is of hydrostatic operation and does not require adjustment. Bleeding, when necessary, is effected from the slave cylinder.

EMISSION AND EVAPORATIVE CONTROL SYSTEM

Every Triumph "Spitfire", 1970 model, entering the American Market incorporates an efficient Emission and Evaporation control system which is designed and built in accordance with stringent Federal Regulations for 1970 model year cars governing the emissions from the crankcase and exhaust. This vehicle is also in compliance with the Federal Regulations for 1971 model year cars governing the loss by evaporation of fuel from the vehicles fuel delivery system.

The system is designed to work within the following requirements:—

Emission	2.2 grammes per mile Hydrocarbons 23 grammes per mile Carbon monoxide
Evaporation	Maximum fuel loss through evaporation is 6 grammes per test. The test is as defined by the Federal Authorities.

EMISSION CONTROL SYSTEM—SERVICING

The importance of servicing at the correct intervals cannot be overstressed as improvements in design and manufacturing techniques count for nothing if the servicing standards are not upheld.

Routine servicing, carried out at the mileage intervals quoted will rectify any deterioration to the system. In addition to normal lubrication and nut tightness checks, those items which should receive attention during routine servicing include: distributor maintenance (page 20), carburettor dash-pot oil replenishment (page 21) and slow running adjustment, spark plugs (page 20) valve rocker clearances (page 19), air cleaner and crankcase ventilation (page 17) and fuel filter (pages 17 and 18).

Ignition Distributor, refer to pages 20 and 29 for ignition timing.

1,000 Mile Free-service

In addition to the operations listed on page 18, check to ensure that the correct ignition timing is maintained at engine idling speed. Repeat this check every 6,000 miles.

EVAPORATION CONTROL SYSTEM—SERVICING

Minimal servicing is required on the evaporation control system apart from changing the filter gauze, in the carbon canister every 12,000 miles (20,000 km) and replacing the canister every 48,000 miles (80,000 km). See page 19.

CARBURETTOR

The single Stromberg CD.SE 150 emission carburettor is the prime component of the emission system and great care is exercised during the manufacture and initial adjustment of this instrument. Because of the precise manufacturing limits involved and the assembly methods adopted to prevent unauthorised tampering during use, the extent of permissible servicing is restricted to the following:

Adjustments (Figs. 77 and 78)

There are only three adjustments that can be made to emission carburettors in the field and these are:

1. **Idling speed:** Ensure that the fast idle screw (1) (Fig. 77) is clear of the cam (2) and the choke lever is against its stop with the facia control pushed fully in. Unscrew the idling screw (3) until the throttle is just closed. Turn the screw 1½ turns to provide a datum setting.
   Start the engine and attain normal running temperature before final adjustment of the idling screw achieves a constant 800 to 850 r.p.m.
2. **Fast idling:** Ensure that the choke lever is fully returned and the facia control knob pushed in. Set the gap "A", between the fast idle screw (1) and the cam (2), at ·035 in. Start the engine and while it is still cold (68–86F) pull the facia control fully out

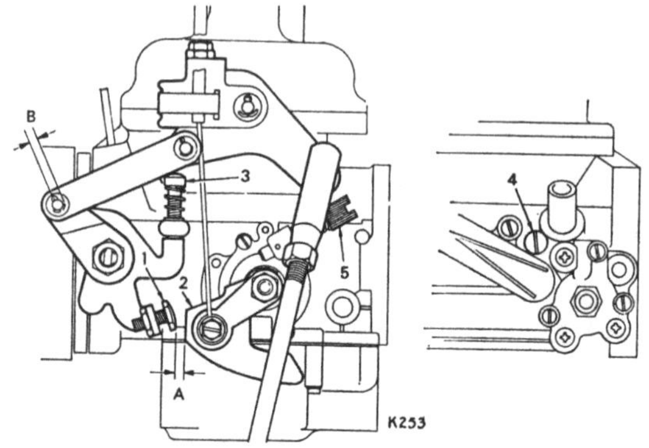

Fig. 77 (*upper*) **Fig. 78** (*lower*)

to check the fast idle speed and, if necessary, adjust to 1,100 r.p.m. with the screw. Tighten the locknut and re-check the fast idling speed.

3. **Idle emission:** An idle trimming screw (4) is provided to give very fine adjustment to compensate for the difference between a new "stiff" engine and one that is "run in". THIS IS NOT AN ORDINARY MIXTURE ADJUSTING SCREW; it

1 Pipe—Rocker cover to carburettor
2 Pipe—Canister purge
3 Activated carbon canister
4 Pipe—Overflow tank to canister
5 Overflow tank
6 Main fuel tank
7 Sealed filler cap
8 Pipe—Filler tube to overflow tank

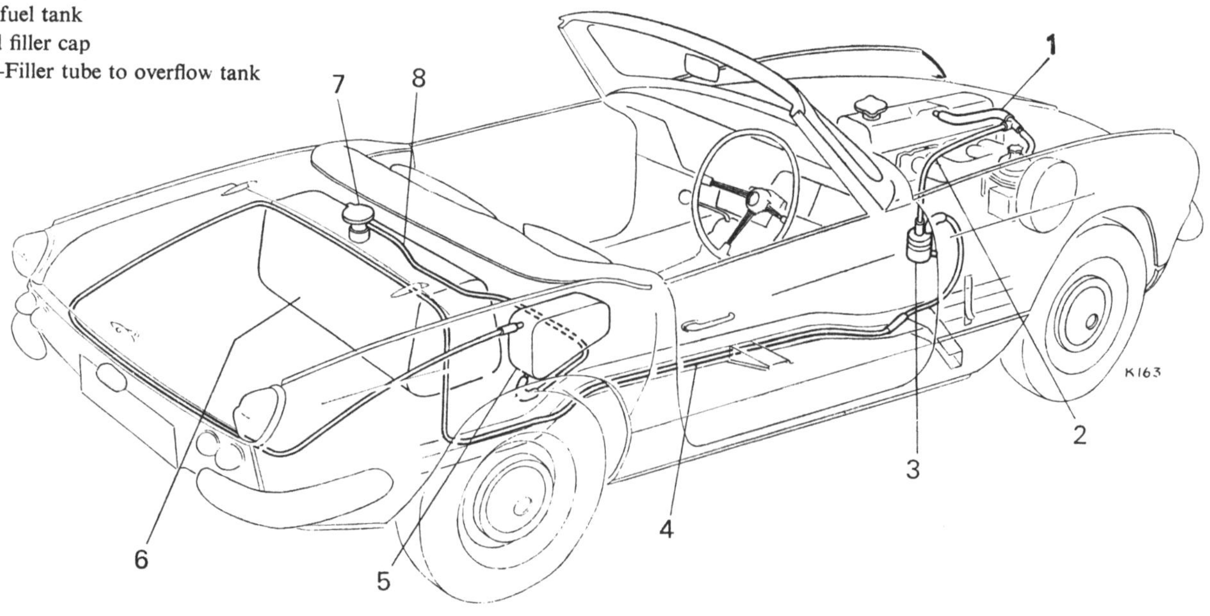

Fig. 79

regulates a limited amount of air that can be introduced into the mixing chamber. It is important to remember that the ear will not detect any difference between the fully "home" and fully "open" position of the screw. The setting should therefore, be checked by means of a C.O. meter or an air/fuel ratio meter to the exhaust pipe. (The correct C.O. level is 1%–3½% and air/fuel ratio is 14·2 : 1–13·2 : 1.

Carburettor Controls

The throttle rod linkage will not require adjustments during normal operation. To ensure complete throttle closure a degree of "lost motion" or slackness is incorporated into the linkage, no attempt must be made to adjust this out "B", Fig. 77.

Occasionally lubricate the linkage and choke cable with thin oil.

Carburettor Servicing Schedules

To maintain the carburettor at peak efficiency, regular servicing at 6,000 miles intervals is essential. This also involves the use of coloured gaskets which give indication that the 24,000 mile service has been carried out. The appropriate servicing operations should be performed by authorised dealers who are trained in the use of the special equipment needed.

RUNNING ADJUSTMENTS

CARBURETTORS — EARLY MODELS

1. Cleaning Float Chambers

Periodically remove the fuel feed pipe, float chamber lid, and float assembly, from each carburettor. Remove all sediment from the float chambers and re-assemble the carburettors.

2. Cleaning Suction Chamber and Piston

At approximate intervals of twelve months, detach the piston unit. Clean the piston and the inside bore of the suction chamber. Re-assemble dry except for a few spots of thin oil on the piston rod.

Replenish the damper reservoir.

3. Float Chamber Fuel Level (Fig. 80)

A practical method of checking the float chamber fuel level may be carried out with the carburettors in situ, as follows:—

1. Remove the dashpot, needle and piston assembly.
2. Operate the fuel pump priming lever to charge the float chamber.
3. Visually check the fuel level in the jet, this should be approximately $\frac{1}{16}$" below the jet bridge.
4. Where the fuel level is incorrect, remove the float chamber lid and bend the prongs of the forked lever towards the lid to raise the fuel level and away from the lid to lower the fuel level. Ensure that the prongs are kept in the same plane.
5. Refit the float chamber and repeat operations (2) and (3).

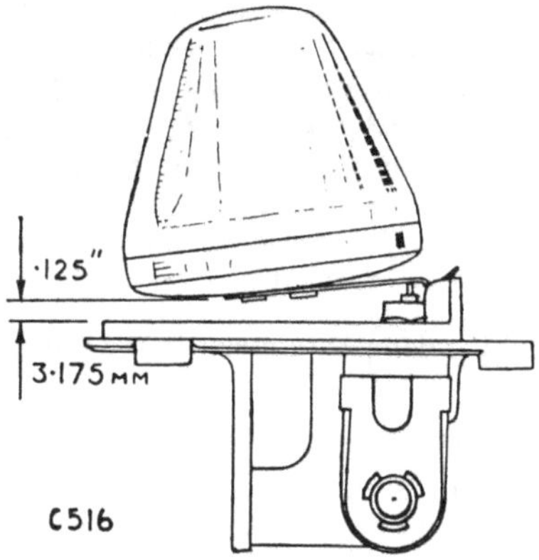

Fig. 80

4. Jet Centralizing (Fig. 81)

The complete jet unit, clamped in position by the gland nut (1) is held in a clearance bore that permits a limited amount of radial float prior to being locked. Therefore, if the gland nut is slackened, the jet assembly can be moved until it is concentric with the needle, thus enabling the piston to move freely to its lower position.

When the suction piston is lifted by hand (air cleaner removed and engine still), it should fall freely and hit the jet bridge with a soft metallic click—that is, with the jet adjusting nut (2) in its topmost position.

If this test is inaudible, but audible when repeated with the jet in the fully lowered position, re-centralize the jet as follows :—

1. Disconnect the rod between the jet lever and jet head (7, Fig. 85).
2. Unscrew the union holding the nylon feed tube into the base of the float chamber and withdraw the tube and the jet together. Unscrew the jet adjusting nut and remove the lock spring. Replace the adjusting nut and screw it to its topmost position, then replace the jet and feed tube.
3. Slacken off the large jet locking screw (1, Fig. 81) until the jet bearing is just free to rotate by finger pressure.
4. With the damper or oil cap removed, and using a pencil on top of the piston rod, gently press the piston and needle down onto the jet bridge.

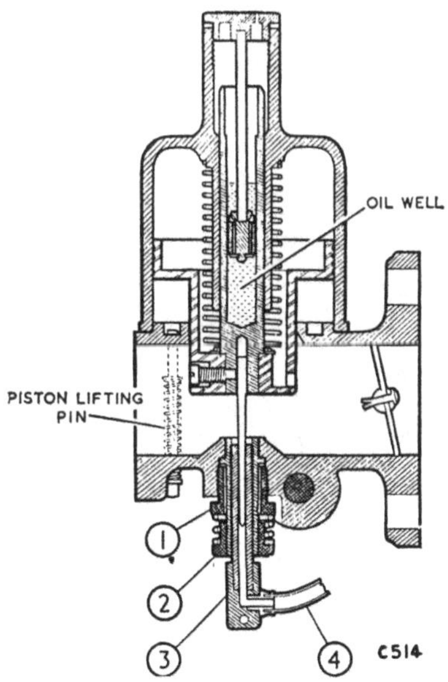

Fig. 81

5. Tighten the jet locking screw, observing that the jet head is still in its correct angular position.
6. Lift the piston and check that it falls freely and evenly, hitting the jet bridge with a soft, metallic click. Then fully lower the jet and re-check to see if there is any difference in the sound of the impact; if there is, and the second test produces a sharper impact sound, the centring operation will have to be repeated until successful.

Refill the damper reservoir in the piston rod.

5. Tuning Carburettors

Twin carburettor installations cannot be successfully tuned unless the general condition of the engine, ignition and the fuel systems is satisfactory.

Remove the air cleaners and run the engine until it reaches normal operating temperature. Slacken the clamping bolts on the throttle spindle connections. Close the throttles fully by unscrewing the idling adjustment screws and then open by screwing down the screws one and a half turns.

Remove the suction chambers and pistons. Screw the jet adjusting nuts until each jet is flush with the bridge of its carburettor, or as near to this as possible. Replace the pistons and suction chamber assemblies and check that the pistons fall freely. Turn down the jet adjusting nuts two complete turns (12 flats).

Restart the engine and adjust the throttle adjusting screws to give the desired idling speed by moving each throttle adjusting screw an equal amount. Using a length of 0·3″ (3 mm.) approx. bore tubing, listen to the hiss in the intakes and adjust the throttle adjusting screws until the intensity of the hiss is similar at both intakes. This will synchronize the throttles.

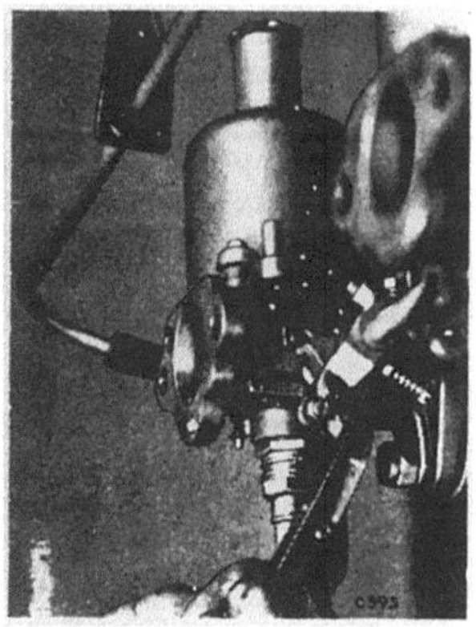

Fig. 73 *(top left)* **Fig. 75** *(above)* **Fig. 74** *(top right)*

When this is satisfactory, adjust the mixture by screwing both jet adjusting nuts up or down by the same amount to give the fastest idling speed consistent with even firing. Press the jets upwards during adjustment to ensure that they are in contact with the adjusting nuts.

Should the engine speed increase as the jets are adjusted, unscrew the throttle adjusting screws a little, each by the same amount, to reduce the speed.

Lift the piston of the front carburettor approximately $\frac{1}{32}$″ (·75 mm.) :

(a) If the engine speed increases, the mixture strength of the front carburettor is too rich ;

(b) If the engine speed immediately decreases, the mixture strength of the front carburettor is too weak ;

(c) If the engine speed momentarily increases very slightly, the mixture strength of the front carburettor is correct.

Repeat the operation at the rear carburettor and, after adjustment, re-check the front carburettor, since the two carburettors are interdependent.

When the mixture is correct the exhaust note should be regular and even. If it is irregular with a splashy type of misfire and with a colourless exhaust, the mixture is too weak. If there is a rhythmical type of misfire in the exhaust beat together with a blackish exhaust the mixture is too rich.

6. Jet and Throttle Interconnection Adjustment (Fig. 85)

With the choke control fully "IN", the engine warm and idling on a closed throttle, adjust the screw (6) to give a clearance of 0·015″ (0·4 mm.) between the end of the screw and rocker lever.

Always check this adjustment when the throttle stop screw (5) is altered.

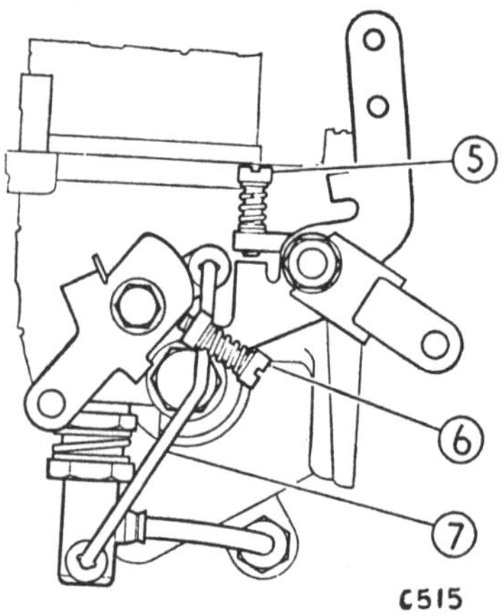

Fig. 85

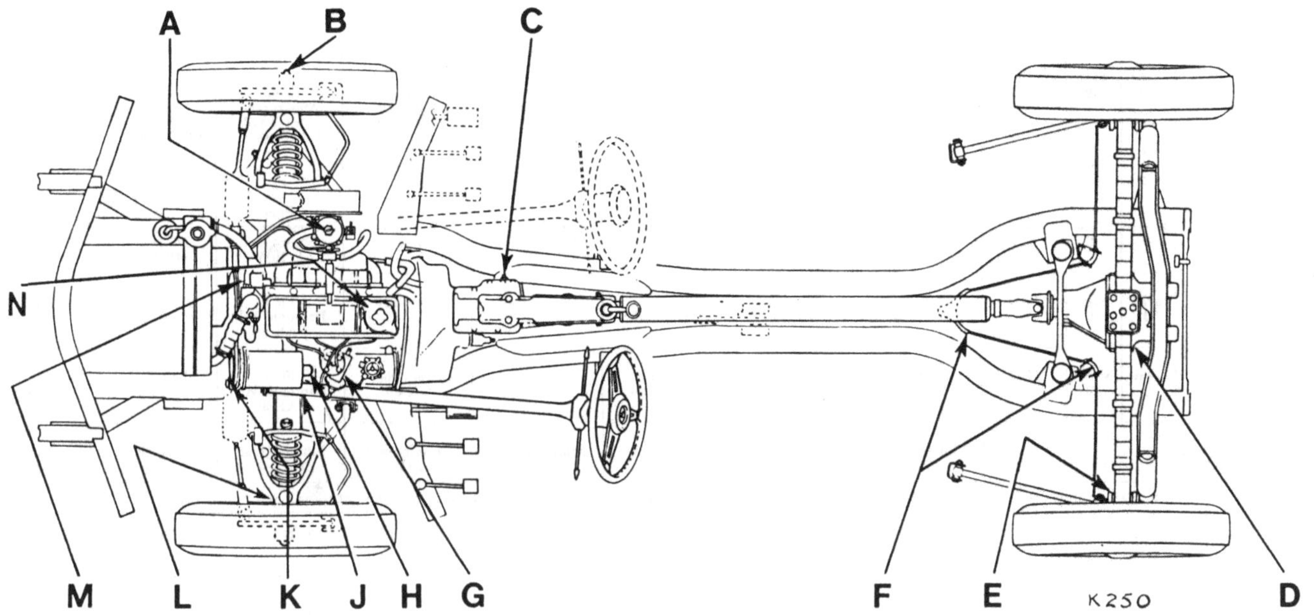

Fig. 80

Chart Ref.	*Items*	*Details*	*Page Ref.*		*Intervals × 100*			
					Miles	*Kms*	*Miles*	*Kms*
A	**Carburettor dampers**	Top up	46	——	6	10		
B	**Front wheel hubs**	Check and adjust	47	——			12	20
C	**Gearbox**	Top up	46	——	6	10		
D	**Rear axle**	Top up	46	——	6	10		
E	**Rear hubs**	Grease	45	——			12	20
F	**Handbrake cable guides**	Grease	49	——			12	20
G	**Distributor**	Oil	44	——	6	10		
H	**Generator rear bearing**	Oil	39	——			12	20
J	**Oil cleaner**	Replace	39	——			12	20
K	**Steering unit**	Grease 5 strokes	42	——			12	20
L	**Lower steering swivels**	Oil	42	——	6	10		
M	**Water pump**	Grease	42	——			12	20
		Top up oil level	36	Daily				
N	**Engine sump**	Drain and refill	39	——	6	10		
		Clean filler cap	39	——	6	10		
	Clutch and brake master cylinders ..	Top up fluid level	37	Monthly				

Renew the seals, cup washers and hoses throughout the braking system, 36,000 miles (60,000 km). **The bonnet catches, all hinges, locks, handbrake and control cables,** should receive occasional attention with the oil can to ensure easy operation.

COMPONENT	AIR TEMP. °C	AIR TEMP. °F	API DESIG-NATION	BP	CASTROL	DUCKHAMS	ESSO	MOBIL	PETROFINA	SHELL	TEXACO CALTEX
ENGINE CARBURETTOR DASHPOTS AND OIL CAN	over 30	over 80	MM or MS	* Super Visco-Static	Castrol XLR or Castrol XL	Q20-50	Esso Extra Motor Oil 20W/40 (Uniflo)	Mobiloil Super or Mobiloil Special 20W/50	Fina Supergrade Motor Oil SAE 20W/50	Shell Super Motor Oil 100	Havoline 20W/40
	0 to 30	30 to 80	MM or MS								
	below 0	below 30	MM or MS		Castrolite	Q5500	Esso Extra Motor Oil 10W/30 (Uniflo)	Mobiloil Super	Fina Supergrade Motor Oil SAE 10W/30	Shell Super Motor Oil 110 or 101	Havoline 10W/30
KING PIN LOWER SWIVELS GEARBOX REAR AXLE O/DRIVE	over 0	over 30	GL4	Gear Oil SAE 90	Castrol Hypoy	Duckhams Hypoid 90	Esso Gear Oil GP 90	Mobilube GX 90	Fina Pontonic MP SAE 90	Shell Spirax 90 EP	Multigear Lubricant EP 90
	below 0	below 30	GL4	Gear Oil SAE 80 EP	Castrol Hypoy 80	Duckhams Hypoid 80	Esso Gear Oil GP 80	Mobilube GX 80	Fina Pontonic MP SAE 80	Shell Spirax 80 EP	Multigear Lubricant EP 80
FRONT HUBS BRAKE CABLES GREASE GUN				Energrease MP	Castrol MP Grease	Duckhams LB 10	Esso Multi-purpose Grease H	Mobilgrease MP	Fina Marson HTL 2	Shell Darina AX	Marfak All Purpose

* Oils marked thus are available in Multigrade forms with viscosity characteristics appropriate to the ambient temperature range in individual markets

CLUTCH & BRAKE RESERVOIRS	CASTROL GIRLING BRAKE & CLUTCH FLUID CRIMSON. WHERE THIS PROPRIETARY BRAND IS NOT AVAILABLE, OTHER FLUIDS WHICH MEET SAE J1703a SPECIFICATION MAY BE USED

APPROVED ANTI-FREEZE SOLUTIONS	Smiths Bluecol	BP Anti-Frost	Castrol Anti-Freeze	Duckhams Anti-Freeze	Esso Anti-Freeze	Mobil Permazone	Fina Termidor	Shell Anti-Freeze	Startex Anti-Freeze

WHERE THESE PROPRIETARY BRANDS ARE NOT AVAILABLE, OTHERS WHICH MEET BSI 3151 OR 3152 SPECIFICATION MAY BE USED

GENERAL SPECIFICATION

Engine

Number of cylinders	4	
Bore of cylinders	2·9 in.	73·7 mm.
Stroke of crankshaft	2·99 in.	76·0 mm.
Cubic capacity	79·2 in.^{3}	1296 c.c.
Piston area	26·5 in.^{2}	171·0 cm.^{2}
Compression ratio	9·0 : 1	
Valve rocker clearances (cold)	0·010 in.	0·25 mm.
Valve timing	Inlet and exhaust valve equally open at T.D.C.	
Maximum power	68 b.h.p. at 5500 r.p.m.	
Maximum torque	880 lb. in. (10·37 kg/m.) at 3000 r.p.m.	
Equivalent to B.M.E.P.	140 p.s.i. (10·1 Kg/cm^2)	
Valves	Stellite faced exhaust valves.	

Lubrication (Engine)

Pump	High capacity eccentric lobe type.
Filter	Replaceable full-flow type.
Pressure at 2000 r.p.m.	40–60 p.s.i. (2·8–4·2 Kg/cm^2)

Cooling System — "No loss" system incorporating plastic overflow reservoir.

Circulation	Pump, driven by "Vee" belt; thermostatically controlled flow.
Fan	Four blade 12·125 in. dia. (308 mm.).

Fuel System — Tank above rear axle. No reserve tap. Separate overflow tank.

Pump	A.C. mechanically operated diaphragm type.
Carburettor (from 1970)	Single 1·50 C.D.S.E. Zenith Stromberg (EMISSION).
(pre-1970)	Twin SU HS2E
Needle size	B5AV. (Zenith Stromberg)
Manifolds	Cast exhaust manifold, cast aluminium inlet manifold water heated.
Air Cleaner	Replaceable paper element.
Crankcase breathing	Positive crankcase ventilation from rocker cover using carburettor depression.
Filter	Replaceable fitted in tank/pipe-line.
Evaporative Emission Control	Sealed fuel filler cap. Emission vented into charcoal canister located in the engine compartment. Canister purged by carburettor depression.

Ignition System

Coil	Lucas HA12.
Distributor—type	Delco-Remy.
contact breaker gap	0·015 in. (0·4 mm.)
rotation—viewed on rotor	Anti-clockwise.
Firing order	1-3-4-2
Sparking plugs—type	Champion UN-12Y.
—gap	0·025 in. (0·64 mm.).
Ignition timing (static)	6° A.T.D.C.
Ignition timing (idle)	2° A.T.D.C.

Electrical System

Voltage	12
Polarity	Negative earth
Fuses	35 amp.
Battery—type	Lucas D9
capacity at 20 hr. rate	40 amp. hour
plates per cell	9
normal charging rate	3·5 amps.
Generator—type	Lucas C40-1.
—output	22 amp.
Control box	Lucas RB340.
Flasher unit	Lucas 8FL 3·6A
Fuel and temperature indication	Smiths bi-metal type—10 volt system.
Oil pressure switch operating pressure	3/5 p.s.i. (0·2/0·35 kg/cm²).

Brake System — Girling hydraulic. Centrally mounted handbrake coupled mechanically to rear wheels only.
Pedal operates on tandem master cylinder to front and rear brakes independently. Pressure differential valve and warning light fitted to indicate partial system failure.

Front	Caliper disc 9 in. dia. (229 mm.).	
Rear	Drum 7 in. dia (178 mm.) × 1¼ in. (31.7 mm.) wide.	
Front lining area	14·8 in.²	95 cm²
Front swept area	150 in.²	962 cm²
Rear lining area	34 in.²	226 cm²
Rear swept area	55 in.²	353 cm²
Total lining area	48·8 in.²	314 cm²
Total swept area	205 in.²	1316 cm²
Maximum retardation	0·98 g. equivalent to stopping from 30 m.p.h. (48 k.p.h.) in 31ft. 9·45 m.).	

Chassis Data

Frame	Double backbone of closed channel section with channel outriggers.	
Wheelbase	6 ft. 11 in.	2110 mm.
Track—Front	4 ft. 1 in.	1245 mm.
—Rear	4 ft.	1220 mm.
Ground clearance	5 in.	125 mm.

Weight (approx.)

Dry (excluding extra equipment)	14 cwt.	712 Kg.
Complete (inc. fuel, oil, water and tools)	14·75 cwt.	748 Kg.
Maximum gross vehicle weight	18·25 cwt.	905 Kg.

Transmission

Clutch	Diaphragm type. 6½ in. dia. (165 mm.)			
Gearbox	4 speed constant mesh. Synchromesh on 2nd, 3rd and Top.			
	Top	3rd	2nd	1st & Rev.
Ratios	1	1·39	2·16	3·75
Overall ratios	4·11	5·73	8·87	15·4
Overdrive (optional)	Laycock de Normanville. Operates on 3rd and Top gears. Ratio 0·801 : 1			
Propellor Shaft	All metal. Needle roller bearings.			
Rear axle	Swing axle. Centre rubber mounted to chassis. Hypoid bevel drive. Ratio 4·11 to 1.			
Wheels	Pressed steel disc type 4½ J rim.			
Tyres	5·20S-13 Tubeless. 1·45SR-13 Radial ply (optional).			

RECOMMENDED TYRE PRESSURES

Tyre Size and Type	FRONT		REAR	
	Psi	Kg/cm²	Psi	Kg/cm²
5·20S—13	18	1·26	24	1·69
1·45SR-13 Radial ply ..	21	1·47	26	1·83

Steering

Turning circle	24 ft.	7·3 m.
Steering unit	Rack and pinion 3¾ turns lock to lock. Energy absorbing non-adjustable steering control system.	

Suspension

Front	Low periodicity independent suspension system with rubber bushed wishbone pivots. Patented screwed bottom bush and top ball joint swivels. Coil springs controlled by telescopic type direct acting dampers and anti-roll bar. Taper roller bearings in hub.
Rear	Swing axle type independent suspension, transverse leaf spring and radius rods. Ball and needle roller bearings in hubs.

Exterior Dimensions

Overall length	12 ft. 3·625 in.	374·5 cm.
Width	4 ft. 9 in.	145 cm.
Height with hood erected (unladen)	3 ft. 11·5 in.	120·5 cm.
Height with hood folded (unladen)	3 ft. 8·25 in.	112·5 cm.

Road speed at 2500 ft./min. (762 m/min.) piston speed r.p.m. top gear 81 m.p.h. 130 k.p.h.

Capacities	U.S.A.	Imperial	Metric
Fuel tank	9·9 galls.	8·25 galls.	37·6 litres
Engine sump	9·6 pints	8 pints	4·5 litres
Gearbox	1·8 pints	1·5 pints	0·85 litres

Road Speed Data

Engine speed at a road speed of	O/D Top	Top	O/D 3rd	3rd	2nd	1st & Rev.
10 m.p.h.	565	635	715	890	1375	2385
10 k.p.h.	350	395	490	550	875	1490

Road speed at 1000 r.p.m.	O/D Top gear	Top gear
	19·6 m.p.h.	15·75 m.p.h.
	31·3 k.p.h.	25·2 k.p.h.

Gearbox and Overdrive	2·85 pints	2·38 pints	1·35 litres
Rear axle	1·2 pints	1 pint	0·57 litres
Cooling system (inc. water bottle)	8·4 pints	7 pints	4·0 litres
Cooling system (inc water bottle) with heater	9·6 pints	8 pints	4·5 litres

CARE OF BODYWORK

Washing

Wash the vehicle frequently with plenty of running water and a clean sponge. Soften, and if possible remove, mud with water before using the sponge. To prevent scratches, do not use a dry cloth to remove dust from the body. When all dirt is removed, dry the vehicle with a clean damp chamois leather.

Chromium Plating

Maintain the attractive appearance of chromium plated parts by frequent washing and thorough drying, particularly in winter when contamination with road salts is likely. Occasionally apply a wax polish.

Removing Grease and Tar

Use de-natured alcohol (methylated spirits) to remove grease and tar. White spirit is also effective, but it must not be applied to rubber.

Glass

To avoid scratching, clean the windows with a damp chamois leather, kept specially for the purpose. Protect glass from contamination by any silicone preparations used in the vicinity. Silicone polishes are extremely difficult to remove and cause the windscreen wipers to smear.

Polishing

After a period of use, the formation of traffic film will cause the paintwork to lose some of its lustre, even though the vehicle may have been carefully and regularly washed. The original brilliance may be restored after washing by using a reputable non-abrasive cleaner and polish.

Wax preparations are recommended for their durability, but if used regularly the old wax must first be removed with a cleaner before further application of new wax. The frequency at which polishing is necessary depends upon local conditions of air pollution.

Interior Care

Brush and clean the inside of your car each time you wash and polish the outside of it. Use a vacuum cleaner where possible and ensure complete removal of all dust from the interior and trim.

Wash the upholstery (and exterior fabric) with luke-warm, non-caustic soapy water. Do not use detergents or household cleaners as these may cause damage. Remove all traces of suds with a clean damp cloth and thoroughly dry the upholstery with a dry duster or towel.

Wipe the facia and instrument panel with a damp cloth only. Wax or other polishes should not be used inside the car.

TRIUMPH SPITFIRE
MK IV & 1500
DRIVER'S HANDBOOK

PART 2

Issued by British Leyland Motor Corporation

FOREWORD

This Handbook provides an introduction to your car together with information on the care and periodic maintenance required to combine trouble-free motoring with minimal running costs.

References to right- or left-hand in this Handbook are made when viewing the car from the rear. In all communications relating to Service or Spares, please quote the Commission Number (Chassis Number), paint and trim numbers. These are stamped on a plate attached to the body adjacent to the left-hand door striker plate. The Commission Number is also stamped on a plate attached to the left-hand windscreen pillar and visible through the screen.

Specification details—the Manufacturers reserve the right to vary their specifications with or without notice, and at such times and in such manner as they think fit. Major as well as minor changes may be involved in accordance with the Manufacturer's policy of constant product improvement.

Whilst every effort is made to ensure the accuracy of the particulars contained in this Handbook, neither the Manufacturer nor the Distributor or Dealer, by whom this Handbook is supplied, shall in any circumstances be held liable for any inaccuracy or the consequence thereof.

Triumph Distributors and Dealers are provided with the latest information concerning special service tools and workshop techniques, enabling them to undertake your service and repairs in the most efficient and economic manner.

Maintenance—The booklet "Passport to Service" is issued with every new car, inside are the list of maintenance schedules and the facility to have a recorded service history. By servicing as recommended and having the booklet stamped, the car will not only be kept in good running order but will have an enhanced re-sale value.

Genuine STANPART and UNIPART service parts are designed and tested for your vehicle and have the full backing of the Triumph Factory Warranty. ONLY WHEN GENUINE STANPART OR UNIPART SERVICE PARTS ARE USED CAN RESPONSIBILITY BE ACCEPTED.

Genuine parts are supplied in cartons bearing either of these symbols:

D1306

L573

List of Sections

[NOTE: Page references in this section refer to the page numbers in the original source. These numbers appear at the bottom of each source-page.]

CONTROLS, INSTRUMENTS AND INDICATORS

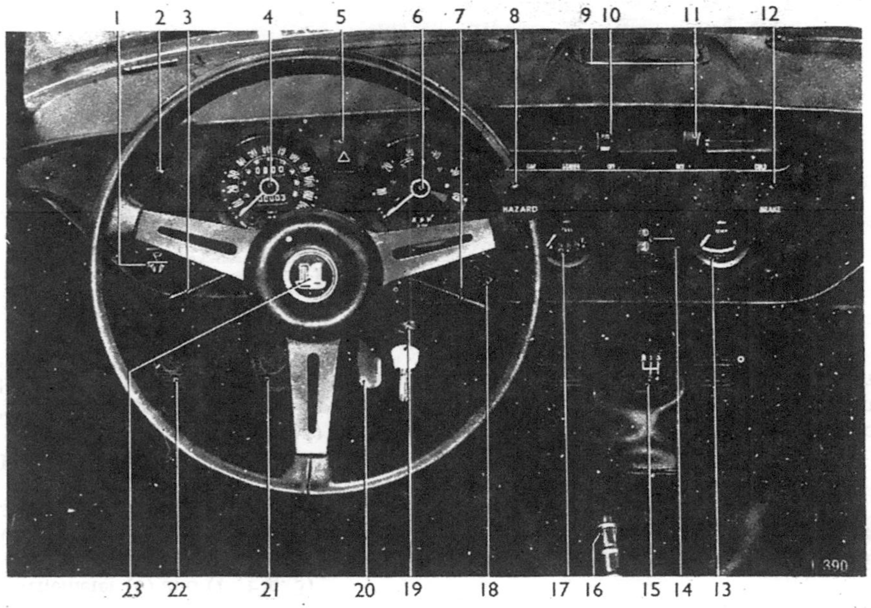

Fig. 2

Key to Fig. 2 (Facia)

1 Windscreen wiper and washer control
2 Turn signal monitor
3 Turn signal control
4 Speedometer
5 Hazard warning switch
6 Tachometer
7 Headlamps dip and flasher switch
8 Hazard warning monitor
9 Ash tray
10 Air distribution control
11 Heater control
12 Brake-line failure indicator
13 Temperature gauge
14 Lighting switch
15 Gear selector lever
16 Handbrake lever
17 Fuel gauge
18 Choke control
19 Ignition and steering lock switch
20 Accelerator pedal
21 Brake pedal
22 Clutch pedal
23 Horn push button

The controls, instruments and indicators shown on Fig. 2 and described in the following pages are positioned within easy reach of the driver to afford maximum ease of operation with minimum distraction. The bracketed figures in the text cross-refer with the key on page 5.

Windscreen Wiper and Washer Control (1)
A dual purpose control operates the two-speed wiper motor and screen washer jets. From the "off" position (full anticlockwise), turn the control clockwise to the next position for slow wiping, or fully clockwise for fast wiping. The wipers park automatically at the base of the windscreen when the control is returned to the "off" position. The wipers operate only when the ignition is switched on.

Clear mud from the screen by repeatedly pushing the control to spray clean fluid as the wiper blades operate. Refer to windscreen washer (page 36).

Turn Signal Monitor (2)
A green monitor at the left-hand side of the facia glows intermittently when the ignition is switched on and the turn signal control operated. Should the monitor fail to operate, this indicates a broken filament in one of the turn signal lamp bulbs or a fault in the monitor bulb.

Turn Signal Control (3)
The turn signal lamps are controlled by moving the lever (3) to either of the arrowed positions. Operation of the turn signal lamps is indicated by the intermittent flashing of a green indicator monitor light (2). Operation of the signal lamps and the signal monitor is cancelled when the steering wheel is returned to the straight ahead position.

Speedometer (4) (See Fig. 3)
Calibrated in miles and kilometres per hour, the speedometer also combines the following indicators and warning devices:

Trip Indicator (A)
When initially set at zero, figures within the aperture above the centre of the dial will record the distance of a particular journey. To zero the trip indicator, turn the knob which is located in the driver's line of view above the ignition/steering lock switch.

Mileage Indicator (B)
Figures within the aperture below the centre of the dial record the total mileage of the vehicle.

Main Beam Warning Light (C)
An indicator at the bottom left-hand side of the dial glows blue when the headlamp main beams are selected and is extinguished when the headlamps are "dipped".

Oil Pressure Warning Light (D)
An indicator at the bottom centre of the dial glows green when the ignition is switched on and is extinguished as the engine speed increases. Should the indicator continue to glow when driving away, stop the engine and check the sump oil level. If this is satisfactory, have the lubrication system checked immediately

Ignition Warning Light (E)
An indicator at the bottom right-hand side glows red when the ignition is switched on and is extinguished as the engine speed increases. Should the indicator continue to glow when driving, this indicates a fault in the charging system which should be rectified without delay.

Hazard Warning Switch (5)
If the vehicle is immobilised and constitutes a hazard to other vehicles, warning may be given by using the "hazard warning system". To operate, depress the lower portion of the switch when all turn-signal lamps and the hazard warning monitor (8) will flash intermittently.

Tachometer (6) and (F. Fig. 3)
The tachometer indicates engine speed in revolutions per minute and is calibrated in increments of 10 × 100, extending to 6,000 r.p.m. The speed range within the coloured segment G is subject to special precautions. These are given on page 24.

Headlamp Dip and Flasher Switch (7)
With the "light switch" (14) at position 3, move the switch lever "down" for "dipped beams" or "up" for "main beams". See main beam warning (4c). "Pulling" the lever towards the steering wheel rim will "flash" the main beams, regardless of the "light switch" position.

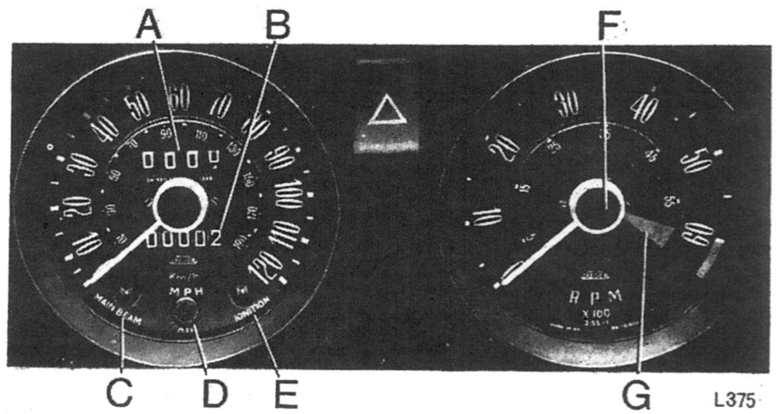

Fig. 3

Hazard Warning Monitor (8)
Refer to "hazard warning switch" (5).

Ash Tray (9)
Raise the flap for access. Lift out the complete unit to clean.

Air Distribution (10) See Page 20

Heater Control (11) See Page 20

Brake-line Failure Indicator (12)
When the ignition is switched on, the "brake-line failure" and "low oil pressure" indicators (4) glow faintly; these are extinguished when the engine is running. Should failure of the front or rear brake lines occur, the indicator (12) will glow brightly.
A broken bulb filament is indicated by the warning light failing to glow when the ignition is switched on, before starting the engine.

Temperature Gauge (13)
When the ignition is switched "on", the pointer moves slowly across the dial, taking up to one minute to reach a steady reading.
Normal operating temperature is reached when the pointer registers in the central sector of the dial. Should the pointer reach the highest mark, stop the engine immediately, allow it to cool and check the level of coolant in the radiator. Refer to page 22.

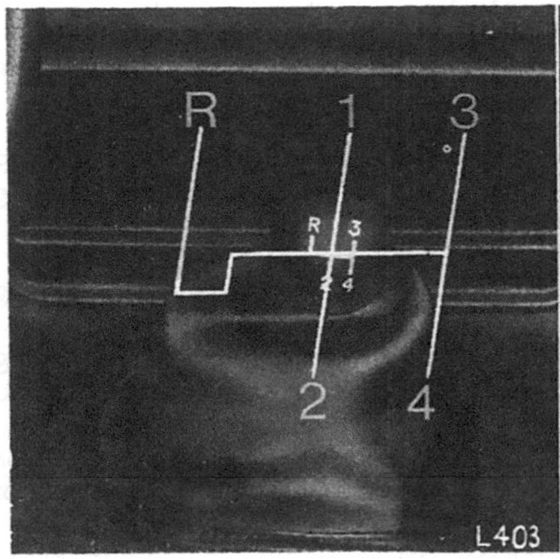

Fig. 4

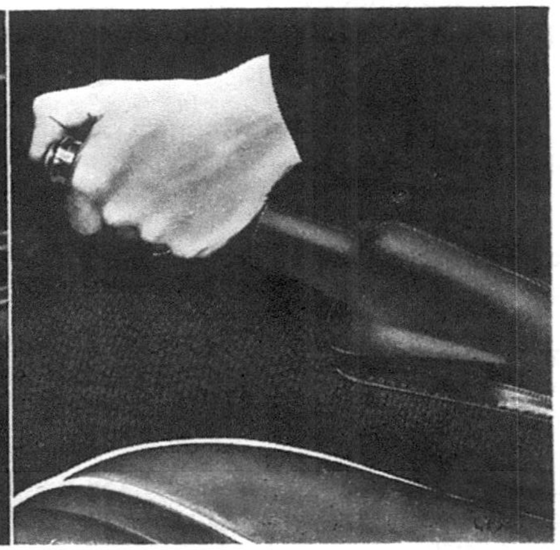

Fig. 5

Light Switch (14)

To operate the three-position light switch:
1. Fully press the top of the switch bar to extinguish all external driving lamps.
2. Move the switch bar to its level position for illuminating the front and rear parking lamps and licence plate.
3. Fully press the bottom of the switch bar to illuminate all external driving lamps.

Gear Selector Lever (15) (See Fig. 4)

Moving the gear selector lever from "neutral", the gear positions are selected as follows:

1st	Move lever left and forward.
2nd synchromesh ..	Move lever left and rearward.
3rd synchromesh ..	Move lever right and forward.
4th (top) synchromesh	Move lever right and rearward.
Reverse	Press the lever downwards, move it to the extreme left and forward. Engage reverse gear only when the vehicle is stationary.

Hand brake lever (16) (See Fig. 5)

To apply the rear wheel brakes, pull the control lever upwards, where it is retained by a rachet and pawl. Release the hand brake by pulling the lever slightly upwards before depressing the button to free the pawl; then allow the lever to move downwards to the "off" position.

Fuel Gauge (17)

The fuel gauge indicates the contents of the fuel tank. When the ignition is switched "on" the needle moves slowly across the scale, taking up to one minute to reach a steady reading which is maintained regardless of vehicle movement, until the ignition is switched "off".

Choke Control (18)

The choke control is used to enrich the petrol-air mixture for easy starting from cold. Pull the control half out for normal cold starting and fully out for extremely cold conditions. Since all engines adopt their own characteristics, experience will determine the best position. Push the control fully home as soon as the engine will run without choke.

Ignition, Starter and Steering Lock Switch (19)

The combined ignition/starter/steering lock switch is operated by a special key.

Incorporated in the switch is an "Audible Alarm System". The alarm system is fitted to encourage the driver to remove the ignition key from the lock before leaving the vehicle.

There are four switch positions which need to be understood for the correct operation of the combined switch (see Fig. 6).

Symbol 0	"OFF" only in this position may the key be inserted or withdrawn (see Audible alarm system).
Symbol I	"Auxiliary", in this position the ignition circuit is isolated to allow the use of a radio when the vehicle is stationary and the ignition is switched off.
Symbol II	"Ignition", all ignition controlled circuits energised.
Symbol III	Start. This key position actuates the starter motor and when released the key returns under spring pressure to Position II.

Audible Alarm System

The Audible Alarm System only functions when the ignition key is positioned in the switch and the driver's door is open. The "warning" denoted by a continuous buzzing sound will terminate when the driver's door is closed or the ignition key is completely removed.

To Disengage Steering Lock and Start Engine

Insert the key and turn in clockwise direction. If difficulty is experienced in turning the key, this can be rectified by simultaneous movement of the steering wheel.

Turn the key clockwise to symbol II (Ignition) the ignition will be switched on.

To start the engine, the key should be turned a little more against spring pressure to symbol III ("start"), as soon as the engine fires release the key which will return automatically under spring pressure to the ignition position (symbol II).

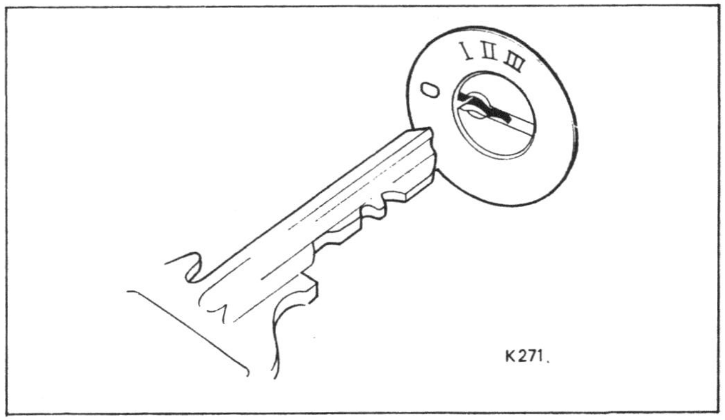

Fig. 6

To Stop Engine and Engage Steering Lock

Turn the key in an anti-clockwise direction from the "ignition" position (symbol II) to the "lock" position (symbol 0). This action stops the engine.

Removal of the key in this position automatically actuates the steering lock mechanism. (See Audible Alarm System.) Slight movement of the steering wheel may be necessary to ensure the lock bolt is fully engaged.

WARNING: If, for any reason, the (ignition) engine is switched off while the car is in motion DO NOT ATTEMPT TO TURN THE KEY INTO THE "LOCK" POSITION O, as this constitutes part of the locking sequence.

Accelerator, Brake and Clutch Pedals (20, 21 and 22)

These are conventional items which should require no further explanation.

Horns (23)

The horns are sounded by pressing the button in the centre of the steering wheel.

Reversing Lamps

Reversing lamps are illuminated by selecting reverse gear, when the ignition is switch on.

Seat Belt Warning Device

On models fitted with a seat belt warning device a lamp lights on the facia and a buzzer sounds if an attempt is made to drive the car without the seat belts in use.

SEATS AND SAFETY HARNESS

Access to the area behind the seats may be gained by lifting lever (1) (Fig. 7) to release a self locking catch, thus enabling the seats to be tilted forward.

Seat Adjustment (Fig. 7)
Fore and aft sliding adjustment is obtained by moving outwards and holding against spring pressure a lever (2) thus permitting the seat to be positioned as required. Releasing the lever allows a locking pawl to engage in the nearest locking notch. To achieve this however, slight additional movement of the seat may be necessary.

To adjust the seat squab for rake, raise and hold the lever (3) whilst using controlled body pressure, move the squab to the desired position. Releasing the lever will lock the squab at the chosen angle. The squab, if unrestrained, will return to the vertical position when the locking lever is raised.

Safety Harness (Fig. 9)
Three anchorage points for each seat are provided for attachment of the safety harness. The lower anchorage (1) is situated on the floor adjacent to the sill to the rear of the seat. The upper anchorage (2) is located on the rear inner wheel arch. The third point (3) incorporates the buckle unit (Fig. 8) and is fixed to the transmission tunnel between the two seats.

Using the Harness
To fasten: Pass the section of the belt whose end is attached to the upper anchorage point over the left shoulder (for driver) and right shoulder (for passenger) and diagonally across the chest. The lower part of the belt pass over the lap. Push the belt tongue into the buckle unit nearest to the wearer until a positive "click" ensures the harness is safely locked.

NOTE: The seat belt warning device will operate if an attempt is made to drive the car without the seat belts in use.

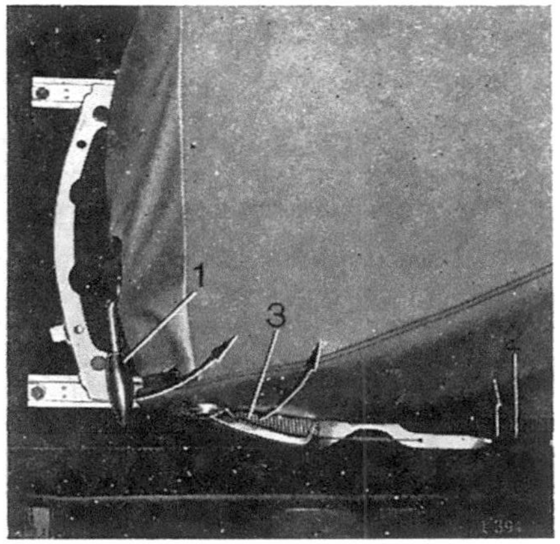

Fig. 7

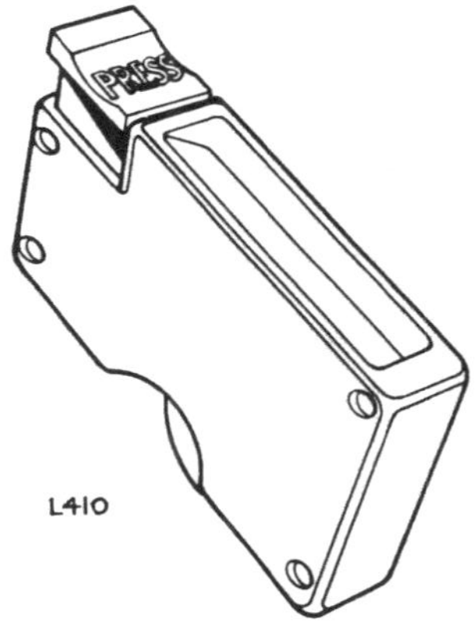

Fig. 8

To release: Press the button on the buckle unit (Fig. 8) nearest the wearer and remove the belt tongue. When the belts are not in use make sure they are stowed on the hook (4) (Fig. 9) so that they cannot become damaged.

Adjustment. The harness should be so adjusted that the hand will just pass between the belt and chest.

To lengthen: Lift the adjuster (5) on the lap part of the harness and lean into the harness until the correct adjustment is obtained.

To shorten: Pull the loose end of the belt through the adjuster.

Cleaning
An occasional wipe with a warm soapy sponge will keep the belts clean. Do not use a bleach or dye as they may affect the efficiency of the material.

IMPORTANT: The belts must not be allowed to rub against sharp surfaces on seats or bodywork. Harness that has been cut, frayed, damaged or used in cars involved in severe accidents should be renewed.

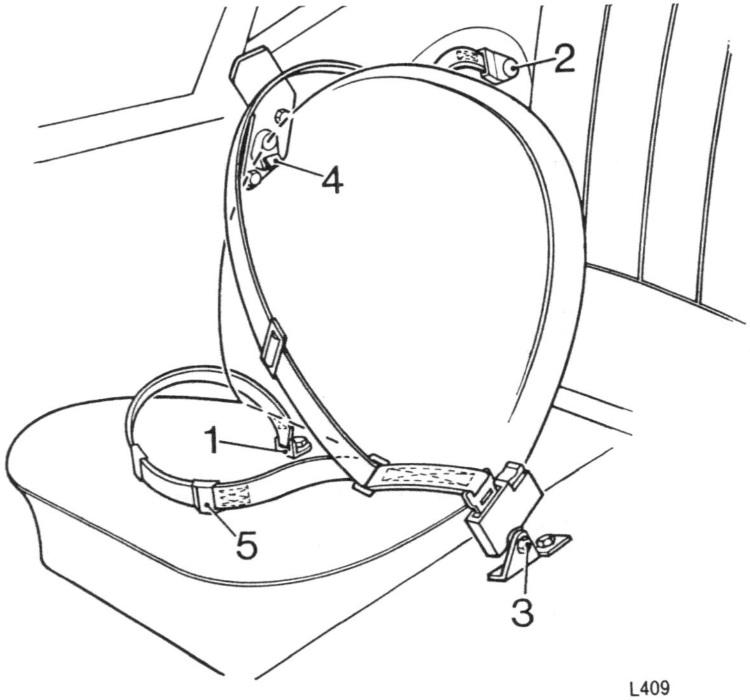

Fig. 9

LOCKS AND KEYS

The car is provided with two sets of keys, comprising:

2. Ignition and steering column lock keys.
2. Door lock keys.
2. Luggage compartment lock keys.

CAUTION: The ignition and steering column lock key is supplied complete with an identification disc. This must not be mislaid since it provides the only record of the key. In the event of loss of keys the disc will have to be submitted to your selling dealer to obtain duplicates.

Door Locks (Figs. 10 and 11)

Both doors have anti-burst locks and are opened from the outside and inside by pulling remote control levers.

To lock the doors from the outside, close the door and insert the appropriate key and turn it a quarter turn towards the front of the car. Turning the key a quarter turn towards the rear of the car unlocks the door. To open the door pull the recessed lever (Fig. 10).

Both doors may be locked from the inside by moving button (A) (Fig. 11) forward. This condition is cancelled by pulling lever (B) to open the door or when the door is unlocked from the outside with the key.

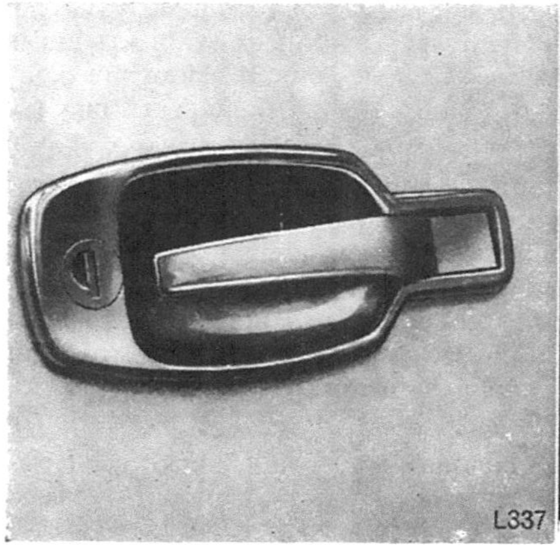

Fig. 10

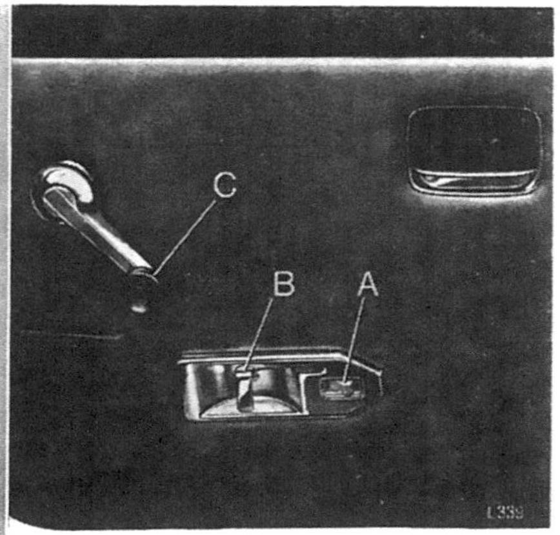

Fig. 11

Luggage Compartment Lock (Fig. 12)

To open the luggage compartment lid, pivot the lock cover plate (A) away as shown to expose the lock, insert the key and turn it clockwise to release the lock. Twist the turn button anti-clockwise and lift the lid to its limit before lowering it on to the telescopic support.

Close the lid by raising it slightly to release the catch in the support, lower the lid and when almost closed push down gently on the lid to engage the catch. Turn the key anti-clockwise to lock remove the key and pivot the cover plate over the lock to prevent ingress of dirt and moisture.

Bonnet Catch (Fig. 13)

The bonnet is opened by lifting a lever on each side to release the catches. Raise the bonnet from the rear until it is fully open and held by a folding strut. To close, pull the centre folding part of the strut while simultaneously supporting and lowering the bonnet. To fasten the catches press each lever flush with the side of the bonnet.

Fuel Filler Cap

The fuel filler cap located forward of the luggage compartment lid is opened by pressing a small lever at the side of the cap. Press the cap to close.

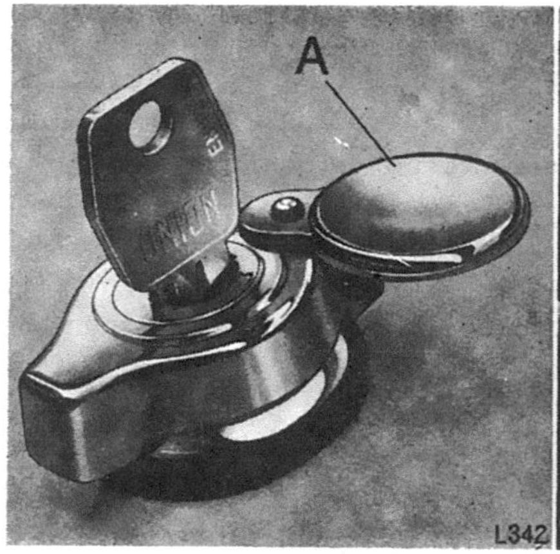

Fig. 12

Fig. 13

SOFT TOP

The soft top is made from P.V.C. material, and is supported by a hinged frame. The assembly folds down into the rear of the car and is protected by a soft top cover.

Lowering the Soft Top

Unclip the fasteners securing the fabric to the centre and rear hoodstick (Fig. 14) and tilt the seats forward. Release the soft top header rail from the windscreen frame by turning the catch levers in the direction shown in Fig. 15 and the fasteners (four each side Fig. 16) securing the edges of the hood to the body.

Push the soft top header rail rearwards and slightly upwards while knocking the second hoodstick forwards, until the assembly begins to fold. **DO NOT PULL the second hoodstick downwards.** Continue lowering the frame and pull the fabric flat over the luggage compartment lid (Fig. 17).

Fold the fabric forward over the hoodsticks and turn the sides inward (Fig. 18) Ensure that the Vybak windows are free from distortion and the hood fabric is not trapped by the hoodsticks.

Soft Top in the Down Position

Retain the soft top in position by fitting the cover (Fig. 19) as follows:

Attach the cover to the fasteners and continue working towards the centre and locate the three hooks (arrowed) on the cover under the bottom hoodstick.

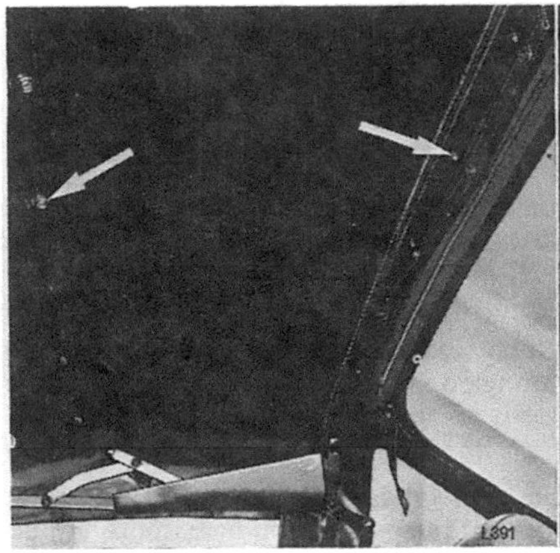

Fig. 14

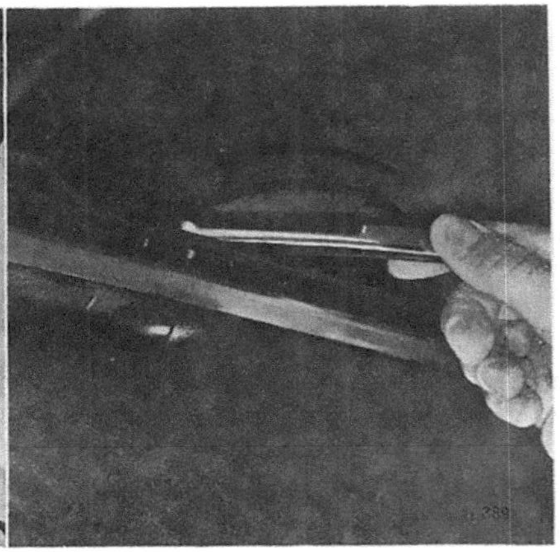

Fig. 15

Tonneau Cover (Optional) (Fig. 20)

The tonneau cover provides weather protection for the vehicle interior when the soft top is lowered. It incorporates press-studs for securing to the car and a zip fastener which permits access to either or both of the front seats.

Raising the Soft Top

Unfasten and remove the hood cover (Fig. 21). Fold the sides of the hood fabric outwards and pull the fabric rearwards over the luggage compartment lid. Lifting the soft top header rail, raise the assembly sufficiently to allow the fabric to lie evenly over the frame.

Fig. 16

Fig. 17

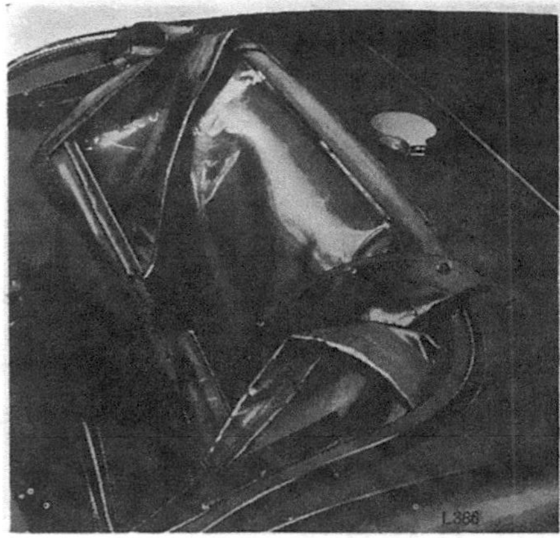

Fig. 18

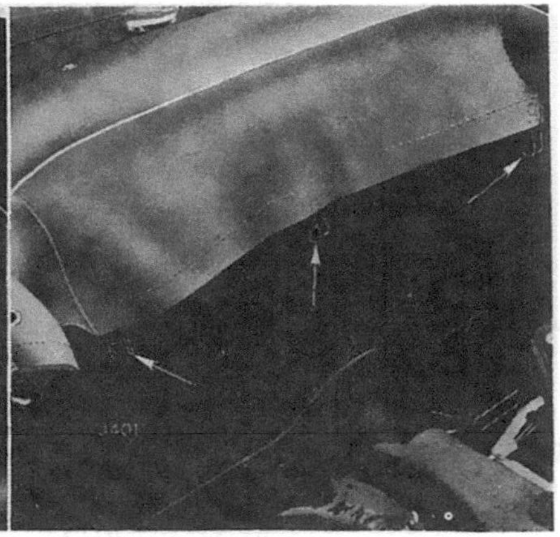

Fig. 19

Secure the fasteners (four each side Fig. 16) to the body. Locate the soft top header rail on the windscreen header rail and turn the levers (Fig. 15) inwards. Knock the centre hoodstick rearwards as far as possible, and secure the fasteners (Fig. 14).

Opening Rear Window (Fig. 22)

To open the rear window, release the zip-fastener and roll the Vybak panel downwards. Retain the rolled panel in position by attaching the straps to the press fasteners located on the back wall of the rear compartment.

Sun Visors (Fig. 20)

Two adjustable sun visors (arrowed) padded to reduce the risk of impact injury, may be unclipped from their inner attachment blocks and swung to eliminate side glare. The passenger's sun visor combines a vanity mirror.

Fig. 20

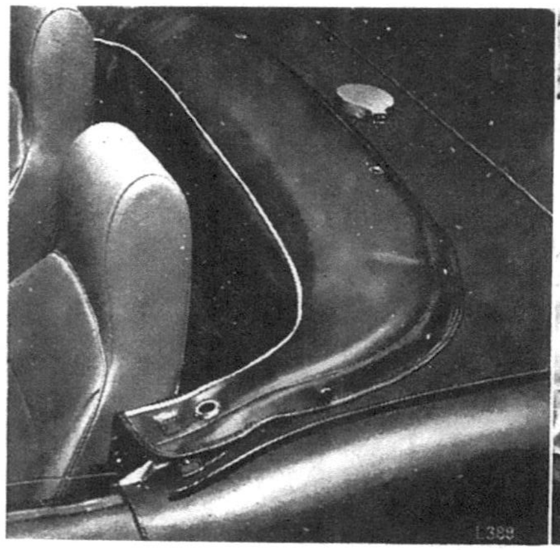

Fig. 21

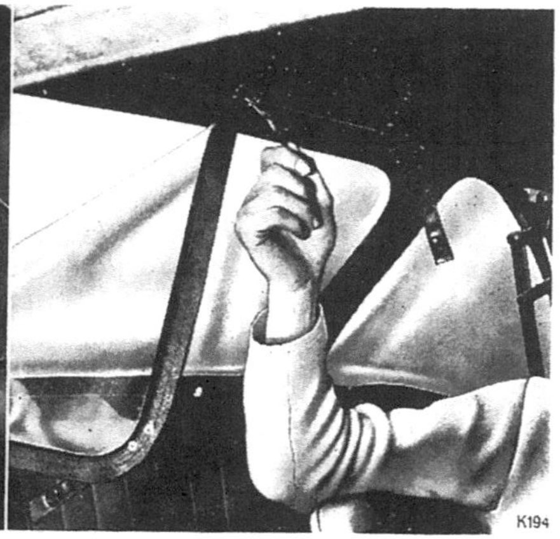

Fig. 22

HARD TOP (Optional)

The vehicle may be used as an open sports car by removing the hard top assembly as follows:

Unscrew the domed-head bolts securing the hard top side brackets to the door pillar brackets (A) (Fig. 23).

Remove the domed-head bolts and washers securing the hard-top to the windscreen header rail (Fig. 24) and the rear deck panel (Fig. 25).

With the aid of an assistant carefully lift off the hard top assembly.

To refit the hard top, reverse the foregoing procedure.

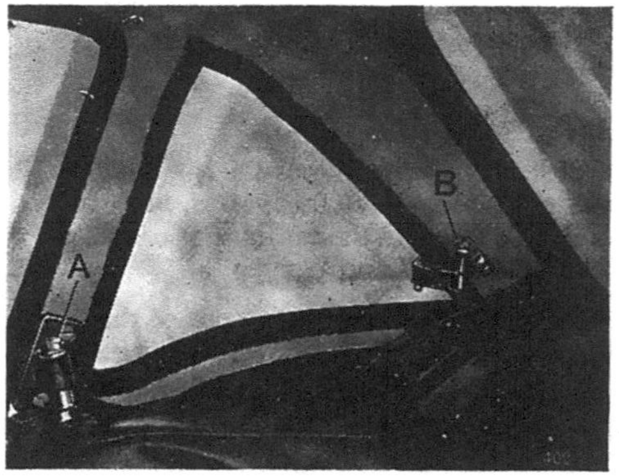

Fig. 23

Fig. 24

Fig. 25

HEATING AND VENTILATING SYSTEM

The heating system consists of a distribution box (mounted between the bulkhead and facia) through which air passes over a water heated element to the interior of the car. The direction, volume and temperature of the air is controlled by remotely operated valves and a two stage booster fan.

Heater Control Panel (Fig. 26)

Movement of control lever (A) determines the volume and direction of air to the interior by operating simultaneously two flap valves, one each side of the distribution box. Thus air may be directed solely to the car interior or screen, or to both by positioning the lever midway between "car" and "screen".

The two speed booster fan may be used to increase the flow of air by pulling the control lever knob out one notch for slow speed and fully out for fast speed. Placing the lever in the "off" position prevents ingress of air to the interior and screen.

Control lever (B) operates a water valve (4) (Fig. 36) which governs the volume of water from the cooling system through the heater element in the distribution box. Placing the lever in the "cold" position closes the water valve causing cold air to enter the car interior. Moving the lever progressively to the "hot" position opens the valve thus warm or hot air is available for direction to interior or screen.

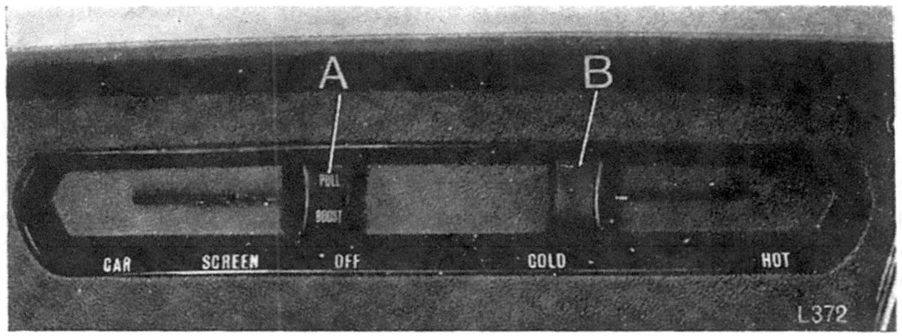

Fig. 26

To Turn Heating System Off

Place lever (A) in the "off" position and lever (B) to "cold".

Cold Air Ventilation

Move lever (A) to "car" and lever (B) to "cold". Operate booster fan to increase air flow if required.

Warm Air Ventilation

Place lever (A) to "car" and lever (B) to "hot" and if necessary operate booster fan.

Windscreen Defrosting

Move lever (A) to "screen" and lever (B) to "hot". Increase air flow by operating booster fan.

Additional ventilation is available by lowering the side windows by turning the handle (C) (Fig. 10) clockwise (drivers side) anti-clockwise (passengers side). When the optional hard top is fitted the rear quarter lights may be opened by pulling the catch (B) (Fig. 23) forward.

COOLING SYSTEM

The pressurised "no loss" cooling system incorporates a translucent plastic overflow reservoir, mounted forward of the radiator, which collects excess coolant from the radiator as the coolant in the system expands with heat. Depression created as the system cools, causes the coolant to flow back from the reservoir into the radiator. The level which is visible through the translucent reservoir should be maintained at least half full when cold. (See Fig. 27.)

Draining

To drain the system, move the heater control (B) Fig. 26, to the hot position, remove the radiator filler cap Fig. 27, open the tap in the bottom of the radiator (Fig. 28) and the tap at the rear right-hand side of the cylinder block (Fig. 29).

CAUTION: If the engine is hot avoid danger from scalding by exercising extreme care when removing the radiator filler cap. Turn it a half-turn and allow pressure to be fully released before completely removing the cap.

Flushing

Efficient cooling is maintained by thoroughly flushing the system once each year before adding anti-freeze. When carrying this out, it is advantageous to remove the drain tap completely and to use plenty of clean running water.

Allowing anti-freeze solution to remain in the system throughout the summer period affords anti-corrosion protection. The solution, however, should be changed at the beginning of each winter period as the solution becomes exhausted.

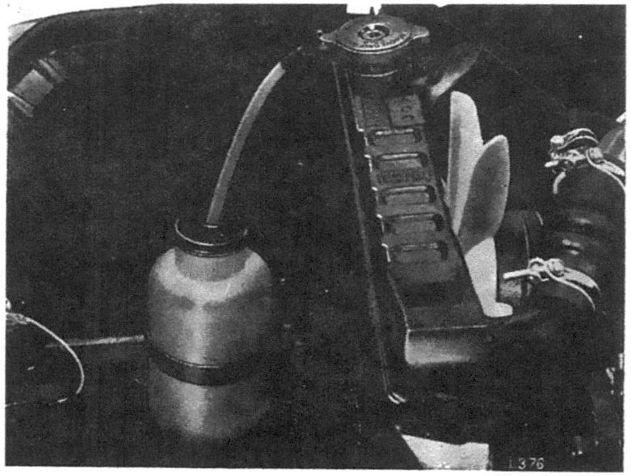

Fig. 27

Filling

Close both drain taps, open the heater control fully and remove the radiator filler cap. Fill the cooling system with clean (soft) water and run the engine at approximately 1,500 r.p.m. for 1 to 2 minutes. Top up the radiator and ıeplace the filler cap. Half fill the plastic overflow reservoir with clean water.

Frost precautions

Since the car heater cannot be completely drained by normal methods, frost damage will not be prevented by merely draining the system.

During freezing conditions the system should be drained, flushed, and refilled with a mixture of soft water and an approved anti-freeze solution (see under "Recommended Lubricants"). Having decided the concentration required (see Chart on page 57) an appropriate mixture should be prepared in a suitable container and the system filled as previously described.

Because of the searching effect of anti-freeze solutions, it is recommended that the cooling system is checked for leaks before filling with anti-freeze mixture.

At certain temperatures anti-freeze and water mixtures adopt a "mushy" state with a viscosity which impairs circulation and can immobilize or damage the water pump. Therefore, consult the concentration chart before adding anti-freeze to ensure adequate protection for temperatures likely to be encountered.

Fig. 28

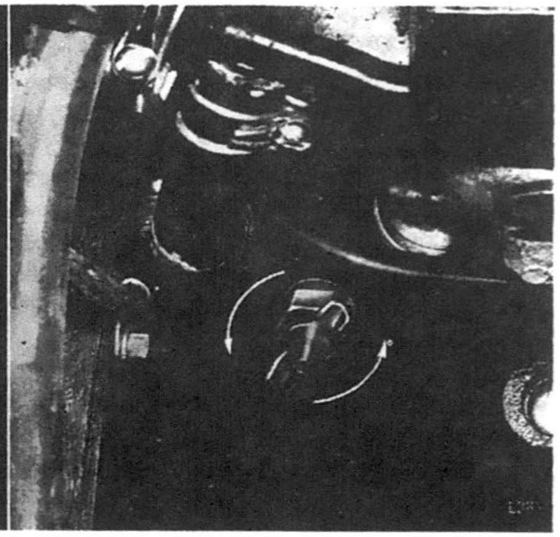

Fig. 29

DRIVING FROM NEW

Starting the Engine from Cold

Check, and if necessary, top up the water and the engine oil levels. If the car has not been used for several days and fuel has evaporated from the carburettor, refill it by. operating the priming lever on the fuel pump (Fig. 30). The slight resistance ceases when the float chamber is full.

Apply the handbrake and ensure that the gear selector lever is in "Neutral". Using the choke as necessary, turn the ignition key to the "start" position and immediately the engine fires, release the key, which will return to the "Ignition" position. Should the engine fail to start at the first attempt, do not re-operate the starter switch until the starter motor has come to rest.

After starting the engine, cylinder wear is minimised if the engine is warmed up quickly by driving away when the oil pressure warning light is extinguished. Do not race the engine to speed up the warming process but, if possible, maintain a moderate speed until the choke may be pushed fully in. Avoid the use of full throttle during the warming-up period. A thermostat incorporated in the cooling system enables the engine to be warmed up quickly from cold.

Starting the Engine Warm or Hot

When re-starting a hot engine, depress the accelerator pedal to about one-third of its travel before operating the starter switch. The choke control should not be used.

Running-in

The importance of correct running-in cannot be too strongly emphasised, since during the first few thousand miles of motoring all working surfaces of the vehicle are bedding down.

Avoid placing heavy loads upon the engine, such as using full throttle at low speeds or when the engine is cold. Running-in should be progressive and no harm will result from the engine being allowed to "rev" fairly fast for short periods provided that it is thoroughly warm and not pulling hard. Always select a lower gear if necessary to relieve the engine load. Full power should not be used until at least 1,000 miles (1600 km.) have been covered and even then, it should be used only for short periods at a time. These periods can be extended as the engine becomes more responsive.

Maximum Engine Speeds

After "running-in" has been completed, owners are advised not to drive the car continuously at engine speeds in excess of 6,000 r.p.m. This speed is indicated by the beginning of an orange coloured segment on the tachometer dial. Speeds within this segment are permissible for short periods, such as when overtaking another vehicle, but under no circumstances must the engine be allowed to rev beyond 6,500 r.p.m., which coincides with the beginning of a red segment. (See (G) Fig. 3.)

Recommended Fuel

The Spitfire IV is capable of performing efficiently on fuels with an octane rating as low as 91 octane (Research Method).

NOTE: The engine is not designed to use unleaded fuel and whilst the occasional tank full will not cause damage, consistent use of unleaded fuels will result in excessive engine wear and the malfunctioning of the emission control system.

Overdrive Unit (Optional fitment)

An overdrive unit serves as a convenient method of providing alternative gear ratios to reduce engine speed which with proper use can effect fuel economy.

Greatest benefit will occur from judicious use of the overdrive, the governing factor being that, when changing into overdrive, the vehicle continues to run easily without sign of engine labouring, combined with the minimum throttle opening necessary to maintain the same road speed. Maximum economies are effected under "motorway" conditions.

The overdrive unit is remotely controlled by a slide switch, mounted on top of the gear selector lever (Fig. 31). To engage the overdrive move the switch rearward which will reveal the word "in". Moving the switch forward will disengage the overdrive, indicated by the word "out". Do not change from overdrive at engine speeds in excess of 4,500 r.p.m. This corresponds with engine speeds within the orange segment of the tachometer, in normal gears. Disengagement of overdrive at higher engine speeds will cause the engine to over-rev and increase the possibility of damage to the engine.

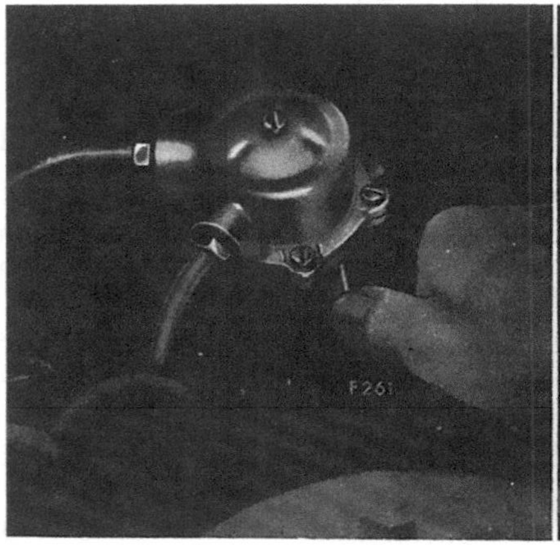

Fig. 30

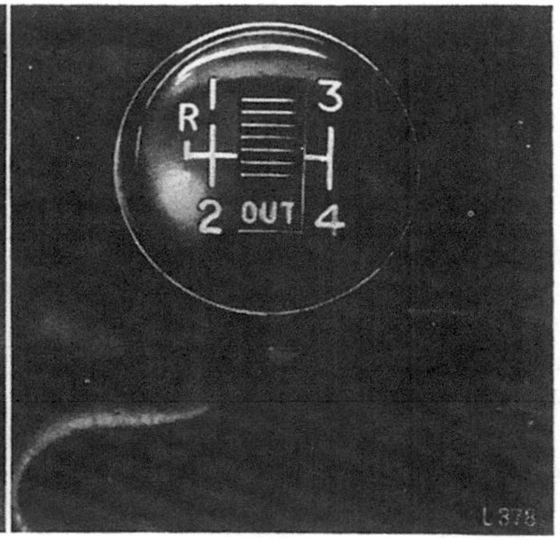

Fig. 31

WHEELS AND TYRES

General

Wheels and tyres, of correct types and pressures, are an integral part of a vehicle's design. Thus the regular maintenance of the tyres contributes not only to the safety but to the designed functioning of the vehicle, as road holding steering and braking are especially vulnerable to the use of incorrectly pressurised, badly fitted or worn tyres.

Radial-ply and Cross-ply Tyres

It is dangerous to use, on the public highways, a vehicle fitted with unsuitable combinations of tyres. The following recommendations should therefore be observed.

1. Do not mix radial-ply and cross-ply tyres on the same axle.
2. Do not fit radial-ply tyres to the front wheels and cross-ply tyres to the rear wheels.
3. With suitable tyre pressure adjustments it may be possible to obtain acceptable handling with cross-ply tyres on the front wheels and radial-ply tyres on the rear wheels, but this combination is not recommended.

Tyre Wear

To prolong tyre life, avoid severe braking, sudden changes of direction at speed, and driving over or against high kerbstones, as this can result in severe damage to the tyre walls. Clean off any soil or grease which may be on the tyres by using cloth moistened in petrol.

Fins on the inside or outside of the pattern ribs indicate excessive toe-in or toe-out respectively. Because sharp pattern edges may also be caused by road camber, it is better to make sure by having the track checked.

"Spotty" tread wear or flats can result from grabbing brakes or unbalanced wheel assemblies. The original degree of balance is not necessarily maintained, but may be affected by uneven tread wear, repairs, tyre removal and refitting, or by wheel damage and eccentricities. The vehicle may also become more sensitive to unbalance due to normal wear of moving parts.

Excessive wear in the centre of the tread results from over inflation. Excessive wear at the outer edges of the tread results from under-inflation, a condition which causes excessive heating and premature tyre failure.

Pressures

Adjust tyre pressures in accordance with the recommendations given below. These pressures are satisfactory for sustained speeds up to 110 m.p.h. (175 km.).

NOTE: Should the vehicle be tuned to increase its maximum speed, or be used for racing, consult the respective tyre company regarding the need for tyres of full racing construction.

Adjust the pressures whilst the tyres are cold, *i.e.* before a run. As tyres warm up, their pressures increase. A warm tyre bled to the recommended pressure will be under-inflated when cold.

Recommended Tyre Pressures

Tyre Size and Type	FRONT		REAR	
	psi.	$kg./cm.^2$	psi.	$kg./cm.^2$
5·20S—13	21	1·47	26	1·38
1·45SR-13 Radial ply SP68	21	1·47	26	1·83

Spare Wheel, Tools and Lifting Jack

A spare wheel, lifting jack and handle together with a tool kit are located in the luggage compartment. The spare wheel is secured by a nut and may be released using the wheel nut box spanner and tommy bar supplied in the tool kit.

Using the Jack (Fig. 32)

In the interests of safety, place the jack under a body mounting bolt at the corner of the chassis nearest to the wheel being lifted. Make sure the jack lifting pad is square on the chassis and the base stands firmly on the ground.

Repairs to Tubeless Tyres

A temporary repair can be made to tubeless tyres using a special kit, provided the puncturing hole is small and confined to the central tread area. The following precautions must, however, be observed.

1. Do not use more than one plug in each hole.
2. Do not use the tyre for high speeds.
3. Ensure that a permanent "cold patch" or vulcanised repair is made at the earliest opportunity.

Wheels Changing Procedure (Pressed Steel Wheels)

1. Firmly apply the handbrake and chock the wheel diagonally opposite the one being lifted.
2. Remove the spare wheel from the luggage compartment and make sure that its tyre pressure is correct.
3. Using the wheel nut spanner slacken, slightly the wheel nuts (Fig. 33).
4. Place the jack in position and lift the wheel clear of the ground. Should it be necessary to lift the vehicle whilst it is on sloping ground, exercise the greatest care.
5. Remove the wheel nuts, hub trim and lift off the wheel.
6. Fit spare wheel together with hub trim and replace wheel nuts. Tighten wheel nuts evenly taking care not to trap the hub trim.
7. Lower the wheel and remove jack.
8. Tighten wheel nuts securely and press hub trim (which should rock on its springs to ensure that the wheel nuts are correctly positioned.

Wheel Changing (Wire Wheels) (Optional)

The procedure for changing wire wheels is the same as for pressed steel wheels except that since no wheel trim is fitted reference to this item should be ignored.

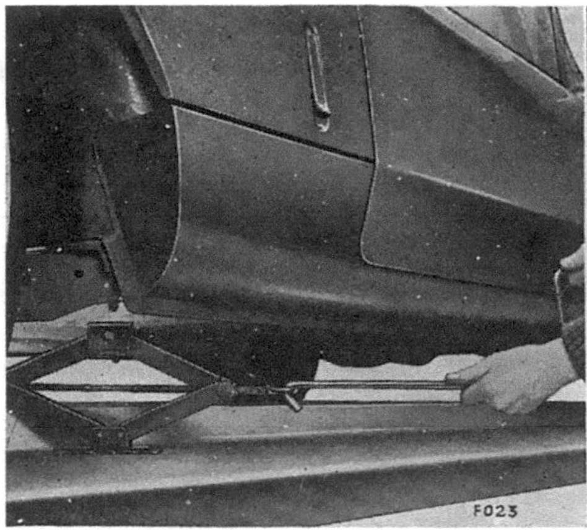

Fig. 32

Fig. 33

ELECTRICAL SYSTEM

BULB CHART

Lamp	Watts	Lucas Part No.	Stanpart No.	
Headlamps—R.H. Dip—U.S.A. ..	50/40	54522231		*
France ..	45/40	411	510219	
Other R.H. Dip markets	60/50	54523079	215735	*
L.H. Dip	60/45	54521872	512231	*
Front parking lamps	5	207	57591	
Front flasher lamps	21	382	502379	
Front marker lamps	4	222	501436	
Rear marker lamps	4	222	501436	
Tail/stop lamps	5/21	380	502287	
Rear flasher lamps	21	382	502379	
Reverse lamp	22	382	502379	
Plate illumination lamp	5	207	57591	
Instrument illumination	2·2	987	59492	
Courtesy light	2·2	987	59492	
Warning lights	2·2	987	59492	
Seat belt warning light	2	281	513000	

* Sealed beam light unit

Wiring Diagram

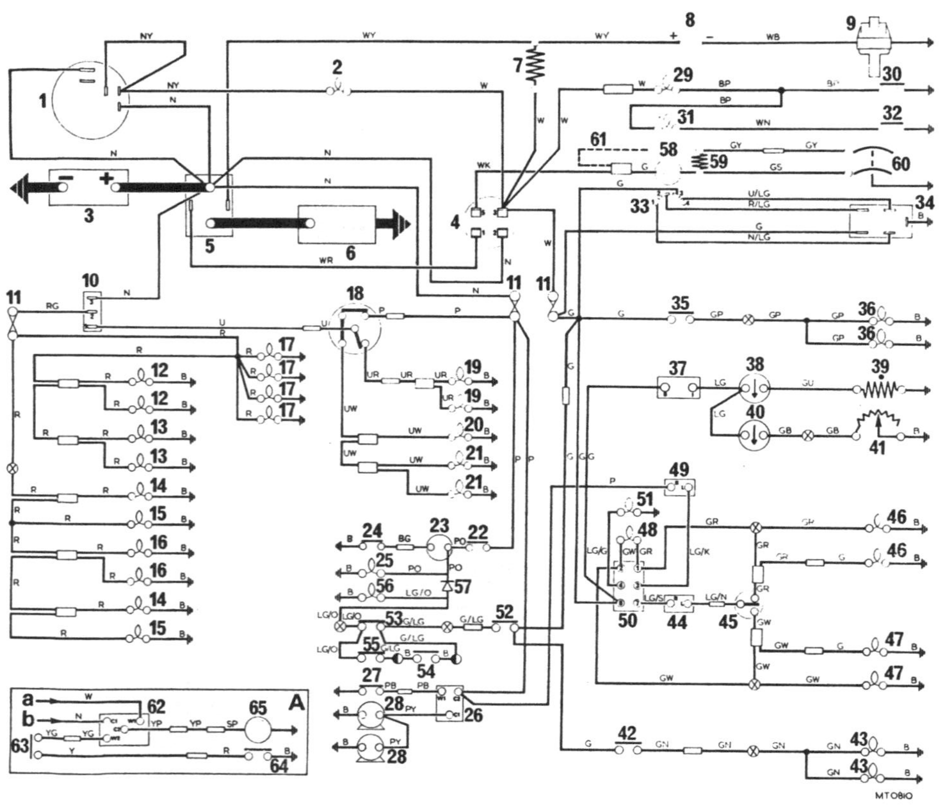

Fig. 34

KEY TO WIRING DIAGRAM—LEFT-HAND STEER

CAUTION: THIS VEHICLE IS FITTED WITH A NEGATIVE EARTH ELECTRICAL SYSTEM ENSURE THAT THE BATTERY EARTH LEAD IS ALWAYS CONNECTED TO THE BATTERY NEGATIVE TERMINAL.

THE ALTERNATOR—AND POSSIBLY SOME ACCESSORIES—CONTAIN POLARITY SENSITIVE COMPONENTS THAT MAY BE IRREPARABLY DAMAGED IF SUBJECTED TO INCORRECT POLARITY.

1 Alternator
2 Ignition warning light
3 Battery
4 Ignition/starter switch
5 Starter solenoid
6 Starter motor
7 Ballast resistor
8 Ignition coil—6 volt
9 Ignition distributor
10 Master light switch
11 Fuse
12 Front parking lamp
13 Front marker lamp
14 Rear marker lamp
15 Tail lamp
16 Plate illumination lamp
17 Instrument illumination
18 Column light switch
19 Dip beam
20 Main beam warning light
21 Main beam
22 Door switch
23 Buzzer
24 Key switch
25 Key light
26 Horn relay
27 Horn push
28 Horn
29 Brake warning light
30 Brake line failure switch
31 Oil pressure warning light
32 Oil pressure switch
33 Windscreen wiper switch
34 Windscreen wiper motor
35 Stop lamp switch
36 Stop lamp
37 Voltage stabilizer
38 Temperature indicator
39 Temperature transmitter
40 Fuel indicator
41 Fuel tank unit
42 Reverse lamp switch
43 Reverse lamp
44 Turn signal flasher unit
45 Turn signal switch
46 L.H. flasher lamp
47 R.H. flasher lamp
48 Turn signal warning light
49 Hazard flasher unit
50 Hazard switch
51 Hazard warning light
52 Belt warning gearbox switch
53 Drivers belt switch
54 Passenger seat switch
55 Passengers belt switch
56 Belt warning light
57 Diode
58 Heater motor
59 Heater rheostat
60 Heater switch
61 Radio faculity

A Overdrive (optional extra)
62 Overdrive relay
63 Overdrive gearbox switch
64 Overdrive gear lever switch
65 Overdrive solenoid

FUSE SYSTEM

The fuse assembly is a component of the main harness. It is secured to an aperture provided on the bulkhead by integral plastic clips. The unit contains three operational fuses and has provision to house two spares. The fuses are protected by a pull-off transparent plastic cover on the bulkhead adjacent to the clutch master cylinder.

Failure of a particular fuse is indicated when all the circuits protected by it become inoperative. If a new fuse fails establish the cause and rectify the fault before fitting a second replacement.

Fuse	Manufacturer	Lucas
	Rating	35 amp.
	Lucas Part No.	188218
	Stanpart No.	58465

Circuits

The fuse fed by a white cable from the ignition/starter switch protects the following circuits:

- Windscreen wiper
- Fuel indication
- Temperature indication
- Stop lamp
- Reverse lamp
- Turn signal
- Seat belt warning

The fuse fed by a red/green cable from the master light switch protects the following circuits:

- Front parking lamp
- Front marker lamp
- Rear marker lamp
- Tail lamp
- Plate illumination lamp
- Instrument illumination

The fuse fed by a brown cable from the battery protects the following circuits:

- Headlamp flasher
- Key warning
- Courtesy light
- Horn
- Hazard warning

RADIO FACILITY

The wiring harnesses include a facility for this optional extra item.

CHARGING SYSTEM

CAUTION: THE ALTERNATOR CONTAINS POLARITY SENSITIVE COMPONENTS THAT MAY BE IRREPARABLY DAMAGED IF SUBJECTED TO INCORRECT POLARITY.

TO PREVENT DAMAGE TO COMPONENTS, DO NOT MAKE OR BREAK ANY CONNECTIONS IN THE CHARGING CIRCUIT —INCLUDING THE BATTERY LEADS—WHILE THE ENGINE IS RUNNING. RUN THE ALTERNATOR ONLY WITH ALL CONNECTIONS MADE OR WITH THE UNIT DISCONNECTED.

Alternator

The Lucas 15ACR alternator—which contains its own control unit—has a nominal output of 28 amps. The unit is driven by a vee belt which should be adjusted as detailed in Maintenance. The rotor runs on two "lubricated for life" ball bearings. No periodic lubrication is required.

BATTERY

The battery is located on the bulkhead. This is normally a Lucas "Pacemaker" unit featuring an "Aqualok" topping up device. Topping up procedure is detailed in Maintenance.

Lucas "Pacemaker" data:

Capacity—at 20-hour rate	.. ..	40 amp.-hour
Plates per cell	.. ..	9
Charge rate	Initial	2·5 amps.
	Normal	3·5 amps.
	Boost	36 amps.

Ensure that the battery top and terminals remain clean and dry. Coat terminals with petroleum jelly (Vaseline) to prevent corrosion.

Check electrolyte level monthly and if required replenish with distilled water as detailed in Maintenance. If electrolyte has been spilled clean the affected area with a cloth moistened with ammonia to neutralise the acid and prevent acid corrosion.

Ensure that the battery is always firmly clamped in position by the retaining assembly. When fitting battery leads do not hammer terminals to terminal posts. Such action may damage battery.

The battery will deteriorate rapidly if left in a discharged condition. If the unit is reduced to a low state of charge it should be recharged at the first opportunity.

HEADLAMPS

NOTE: IF BEAM AIMING IS NOT INTENDED DO NOT DISTURB BEAM AIMING SCREWS "A" AND "B" (Fig. 35).

Beam Aiming

It should be possible to adjust the beams without removing the headlamp surround. Screw "A" positions the beam in the horizontal plane. Screw "B" controls beam height.

Beam aiming can best be accomplished using equipment such as Lucas "Beamsetter" or "Lev-L-Lite". This service is available at Triumph distributors or dealers and will ensure maximum road illumination with minimum discomfort to other road users.

Filament Failure (Fig. 35)

Remove two bolts (1 and 2). Remove nut (3) and lift headlamp surround from upper retainers. Remove three screws (4, 5 and 6) to release retaining rim and light unit. Pull connector from sealed beam light unit or bulb. Renew light unit or bulb and reassemble.

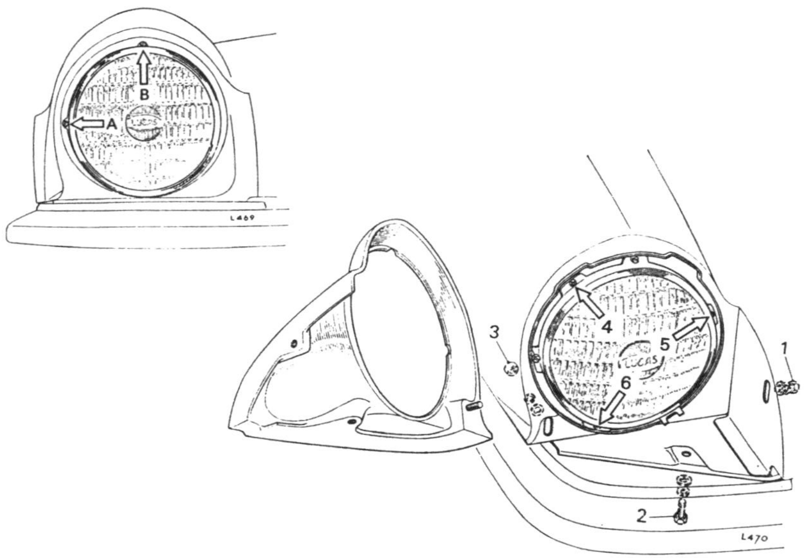

Fig. 35

ROUTINE SERVICING

Braking System

In addition to adjustment, examination and renewal of shoes and pads at the intervals recommended in the following pages, it is strongly recommended that the brake system is overhauled and the brake fluid renewed every 36,000 miles (60,000 km.) or 3 years (whichever is sooner). This involves dismantling, examining and renewing all seals, flexible hoses and defective items.

Owners are urged to seek the assistance of a Triumph Dealer who will be pleased to estimate for the work which, for reasons of safety, should be entrusted only to skilled workshop personnel.

Lubricants

When carrying out the following maintenance work, use high grade lubricants only. The lubricants listed on page 56 have maintained a high standard of quality for many years and are approved only after extensive tests in collaboration with the oil companies concerned. In countries where these oils are unobtainable, use similar oils having the same characteristics.

Engine

When a new car is delivered the engine sump contains a quantity of special running-in oil, sufficient for the running-in period. Provided that the level does not fall below the low mark on the dipstick, topping-up is unnecessary. If replenishment is required, top-up with one of the engine oils listed on page 56.

At the "Free Service", or after completing the first 1,000 miles (1,600 km.), drain the running-in oil from the sump and replenish with new oil, of the appropriate grade, to the high mark on the dipstick.

Transmission, Overdrive and Final Drive

Final Drive, transmission and overdrive units fitted to new cars are filled with a special oil, formulated to give all necessary protection to new gears. This oil should not be drained but may be topped up with any of the approved oils.

PERIODIC CHECKS—WEEKLY

Engine (Fig. 36)

Once a week or prior to starting out on a long run, or every 250 miles (400 km.), check the level of oil in the engine sump, first making sure that the car is standing on level ground. If the engine has been running, wait a few minutes to allow the oil to drain back into the sump.

Withdraw the dipstick (1), wipe it clean and push it fully home before again withdrawing it for reading. Add oil via the filler cap (2) until the level reaches the high mark on the dipstick. **Do not overfill,** and ensure that the filler cap and dipstick are replaced.

Radiator Water Level

The level of water, visible through the translucent plastic reservoir, should be maintained at least "half full" by adding soft water, when required, via the screwed cap. Refer to "filling" on page 23.

Windscreen Washer (3) (Fig. 36)

Examine the water level in the plastic windscreen washer container. To refill the reservoir, hold the plastic tubing while unscrewing the cap. Fill with clean soft water to the bottom of the neck and replace the cap. If the wiper blades tend to smear, a little household detergent added to the water in the reservoir will help to prevent this.

To avoid frost damage during cold weather, either empty the reservoir or fill with a mixture of 1 part de-natured alcohol (methylated spirits) to 2 parts water. Do not use glycol anti-freeze.

Fig. 36

Brake Master Cylinder (1) (Fig. 37)

Every week check the level of fluid in the brake master cylinder reservoir. The fluid level is visible through the translucent casing of the reservoir, **do not remove the cap.** A gradual lowering of the level over a long period is caused by brake pad wear and does not require topping-up. A sudden appreciable drop in the level must be investigated, the cause ascertained and rectified immediately.

Do not allow the level to drop below the danger line on the side of the casing.

To avoid dirt entering the system ensure that the reservoir is clean externally before removing the cap. Use only new fluid taken from a sealed container and re-seal the container after use. Replace the reservoir cap immediately after filling.

Clutch Master Cylinder (2) (Fig. 37)

Every week, check the level of fluid in the clutch master cylinder. To prevent dirt entering the system, clean the cap and surrounding area prior to removing the cap. Top-up the fluid until it is level with the line on the side of the reservoir.

Tyres

The maintenance of correct tyre pressures, tread thickness and freedom from cuts and other tyre damage, are important safety factors which affect the behaviour of steering and suspension. It is, therefore, expedient to examine the tyres once a week and to ensure that pressures are maintained in accordance with the figures given on page 27.

Fig. 37

PERIODIC CHECKS—MONTHLY

Battery (Figs. 38 and 39)

At periods not exceeding four weeks, examine the level of the electrolyte in the cells and top up with distilled water if necessary. The electrolyte level should just cover the separators. More frequent checks should be made during hot weather and if the car is subjected to long journeys.

IMPORTANT: Never use a naked flame when examining the battery since the mixture of oxygen and hydrogen given off by the battery is dangerously explosive.

When a Lucas "Pacemaker" battery is fitted the following topping-up procedure should be followed. Make sure the car is standing on level ground and lift and tilt the battery cover. If the electrolyte level is below the separators, add distilled water until the filling tubes are full and the trough is just covered. Replace the cover immediately.

Key to Fig. 39

1. Cover
2. Trough
3. Vent holes
4. Filling tubes
5. Separators
6. Vent cover raised, air lock created, filling tube full
7. Vent cover replaced, air lock broken, correct level indicated

Fig. 38

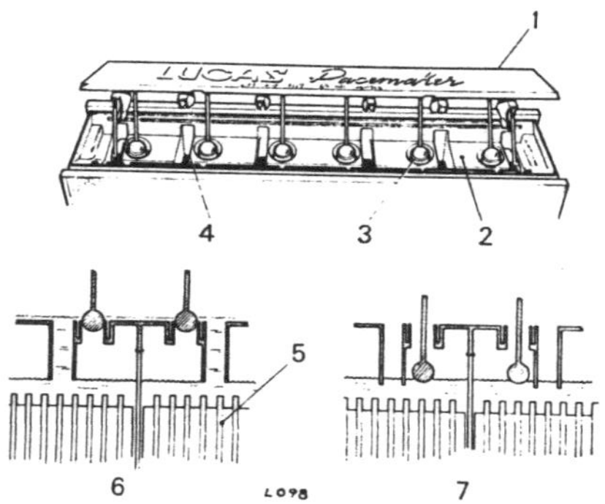

Fig. 39

Wheels

In addition to checking tyres each week, once a month examine the wheel stud threads for damage. Before replacing the nuts smear grease on the stud threads to resist wear and corrosion. Check nuts for tightness.

Lights

Check operation of headlamps and front and rear parking lights including licence plate illumination lamp.

1,000 MILES (1,600 KM.) FREE SERVICE

At the completion of 1,000 miles (1,600 km.) or as near to this figure as possible, it is important that the car is returned to the selling Distributor or Dealer, who will complete the work listed in the Passport to Service (U.S.A. Edition), free of charge except for the cost of oil and grease.

The importance os carrying out this free service cannot be over emphasised, as it corresponds with the end of the running-in period when many of the components will have settled down, thus necessitating re-adjustment. Furthermore, the special running-in engine oil, used for the initial fill, is unsuitable for normal motoring and must, therefore, be replaced by the correct grade of oil.

ROUTINE SERVICING

In addition to the periodic checks listed on pages 36, 37 and 38, perform the following operations at the intervals recommended in the "Passport to Service" (U.S.A. Edition).

Air Cleaner (Fig. 41)

(a) Remove and clean the paper element (4). Do this more frequently where conditions of extreme dust prevail.

To gain access remove bolts (1) securing the unit to the carburettor intake flange. Detach the closing plate (5), lift out the element (4) and, using a low pressure air line or a soft brush, clean between the folds of the paper element.

Adopt the reverse procedure to refit, ensuring that the gasket (6) is in good condition and that the slots in the closing plate (5) and the gasket (6) align with those in the carburettor flange. Replace or renew the sealing rings (3).

(b) Remove and dismantle air cleaner as described and renew paper element.

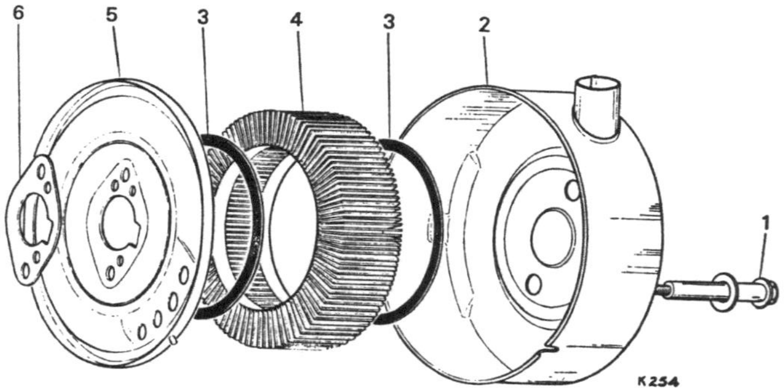

Fig. 41

Engine Oil Change (Fig. 42)

Changing the oil in the sump is best carried out when the engine is warm. Place a suitable receptacle under the drain plug (1) and unscrew the plug slowly until oil begins to escape. When the rate of flow lessens remove the plug completely and allow the oil to drain for a few minutes. Clean and refit the drain plug and refill the sump with the appropriate grade of oil see page 56. More frequent oil changes are recommended if the car is used for short journeys during cold weather when the choke is used extensively.

Gearbox (Fig. 42) (and overdrive when fitted)

With the vehicle standing on level ground, remove the oil filter plug (2) and top up the gearbox until the oil is level with the bottom of the filler plug threads. Allow surplus oil to drain away before refitting the plug and wiping clean. An oil transfer hole between the gearbox and overdrive unit (when fitted) provides a common oil level. Maintenance of the overdrive unit is thus limited to ensuring that the correct oil level is maintained in the gearbox.

Final Drive (Fig. 43)

Remove the oil level plug (shown arrowed) and top up the final drive initl the oil is level with the bottom of the filler plug threads. Allow surplus oil to drain before refitting the plug and wiping clean.

Fig. 42

Fig. 43

ROUTINE SERVICING

Ignition Distributor (Fig. 44)

(a) Release the clips and remove the distributor cap and rotor arm. Apply a few drops of thin oil to points (3) and (4). Lightly grease the cam surface (6) and inject approximately 5 c.c. (one teaspoonful) of engine oil thıough the hole (5).

Adjustment

Turn the engine until the contact breaker lever is operating on the highest point of the cam lobe, *i.e.* gap at its widest. Slacken the fixed contact screw (2) and using a feeler gauge between the contacts (1), set the gap to 0.015 in. (0.4 mm.). Tighten screw (2).

Renew worn or damaged points when required.

If the distributor has been removed from the engine, use a static ignition timing of 8° BTDC for starting the engine only. As this setting cannot achieve the extreme accuracy required for the proper functioning of the emission control system, it is vitally important that the final ignition setting is made dynamically to achieve a precise setting of 2° ATDC at the engine idling speed of 800/850 r.p.m.

(b) Renew contact breaker points and adjust gap as decribed above.

Timing Marks (Fig. 46)

Marks are engraved in four degree increments at both sides of zero. Numbers one and four pistons are at top dead centre when the zero mark on the pulley is turned to coincide with the pointer on the timing cover.

Carburettor Damper (Fig. 45)

Unscrew the plug from the top of the carburettor and withdraw the plug and damper assembly. Top-up the damper chamber with engine oil. The oil level is correct when utilising the damper as a dipstick, its threaded plug is ¼ in. above the dash-pot when resistance is felt. Refit the damper and hexagonal plug. Using an oil can, apply oil to the throttle and choke control linkages, and accelerator pedal pivot. Check the carburettor slow running as described on page 53.

Fig. 44

Fig. 45

Sparking Plugs

(a) Remove the sparking plugs for cleaning and reset the gaps to 0.025 in. (0.63 mm.). Clean the ceramic insulators and examine them for cracks or other damage likely to cause "H.T." tracking. Test the plugs and renew those which are suspect.

(b) Remove and discard the sparking plugs. When fitting replacements, ensure that they are of the correct type and that the gaps are set to 0.025 in. (0.6 mm.). Reconnect the leads in the order shown in Fig. 47.

Plug Lead Position (Fig. 47)

Ensure that the plug leads are attached to the sparking plugs as shown. Firing order is 1, 3, 4, 2, taken in anti-clockwise order, as viewed from the top of the rotor arm.

High Tension Cables

All high tension cables fitted to the ignition system are made from carbon impregnated nylon or cotton cords encased in rubber or neoprene to form a high resistive conductor. Replacement cables :hould be obtained from a Triumph Dealer and must be the same length and type as the original.

Valve Seat Attention

Have the compression pressures checked by your Triumph dealer; providing that the engine is functioning satisfactorily, and the compression pressures of all the cylinders are equal, you are advised not to disturb the engine.

The need for decarbonising arises when the build-up of carbon, a product of combustion, becomes excessive. If premium grade fuels and high quality lubricants are used, carbon deposit is so minimised that frequent decarbonising is unnecessary. Carbon removal may, therefore, be restricted to occasions when the cylinder head is removed for attention to the valves and seats.

Fig. 46

Fig. 47

Hydraulic Pipes

Trace the run of all hydraulic pipes and make sure they are not rubbing against the body or chassis. Check for leaks at the joints and examine the flexible hoses. Check that when the front wheels are turned from lock to lock the flexible hoses are not strained, kinked or rubbing. It is recommended that this check is carried out by your Distributor or Dealer.

Rear Brakes

Jack up the car, remove road wheel and brake drum. Examine brakes shoes for wear. Replace worn or oil contaminated shoes. Examine wheel cylinders for leaks and perished rubber boots. Using a high pressure air line or brush remove loose dust from the brake mechanism and use petrol to remove oil or grease from the back plate. Wipe clean the inside of the brake drum and avoid touching the braking surface with greasy hands. Adjust brakes if necessary see page 49.

Front Brakes

Jack up the front of the car, remove the road wheel and examine the brake pads for wear. Renewal is necessaiy when the lining material bonded to the steel backing plate is reduced to a thickness of $\frac{1}{8}$ in. (3 mm.). (Refer to page 49 for procedure).

Fan Belt Adjustment (Fig. 48)

The fan belt should be sufficiently tight to drive the water pump and alternator without unduly loading the bearings.

To adjust the belt, slacken the pivot bolt nut (1) and the adjustment bracket bolt (2). Pivot the alternator away from the engine until the belt can be moved $\frac{3}{4}$ to 1 in. (19 to 25 mm.) at the mid-point of its longest run. Maintaining the alternator in this position, tighten the bolt (1) and nut (2).

Fig. 48

Fig. 49

Rocker Clearances (Fig. 49)
Adjust the valve rocker clearances, at valve stem tip, to 0.010 in. (0.25 mm.) when cold, in the following sequence.

Adjust Nos. 1 and 3 valves with Nos. 8 and 6 valves open.
,, ,, 5 ,, 2 ,, ,, ,, 4 ,, 7 ,, ,,
,, ,, 8 ,, 6 ,, ,, ,, 1 ,, 3 ,, ,,
,, ,, 4 ,, 7 ,, ,, ,, 5 ,, 2 ,, ,,

Fuel Pump (Fig. 50)
Access to the petrol pump bowl and filter is gained by unscrewing the bolt (1) and removing the domed cover (2). Lift the filter gauze (3) from its seating and wash it in petrol.

Using a small screwdriver, loosen the sediment in the bowl and remove it with compressed air. A foot-pump used for tyre inflation is ideal for the purpose.

IMPORTANT: Do not damage the non-return valve (5) during this process. Renew the cork gasket (4) if this has hardened or broken. Assemble the filter gauze into its seating, taking care to place the gauze face downwards so that it can be removed easily when required.

Evaporation System—Carbon Canister (Fig. 51)
Replace the filter in the carbon canister as follows:

Remove the inlet and purge tubes from the top of the canister (1). Slacken the clip securing the canister to its mounting bracket and remove the canister.

Unscrew the base cover (3) and remove the filter gauze (2). Clean the base cover, fit a new gauze, replace cover and refit the canister.

Ensure piping is not chafing and free from kinks.

Every 48,000 miles replace the canister adopting the method described above.

Crankcase Ventilation
Remove the piping connecting the rocker cover to the evaporation canister and carburettor. Wash the internal bore of the pipes in petrol and allow to dry before refitting. Remove the rocker cover and wash in petrol and refit when dry.

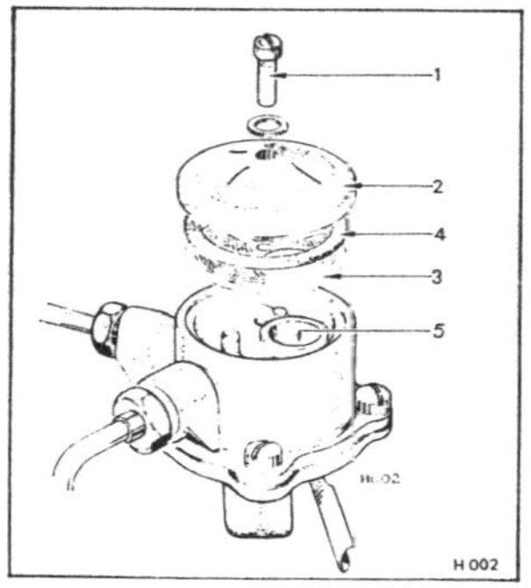

Fig. 50

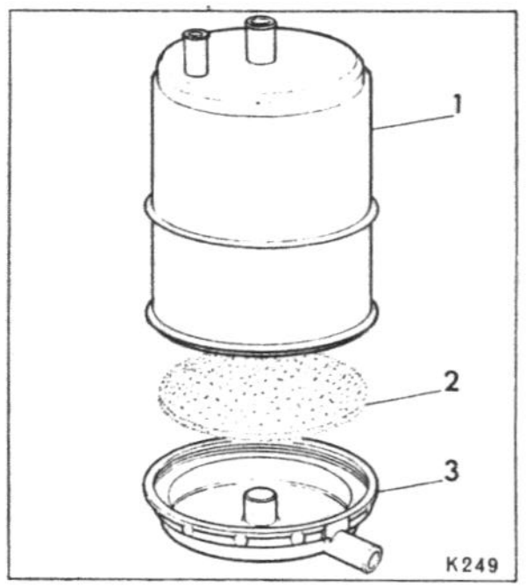

Fig. 51

Oil Filter (Fig. 52)
Unscrew the filter from the cylinder block and replace it with a new one, fitting a new "O" ring. Ensure an oil-tight seal by cleaning and smearing the joint faces with oil before screwing the new filter unit tightly home.

Fuel Line Filter (Fig. 53 Early Models)
The fuel line filter is a sealed unit and cannot be dismantled for cleaning. Remove the filter from the attachment bracket and fit a new unit.

Exhaust System
To prevent injury to health, check the exhaust system for leaks and deterioration, and rectify as necessary.

Universal Coupling Bolts
Check for tightness th e unversal coupling bolts at both ends of the propellor shaft linking the gearbox and finaldrive and coupling bolts on the left and right hand drive shafts.

Rear Hubs
Remove the plug (1) (Fig. 54) and fit a grease nipple. Using a grease gun, pump, until grease exudes from the bearing. Remove nipple, replace the plug and remove surplus grease.

Cylinder Head Nuts (Fig. 59)
Tighten the cylinder head nuts in the order shown. Slacken them by reversing the sequence.

Fig. 52

Fig. 53

Front Hubs (Fig. 56)

Jack up each front wheel in turn, remove the road wheel and check the hub bearings for end float. If excessive, adjust as follows:

1. Prise off the grease cap (11) and withdraw the split pin (12).
2. Whilst spinning the hub tighten the slotted nut (10) to 5 lb./ft. and unscrew the nut one flat to give 0·002 in. to 0·008 in. (0·05 mm. to 0·2 mm) hub end float.
3. Fit a new split pin, replace the grease cap and road wheel. Repeat operation on opposite hub.

At major overhaul periods or every 12,000 miles if the car is being used for competition work, re-pack the front hubs with grease as follows:

1. Jack up the front of the car and remove road wheel.
2. Without disturbing the hydraulic pipe unions, unscrew the caliper securing bolts A and B (Fig. 55) and lift the caliper from the disc, tying it to a convenient point to prevent it hanging by the attached hydraulic pipe. Note the number of shims fitted between the caliper and the vertical link.
3. Prise off grease cap (11), withdraw the split pin (12) and unscrew the slotted nut (10). Remove the "D" washer and pull the hub assembly from the stub axle.
4. Wash all grease from the hub and bearings. Pack the hub and bearings with grease, working it well into the rollers and re-assemble.
5. Adjust the end float as described above.
6. Re-assemble the brake caliper unit, refitting any shims removed during dismantling.
7. Refit the road wheel, lower the jack, and repeat the above operations with the opposite wheel hub.

Lower Steering Swivel (Fig. 55)

Remove the plug (C) and fit a grease nipple. Using a grease gun charged with a recommended E.P. oil (see page 56) pump the gun until surplus oil exudes from the bearing. Remove the nipple and refit the plug.

Fig. 54

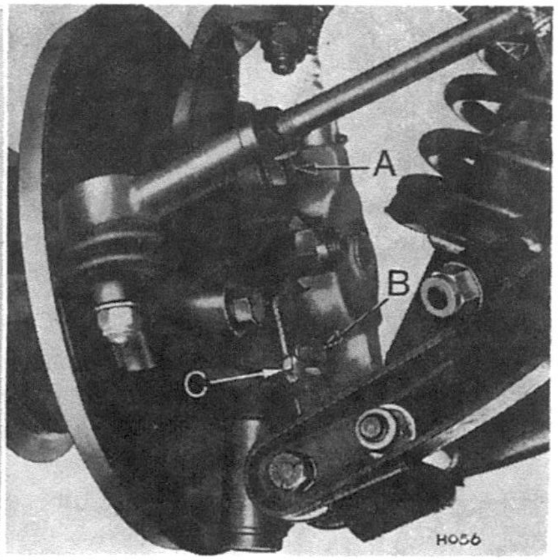

Fig. 55

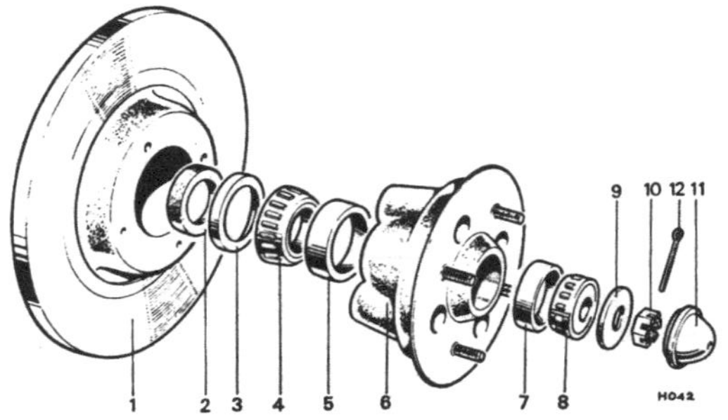

Fig. 56

Key to Fig. 56

1. Brake disc
2. Oil seal
3. Oil seal cage
4. Taper bearing (inner)
5. Bearing track (inner)
6. Hub casting
7. Bearing track (outer)
8. Taper bearing (outer)
9. "D" washer
10. Slotted nut
11. Grease cap
12. Split pin

Steering Unit and Suspension (Fig. 57)

Remove the plug from the top of the unit (shown arrowed) and fit a grease nipple. Apply the grease gun and give five strokes only. Remove the nipple and refit the plug. Over greasing can cause damage to the rubber bellows.

Have your Triumph Distributor or Dealer check the steering unit for excessive backlash and check the steering unit attachment bolts for tightness. Check also tie rods and levers and suspension nuts and bolts.

Handbrake Cable Guides (Fig. 58)

Smear grease around the handbrake cable guides shown arrowed and clevis pins.

Fig. 57

Fig. 58

Electrical Equipment
Check, and if necessary rectify any defects that may exist with the lights, heater, and warning equipment. Check headlamp beam alignment see page 34.

Body
Lubricate and check operation of hinges, door strikers and locks. Make sure surplus oil is removed to prevent damage to clothing.
Check safety harness bolts for tightness and examine the harness for wear and damage. Examine and if necessary replace worn pedal rubbers.

General
Check all nut and bolt torques and road test car.

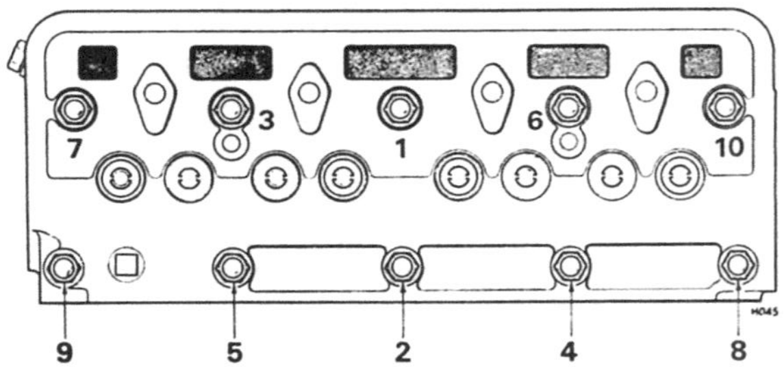

Fig. 59

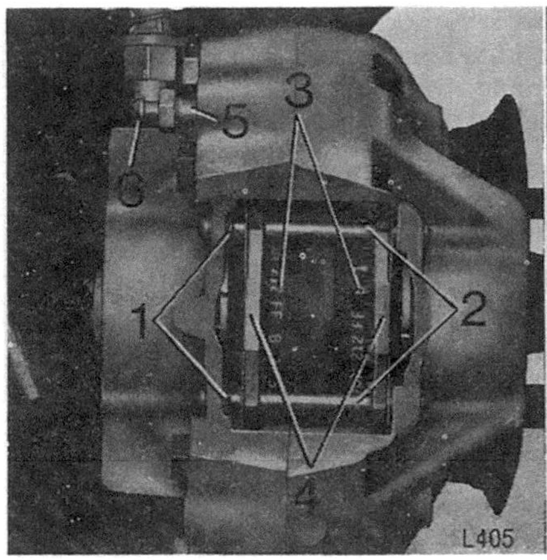

Fig. 60

Fig. 61

BRAKE AND HYDRAULIC SYSTEM

BRAKES

The mechanical design of the braking system employs self adjusting disc brakes for the front and leading and trailing manually adjusted drum brakes for the rear wheels. Handbrake operation is by cables on the rear brakes only.

Front Brake Pads (Fig. 60)
Inspect the friction linings of the front disc brake pads for wear at mileage intervals previously stated and when they are reduced to a thickness of $\frac{1}{8}$ in. (3 mm.) renew them as follows:

1. Apply the handbrake, jack up the front of the car and remove the front road wheels.
2. Release the retaining clips (1) and remove the pad retaining pins (2).
3. Lift the friction pads (3) and the anti-squeal plates (4) from the caliper.

   IMPORTANT: Do not depress the brake pedal while the pads are removed.
4. Clean the exposed faces of the pistons and the recesses into which the pads fit. Open the bleed screw (5) one turn and carefully push the pistons back into the calipers. A small amount of brake fluid will be discharged from the bled screw nipple.
5. Tighten the bled screw and fit the new brake pads. Replace the anti-squeal plates, ensuring that the arrows on the plates are pointing in the direction of the wheel rotation.
6. Insert the pad retaining pins and secure them with the spring clips.
7. Pump the brake pedal several times to re-position the pads against the disc.
8. Top-up master cylinder if necessary.
9. Replace front wheels and remove the jack.

Rear Brakes—Adjustment (Fig. 54)
Excessive foot pedal and handbrake travel indicates the need for rear brake adjustment.
Each rear brake is provided with an adjuster on the rear of the backing plate. To adjust the shoes, jack up the rear wheels and turn the adjuster (2) clockwise until the shoes are hard against the drum; then slacken the adjuster by one notch increments until the drum is free to rotate.

NOTE: There is a constant drag on the rear wheels caused by the friction of the differential and axle oil. Do not confuse this with brake drag.

Handbrake Adjustment
Handbrake "free-travel" is reduced to normal when the rear brake shoes are correctly adjusted.

HYDRAULIC SYSTEM

A split hydraulic system employs a tandem master cylinder to pressurise independent front and rear brake lines.

A brake line failure in either line will be indicated to the driver by illumination of the brake line failure warning light mounted on the facia panel (12) (Fig. 2). The dangerous possibility of continual motoring with a brake line failure, therefore, is prevented.

The warning system is actuated by a pressure differential warning actuator.

The system circuit is amalgamated with the oil pressure indication circuit—see Wiring diagram—so that when the ignition circuits are energised both warning lights will illuminate faintly to indicate no bulb filament failure.

When the engine is started the oil pressure will rise causing the oil pressure switch to break the earth contact. Both warning lights will extinguish.

Should pressure loss occur in either brake line the pressure differential warning actuator switch will operate. The "BRAKE" warning light will glow brightly.

Should the oil pressure fall below the safe operating pressure while the engine is running the oil pressure switch will actuate. Both the "OIL" and the "BRAKE" warning lights will glow faintly.

Bleeding the Hydraulic Braking System

General

If air has entered one of the hydraulic braking systems then only the system affected need be bled. During bleeding, exercise care, as described in the following procedure, to avoid moving the shuttle from its mid-position. However, if the shuttle has moved during bleeding or subsequent to a fault condition, centralise the shuttle by performing operations 1 to 5 page 51.

Preparation for Bleeding

Before commencing to bleed the brakes ensure that all the bleed nipples (6) (Fig. 60) are clean and, taking care to avoid dirt entering the fluid reservoir, remove its filler cap and top-up with new hydraulic fluid. During the bleeding operation keep the level of fluid above the dividing partition in the reservoir.

NOTE: Hydraulic fluid that has been exposed to the atmosphere or expelled from the system during bleeding must never be used for topping-up and should be discarded. Use new fluid from a sealed container, resealing the container after use.

Procedure

Commence with the brake, of the pair being bled, farthest from the master cylinder. If both systems are to be bled, bleed the rear brakes first. When bleeding the rear brakes, release the handbrake and turn the brake adjusters to lock the shoes against the drums. When the bleeding is completed adjust the brakes as detailed on page 49.

1. Attach a rubber tube of approx. ¼ in. (6 mm.) bore to the brake bleed nipple allowing the other end of the tube to hang submerged in a jar containing a quantity of clean brake fluid.
2. Unscrew the bleed-screw enough to allow the fluid to be pumped out (half a turn is normally sufficient).
3. Depress the brake pedal and allow it to return slowly noting that only a LIGHT pedal effort is required and the pedal must NOT be pushed through at the end of the stroke. (In addition, never "try" the pedal until all air has been dispelled and the system fully bled, as either action will cause the shuttle to move and actuate the switch). Pausing between each depression of the pedal, continue pumping until all air has been dispelled from the bleed-screw nipple (denoted by the absence of bubbles in the fluid being pumped into the jar).
4. With the pedal depressed, close the bleed-screw nipple and repeat the operation on the other brake.

Procedure for Re-Centralising the P.D.W.A. Shuttle*

If, for reasons previously described, the P.D.W.A. shuttle requires to be re-centralised, adopt the following procedure.

1. Fit a rubber tube, as described in 1 above, to a brake bleed-screw nipple at the opposite end of the car to that which has just been bled.
2. Open the bleed-screw.
3. Switch the ignition on but DO NOT START THE ENGINE. (The brake warning light will glow but the oil warning light will remain extinguished.)
4. Exert a steady pressure on the brake pedal until the brake light dims and the oil light glows. (A click should be felt on the pedal as the shuttle returns to its mid-position.)
5. Tighten the bleed-screw.

NOTE: If the pedal has been pushed too hard the shuttle will move to the other side of the valve, thus requiring the procedure to be repeated on a brake at the opposite end of the car.

*Pressure Differential Warning Actuator

Clutch

The clutch is of hydrostatic operation and does not require adjustment. Bleeding, when necessary, is effected from the slave cylinder (3) (Fig. 42).

EMISSION AND EVAPORATIVE CONTROL SYSTEM

Each Spitfire Mk. IV imported into the United States and Canada incorporates an efficient Emission and Evaporation control system. The systems are designed to limit the emission of Nitrogen, Hydro carbon and Carbon Monoxide from the crankcase and exhaust systems, and the emission by evaporation of fuel from the fuel delivery system.

Emission Control System—Servicing

The servicing of the emission and evaporation system is vital to the correct functioning of the system and the Warranty can be invalidated if the schedules detailed in the Passport to Service are not adhered to. Please see also the Emission Control System Maintenance and Warranty Information booklet which is supplied with the new car and should be passed, together with the Passport to Service and this handbook to all subsequent owners.

CARBURETTOR

The single Stromberg CD.SE 150 emission carburettor is the prime component of the emission system and great care is exercised during the manufacture and initial adjustment of this instrument. Because of the precise manufacturing limits involved and the assembly methods adopted to prevent unauthorised tampering during use, the extent of permissible servicing is restricted to the following:

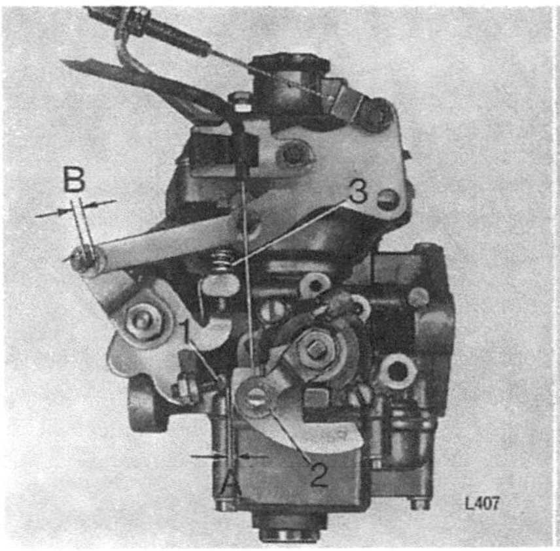

Fig. 62

Fig. 63

Adjustments (Figs. 62 and 63)

There are only three adjustments that can be made to emission carburettors in the field and these are:

1. **Idling speed:** Ensure that the fast idle screw (1) (Fig. 62) is clear of the cam (2) and the choke lever is against its stop with the facia control pushed fully in. Unscrew the idling screw (3) until the throttle is just closed. Turn the screw 1½ turns to provide a datum setting.
   Start the engine and attain normal running temperature before final adjustment of the idling screw achieves a constant 800 to 850 r.p.m.

2. **Fast idling:** Ensure that the choke lever is fully returned and the facia choke control knob pushed in. Set the gap "A", between the fast idle screw (1) and cam (2), at ·045 in. Start the engine and while it is still cold (68–86°F) pull the facia choke control fully out to check the fast idle speed and, if necessary, adjust to 1,100 r.p.m. with the screw. Tighten the locknut and re-check the fast idling speed.

3. **Idle emission:** An idle triming screw (4) (Fig. 63) is provided to give very fine adjustment to compensate for the difference between a new "stiff" engine and one that is "run-in". THIS IS NOT AN ORDINARY MIXTURE ADJUSTING SCREW; it regulates a limited amount of air that can be introduced into the mixing chamber. It is important to remember that the ear will not detect any difference between the fully "home" and fully "open" position of the screw. The setting should therefore, be checked by means of a C.O. meter or an air/fuel ratio meter to the exhaust pipe. (The correct C.O. level is ½%–2½% and air/fuel ratio is 14·4:1–13·6:1.)

Carburettor Controls

The throttle rod linkage will not require adjustments during normal operation. To ensure complete throttle closure a degree of "lost motion" or slackness is incorporated into the linkage, no attempt must be made to adjust this out "B", Fig. 62.

Occasionally lubricate the linkage and choke cable with thin oil.

Carburettor Servicing Schedules

To maintain the carburettor at peak efficiency, regular servicing at the intervals recommended in the Passport to Service is essential. The appropriate servicing operations should be performed by authorised dealers who are trained in the use of the special equipment needed.

EVAPORATION CONTROL SYSTEM—SERVICING

Minimal servicing is required on the evaporation control system apart from changing the filter gauze, in the carbon canister every 12,000 miles (10,000 km.) and replacing the canister every 48,000 miles (80,000 km.). See page 44.

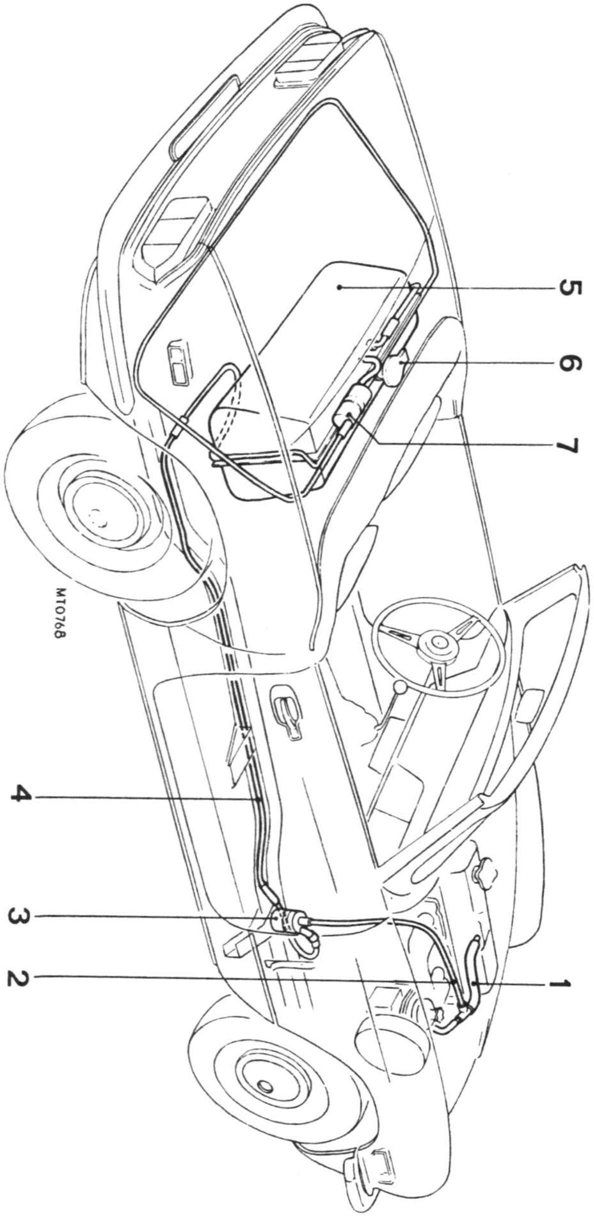

KEY TO FIG. 64—1 Pipe—Rocker cover to carburettor
2 Pipe—Canister purge
3 Activated carbon canister
4 Pipe—Seperator tank to canister
5 Main fuel tank
6 Sealed filler cap
7 Seperator tank
8 Pipe—Filler tube to overflow tank

Fig. 64

CARE OF BODYWORK

To preserve the "new look" of the body, the following suggestions are offered.

Washing

Avoid using a dry cloth to wipe dust from the paintwork and chromium surfaces. Dust is an abrasive and if removed in this way it will scratch the polished surfaces. Wash the car frequently with plenty of running water and a clean, soft sponge. Soften, and if possible remove the mud with water before using the sponge. When all dirt is removed, sponge off and dry with a clean, damp chamois-leather. Never wash or polish the car under a hot sun.

Removing Grease and Tar

Remove grease or tar with methylated spirits. White spirit is also effective, but this must not be applied to rubber, particularly the windscreen wiper blades.

Glass Surfaces

Glass is easily scratched. This can be avoided by always using a damp chamois-leather which is especially reserved for use on glass only. If silicone polishes have been used on the body, take care that the polish does not come into contact with the glass. It is extremely difficult to remove and causes the windscreen to smear.

Chromium Plating and Stainless Steel

Frequent washing and thorough drying is recommended, especially during the winter months when there is likelihood of corrosion through contamination with road salt.

Polishing

After a period of use, the formation of traffic film will cause the paintwork to lose some of its lustre, even though the car has been carefully and regularly washed. The original brilliance may be restored after washing by using a reputable non-abrasive cleaner and polish.

Being the most durable, wax preparations are preferable, but where these are used regularly the old wax must first be removed with a cleaner before further application of new wax. The frequency at which polishing is necessary will depend upon local conditions of air pollution.

Interior of Car and Soft Top

Brush and clean the inside of your car each time you wash and polish the outside. Use a vacuum cleaner where possible and ensure complete removal of all dust from the interior and trim.

Wash the upholstery and soft top with luke-warm, non-caustic, soapy water. Do not use detergents or household cleaners as these may cause damage.

Remove all traces of suds with a clean, damp cloth and thoroughly dry the upholstery with a dry duster or towel.

Wipe the facia and instrument panel with a damp cloth only. Wax or other polishes should not be used inside the car.

ANTI-FREEZE CONCENTRATION CHART

ANTI-FREEZE CONCENTRATION	25%	30%	35%
Complete Protection: Car may be driven away immediately from cold.	10°F (−12°C)	3°F (−16°C)	−4°F (−20°C)
Safe Limit: Coolant in mushy state. Engine may be started and car driven away after short warm-up period.	0°F (−17°C)	−8°F (−22°C)	−18°F (−28°C)
Lower Protection: Prevents frost damage to cylinder head, block and radiator. Engine should NOT be started until thawed)	−14°F (−27°C)	−22°F (−30°C)	−28°F (−33°C)

RECOMMENDED LUBRICANTS AND ANTIFREEZE SOLUTIONS

COMPONENT	Air temp. °C	Air temp. °F	API Designation	BP	CASTROL	DUCKHAMS	ESSO	MOBIL	PETROFINA	SHELL	TEXACO
ENGINE	over 30	over 80	MM or MS	*BP Super Visco-Static	Castrol XLR or Castrol XL	Q20-50	Esso Extra Motor Oil 20W/40	Mobiloil Super 10W/50 Mobiloil Special 20W/50.	Fina Supergrade Motor Oil 20W/50	Shell Super Motor Oil 20W/50	Havoline 20W/40
CARB. DASHPOTS	30 to 0	80 to 30	MM or MS								
OIL CAN	0 to −20	30 to −4	MM or MS		Castrolite	Q5500	Esso Extra Motor Oil 10W/30	Mobiloil Super 10W/50	Fina Supergrade Motor Oil 10W/30	Shell Super Motor Oil 10W/30	Havoline 10W/30
	below −20	below −4	MM or MS		Castrol 5W/20	Q5W/30	Esso Extra Motor Oil 5W/20	Mobiloil 5W/20	Fina Supergrade 5W/20	Shell Winter Oil 5W/20	Havoline 5W/20
GEARBOX AND OVERDRIVE REAR AXLE	over 0	over 30	GL4	BP Gear Oil SAE 90 EP	Castrol Hypoy	Duckhams Hypoid 90	Esso Gear Oil GX 90	Mobilube GX 90	Fina PONTONIC MP SAE 90	Shell Spirax 90 EP	Multigear Lubricant EP 90
LOWER STEERING SWIVELS	below 0	below 30	GL4	BP Gear Oil SAE 80 EP	Castrol Hypoy 80	Duckhams Hypoid 80	Esso Gear Oil GX 80	Mobilube GX 80	Fina PONTONIC SAE 80	Shell Spirax 80 EP	Multigear Lubricant EP 80
FRONT AND REAR HUBS BRAKE CABLES GREASE GUN				Energrease MP	Castrol MP Grease	Duckhams LB 10	Esso Multi-purpose Grease H	Mobilgrease MP	Fina Marson HTL 2	Shell Darina AX	Marfax All-purpose

* OILS MARKED THUS ARE AVAILABLE IN MULTIGRADE FORMS WITH VISCOSITY CHARACTERISTICS APPROPRIATE TO THE AMBIENT TEMPERATURE RANGE IN INDIVIDUAL MARKETS

CLUTCH AND BRAKE RESERVOIRS	CASTROL GIRLING BRAKE & CLUTCH FLUID CRIMSON. WHERE THIS PROPRIETARY BRAND IS AVAILABLE, OTHER FLUIDS WHICH MEET THE SAE J1703 A SPECIFICATION MAY BE USED								
APPROVED ANTI-FREEZE SOLUTIONS	Smiths Bluecol	BP Anti-freeze	Castrol Anti-freeze	Duckhams Anti-freeze	Esso Anti-freeze	Mobil Permazone	Fina Thermidor	Shell Anti-freeze	Startex

WHERE THESE PROPRIETARY SOLUTIONS ARE NOT AVAILABLE, OTHERS WHICH MEET BSI 3151 OR 3152 SPECIFICATION MAY BE USED

GENERAL SPECIFICATION

Engine

Number of cylinders	4.
Bore of cylinders	2·9 in. 73·7 mm — 1500 model: 3·44 in. 87·5 mm
Stroke of crankshaft	2·99 in. 76·0 mm
Cubic capacity	79·2 in.^{3} 1296 c.c.
Piston area	26·5 in.^{2} 171·0 cm.^{2}
Compression ratio	8·0:1 — 1500 model: 7·5:1
Valve rocker clearances (cold)	0·010 in. 0·25 mm.
Valve timing	Inlet and exhaust valve equally open at T.D.C.
Maximum power (1300)	48 b.h.p. nett at 5,500 r.p.m. (S.A.E. nett)
Maximum torque (1300)	730 lb. ins. at 2,900 r.p.m. (S.A.E. nett)
Equivalent to B.M.E.P.	116 lbs./sq. in. b.m.e.p.
Maximum power (1500)	57 b.h.p. nett at 5,000 r.p.m. (S.A.E. nett)
Maximum torque (1500)	885 lb. ins. at 3,000 r.p.m. (S.A.E. nett)
Equivalent to B.M.E.P.	122 lbs./sq. in. b.m.e.p.

Lubrication (Engine)

Pump	High capacity eccentric lobe type.
Filter	Full flow type. Replaceable element.

Cooling System

Cooling System	"No loss" system incorporating plastic overflow reservoir.
Circulation	Pump, driven by "Vee" belt; thermostatically controlled flow.
Fan	Seven blade 11½ in. dia. (292 mm.).

Fuel System

Fuel System	Tank above rear axle. No reserve tap.
Pump	A.C. mechanically operated diaphragm type.
Carburettor	Single 1·50 C.D.S.E. Zenith Stromberg (EMISSION). Side draught.
Needle size	B5 CH.
Manifolds	Cast iron exhaust manifold, cast aluminium inlet manifold water heated.
Air Cleaner	Replaceable paper element.
Crankcase breathing	Positive crankcase ventilation valveless closed circuit breathing from rocker cover using carburettor depression.
Filter	Replaceable fitted in tank/pipe-line.
Evaporative emission control	Sealed fuel filler cap. Emission vented into charcoal canister located in the engine compartment. Canister purged by carburettor depression.

Ignition System

Coil	Lucas 16C6 6 volt with ballast resistor coil.
Distributor—type	Delco-Remy.
contact breaker gap	0·015 in. 0·4 mm.
rotation—viewed on rotor	Anti-clockwise.
Firing order	1-3-4-2.
Sparking plugs—type	Champion UN-12Y.
—gap	0·025 in. 0·64 mm.
Ignition timing (static)	8° B.T.D.C.
Ignition timing (dynamic)	2° A.T.D.C. @ 800–850 r.p.m.

Electrical System

Voltage	12.
Polarity	Negative earth.
Fuses	35 amp.
Battery—type	12 volt 40 amp. hr. @ 20 hr. rate
Alternator—type	Lucas 15ACR.
—output	28 amps.
Oil pressure switch operating pressure	3/5 p.s.i. 0·2/0·35 kg./cm.^{2}
Starter motor	Lucas M35J inertia type.

Transmission

Clutch	Diaphragm type. 6½ in. dia. (165 mm.)
Gearbox	4 speed constant mesh. Synchromesh on all forward ratios.

	Top	3rd	2nd	1st	Rev.
Ratios	1	1·39	2·16	3·50	3·99
Overall ratios	4·11	5·71	8·87	14·40	16·39

Overdrive (optional)	Laycock de Normanville. Operates on 3rd and Top gears. Ratio 0·802:1.
Propellor shaft	All metal. Needle roller bearings.
Rear axle	Hypoid bevel drive. Ratio 4·11
Wheels	Pressed steel disc type 4½ J rim. Bolt on wire wheels (optional).
Tyres	5·20S-13 tubeless. Cross ply 1·45SR-13 Radial ply (optional).

Steering

Turning circle	24 ft. 7·3 m.
Steering unit	Rack and pinion 3¾ turns lock to lock. Impact absorbing steering column with anti-theft locking device.

Suspension

Front	Independent double wishbone type, with rubber bushed wishbone pivots. Patented screwed bottom bush and top ball joint swivels. Coil springs controlled by telescopic type direct acting dampers and anti-roll bar. Taper roller bearings in hub.
Rear	Swing axle type independent suspension, pivoted transverse leaf spring and radius rods direct acting telescopic dampers. Ball and needle roller bearings in hubs.

Brake System — Girling hydraulic. Centrally mounted hand-brake coupled mechanically to rear wheels only.

Pedal operates on tandem master cylinder to front and rear brakes independently. Pressure differential valve and warning light fitted to indicate partial system failure.

Front	Caliper disc 9 in. dia. (229 mm.).	
Rear	Drum 7 in. dia. (178 mm.) × 1¼ in. (32 mm.). wide. Leading and trailing shoe.	
Front lining area	14·8 in.²	95 cm.²
Front swept area	150 in.²	967 cm.²
Rear lining area	34 in.²	220 cm.²
Rear swept area	55 in.²	355 cm.²
Total lining area	48·8 in.²	315 cm.²
Total swept area	205 in.²	1322 cm.²

Exterior Dimensions

Overall length	12 ft. 5 ins.	3785 mm.
Width	4 ft. 10½ ins.	1488 mm.
Height with hood erected (unladen)	3 ft. 11½ in.	1205 mm.
Height with hood folded (unladen) to top of screen	3 ft. 8¼ in.	1125 mm.

Weight (approx.)

Dry (excluding extra equipment)	1620 lbs.	735 kg.
Complete (inc. fuel, oil, water and tools)	1708 lbs.	775 kg.
Maximum gross vehicle weight	2230 lbs.	1011 kg.

Chassis Data

Frame	Double backbone of closed channel section with channel outriggers.	
Wheelbase	6 ft. 11 in.	2110 mm.
Track—Front	4 ft. 1 in.	1245 mm.
—Rear	4 ft.	1220 mm.
Ground clearance	5 in.	127 mm.

Capacities	U.S.A.	Imperial	Metric
Fuel tank	9·9 galls.	8·25 galls.	37·6 litres
Engine sump	9·6 pints	8 pints	4·5 litres
Engine sump Drain and refil	8·4 pints	7 pints	4·0 litres
Gearbox	1·8 pints	1·5 pints	0·85 litres
Gearbox and overdrive	3·0 pints	2·5 pints	1·4 litres
Rear axle	1·2 pints	1 pint	0·57 litres
Cooling system (inc. water bottle)	8·4 pints	7 pints	4·0 litres
Cooling system (inc. water bottle) with heater	9·6 pints	8 pints	4·5 litres

Road Speed Data

	Engine Speed r.p.m.				
	Top	**3rd**	**2nd**	**1st**	**Rev.**
At road speed of 10 m.p.h.	628	872	1356	2202	2512
At road speed of 10 km.p.h.	390	542	842	1369	1560
Road speed at 1,000 r.p.m. in top gear	=15·9 m.p.h.		=26 km.p.h.		
Road speed at 2,500 ft./min. in top gear	=79·8 m.p.h.		=129 km.p.h.		

Overdrive

Overall ratio	Top gear 3·29	3rd gear 4·58.
	Engine Speed r.p.m.	
At road speed of 10 m.p.h.	Top gear 503	3rd gear 701
At road speed of 10 k.p.h.	313	436
Road speed at 1,000 r.p.m. in top gear	19·9 m.p.h.	32 km.p.h.

TRIUMPH SPITFIRE
MK III, MK IV & 1500
WORKSHOP MANUAL

PART 3

Purchasers are advised that the specification details set out in this Manual apply to a range of vehicles and not to any one. For the specification of a particular vehicle, purchasers should consult their Distributor or Dealer.

The Manufacturers reserve the right to vary their specifications with or without notice, and at such times and in such manner as they think fit. Major as well as minor changes may be involved in accordance with the Manufacturer's policy of constant product improvement.

Whilst every effort is made to ensure the accuracy of the particulars contained in this Manual, neither the Manufacturer nor the Distributor or Dealer, by whom this Manual is supplied, shall in any circumstances be held liable for any inaccuracy or the consequences thereof.

[NOTE: Any material added to this *Workshop Manual* by Robert Bentley, Inc. has been indicated by the use of brackets ([ ]). Material from other British Leyland publications, which has been added to include information pertaining to the MK. III model and the 1500 model, has been indicated by the addition of a source reference at the bottom of the appropriate page.]

CONTENTS

[NOTE: From page 65 to page 470 of this Manual, when a reference on one page is made to information found on another page, numbers (usually six) are used. These numbers are found in the lower outside margin of each page.

For example, on page 93 (numbered 10.00.09 in the lower right-hand margin) we are referred at the top of the page to 86.35.14. This number reference is found in the lower left-hand margin of page 424.]

INTRODUCTION

The purpose of this manual is to assist skilled mechanics in the efficient repair and maintenance of British Leyland vehicles. Using the appropriate service tools and carrying out the procedures as detailed will enable the operations to be completed in the time stated in the 'Repair Operation Times'.

Indexing

For convenience, the manual is divided into a number of divisions.

A list of the operations within each division appears in alphabetical order on the page preceding each division.

Operation Numbering

A master index of numbered operations has been compiled for universal application to all vehicles manufactured by the British Leyland Motor Corporation and, therefore, because of the different specifications of various models, continuity of the numbering sequence cannot be maintained throughout this manual.

Each operation described in the manual is allocated a number from the master index and cross-refers with an identical number in the 'Repair Operation Times'. The number consists of six digits arranged in three pairs.

Each instruction within an operation has a sequence number and, to complete the operation in the minimum time, it is essential that the instructions are perfórmed in numerical sequence commencing at 1 unless otherwise stated. Where applicable, the sequence numbers identify the relevant components in the appropriate illustration.

Emission Control

With the exception of Section 17, all remaining sections of this manual relate to basic vehicles not fitted with *anti-pollution* equipment. Where an operation is affected by the presence of this equipment, refer also to Anti-pollution (Section 17) as appropriate.

Service Tools

Where performance of an operation requires the use of a service tool, the tool number is quoted under the operation heading and is repeated in, or following, the instruction involving its use. An illustrated list of all necessary tools is included in section 99.

References

References to the left- or right-hand side in the manual are made when viewing from the rear. With the engine and gearbox assembly removed, the 'timing cover' end of the engine is referred to as the front. A key to abbreviations and symbols is given on page 01-5.

REPAIRS AND REPLACEMENTS

When service parts are required it is essential that only genuine British Leyland Stanpart or Unipart replacements are used.

Attention is particularly drawn to the following points concerning repairs and the fitting of replacement parts and accessories.

Safety features embodied in the car may be impaired if other than genuine parts are fitted. In certain territories legislation prohibits the fitting of parts not to the vehicle manufacturer's specification. Torque wrench setting figures given in the Repair Operation Manual must be strictly adhered to. Locking devices, where specified, must be fitted. If the efficiency of a locking device is impaired during removal it must be renewed. Owners purchasing accessories while travelling abroad should ensure that the accessory and its fitted location on the car conform to mandatory requirements in their country of origin.

The car warranty may be invalidated by the fitting of other than genuine British Leyland parts. All British Leyland Stanpart or Unipart replacements have the full backing of the factory warranty.

British Leyland Distributors and Dealers are obliged to supply only genuine service parts.

ABBREVIATIONS AND SYMBOLS

Across flats (bolt size)	A.F.
After bottom dead centre	A.B.D.C.
After top dead centre	A.T.D.C.
Alternating current	a.c.
Amperes	amp
Ampere-hour	Ah
Before bottom dead centre	B.B.D.C.
Before top dead centre	B.T.D.C.
Bottom dead centre	B.D.C.
Brake horse-power	b.h.p.
Brake mean effective pressure	b.m.e.p.
British Standards	B.S.
Carbon monoxide	CO
Centigrade (Celsius)	C
Centimetres	cm
Cubic centimetres	cm^3
Cubic inches	in^3
Cycles per minute	c/min
Degree (angle)	deg. or °
Degree (temperature)	deg. or °
Diameter	dia.
Direct current	d.c.
Fahrenheit	F
Feet	ft
Feet per minute	ft/min
Fifth	5th
Figure (illustration)	Fig.
First	1st
Fourth	4th
Gallons (Imperial)	gal
Gallons (U.S.)	U.S. gal
Grammes	g
High compression	h.c.
High tension (electrical)	h.t.
Horse-power	hp
Hundredweight	cwt
Inches	in
Inches of mercury	inHg
Independent front suspension	i.f.s.
Internal diameter	i.dia.
Kilogrammes (force)	kgf
Kilogrammes (mass)	kg
Kilogramme centimetre	kgf cm
Kilogramme metres	kgf m
Kilogrammes per square centimetre	kg/cm^2
Kilometres	km
Kilometres per hour	km/h
Kilovolts	kV
King pin inclination	k.p.i.
Left-hand	L.H.
Left-hand steering	L.H.Stg.
Left-hand thread	L.H.Thd.
Low compression	l.c.
Low tension	l.t.
Maximum	max.
Metres	m
Microfarad	mfd
Midget Edison Screw	MES
Miles per gallon	m.p.g.
Miles per hour	m.p.h.
Millimetres	mm
Millimetres of mercury	mmHg
Minimum	min.
Minus (of tolerance)	–
Minute (of angle)	
Negative (electrical)	–
Number	No.
Ohms	ohm
Ounces (force)	ozf
Ounces (mass)	oz
Ounce inch (torque)	ozf in
Outside diameter	o.dia.
Overdrive	O/D
Paragraphs	para.
Part Number	Part No.
Percentage	%
Pints (Imperial)	pt
Pints (U.S.)	U.S. pt
Plus or minus	±
Plus (tolerance)	+
Positive (electrical)	+
Pounds (force)	lbf
Pounds (mass)	lb
Pounds feet (torque)	lbf ft
Pounds inches (torque)	lbf in
Pounds per square inch	lb/in^2
Radius	r
Ratio	:
Reference	ref.
Revolutions per minute	rev/min
Right-hand	R.H.
Right-hand steering	R.H.Stg.
Second (angle)	"
Second (numerical order)	2nd
Single carburetter	SC
Society of Automobile Engineers	S.A.E.
Specific gravity	sp. gr.
Square centimetres	cm^2
Square inches	in^2
Standard	std
Standard wire gauge	s.w.g.
Synchronizer/synchromesh	synchro.
Third	3rd
Top dead centre	T.D.C.
Twin carburetters	TC
United Kingdom	UK
Volts	V
Watts	W

Screw threads

American Standard Taper Pipe	N.P.T.F.
British Association	B.A.
British Standard Fine	B.S.F.
British Standard Pipe	B.S.P.
British Standard Whitworth	B.S.W.
Unified Coarse	U.N.C.
Unified Fine	U.N.F.

LOCATION OF COMMISSION AND UNIT NUMBERS

THE COMMISSION NUMBER is the identification number which is required for registration and other purposes. It is stamped on a plate attached to the left hand bulkhead panel (not U.S.A.) and is visable when the bonnet is raised. On vehicles for the U.S.A. type markets this plate is attached to the body adjacent to the left hand door striker plate and the Commission Number is also stamped on a small plate visable through the left hand side of the windscreen.

The significance of the Commission Numbers and suffix is as follows:

FH – this prefix denotes 'Spitfire Mk IV' model range.
FK – is an alternative prefix for U.S.A. market only.
FL – is an alternative prefix for Swedish market (pre 1972 condition only).
1234 – is the accumulated total build of this model.
L – denotes left hand steering
(No letter is given to right hand steering models).
U – denotes U.S.A. type markets 1972 condition.

The Commission Number plate also bears code symbols for identification of the vehicle's exterior colour, trim material and trim colour. Refer to page 04-6.

Mk. IV and 1500

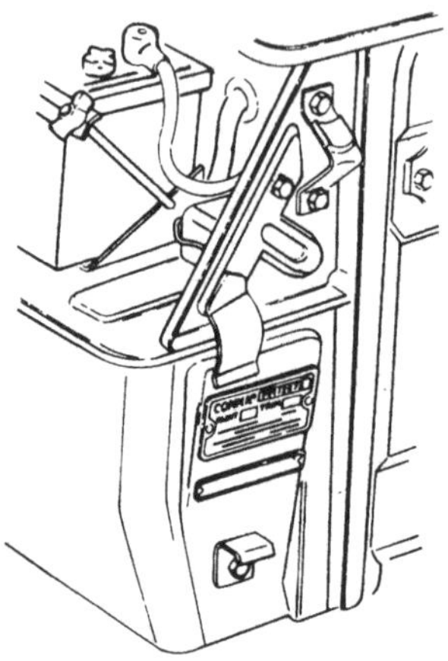

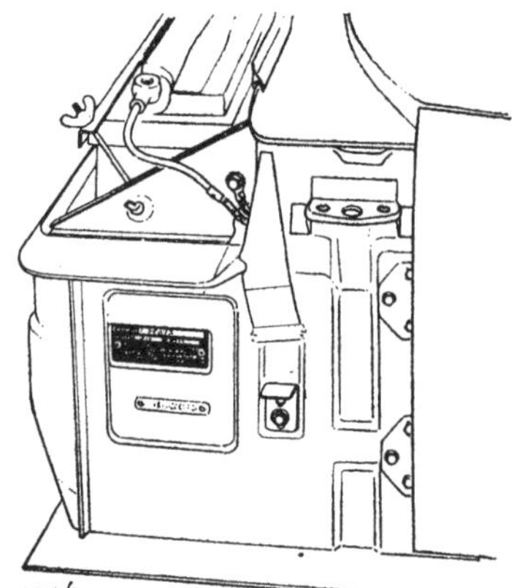

J191/1

SPITFIRE MK. III

Commission Numbers	FD1 and future — built from January 1967
Engine Numbers	FD1 or FE1 .... HE or LE
Gearbox Numbers	FD1
Rear Axle Numbers	FC
Body Numbers	81311FC to 81732FC; 422FD and future

On Mk. III models the Commission Number (Chassis Number), paint and trim code numbers are stamped on the plate attached to the dash side panel. The Body Number is stamped on a separate plate mounted below the Commission Number plate.

Mk. III, Mk. IV & 1500

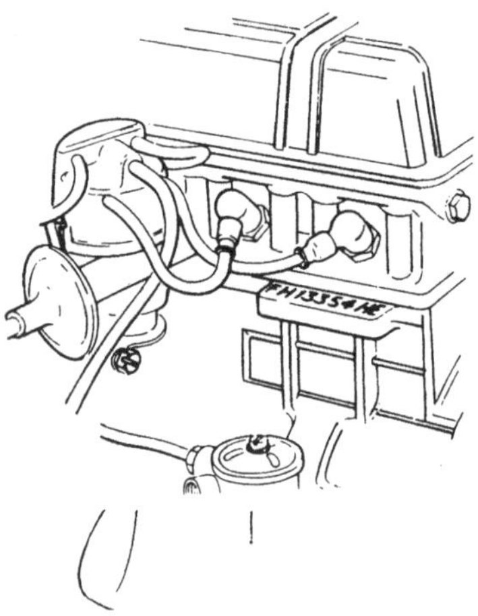

THE ENGINE NUMBER is stamped on a machined flange on the left hand side of the cylinder block. The significance of the Engine Numbers and suffix is as follows:

FH – this prefix denotes model range.
FK – an alternative prefix for U.S.A. market only.
FL – an alternative prefix for Swedish market (pre 1972 condition only).
1234 – is the accumulated total build of this type.
H – denotes High Compression. Alternatively.
L – denotes Low Compression. Alternatively.
U – denotes U.S.A. type markets 1972 condition.
E – denotes engine unit.

Triumph Spitfire Mk IV Manual. Part No. 545254. Issue 1.
Part No. 511243 page 0.108. Issue 1.

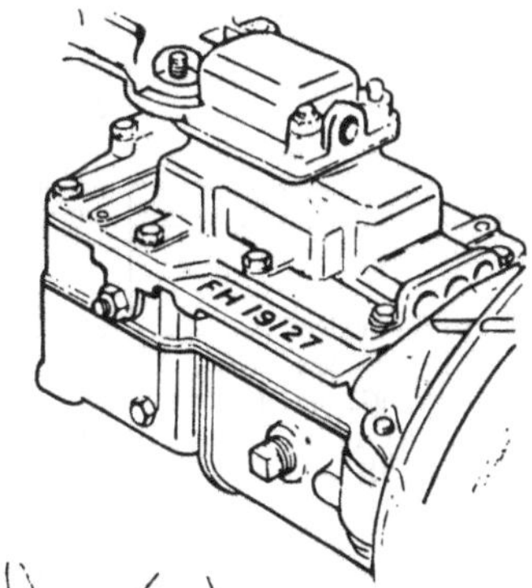

THE GEARBOX NUMBER is stamped on the right hand side of the gearbox casing. The significance of the Gearbox Numbers is as follows:

FH – this prefix denotes model range.
1234 – is the accumulated build of this type. There are no suffix letters.

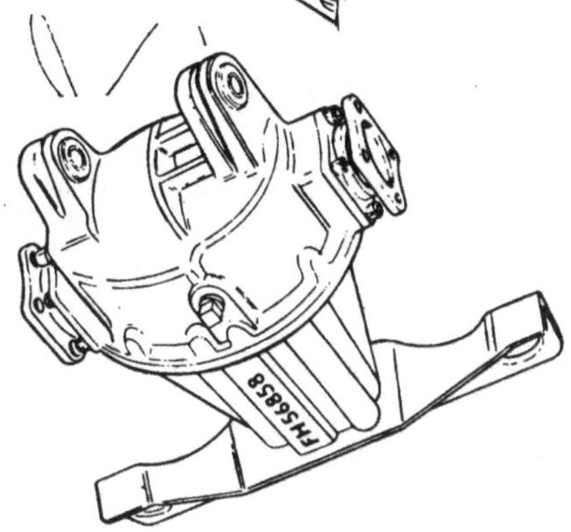

THE REAR AXLE NUMBER is stamped on the bottom flange of the axle housing. The significance of the Axle Numbers is as follows:

FH – this prefix denotes model range.
FK – an alternative prefix for U.S.A. market only 1972 condition
1234 – is the accumulated total build of this type. There are no suffix numbers.

IMPORTANT: In all communications relating to Service and Spares it is essential to quote Commission Number, paint and trim codes and unit numbers (if applicable).

		U.S.A. market differences
ENGINE		
Number of cylinders	4 in line	—
Bore of cylinders	2·9 in (73·7 mm)	—
Stroke of crankshaft	2·992 in (76 mm)	3·44 in (87·5 mm)— 1500 models only
Capacity	79·2 in³ (1296 cm³)	91 in³ (1493 cc) — 1500 models only
Compression ratio	9·0:1	7·5:1 — 1500 models only 8·0:1 — 1972 models only 8·5:1 — Mk. III from Jan. 1, 1970
Maximum Power — 1973 models	61 bhp nett at 5500 rev/min	57 bhp SAE nett at 5000 rev/min
— 1972 models	61 bhp nett at 5500 rev/min	48 bhp SAE nett at 5500 rev/min
— pre 1972 models Mk. IV	63 bhp nett at 6000 rev/min	58 bhp nett at 5200 rev/min
— Mk. III	75 bhp nett at 6000 rev/min	68 bhp nett at 5500 rev/min — from Jan. 1, 1970
Maximum Torque — 1973 models	818 lbf/in at 2900 rev/min equivalent to 130 lb/in² bmep	885 lbf/in SAE nett at 3000 rev/min equivalent to 122 lb/in² bmep
— 1972 models	818 lbf/in at 2900 rev/min equivalent to 130 lb/in² bmep	730 lbf/in SAE nett at 2900 rev/min equivalent to 116 lb/in² bmep
— pre 1972 model Mk. IV	837 lbf/in at 3500 rev/min equivalent to 133 lb/in² bmep	*860 lbf/in at 3000 rev/min equivalent to 137 lb/in² bmep
— Mk. III	900 lbf/in at 4000 rev/min equivalent to 144 lb/in² bmep	880 lbf/in at 5500 rev/min equivalent to 140 lb/in² bmep — from Jan. 1, 1970
LUBRICATION		
Oil pump	High capacity, eccentric lobe type	—
Oil filter	Full flow type, replaceable element	—
Oil warning light	Extinguishes at 3 to 5 lb/in² (0·21 to 0·35 kg/cm²) oil pressure	—
COOLING SYSTEM		
Type	Water, "No Loss" system	—
Circulation	By impellor type pump, Vee belt drive	—
Pressure	13 lb/in² (0·91 kg/cm²)	—
Thermostat	Opens at 82°C (180°F) normal 88°C (190°F) cold climate	— U.S.A. market only from engine number FK 33745UE — 88°C (190°F) cold climate
Fan — Mk. IV and 1500	7 blades, 11½ in diameter (292 mm)	—
— Mk. III	4 blades, 12⅛ in diameter (308 mm)	—
FUEL SYSTEM		
Tank	Tank at rear	Tank at rear with separate overflow tank — from Jan. 1, 1970
Pump	Mechanically operated, diaphragm type	—
Carburetter — 1973 model	Two, SU HS2E side draught exhaust emission controlled	Single, Stromberg 1·50 C.D.S.E.V. exhaust emission controlled
— 1972 model	Two, SU HS2E side draught exhaust emission controlled	Single, Stromberg 1·50 C.D.S.E. exhaust emission controlled
— pre 1972 model Mk. IV	Two, SU HS2E side draught	Single, Stromberg 1·50 C.D.S.E. exhaust emission controlled
— Mk. III	Two, SU HS2 side draught	Two, SU HS2E side draught
Air cleaner	Combined air cleaner and silencer with replaceable paper elements	—
Crankcase ventilation — from Jan. 1, 1970	Closed circuit breathing from rocker cover to constant depression side of carburetters	—
— through Dec. 1969	Open P.C.V. with diaphragm-type valve	—
Evaporative emission control — from Jan. 1, 1970	—	Sealed tank filler cap. Vapour emissions from the tank are vented via a separator tank to a charcoal canister located in the engine compartment. Canister is purged by carburetter depression.
CLUTCH		
Make/type	Borg and Beck single dry plate, diaphragm spring type	—
Release mechanism	Hydraulically operated	—
Plate diameter	6½ in (165 mm)	7¼ in (184 mm) — 1500 models only
Facing material	Mintex M19	H-K Porter 11046 — 1500 models only

*Also applicable to Swedish Market pre-1972 condition

		U.S.A. market differences
GEARBOX Manual		
Synchromesh — Mk. IV and 1500	On all forward gears	—
— Mk. III	On top three forward gears only	—

	O/D Top	Top	O/D 3rd	3rd	2nd	1st	Rev.
Gear ratios — Mk. IV and 1500	—	1·00	—	1·39	2·16	3·50	3·99
— Mk. III	—	1·00	—	1·39	2·16	3·75	3·75
Overall ratios — U.S.A. 1972 only	3·29	4·11	4·58	5·71	8·87	14·40	16·39
— U.S.A. pre 1972 & other markets	3·12	3·89	4·34	5·41	8·41	13·65	15·55
— Mk. III	3·12	3·89	4·34	5·41	8·41	15·42	15·42

		U.S.A. market differences
Overdrive (optional)		
Make/type	Laycock, Type D	—
Operative on	Top & 3rd gears	—
Overall ratio	0·802:1	—
FINAL DRIVE		
Type	Hypoid bevel gears in rear axle	—
Ratio — Mk. IV and 1500	3·89:1	4·11:1 — 1972 condition only
— Mk. III	4·11:1	—

EFFECTIVE GEARING (Approximate values depending on type and conditon of tyre fitted)

Engine speeds (rev/min) at road speeds of:

	O/D Top	Top	O/D 3rd	3rd	2nd	1st	Rev.
10 m.p.h. — U.S.A. 1972 only	503	628	701	872	1356	2202	2511
— Mk. IV U.S.A. pre 1972 and other markets	482	599	669	834	1295	2100	2395
— Mk. III	565	635	715	890	1375	2385	2385
10 km/h — U.S.A. 1972 only	313	390	436	542	842	1360	1555
— Mk. IV U.S.A. pre 1972 and other markets	301	372	417	518	803	1304	1488
— Mk. III	350	395	490	550	875	1490	1490

ROAD SPEED DATA (Approximate values depending on type and condition of tyre fitted)

		U.S.A. market differences
Road speed at 1,000 rev/min engine speed:		
O/D top — Mk. III	19·6 m.p.h. (31·3 km/h)	—
— 1972 only	—	19·9 m.p.h. (32·0 km/h)
— Mk. IV pre 1972	—	20·8 m.p.h. (33·5 km/h) — and other markets
Top gear — Mk. III	15·75 m.p.h. (25·2 km/h)	—
— 1972 only	—	15·9 m.p.h. (26·0 km/h)
— Mk. IV pre 1972	—	16·7 m.p.h. (27·0 km/h) — and other markets
Road speed at 2,500 ft/min piston speed:		
Top gear — 1500 models	72·7 m.p.h. (116;3 km/h)	—
— 1972 only	—	79·8 m.p.h. (129·0 km/h)
— Mk. III	79 m.p.h. (126·6 km/h)	—
— Mk. IV pre 1972	—	83·6 m.p.h. (134·0 km/h) — and other markets
STEERING		
Make/type	Alford and Alder, Rack and pinion	—
Turning circle	24 feet (7·3 metres)	—
Steering wheel diameter — 1973 models	14·5 in (368 mm) — turns, lock to lock 3¾	—
— pre 1973 models	15 in (381 mm) — turns, lock to lock 3¾	—
BRAKE SYSTEM		
Foot pedal operation	Hydraulic on all four wheels	Tandem master cylinder operating front and rear brakes independently
Handbrake operation	Mechanical on rear wheels only	—
Front		
Type	Caliper disc	—
Dimensions	Disc diameter 9 in (229 mm)	—
Lining area	14·8 in² (95 cm²)	—
Swept area	150·0 in² (967 cm²)	—
Rear		
Type	Drum with leading and trailing shoes	—
Dimensions	7 in dia. x 1¼ in wide (178 mm x 32 mm)	—
Lining area	34·0 in² (220 cm²)	—
Swept area	55·0 in² (355 cm²)	—

		U.S.A. market differences
WHEELS AND TYRES		
Wheels	Steel disc type, 4½J rims (Wire wheels optional)	—
Tyres	155SR - 13 radial ply tyres standard from Commission No. FH 38271	(optional on U.S. 1500 models)
	145SR - 13 SP68 radial ply tyres optional	5·20S - 13 cross ply tubeless with narrow white side band
Tyre pressures (cross ply and radial ply)		
— front	21 lb/in² (1·476 kg/cm²)	—
— rear	26 lb/in² (1·828 kg/cm²)	—
CHASSIS DATA		
Wheelbase	6 ft 11 in (2108 mm)	—
Track: (2 up condition)	Disc Wheels / Wire wheels	
front	4 ft 1 in (1244 mm) / 4 ft 1½ in (1257 mm)	
— 1500 model	4 ft 1 in (1244 mm) / 4 ft 1¼ in (1251 mm)	
rear	4 ft 0 in (1220 mm) / 4 ft ½ in (1232 mm)	
— 1500 model	4 ft 2 in (1270 mm) / 4 ft 2¼ in (1276 mm)	
Wheel alignment: (2 up condition) front	0 to 1/16 in toe-in (0 to 1·5875 mm)	—
rear	0 to 1/16 in toe-in (0 to 1·5875 mm)	—
Ground clearance: (2 up condition)	5 in (127 mm)	—
Camber: (2 up condition) front	2° positive ± ½°	—
rear	3¾° negative ± 1°	—
— Mk. III	3° negative	—
Caster: (2 up condition)	4½° ± ½°	—
— Mk. III	4° positive	—
King pin inclination (2 up condition)	6¾° ± ¾°	—
ELECTRICAL EQUIPMENT		
Electrical system	12 volt negative earth	—
Battery capacity	40 amp hour at 20 hour rate	—
— Mk. III	43 amp hour at 20 hour rate	—
Alternator: type	Lucas 15 ACR	—
— 1500 model	Lucas 16 ACR	—
(Generator) — Mk. III	Lucas C40-1	—
output	28 amps	—
— 1500 model	34 amps	—
(Generator) — Mk. III	22 amps	—
Starter motor	M 35 J inertia type	—
— early Mk. III	M 35 G inertia type	—
OVER £LL DIMENSIONS		
Length	12 ft 5 in (3785 mm)	—
— Mk. III	12 ft 2½ in (3721 mm)	—
Width	4 ft 10½ in (1488 mm)	—
— Mk. III	4 ft 9 in (1448 mm)	—
Height (unladen) to top of windscreen	3 ft 8¼ in (1125 mm)	—
Soft top, hood erect	3 ft 11½ in (1205 mm)	—
WEIGHTS (Approximate)		
Dry (excluding extra equipment)	14½ cwt (734 kg)	—
— Mk. III	14 cwt (712 kg)	—
Basic kerb (including water, oil, fuel & tools)	15¼ cwt (775 kg)	—
— Mk. III	14¾ cwt (748 kg)	—
Kerb (including optional extras, etc.)	16¼ cwt (822 kg)	—
Maximum gross vehicle weight	20 cwt (1011 kg)	—
— Mk. III	17¾ cwt (900 kg)	—
TOWING INFORMATION (Not applicable to Mk. III)		
Maximum recommended trailer weight	12 cwt (610 kg) when the trailer being towed is equipped with brakes.	—
	3·94 cwt (200 kg) when the trailer being towed is not equipped with brakes — providing that the total car and trailer laden weights do not exceed 20·89 cwt (1112 kg).	—
Maximum starting gradient (fully laden car and trailer)	5·5 with car engine in peak condition	—
Maximum climable gradient (fully laden car and trailer)	4·3 with car engine in peak condition	—

Triumph Spitfire Mk IV Manual. Part No. 545254. Issue 1.
Part No. 511243 page 0.104. Issue 1.

VEHICLE DIMENSIONS — MK. III

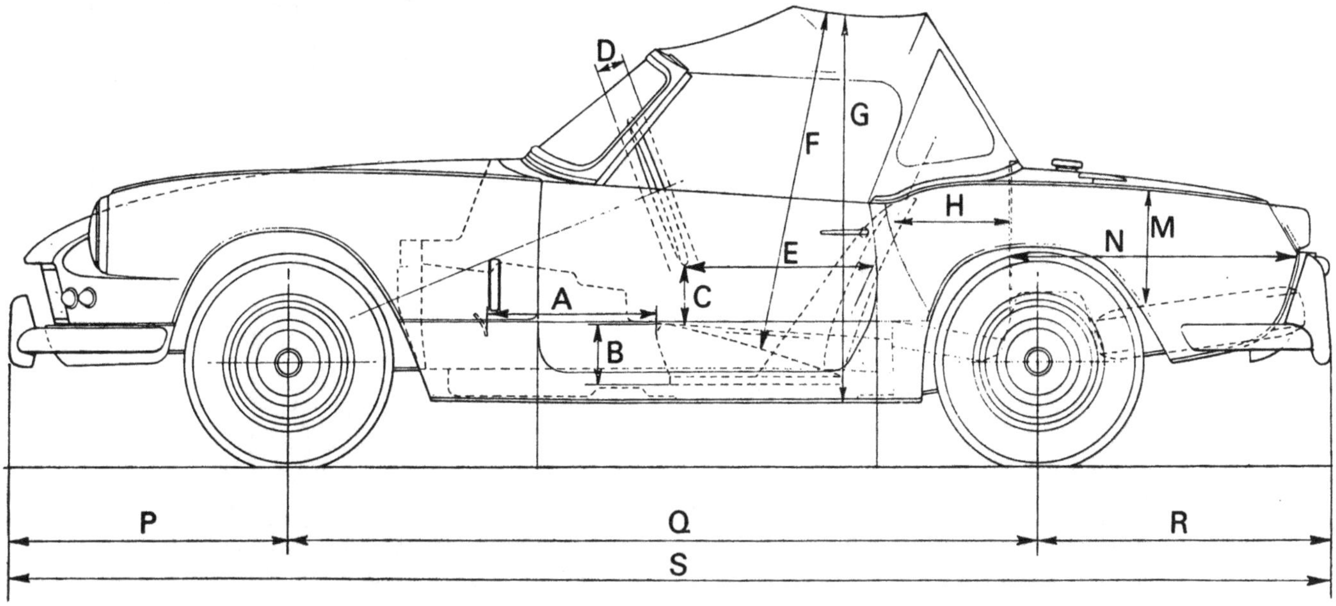

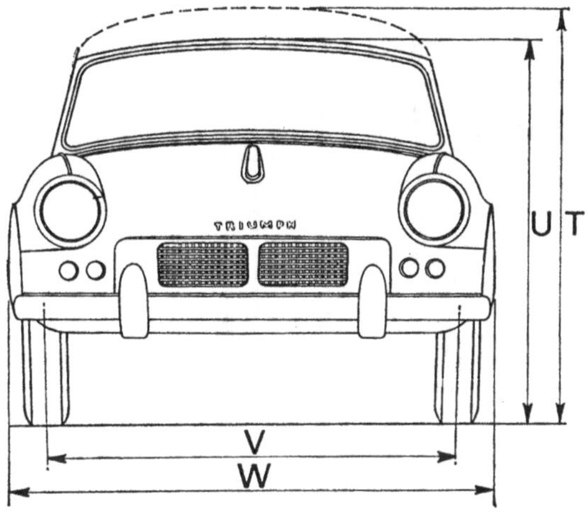

	SPITFIRE MK. III (above)	
A Max.	2′ 2½″	67·00 cm.
Min.	1′ 7½″	49·53 cm.
B	7″	17·78 cm.
C	7″	17·78 cm.
D	4″	10·16 cm.
E Max.	1′ 10″	55·88 cm.
Min.	1′ 2½″	38·83 cm.

	SPITFIRE MK. III (continued)	
F	2′ 11″	88·90 cm.
G	3′ 8″	111·76 cm.
H Max.	1′ 4¾″	42·54 cm.
Min.	9½″	24·15 cm.
N	2′ 11″	88·90 cm.
M	10″	25·40 cm.
P	2′ 7″	78·74 cm.
Q	6′ 11″	210·82 cm.
R	2′ 7½″	80·01 cm.
S	12′ 1″	368·30 cm.
T	3′ 11½″	120·65 cm.
U	3′ 8¼″	111·40 cm.
V	4′ 1″	124·46 cm.
Rear Track	4′ 0″	121·90 cm.
W	4′ 9″	144·78 cm.

VEHICLE DIMENSIONS — MK. IV and 1500

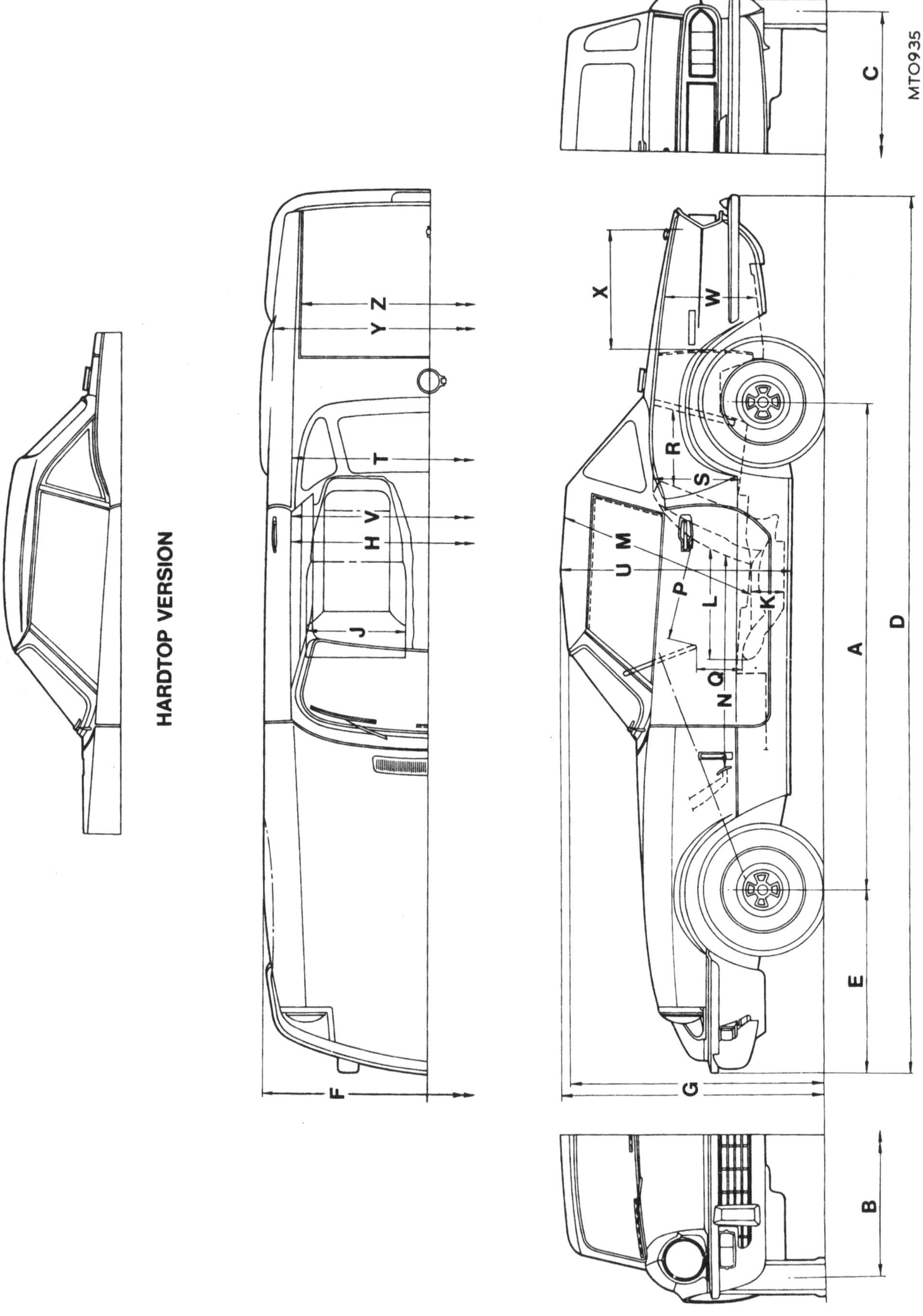

Triumph Spitfire Mk IV Manual. Part No. 545254. Issue 1.

VEHICLE DIMENSIONS — MK. IV and 1500

Dim.	Description	inches	mm
A	Wheelbase	83·00	2108
B	Front track:		
	Disc wheels	49·00	1244
	Wire wheels	49·50	1257
C	Rear track:		
	Disc wheels	48·00	1220
	Wire wheels	48·50	1232
D	Overall length	149·00	3785
E	Front overhang	31·06	788
F	Overall width	58.50	1488
G	Height:		
	Soft top hood erect	47·50	1205
	to top of windscreen	44·25	1125
H	Width – door to door (shoulder height)	45·50	1156
J	Seat width	18·00	457
K	Seat height – floor to cushion	7·50	191
L	Seat depth	20·00	508
M	Headroom from seat cushion	35·00	889
N	Seat squab to clutch pedal:		
	Max.	41·50	1055
	Min.	35·50	902

Dim.	Description	inches	mm
P	Seat squab to steering wheel:		
	Max.	18·00	457
	Min.	12·00	305
Q	Seat cushion to steering wheel	7·00	178
R	Length of luggage space behind seats		
	Max.	18·00	457
	Min.	12·00	305
S	Height-floor to top of seat squab	16·00	406
T	Width between wheel arches	35·50	902
U	Maximum interior height	40·00	1016
V	Maximum interior width at hip point	47·50	1208
W	Luggage compartment height:		
	Max.	17·00	432
	Min.	7·00	178
X	Luggage compartment depth:		
	Max.	22·00	560
	Min.	17·00	432
Y	Luggage compartment width:		
	(between wheel arches) Max.	48·00	1220
	Min.	35·50	902
Z	Luggage compartment effective opening width	42·00	1068

PAINT AND TRIM CODING SYSTEM

The commission number plate affixed to the front wing valance panel bears code symbols for identification of the vehicle's exterior colour, trim material and trim colour.

Colour Code

Nine basic colours are allocated a number as shown in the table. Shades of these colours are classified as 1st shade, 2nd shade, 3rd shade, etc. The number of each shade change prefixes the basic colour to indicate the shade colour. Dual colours are identified by two code numbers separated by a stroke, e.g. 19/26 denotes 'White and 'Wedgwood', the predominant colour being white, this symbol being quoted first.

The main trim material is identified by prefixing the colour code number with a letter, e.g.:

Leathercloth – No prefix letter
Leather – Prefix letter H
Cloth – Prefix letter C

Basic colour	Basic colour number	1st shade	2nd shade	3rd shade	4th shade	5th shade	6th shade	7th shade	8th shade	9th shade	10th shade	11th shade	12th shade
Black	01	11											
Red	02	12 Matador	22 Cherry	32 Signal	42 Burgundy	52 Scarlet	62 Inca Red	72 Pimiento	82 Carmine	92 Magenta			
Brown	03	13 Light Tan	23 Sienna	33 New Tan	43 Saddle Tan	53 Dark Brown	63 Chestnut						
Yellow	04	14 Jonquil	24 Wimpey	34 Jasmine	** 44 Beige	54 Saffron	64 Mimosa						
Green	05	15 Cactus	25 Confer	35 Olive	45 Lichfield	55 Laurel	65 Emerald						
Blue	06	16 Midnight	26 Wedg-wood	36 Dark Blue	46 Renoir	56 Royal	66 Valencia	76 Print Blue	86 Navy	96 Sapphire	106 Mallard	116 Ice	126 French
Purple	07	17 Damson	27 Shadow Blue										
Grey	08	18 Gunmetal	28 Dark Grey	38 Phantom	48 Dolphin	58 Shadow Blue	68 Slate	78 Grey					
White	09	19 White	29 Sebring White	** 39 Honey-suckle									

Thus: Paint 19/26 Trim 16 denotes that the vehicle is painted 'White' and 'Wedgwood', and trimmed in leathercloth coloured Midnight Blue.

NOTE: Acrylic paints bear the suffix 'A'.

ENGINE

Firing order	1–3–4–2			
No. 1 cylinder	at front			
Idle speed	800 to 850 rev/min U.S.A. market only, 750 to 800 rev/min other markets			
Fast idle speed	1100 to 1300 rev/min all markets			
Valve clearance (cold)	0·010 in (0·25 mm) inlet and exhaust			
Valve clearance adjustment	screw and locking nut on rocker			
Location of timing marks	scale on front engine cover, mark on crankshaft pulley.			
Valve timing	Engine No. 25001 and upwards		Below Engine No. 25001	
	U.S.A. market	other markets	U.S.A. markets	other markets
Inlet opens	18°B.T.D.C.	18°B.T.D.C.	10°B.T.D.C.	25°B.T.D.C.
closes	58°A.B.D.C.	58°A.B.D.C.	50°A.B.D.C.	65°A.B.D.C.
Exhaust opens	58°B.B.D.C.	58°B.B.D.C.	50°B.B.D.C.	65°B.B.D.C.
closes	18°A.T.D.C.	18°A.T.D.C.	10°A.T.D.C.	25°A.T.D.C.
Ignition timing:				
static	8°B.T.D.C. Up to Eng. No. F.M.I. 10° B.T.D.C. Eng. No. F.M. 1 onwards	8°B.T.D.C.	6°A.T.D.C.	6°B.T.D.C.
dynamic	See 86.35.00			

CARBURETTER

Make/type	Single Stromberg 1·50 C.D.S.E. Up to 1973 condition Single Stromberg 1·50 C.D.S.E.(V) 1973 condition	Twin S.U. HS2E	Single Stromberg 1·50 C.D.S.E.	Twin S.U. HS2
Main jet	0·090 in	0·090 in	0·090 in	0·090 in
Needle	B 5 C H	AAN	B 5 A V	AAN
Float height	16 to 17 mm	–	16 to 17 mm	–

IGNITION COIL

	MK. III	*Mk. IV*	*1500*
Make/type	[Lucas HA12]	Lucas 16C6	Lucas 15C6
Primary winding resistance	[3·1 to 3·5 ohms]	1·43 to 1·58 ohms	1·30 to 1·45 ohms

BALLAST RESISTOR (Fitted up to 1973 condition only. 1973 models have a resistance wire built into the harness as an alternative to the ballast resistor) **[Mk. III models have no ballast resistance.]**

Make/type	Lucas 3BR
Resistance	1·3 to 1·4 ohms

IGNITION DISTRIBUTOR

	Mk. III	*Mk. IV and 1500*
Make/type	[A.C. Delco, D200 (different to each model)]	A.C. Delco, D 204 (different to each model)
Rotation viewed on rotor	anti-clockwise	
Dwell angle	38° to 40°	
Capacitor capacitance	0·18 to 0·23 mfd	
Contact breaker gap	0·014 in to 0·016 in (0·35 to 0·40 mm)	
Centrifugal advance	See 86.35.00	
Vacuum advance	See 86.35.00	

SPARKING PLUGS

Make/type	Champion N12Y	U.S.A. market only	Eng. No. FM1349 onwards
	Champion UN12Y	U.S.A. market only	Up to Eng. No. FM1349
	Champion N9Y	other markets	
Gap	0·025 in (0·64 mm)		

SPITFIRE MkIV TIGHTENING TORQUES

Operation ENGINE	Description	Specified Torque (lbf. ft.)	(kgf. m.)
Air cleaner to backplate	¼" UNF X ½" setscrew	8	1·1
Air cleaner to carburettor	5/16" UNC X 2⅛" bolt	8	1·1
Alternator [or generator] to mounting bracket & front engine plate	5/16" UNF X 4⅞" bolt	22	3·0
Alternator [or generator] to adjusting link	5/16" UNC X 1" setscrew	20	2·8
Clutch attachment to flywheel	5/16" UNF X ⅝" setscrew	22	3·0
Connecting rod bolt – Colour Dyed	⅜" UNF X 1·65" bolt	50	6·9
Connecting rod bolt – Phosphated	⅜" UNF X 1·65" bolt	46	6·4
Chainwheel to camshaft	5/16" UNF setscrew	24	3·3
Crankshaft pulley nut	1" X 16 T.P.I.	150	20·7
Cylinder block drain plug [where fitted]	½" X 20 NPT	35	4·8
Cylinder head to block	⅜" UNF stud	46	6·4
Distributor to pedestal	5/16" UNF X 2" bolt	20	2·8
Fan attachment	¼" UNF X ¾" setscrew	9	1·2
Flywheel to crankshaft	⅜" UNF X 1·03" bolt		
	Cad. Plated	40	5·5
	Parkarised	45	6·2
Fuel pump to cylinder block	5/16" X 1·16" stud	14	1·9
Gearbox & rear engine plate to block	5/16" UNF X 1·56" stud	14	1·9
Manifold inlet to exhaust	5/16" UNF stud	14	1·9
Manifold to head	⅜" UNF stud	25	3·5
Main bearing cap bolts	7/16" UNF X 3" bolt	65	9·0
Oil sump drain plug	⅜" Dryseal taper plug	25	3·5
Oil sump to block	5/16" UNF X ⅝" setscrew	20	2·8
Oil pressure switch plug to cylinder head	⅜" X 18 NP Tapered plug	14	1·9
Oil seal block attachment	5/16" UNF X 0·94" screw	14	1·9
Rocker cover to head	5/16" UNF X 4·13" stud	2	0·3
Rocker pedestal to cylinder head	⅜" UNF X 3·09" stud	34	4·7
Rear crankshaft seal	5/16" UNF X 1⅛" setecrew	20	2·8
Rear engine mounting platform on frame	5/16" UNF X ⅝" setscrew	20	2·8
Sealing block to engine plate	5/16" UNF X ⅝" setscrew	20	2·8
Spark plug to head	14mm X ¾" long thread	20	2·8
Starter motor attachment	⅜" UNF X 2⅛" bolt	34	4·7
Timing cover to front engine plate	5/16" UNF X ⅜" setscrew	10	1·4
	5/16" UNF X ⅞" setscrew	20	2·8
	5/16" UNF X 1·16" stud	16	2·2
Water elbow to water pump	5/16" UNF X 1¾" bolt	20	2·8
	5/16" UNF X 1" setscrew	20	2·8
Water pump bearing housing to pump	5/16" UNF X 1·16 stud	14	1·9
Water pump to cylinder head	5/16" UNF X 2¾" bolt	20	2·8
	5/16" UNF X 2¼" bolt	20	2·8
	5/16" UNF X 2⅜" bolt	20	2·8

Operation	Description	Specific Torque (lbf. ft.)	(kgf. m.)
GEARBOX			
Drain plug	3/8" NP Taper	25	3·5
Extension to gearbox	5/16" UNF X 7/8" setscrew	20	2·8
Filler Plug	3/8" NP Taper	25	3·5
Flange to mainshaft	5/8" UNF	[100]	[16·6]
[Flange to mainshaft — Mk. III	**5/8" UNF**	**80**	**11·1]**
Gearbox case to clutch housing	3/8" UNF X 1" bolt	32	4·4
Hand lever to operating shaft	1/4" UNF X 1" bolt	9	1·2
Slave cylinder to boss	5/16" UNF X 2 1/4" bolt	14	1·9
Selector fork attachment	5/16" UNF setscrew	10	1·4
Top cover to gearbox	1/4" UNF X 7/8" setscrew	10	1·4
	1/4" UNF X 1 3/8" bolt	10	1·4
OVERDRIVE			
Top cover attachment	1/4" UNF X 7/8" setscrew 1/4" UNF X 1" bolt	9	1·2
Support bracket attachment	5/16" UNF X 5/8" setscrew	20	2·8
Support bracket to overdrive	5/16" UNF X 7/8" setscrew	20	2·8
PROPELLOR SHAFT AND REAR AXLE			
Axle mounting plate to hypoid housing	3/8" UNF	38	5·2
Back plate attachment (axle shaft & hub)	5/16" UNF bolt	20	2·8
Bearing cap to housing	3/8" UNF X 1·84" bolt	38	5·2
Crown wheel to differential unit	3/8" UNF X 0·72" bolt	46	6·4
Front and rear attachment of prop. shaft	3/8" UNF X 1" bolt	34	4·7
Hub to axle shaft	5/8" UNF	120	16·6
Hypoid housing to rear axle housing	5/16" UNF X 1" setscrew	20	2·8
Hypoid flange to pinion	5/8" UNF	120	16·6
[Hypoid flange to piston — Mk. III	**9/16" UNF**	**85**	**11·8]**
Plugs, drain and filler	3/8" X 18 dryseal	25	3·5
Rear axle to frame	7/16" UNF X 7·4" bolt	45	6·2
Rear axle nose mounting to frame	3/8" UNF Special stud	34	4·7
Road spring attachment	3/8" UNF stud	34	4·7
Shaft joint to inner axle shaft	3/8"UNF stud	46	6·4
[Shaft point to inner axle shaft — Mk. III	**5/16" UNF bolt**	**28**	**3·6]**
FRONT SUSPENSION			
Anti-roll bar attachment to chassis	5/16" UNF 'U' bolt	4	0·6
Anti-roll bar link assembly	7/16" UNF	45	6·2
Anti-roll bar stud	3/8" UNF stud	14	1·9
Ball assembly to upper wishbone	5/16" UNF X 2 1/4" bolt	20	2·8
Ball pin to vertical link	7/16" UNF ball pin	38	5·2
Brake caliper to mounting plate	7/16" UNF X 1·31" bolts	65	9·0
Brake disc	3/8" UNF X 1" bolt	34	4·7
Damper attachment	7/16" UNF bolt	46	6·4
Front suspension and engine mounting bracket to frame	3/8" UNF X 1" setscrew 3/8" UNF X 1 1/8" bolt	32	4·4
Fulcrum bracket to lower wishbone	3/8" UNF X 2 3/8" bolt	32	4·4
Fulcrum bracket to upper wishbone	3/8" UNF X 2 1/4" bolt 3/8" UNF X 2 1/2" bolt	32	4·4
Stub axle to vertical link	1/2" UNF on stub shaft	65	9·0
Tie rod end ball joint assembly	3/8" UNF	32	4·4
Tie rod lever & dust cover to vertical link	7/16" UNF X 1·88" bolt	65	9·0
[Tie rod lever & dust cover to vertical link — Mk. III	**3/8" x 1.88" bolt**	**35**	**4·9]**
Trunion to wishbone	7/16" UNF X 2 1/2" bolt	45	6·2
Wishbone assembly to frame	3/8" UNF stud	25	3·5

Operation	Description	Specific Torque (lbf. ft.)	(kgf. m.)
REAR SUSPENSION			
Damper fulcrum pin	½" UNF fulcrum pin	48	6·6
Damper lower attachment	7/16" UNF	38	5·2
Pivot bracket to body floor	3/8" UNF X 1 7/8" setscrew	32	4·4
Radius arm attachment to pivot brackets and links	3/8" UNF X 2 1/4" bolt	32	4·4
Spring to pivot bracket **[except Mk. III]**	5/16" UNF X 2 5/8" bolt	20	2·8
[Spring to axle unit — Mk. III only	**3/8" UNF x 2 5/8" stud**	**30**	**4·2]**
Spring ends to vertical links	7/16" UNF bolt	48	6·6
Vertical link plates to rear hub inner	7/16" UNF X 3 3/8" bolt	48	6·6
STEERING			
Ball joint to tie rod lock nut	½" UNF on tie rod	38	5·2
Column Lower assembly to bracket	¼" UNF X ½" setscrew	9	1·2
Coupling Lower to upper clamp	¼" UNF bolt	9	1·2
Coupling pinch bolt	5/16" UNF X 1¼" bolt	14	2·0
Rack mounting to frame	5/16" UNF 'U' bolt	16	2·2
Road wheel to hub	3/8" UNF stud	48	6·6
Safety clamp socket setscrew	7/16" UNF setscrew	20	2·8
Steering column lock attachment	5/16" UNF torque head screw	Tighten to shear	
Steering column safety clamp	¼" UNF bolt	9	1·2
Steering wheel to column	9/16" UNF on column	34	4·7
BODY			
Accelerator mounting bracket	5/16" UNF X 1½" bolt	20	2·8
Accelerator mounting bracket to dash	¼" UNF X ½" setscrew	9	1·2
Bonnet hinge pivot to support bracket	3/8" UNF X 1 7/8" setscrew	32	4·4
Bumper to rear body wing sides	5/16"UNF X ¾" setscrew	14	1·9
Bumper assembly to rear support and support attachment	3/8" UNF setscrew	32	4·4
Clutch and brake pedal mounting bracket	¼" UNF X 5/8" setscrew	9	1·2
Clutch and brake master cylinder attachment	5/16" UNF X 7/8" setscrew	20	2·8
Door hinge to body	5/16" UNF X 7/8" setscrew	20	2·8
Door hinge to door	5/16" UNF X ¾" setscrew	14	1·9
Seat slide attachment	¼" UNF X 7/8" setscrew	9	1·2
Seat slide to floor	¼" UNF X ¾" setscrew	9	1·2
Seat belt attachment to tunnel & sill	7/16" UNF special bolt	32	4·4
Seat belt attachment to rear wheel arch	7/16" UNF special bolt	32	4·4

RECOMMENDED LUBRICANTS – BRITISH ISLES

(The products recommended are not listed in order of preference)

COMPONENT	BP	CASTROL	DUCKHAMS	ESSO	MOBIL	PETROFINA	REGENT	SHELL
**ENGINE **CARB. DASHPOTS AND ** OIL CAN	Super Visco-static 20W-50	Castrol GTX	Duckhams Q20-50	Uniflo	Mobiloil Super 10W/50 or Mobiloil Special 20W/50	Fina Super Grade Motor Oil SAE 20W/50	Havoline Motor Oil 20W-50	** Shell Super Multigrade **
GEARBOX OVERDRIVE AND REAR AXLE LOWER STEERING SWIVELS	BP Gear Oil SAE 90 EP	Castrol Hypoy	Duckhams Hypoid 90	Esso Gear Oil GX 90/140	Mobilube HD 90	** Fina Pontonic XP 90-140 **	Multigear Lubricant EP 90	Shell Spirax 90 EP
FRONT & REAR HUBS BRAKE CABLES GREASE GUN	** BP Energrease L2 **	Castrol LM Grease	Duckhams LB 10	Esso Multi-purpose Grease H	Mobilgrease MP	Fina Marson HTL 2	Marfak All purpose	Shell Retinex A

RECOMMENDED LUBRICANTS – OVERSEAS

(The products recommended are not listed in order of preference)

COMPONENT	Air temp. °C	Air temp. °F	API Designation	BP	CASTROL	DUCKHAMS	ESSO	MOBIL	PETROFINA	SHELL	TEXACO
**ENGINE CARB. DASHPOTS ** AND ** OIL CAN	over 30	over 80	**SD or SE	**	Castrol GTX Castrol Super 20W/50 or XLR (USA only)	Q20/50 / Q10-50	Esso Extra Motor Oil 20W/50 / Uniflow	Mobiloil Super 10W/50 Mobiloil Special 20W/50	Fina Supergrade Motor Oil 20W/50	**	Havoline 20W/50
	30 to 0	80 to 30	SD or SE	BP Super Visco-Static							
	0 to −20	30 to −4	SD or SE		Castrolite or Castrol GTZ **	** Q10-40 **	Esso Extra Motor Oil 10W/30	Mobiloil Super 10W/50	**Fina Supergrade Motor Oil 10W/40	Shell Super Motor Oil	Havoline 10W/30
	below −20	below −4	SD or SE **	**	Castrol 5W/20	** Q5-30 **	Esso Extra Motor Oil 5W/20	Mobiloil 5W/20	Fina Supergrade 5W/30**	**	Havoline 5W/20
GEARBOX OVERDRIVE AND REAR AXLE	over 0	over 30	GL4	BP Gear Oil SAE 90 EP	Castrol Hypoy	Duckhams Hypoid 90	Esso Gear Oil GX 90	Mobilube HD 90	Fina PONTONIC MP SAE 90	Shell Sprìax 90 EP	Multigear Lubricant EP 90
LOWER STEERING SWIVELS	below 0	below 30	GL4	BP Gear Oil SAF 80 EP	Castrol Hypoy 80	Duckhams Hypoid 80	Esso Gear Oil GX 80	Mobilube HD 80	Fina PONTONIC MP SAE 80	Shell Spirax 80 EP	Multigear Lubricant EP 80
FRONT AND REAR HUBS BRAKE CABLES SUSPENSION BALL JOINTS (SWEDISH MARKET ONLY) GREASE GUN				** BP Energrease MP **	Castrol MP Grease **	Duckhams LB 10	Esso Multi-purpose Grease H	Mobilgrease MP	Fina Marson HTL 2	Shell Retinax A or ** Darina AX (USA only) **	Marfax All-purpose

* OILS MARKED THUS ARE AVAILABLE IN MULTIGRADE FORMS WITH VISCOSITY CHARACTERISTICS APPROPRIATE TO THE AMBIENT TEMPERATURE RANGE IN INDIVIDUAL MARKETS.

** WHERE CIRCUIT RACING OR OTHER SEVERE COMPETITIVE EVENTS ARE CONTEMPLATED IT IS ADVISABLE, IN VIEW OF THE INCREASED OIL TEMPERATURE ENCOUNTERED, TO USE OILS OF HIGH VISCOSITY.

RECOMMENDED HYDRAULIC FLUIDS

Clutch and Brake Reservoirs: Castrol Girling Brake and Clutch Fluid – Crimson or Unipart 550 Brake Fluid.
Where these proprietary brands are not available, other fluids which meet the S.A.E. J.1703 specification may be used.

RECOMMENDED FUEL

The Triumph Spitfire engine is designed to operate on fuel having a minimum octane rating of 97 (High compression engines) OR 91 (Lower compression engines): this is equivalent to the British 4 star and 2 star ratings respectively.
Where such fuels are not available and it is necessary to use fuels of lower or unknown rating, the ignition timing must be retarded from the specified setting, just sufficiently to prevent audible detonation (pinking) under all operating conditions, otherwise damage to the engine may occur.

IMPORTANT: When cars for the U.S.A. market enter the "United States" the ignition timing must be set to suit the use of the recommended grade of fuel AND TO COMPLY WITH REGULATIONS ON EMISSIONS FROM THE CRANKCASE AND EXHAUST.

ANTI-FREEZE SOLUTIONS

Only solutions which meet B.S.I. 3151 or 3152 specifications may be used.

ANTI-FREEZE CONCENTRATION		25%	30%	35%	50%
SPECIFIC GRAVITY OF COOLANT AT 15·5°C (60°F)		1·039	1·048	1·054	1·076
ANTI-FREEZE QUANTITY (Cooling system with heater)	PINTS IMP.	2·0	2·4	2·8	4·0
	PINTS U.S.A.	2·4	2·9	3·4	4·8
	LITRES	1·1	1·4	1·6	2·3
DEGREE OF PROTECTION	**Complete** Car may be driven away immediately from cold	−12°C 10°F	−16°C 3°F	−20°C −4°F	−36°C −33°F
	Safe Limit Coolant in mushy state. Engine may be started and driven away after short warm-up period	−18°C 0°F	−22°C −8°F	−28°C −18°F	−41°C −42°F
	Lower Protection Prevents frost damage to cylinder head, block and radiator. Thaw out before starting engine.	−26°C −15°F	−32°C −26°F	−37°C −35°F	−47°C −53°F

CAPACITIES

Fuel tank . . . . . pre 1972 condition	8¼ gal. (9·9 U.S. gal.) (37·6 litres)
. . . . 1973/1972 condition	7¼ gal. (8·7 U.S. gal.) (33·0 litres)
Engine sump and oil filter	8 pints (9·6 U.S. pints) (4·5 litres)
Engine sump (drain and refill	7 pints (8·4 U.S. pints (4·0 litres)
Gearbox (from dry)	1½ pints (1·8 U.S. pints) (0·85 litres)
Gearbox and overdrive (from dry)	2½ pints (2·8 U.S. pints) (1·4 litres)
Rear axle (from dry)	1 pint (1·2 U.S. pints) (0·57 litres)
Cooling system (including heater)	8 pints (9·6 U.S. pints) (4·5 litres)
Heater	1 pint (1·2 U.S. pints) (0·57 litres)

MAINTENANCE OPERATIONS

LUBRICATION CHART

Weekly or before a long journey

1. Check/top up cooling system level.
2. Check/top up engine oil level.

Every 6,000 miles (10,000 km)

1. Check/top up cooling system level.
2. Change engine oil.
3. Lubricate steering rack and pinion.
4. Using OIL lubricate lower steering swivels.
   Swedish market only: Grease suspension/steering balls joints.
5. Check/top up carburetter piston damper(s) and lubricate throttle linkage.
6. Check/top up brake and clutch fluid reservoirs.
7. Lubricate accelerator, brake and clutch pedal pivots.
8. Check/top up gearbox oil level.
9. Check/top up rear axle oil level.
11. Lubricate handbrake linkage and cable.
12. Lubricate all door, bonnet and boot locks and hinges.
13. Lubricate battery terminals (petroleum jelly).
14. Lubricate distributor.

Every 3,000 miles (5,000 km)

1. Check/top up cooling system level.
2. Check/top up engine oil level.
6. Check/top up brake and clutch fluid reservoirs.

Every 12,000 miles (20,000 km)

1. Check/top up cooling system level.
2. Change engine oil.
3. Lubricate steering rack and pinion.
4. Using OIL lubricate lower steering swivels.
   Swedish market only: Grease suspension/steering ball. joints.
5. Check/top up carburetter piston damper(s) and lubricate throttle linkage.
6. Check/top up brake and clutch fluid reservoirs.
7. Lubricate accelerator, brake and clutch pedal pivots.
8. Check/top up gearbox oil level.
9. Check/top up rear axle oil.
10. Lubricate rear hubs.
11. Lubricate handbrake linkage and cable.
12. Lubricate all door, bonnet and boot locks and hinges.
13. Lubricate battery terminals (petroleum jelly).
14. Lubricate distributor.
15. Renew oil filter element.

[NOTE: See Part 1, *Driver's Handbook* (page 27) for Mk. III data.]

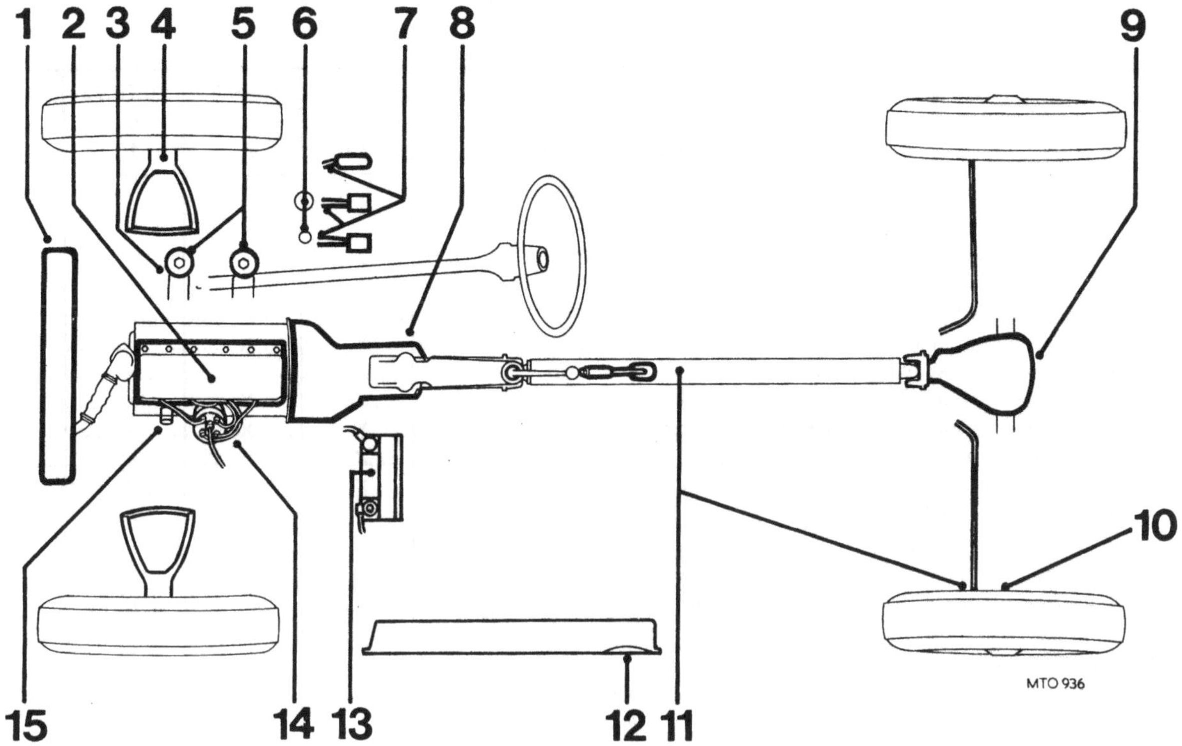
1
2
3
4
5
6
7
8
9
10
11
12
13
14
15
MTO 936

SUMMARY CHART

[NOTE: See Part 1, *Driver's Handbook* (beginning on page 16) for Mk. III data.]

The Summary Chart below lists general recommendations for Service Operations and Intervals. Overseas Service Engineers are advised to consult the "Passport to Service" booklet supplied with the car for amendments to these recommendations that may be specially applicable to their local operating conditions OR that may be obligitory to meet Regulations for a specific Country.

Operation Number Interval in miles x 1,000 Interval in Kilometres x 1,000	10.10.03 1 1·6	10.10.06 3 5	10.10.12 6 10	10.10.24 12 20
Operation Description				
ENGINE COMPARTMENT				
1. Check/top up engine oil level (E)		X		
2. Check/top up cooling system (E)	X	X	X	X
3. Check/top up brake fluid reservoir	X	X	X	X
4. Check/top up clutch fluid reservoir	X	X	X	X
5. Check/top up windscreen washer fluid reservoir	X	X	X	X
6. Check/top up battery	X	X	X	X
7. Check/top up carburetter piston(s) damper(s) (E)	X		X	X
8. Drain engine oil and refill (E)	X		X	X
9. Renew oil filter element (E)				X
10. Clean fuel pump sediment bowl	X			X
11. Lubricate distributor and check automatic advance (E)	X		X	X
12. Check/adjust/report condition of distributor points (E)	X		X	
13. Distributor points – renew (E)				X
14. Check/adjust ignition timing using electronic equipment (E)	X		X	X
15. Check/report ignition wiring for fraying, chafing and deterioration (E)	X		X	X
16. Condensor and coil check for breakdown on oscilascope tune (E)			X	X
17. Clean/adjust sparking plugs (E)			X	
18. Renew sparking plugs (E)				X
19. Check/adjust torque of cylinder head nuts/bolts (E)	X			
20. Check/report cylinder compression (E)			X	X
21. Check/adjust valve rocker clearances (E)	X			X
22. Clean engine oil filler cap (E)				X
23. Clean carburetter air cleaner elements (E)			X	
24. Renew carburetter air cleaner elements (E)				X
25. Check/adjust/report condition of all driving belts (E)	X	X	X	X
26. Check security of starter motor and alternator retaining bolts	X			
27. Check security of engine mountings	X			
28. Check/adjust carburetter settings (E)	X		X	X
29. Carburetter – overhaul – at 24,000 miles (E)				
30. Fuel filter – change (E)				X
31. Fuel system – check for leaks (E)	X			X
32. Lubricate accelerator linkage/pedal fulcrum and check operation	X		X	X
33. Check battery condition: clean and grease connections			X	X
34. Check/report for oil/fuel/fluid leaks (general) (E)	X	X	X	X
35. Check/report leaks from cooling and heater systems (E)	X	X	X	X
36. Evapourative and crankcase ventilations systems – check hoses and restrictors for blockage, security and deterioration (E)			X	X
37. Carbon cannister – renew filter (E)				X
38. Carbon cannister – renew 48,000 miles (E)				

Operation Number Interval in miles x 1,000 Interval in kilometers x 1,000	10.10.03 1 1·6	10.10.06 3 5	10.10.12 6 10	10.10.24 12 20
Operation Description				
UNDERBODY				
39. Check/top up level of gearbox and overdrive oil	X		X	X
40. Check/top up level of final drive unit oil	X		X	X
41. Lubricate lower steering swivel	X		X	X
42. Lubricate all grease points except hubs	X		X	X
43. Lubricate steering rack and pinion			X	X
44. Lubricate rear hubs				X
45. Lubricate handbrake linkage and cable guides			X	X
46. Check transmission, engine, final drive, suspension and steering unit for oil leaks and report	X	X	X	X
47. Check visually brake, fuel and clutch pipes, hoses and unions for chafing, leaks and corrosion and report	X	X	X	X
48. Check/report exhaust system for leakage and security (E)		X	X	X
49. Check security of suspension fixings, tie-rod levers, steering unit attachments and steering universal joint coupling bolts	X			X
50. Check security of propeller shaft and drive shaft universal coupling bolts				X
51. Check security of sub-frame or body mountings	X			X
52. Check/report condition of steering unit/joints for security, backlash and gaiter condition	X	X	X	X
EXTERIOR				
53. Adjust front hubs				X
54. Check/adjust front and rear wheel alignment with tracking equipment	X			
55. Check/report front and rear wheel alignment with tracking equipment			X	X
56. Inspect brake pads for wear, and discs for condition		X	X	X
57. Inspect and report brake linings for wear and drums for condition				X
58. Check security of road wheel fastenings	X	X	X	X
59. *Check that tyres are in accordance with manufacturers specification		X	X	X
60. *Check visually and report depth of tread, cuts in tyre fabric, exposure of ply or cord structure, lumps or bulges	X	X	X	X
61. Check/adjust tyre pressures (including spare wheel)	X	X	X	X
62. Check/adjust/report headlamp alignment	X	X	X	X
63. Check, if necessary replace windscreen wiper blades		X	X	X
64. Fuel tank filler cap – check seal for security (E)	X		X	X
INTERIOR				
65. Check brake pedal travel and hand brake operation adjust if necessary	X			
66. Check/report brake pedal travel and handbrake operation		X	X	X
67. Check operation of window controls, locks and bonnet release	X			
68. Check function of all electrical systems and windscreen washer	X	X	X	X
69. Lubricate clutch and brake pedal pivots			X	X
70. Lubricate all locks, door hinges, strikers and bonnet release	X		X	X
71. Check/report condition and security of seats and seat belts		X	X	X
72. Check/report rear view mirrors for looseness, cracks and crazing		X	X	X
ROAD TEST				
73. Road/roller test and report additional work required	X		X	X
74. Ensure cleanliness of controls, door handles, steering wheels etc	X	X	X	X

***Important** – If the tyres do not conform with legal requirements report to the owner.

Items marked (E) are particularly relevant to the emmission and evaporative control systems and must receive attention at the recommended intervals to keep these systems in good order.

[NOTE: See Part 1, *Driver's Handbook* (beginning on page 16) for Mk. III data.]

The maintenance summary list on pages 10.00.02 and 10.00.03 gives details of mile and kilometer intervals for the following operations. The figure in parenthesis to the left of each heading refers to the item number on the summary list.

(1) Check/top up engine oil level

NOTE: Allow time for oil to drain back into sump after running engine.
Stand vehicle on level ground.

1. Withdraw dipstick, wipe it clean and replace in position.
2. Withdraw dipstick again and note oil level.
3. Wipe dipstick clean and replace in position.
   If topping up is necessary:-
4. Remove oil filler cap.
5. Add recommended grade of oil, via filler cap, to bring level just below high mark on dipstick.
   DO NOT OVERFILL
6. Replace filler cap.
7. Allow time for added oil to drain into sump, then check final oil level using the procedure in 1 to 3 above.

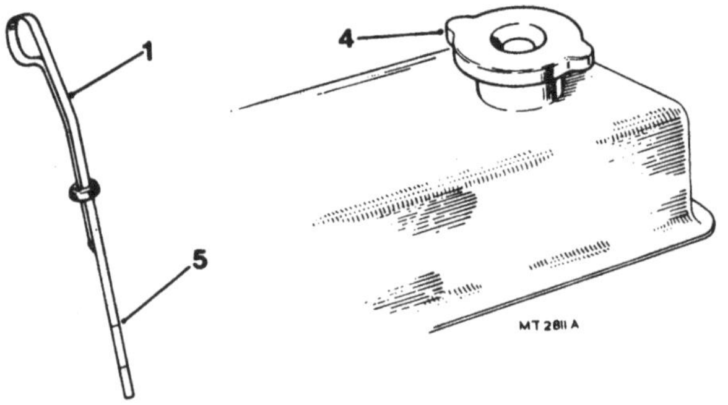

(2) Check/top up cooling system

WARNING: Do NOT remove cooling system filler caps or plugs when engine is hot.

1. Remove radiator expansion tank cap.
2. If necessary, top up expansion tank with soft water to maintain level at approximately half full.
3. Replace cap.

   If the expansion tank is empty:–
4. Remove the cooling system filler cap.
5. Add soft water, via filler cap, until the system is full.
6. Replace filler cap.
7. Half fill expansion tank with soft water using the procedure in 1 to 3 above.
8. Run the engine until normal operating temperature is reached, allow engine to cool and re-check cooling system level.

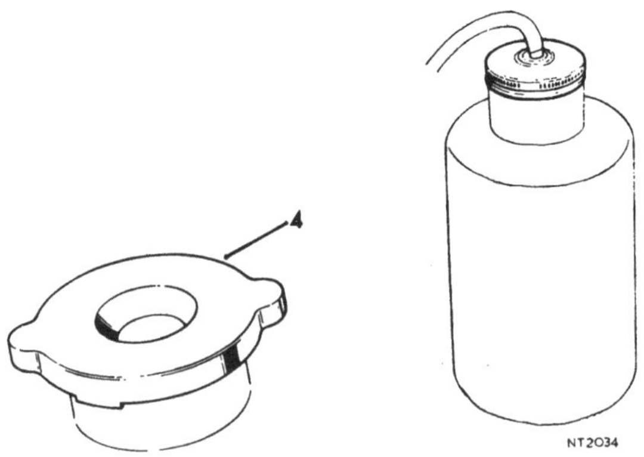

(3) Check/top up brake fluid reservoir

1. Wipe clean the reservoir cap and surrounding area.
2. Remove the reservoir cap.
3. Check fluid level against mark on side of reservoir
4. If necessary, add fluid to bring level up to mark on side of reservoir.

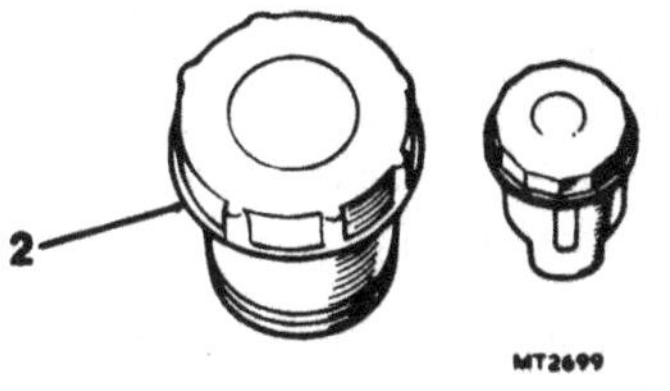

WARNING: Use only new fluid of the correct specification. Do NOT use fluid of unknown origin, or fluid that has been exposed to the atmosphere, or fluid that has been discharged during bleeding operations.

5. Replace reservoir cap.
6. Remove any spilled fluid with a clean cloth.

CAUTION: Paintwork can be damaged by direct contact with brake fluid.

(4) Check/top up clutch fluid reservoir

1. Wipe clean the reservoir cap and surrounding area.
2. Remove the reservoir cap.
3. Check fluid level against mark on side of reservoir.
4. If necessary, add fluid to bring level up to mark on side of reservoir.

WARNING: Use only new fluid of the correct specification. Do NOT use fluid of unknown origin, or fluid that has been exposed to the atmosphere, or fluid that has been discharged during bleeding operations.

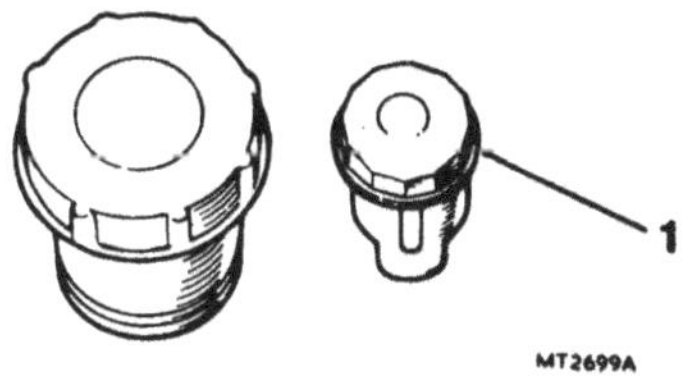

5. Replace reservoir cap.
6. Remove any spilled fluid with a clean cloth.

CAUTION: Paintwork can be damaged by direct contact with clutch fluid.

(5) **Check/top up windscreen washer fluid level**

1. Check fluid level in translucent reservoir.

   If topping up is necessary:–

2. Wipe clean the reservoir cap and surrounding area.
3. Remove the reservoir cap.
4. Add soft water to bring level up to approximately 1 in (25·4 mm) from top of reservoir.
5. Replace reservoir cap.

CAUTION: As a precaution against freezing conditions, fill the reservoir with a mixture of one part methylated spirits and two parts water.
Do NOT use glycol anti-freeze solutions in the washer reservoir, as these may discolour paintwork and damage wiper blades and sealing rubbers.

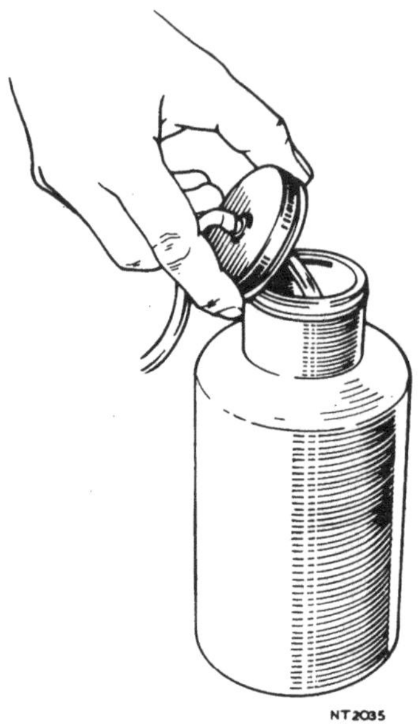

(6) **Check/top up battery**

NOTE: Alternative procedures are given for each of the two battery types that may be fitted.

1. Lift and tilt battery cover.
2. Check electrolyte level, which if correct should just cover the seperators.
   If topping up is necessary:–
3. Add DISTILLED WATER until the filler tubes are full and the trough is just covered.
4. Replace battery cover.

Alternatively:

1. Remove battery filler plugs.
2. Check electrolyte level, which if correct should just cover the seperators.
   If topping us is necessary:–
3. Add DISTILLED WATER until the seperators are just covered. DO NOT OVERFILL.
4. Replace filler plugs.

CAUTION: Paintwork can be damaged by direct contact with the base of filler plugs.

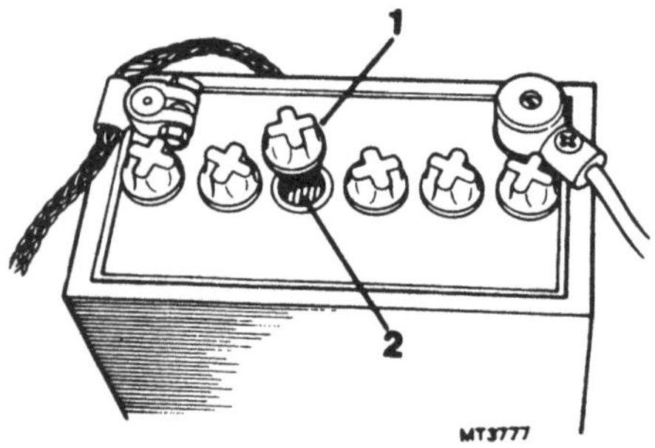

(7) Check/top up carburetter piston dampers

Prodedure for S. U. Carburetters.

1. Unscrew hexagon plug from top of carburetter.
2. Withdraw plug and damper assembly from carburetter.
3. Check dashpot oil level, which if correct will be ½ in (13 mm) above the top of the hollow piston rod.
4. If necessary, add a recommended engine oil, using an oil can, until the level is correct.
5. Replace plug and damper assembly.
6. Screw hexagon plug firmly in position.

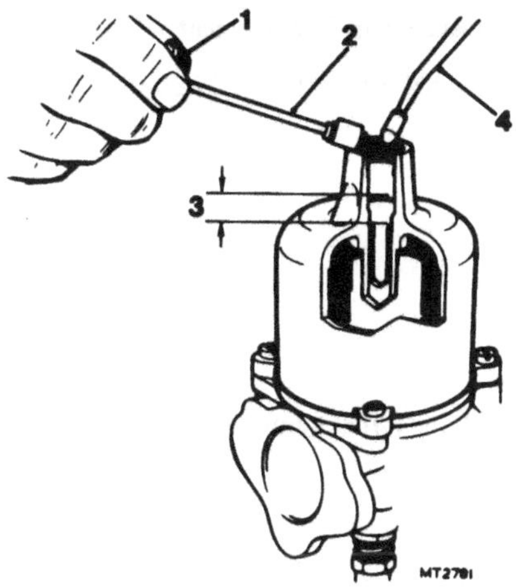

Procedure for Stomberg Carburetters

1. Unscrew hexagon plug from top of carburetter.
2. Withdraw plug and damper assembly from carburetter.
3. Replace plug and damper assembly to check oil level, which if correct will offer resistance to the assembly when the bottom of the plug threads are ¼ in (6 mm) above the rim of the dashpot.
4. If necessary, again withdraw plug and damper assembly and add a recommended engine oil, using an oil can, until the oil level is correct.
5. Replace plug and damper assembly.
6. Screw hexagon plug firmly in position.

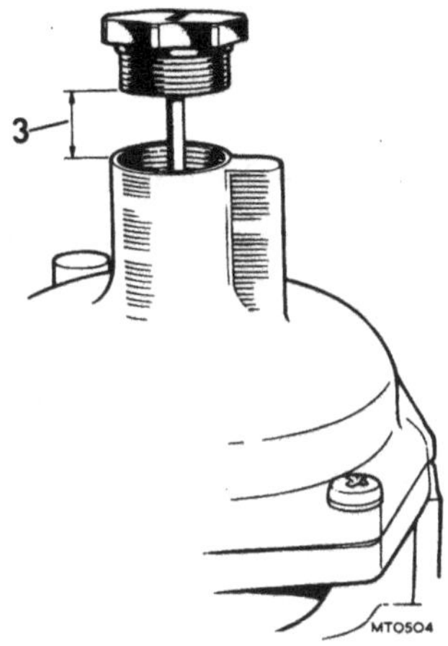

(8) Drian engine oil and refill

NOTE: This operation is best carried out when the engine is warm and with the vehicle standing level on a ramp or over a pit.

1. Wipe clean the engine drain plug and surrounding area.
2. Place a suitable receptical under the drain plug.
3. Unscrew the drain plug slowly until oil begins to escape.
4. When the rate of oil flow lessens, remove drain plug from sump and allow oil to drain completely.
5. Wipe clean the drain plug and replace it in sump.
6. Tighten drain plug to 20 to 25 lbf ft. (2·8 to 3·5 kgf m).
7. Remove oil filler cap.
8. Add a recommended engine oil, via filler cap, to bring level just below high mark on dipstick. **DO NOT OVERFILL.**
9. Replace oil filler cap.
10. Allow time for added oil to drain into sump, then check final oil level on dipstick.

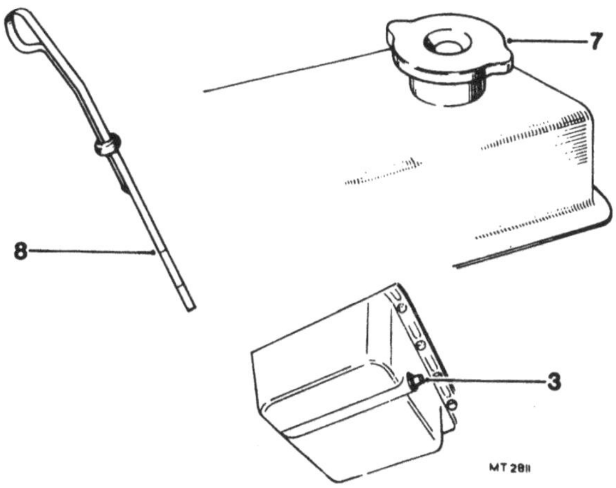

(9) **Renew oil filter element**

See 12.60.01 and 12.60.08.

(10) **Clean fuel pump sediment bowl**

See 19.45.05

(11) **Lubricate distributor and check automatic advance**

Lubricate distributor – See 86.35.18

Check automatic advance

1. Fit a strobe Timing Light in accordance with the Timing Light manufacturers instructions.
2. Disconnect vacuum pipe between distributor and induction side of engine.
3. Start engine.

Check centrifugal advance.

4. Using a second operator to vary engine speed, check apparent movement of timing marks under strobe light.
5. Reconnect vacuum pipe.

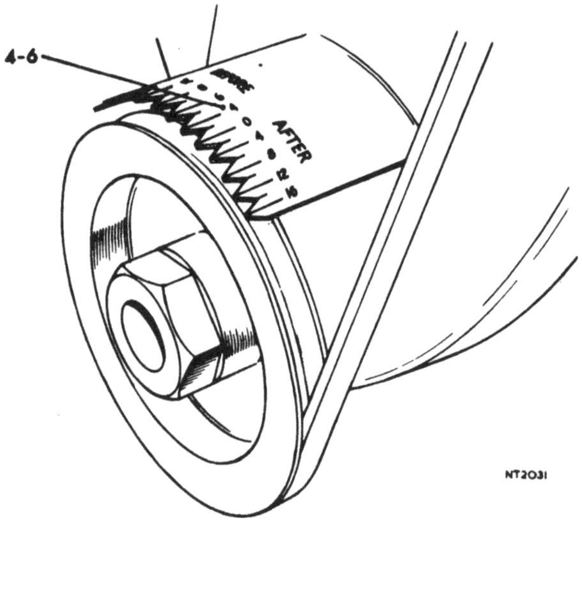

NOTE: If more accurate results are required electronic tuning equipment may be used in conjuction with the data on page 86.35.00. This is extra to normal service requirements.

(12) Check/adjust/report condition of distributor points

See 86.35.14.

(13) Renew distributor points

See 86.35.13

(14) Check/adjust ignition timing

See 86.35.16

(15) Check/report ignition wiring for fraying, chaffing and deterioation

Low tension circuit.

1. Check exposed wiring between coil and ignition switch.
2. Check ignition coil connections.
3. Check wiring between coil and distributor.
4. Check distributor external connections.
5. Remove distributor cap and check internal wiring.
6. Check internal distributor connections.
7. Replace distributor cap.

High tension circuit.

8. Check lead between coil and distributor.
9. For each sparking plug in turn:– Check lead between plug and distributor.
10. Check high tension lead connections.
11. Report wiring condition.

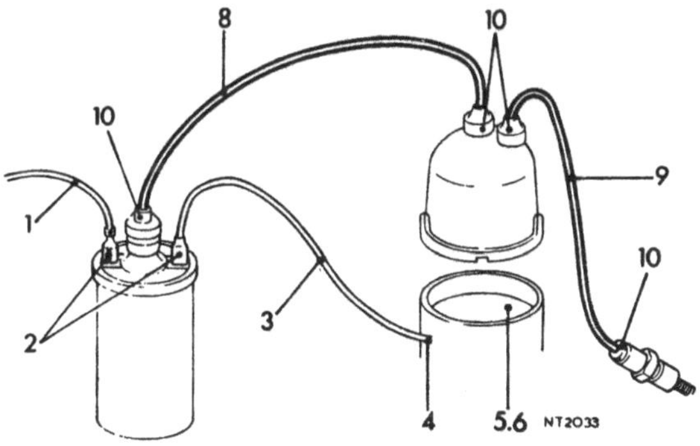

(16) Check condensor and coil for breakdown on oscilascope oscilascope

Using proprietory electronic testing equipment

1. Check distributor condensor performance.
2. Check ignition coil performance.

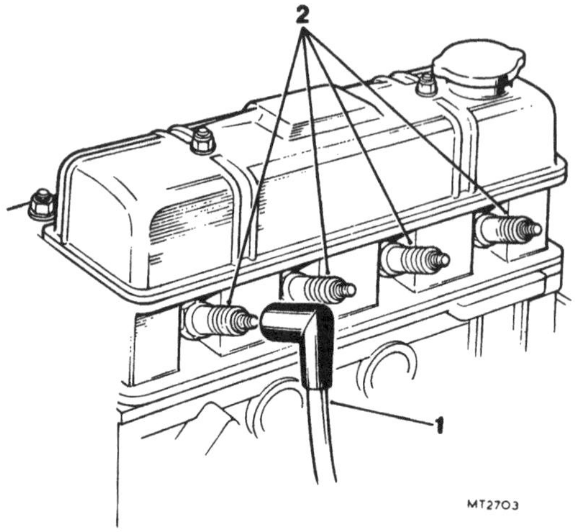

(17) Clean/adjust sparking plugs

For each sparking plug in turn

1. Remove ignition high tension lead from plug.
2. Unscrew plug from engine using a special plug spanner or a box type spanner.
3. Wipe clean ceramic body of plug.
4. Visually check plug body for cracks, and renew plug if cracks are present.
5. Unscrew end terminal cap from plug.
6. Clean plug terminal threads with a wire brush.
7. Clean cap threads using a low pressure air line.
8. Screw end terminal cap firmly into position on plug.
9. Clean electrode area and plug threads with a wire brush or sand blasting machine.

[NOTE: If you use a wire brush to clean the plugs, be careful not to scrub the insulator nose. Doing so may leave conductive "pencil marks" which can short the plug. Motor-powered wire brushes should not be used since they are difficult to control and may deform the plug threads. After sandblast cleaning, or if the plug electrodes appear rounded, file the spark plug electrodes to the sharp, square profiles which are typical of new spark plug electrodes.]

10. Visually check electrode surfaces for damage, and renew plug if damage is present.
11. Check electrode gap, which if correct will just allow a 0·025 in (0·64 mm) feeler gauge to slide slowly between the electrodes under light pressure.

If adjustment is necessary.

12. *(a)* Using a suitable tool, carefully move the side electrode.
    (b) Recheck the gap and repeat this procedure until the gap is correct.
13. Check sealing washer for cracks and distortion, and renew washer if necessary.
14. Refit sparking plug to engine.
15. Tighten plug to 14 to 20 lbf ft (1·9 to 2·8 kgf m).
16. Refit high tension lead to plug.

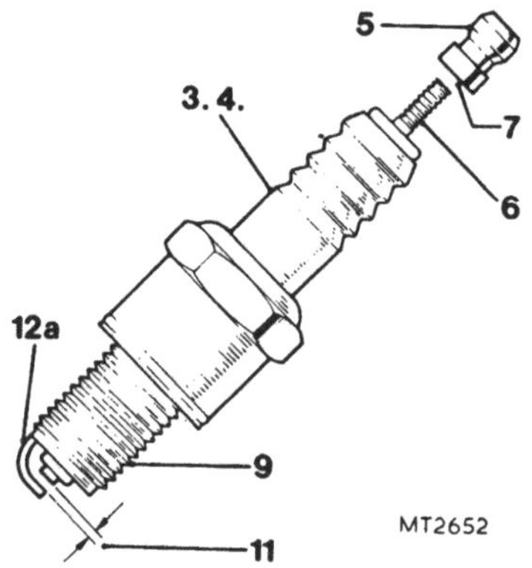

(18) **Renew sparking plugs**

For each sparking plug in turn

1. Remove ignition high tension lead from plug.
2. Unscrew plug from engine using a special plug spanner or a box type spanner.
3. Discard plug.
4. Visually check new plug for damage to body and electrodes, discard plug if damage is present.
5. Check electrode gap on new plug, which if correct will just allow a 0·025 in (0·64 mm) feeler gauge to slide slowly between the electrodes under light pressure.

If adjustment is necessary.

6. *(a)* Using a suitable tool. carefully move the side electrode.
   (b) Recheck the gap and repeat this procedure until the gap is correct.

7. Check sealing washer for cracks and distortion, and renew washer if necessary.
8. Fit new sparking plug to engine.
9. Tighten plug to 14 to 20 lbf ft (1·9 to 2·8 kgf m).
10. Refit high tension lead to plug.

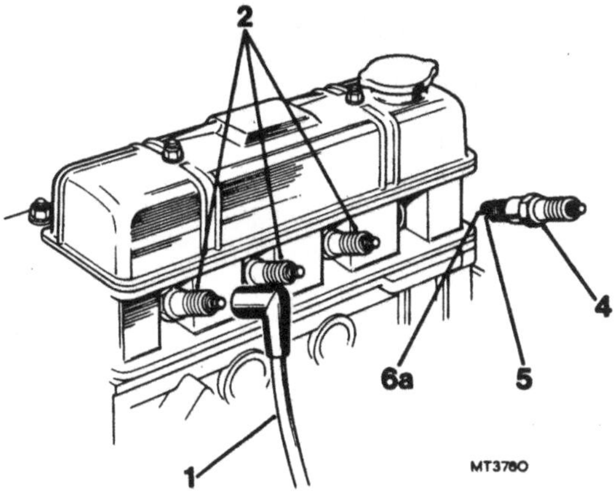

(19) **Check/adjust torque of cylinder head nuts/bolts**

1. Remove rocker cover – See 12.29.42.
2. Using the sequence shown, tighten cylinder head nuts to 38 to 46 lbf ft (5·2 to 6·4 kgf m).
3. Check/adjust valve rocker clearances – See 12.29.48
4. Check rocker cover gasket for damage, and renew if necessary.
5. Refit rocker cover – See 12.29.42.
6. With gears in neutral, handbrake on, start engine and check for leaks from rocker cover gasket.

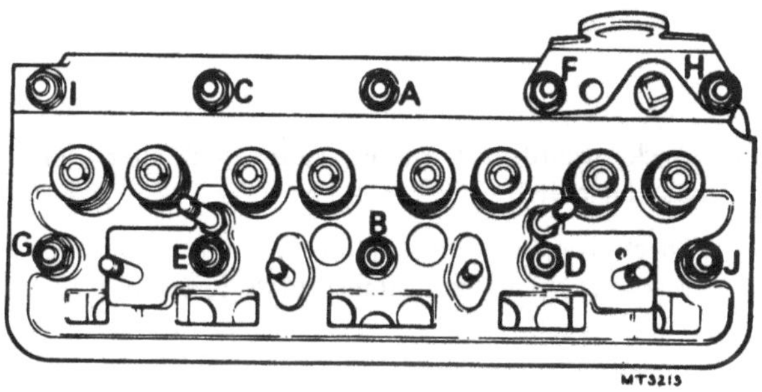

(20) **Check/report cylinder compression**

See 12.25.01

(21) **Check/adjust valve rocker clearances**

See 12.29.48

(22) **Clean engine oil filler cap**

1. Remove filler cap.
2. Clean cap with clean petrol.
3. Allow to dry.
4. Refit filler cap.

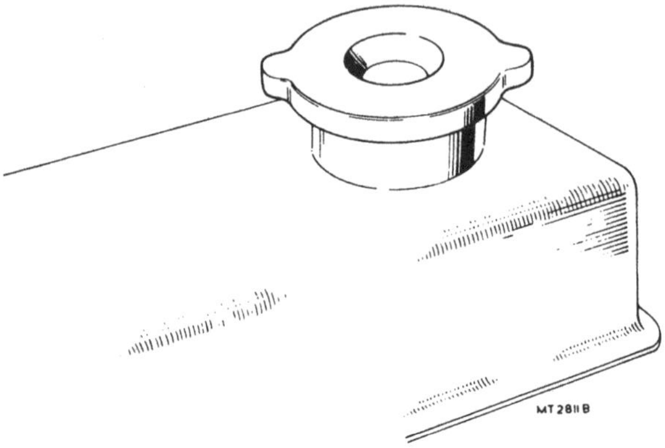

(23) **Clean carburettor air cleaner elements**

See 19.10.08

(24) **Renew carburettor air cleaner elements**

See 19.10.08

(25) **Check/adjust/report condition of driving belts**

1. Check and adjust – See 26.20.01.
2. Report condition where belt is visably
   (a) worn or
   (b) damaged

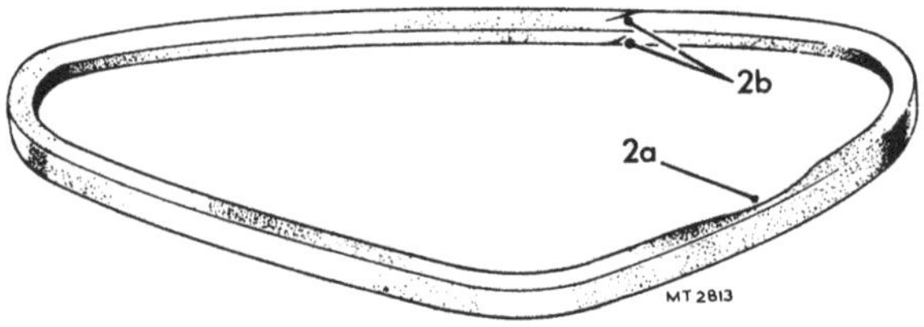

(26) **Check security of starter motor and alternator retaining bolts**

1. Check security of starter motor retaining bolts, which if correct should be tightened to 26 to 34 lbf ft (3·6 to 4·7 kgf m).
2. Check security of alternator to adjusting link bolt, which if correct should be tightened to 15 to 20 lbf ft (2·1 to 2·8 kgf m).
3. Check security of alternator mounting bracket bolt, which if correct should be tightened to 16 to 22 lbf ft (2·2 to 3·0 kgf m).

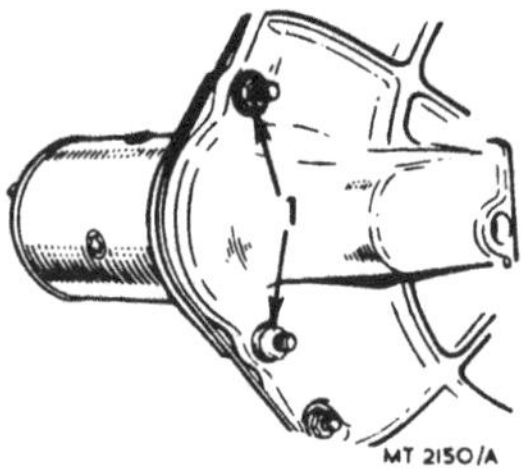

(27) **Check security of engine mountings**

1 Check security of front engine mountings, which if correct should be tightened to 24 to 32 lbf ft (3·3 to 4·4 kgf m).
2. Check security of rear engine mountings, which if correct should be tightened to 15 to 20 lbf ft (2·1 to 2·8 kgf m).

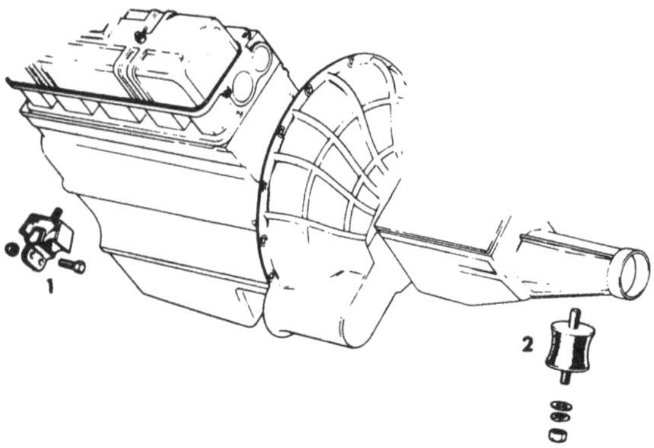

(28) Check/adjust carburettor settings

See 19.15.02

(29) Overhaul carburetter

See 19.15.18

(30) Change fuel filter

See 19.25.01

(31) Check fuel system for leaks

1. Check for leaks from fuel system connections.
2. Check fuel pipes for fractures and damage.
3. Check for leaks from fuel tank(s), pump and carburetter(s).
   On vehicles fitted with an evaporative control system, additional checks are given under 17.15.01.

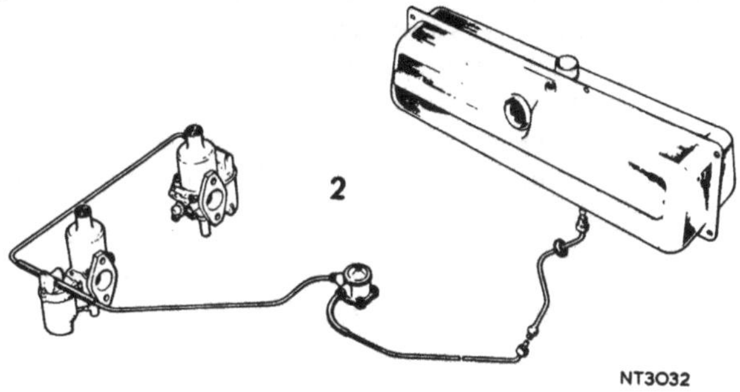

(32) Lubricate accelerator linkage/pedal fulcrum and check operation

1. Lubricate accelerator linkage on carburetter(s), using an oil can.
2. Wipe away surplus oil from linkage.
3. Check for roughness in linkage operation.
4. Lubricate accelerator pedal fulcrum, using an oil can.
5. Wipe away surplus oil from pedal fulcrum.

CAUTION: Surplus oil on the pedal fulcrum can cause staining of the carpet.

6. Check carburetter throttle response to initial movement of the accelerator pedal.

   If adjustment is necessary – see 19.20.05

7. Check carburetter throttle position with accelerator pedal fully depressed.

   If adjustment is necessary – See 19.20.05.

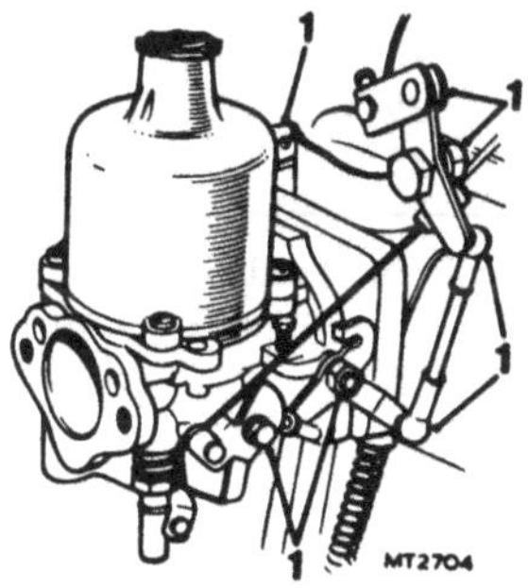

(33) Check battery condition: clean and grease connections

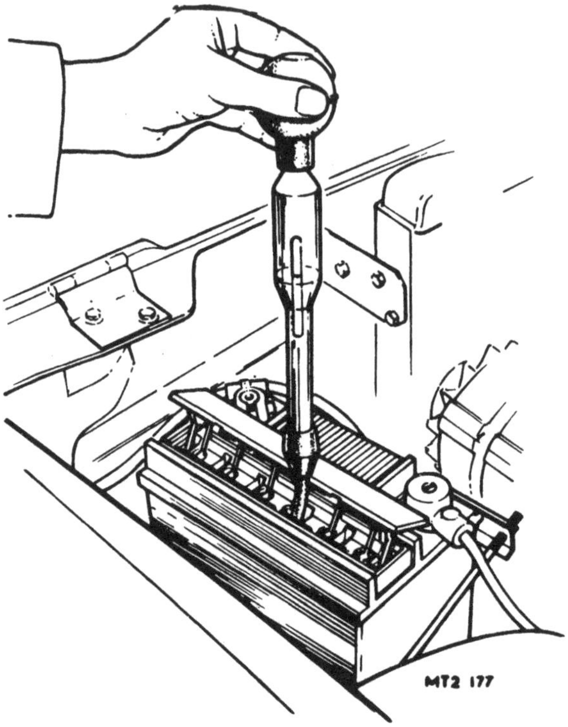

With battery in location

1. Check battery and surrounding area for corrosion from battery chemicals.
2. Clean off any corrosion found.
3. Check visually for cracks in battery case.
4. Report any case cracks found.
5. Check security of terminal connections.
6. Coat terminals with petroleum jelly.

For each cell in turn:-

7. Check electrolyte specific gravity, using an hydrometer, which if correct will approximate to the tabled readings below.

NOTE *(a)* Do NOT check S.G. immediately after adding distilled water as a false reading may be obtained.
(b) S.G. readings approximately equal for each cell indicate a battery in good condition. Conversely, if one or more cells show a reading lower than the others the battery is approaching the end of its usefull life.

Charge condition of cell – temperate climate			
Ambient Temperature °C	**Specific Gravity of Electrolyte**		
	Charged	**Half-Charged**	**Discharged**
5	1·287	1·207	1·117
15	1·280	1·200	1·110
25	1·273	1·193	1·103
35	1·226	1·186	1·096
Charge condition of cell – tropical climate			
15	1·250	1·180	1·100
25	1·243	1·173	1·093
35	1·236	1·166	1·086
52	1·224	1·154	1·074

8. Check voltage, using a heavy discharge tester, which if correct will give approximately equal readings for each cell.
CAUTION: This check should **NOT** be made on a battery in a low state of charge as shown by procedure 7 as damage to the battery can result.

NOTE *(a)* Before making this check on a battery that has just completed an operational journey, the headlamps should be switched on for 2 or 3 minutes to remove any surface charge. *(b)* Voltage readings approximately equal for each cell indicate a battery in good condition. Conversely, if one or more cells show a reading lower than the others, or a reading that falls during the test, the battery is approaching the end of its usefull life.

(34) Check/report oil/fuel/ fluid leaks

1. Check for oil leaks from engine and transmission.
2. Check for fuel leaks from pump, carburetter, pipes, joints and unions.
3. Check for fluid leaks from brake master cylinder, pipes, joints and unions.
4. Check for fluid leaks from clutch master cylinder, pipes, joints and unions.
5. Report any leaks found.

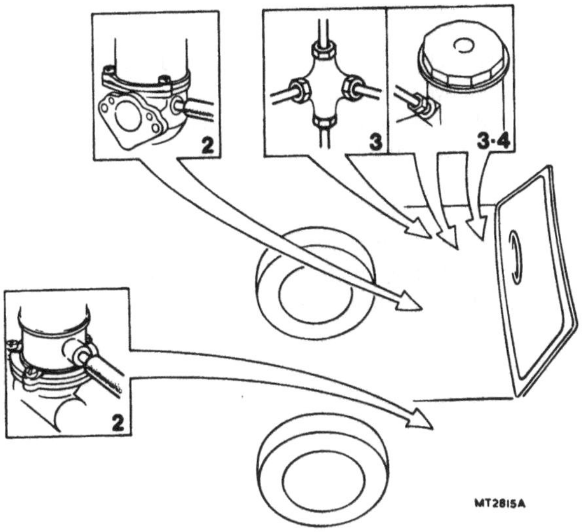

(35) Check/report leaks from cooling and heater systems.

1. Check for leaks from engine and radiator drain taps/plugs.
2. Check for leaks from water hose joints.
3. Check for leaks from water hoses through damage or porosity.
4. Check for leaks from water pump, thermostat housing, radiator and heater unit.
5. Report any leaks found.

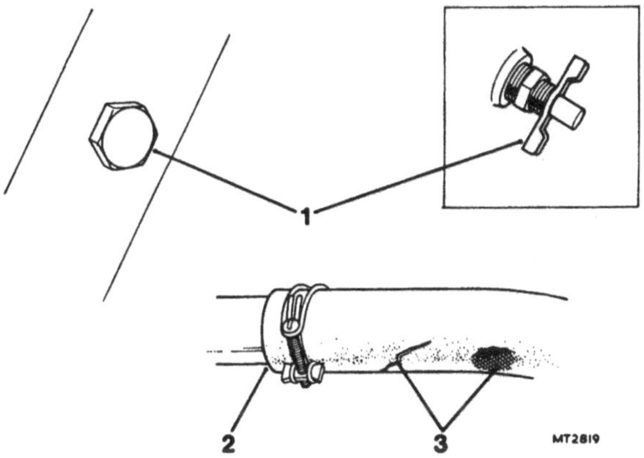

(36) Evaporative and crankcase ventilation systems-check hoses and restrictors for blockage, security and deterioration.

See 17.15.01 and 17.15.36

(37) Carbon cannister – renew filter

See 17.15.07

(38) Carbon cannister – renew at 48,000 miles

See 17.15.13

NOTE: OPERATIONS 39 to 52 ARE BEST CARRIED OUT WITH THE CAR ON A RAMP OR OVER A PIT.

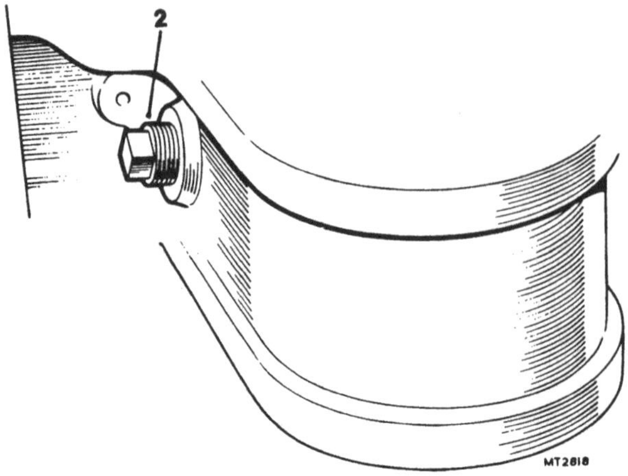

(39) Check/top up level of gearbox and overdrive oil

With vehicle standing level

1. Wipe clean gearbox filler plug and surrounding area.
2. Remove filler plug.
3. Add new oil of the recommended grade, via the filler plug hole, until the oil level reaches the bottom of the hole.
4. Allow surplus oil to drain.
5. Replace filler plug.
6. Tighten plug to 20 to 25 lbf ft (2·8 to 3·5 kgf m).
7. Wipe away surplus oil.

(40) Check/top up level of final drive unit oil

With vehicle standing level

1. Wipe clean final drive unit filler plug and surrounding area.
2. Remove filler plug.
3. Add new oil of the recommended grade, via the filler plug hole, until the oil level reaches the bottom of the hole.
4. Allow surplus oil to drain.
5. Replace filler plug.
6. Tighten plug to 20 to 25 lbf ft (2·8 to 3·5 kgf m).
7. Wipe away surplus oil.

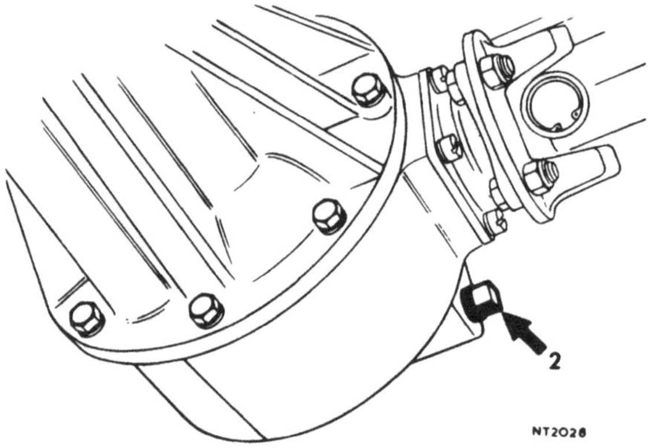

(41) Lubricate lower steering swivel

WARNING: OIL must be used for the operation. Do NOT use grease.

1. Wipe clean the plug and surrounding area.
2. Remove the plug.
3. Fit a suitable grease nipple to the plug hole.
4. Using a grease gun, **CHARGED WITH A RECOMMENDED OIL,** lubricate the lower steering swivel, via the grease nipple, until oil exudes from the bearing.
5. Remove grease nipple
6. Refit plug.
7. Wipe away surplus oil.

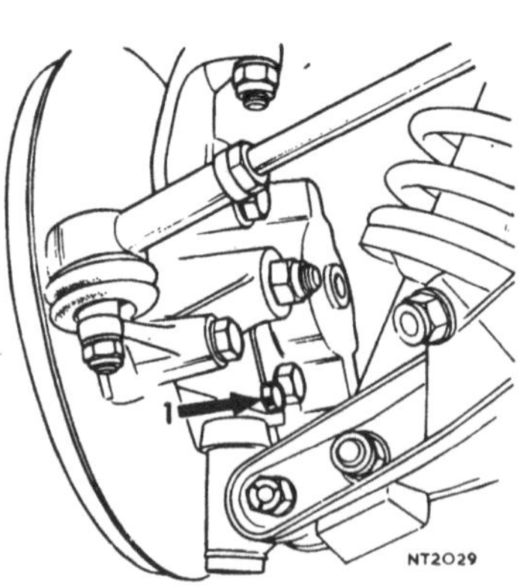

(42) Lubricate all grease points except hubs

Suspension/steering ball joints – Swedish Market Only.

1. Wipe clean the grease nipples and surrounding areas.
2. Apply a grease gun to nipples and stroke until grease exudes from joint.
3. Wipe away surplus grease.

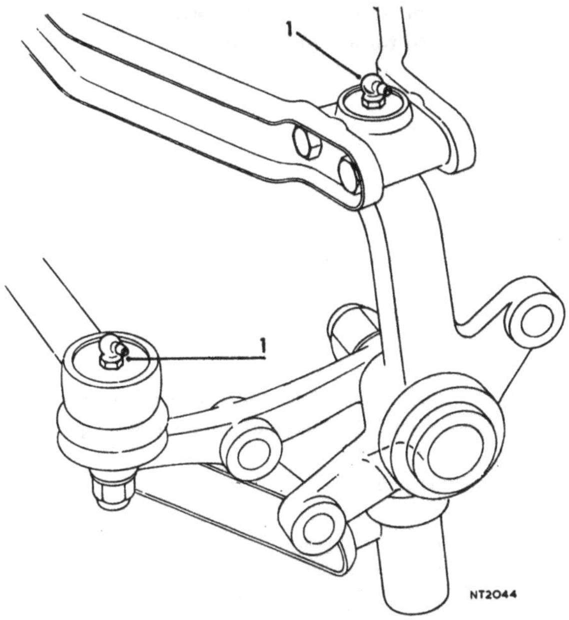

(43) Lubricate steering rack and pinion

1. Wipe clean the plug and surrounding area.
2. Remove the plug.
3. Fit a suitable grease nipple to the plug hole.
4. Apply a grease gun to nipple, and stroke for 5 times only.

CAUTION: Over greasing can cause damage to the rubber bellows.

5. Remove grease nipple.
6. Refit plug.
7. Wipe away surplus grease.

(44) Lubricate rear hubs

1. Wipe clean the plug and surrounding area.
2. Remove the plug.
3. Fit a suitable grease nipple to the plug hole.
4. Apply a grease gun to nipple and stroke until grease exudes from the bearing.
5. Remove grease nipple.
6. Refit plug.
7. Wipe away surplus grease.

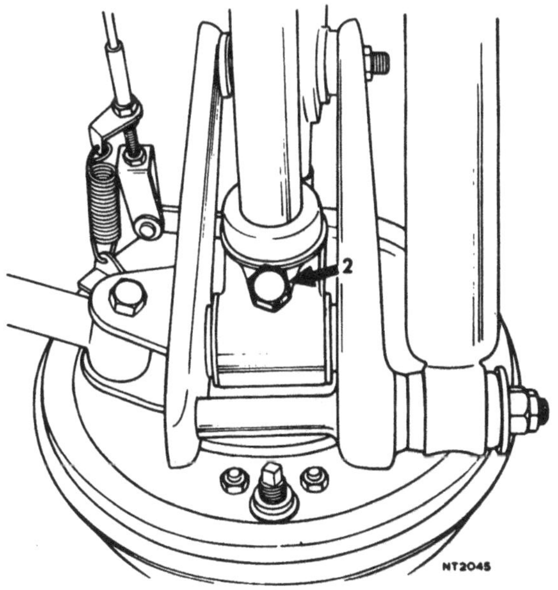

(45) Lubricate handbrake linkage and cables

1. Lubricate handbrake pivot.
2. Smear grease around handbrake lever cable connections, working it well into the clevis.
3. Smear grease around brake drum cable connections, working it well into the clevis pin.
4. Grease exposed sections of inner cable to resist corrosion.

(46) Check engine, transmission, final drive, suspension and steering unit for oil leaks and report.

(47) Check visually brake, fuel and clutch pipes, hoses and unions for chaffing leaks and corrosion and report.

Check visually

1. Brake and clutch pipes,
2. Brake and clutch hoses,
3. Brake and clutch pipe and hose unions,
4. Fuel pipes,
5. Fuel pipe unions,

   for chaffing leaks and corrosion.
6. Report any defects found.

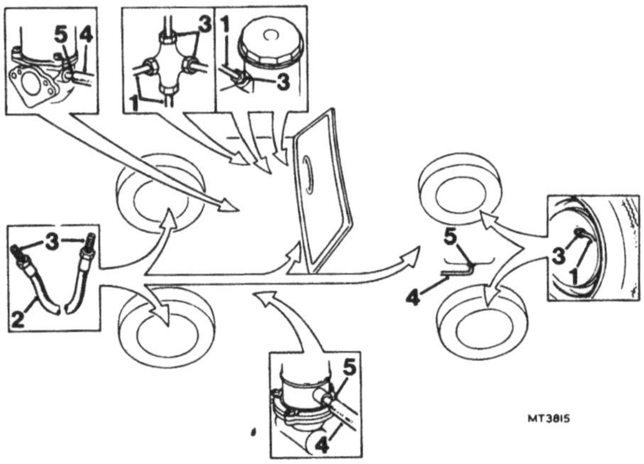

(48) Check/report exhaust system for leakage and security

1. Place car on ramp or over a pit.
2. Check security of exhaust pipe to manifold nuts, which if correct should be tightened to 12 to 14 lbf ft (1·7 to 1·9 kgf m).
3. Check security of exhaust pipe joint clips.
4. Check security of exhaust system mounting bolts.
5. Using a second operator, run engine at fast idle speed.
6. Check exhaust system joints for leaks.
7. Check exhaust pipes for leaks arising from damage or deterioration.
8. Check exhaust silencers for leaks arising from damage or deterioration.
9. Stop engine.
10. Report any defects found.

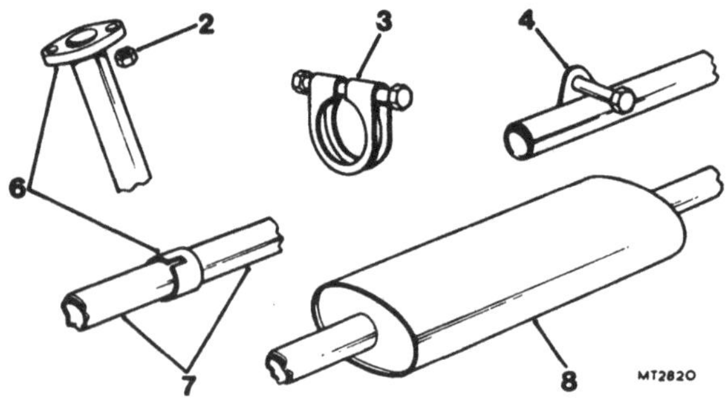

(49) Check security of suspension fixings, tie-rod levers, steering unit attachment and steering universal joint coupling bolts.

Check security of

1. Suspension fixings,
2. Tie-rod levers,
3. Steering unit attachment,
4. Steering universal joint coupling bolts.

(50) Check security of propeller shaft and drive shaft universal coupling bolts

1. Check security of propeller shaft coupling bolts, which if correct should be tightened to 26 to 34 lbf ft (3·6 to 4·7 kgf m).
2. Check security of half shaft to final drive unit coupling bolts, which if correct should be tightened to 38 to 46 lbf ft (5·2 to 6·4 kgf m).

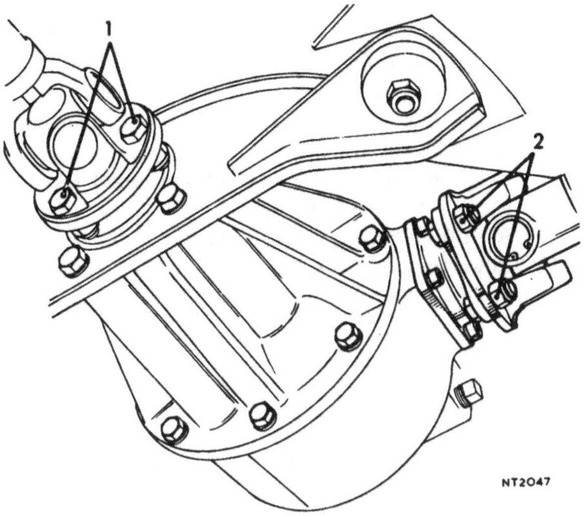

NT2047

(51) Check security of sub-frame to body mountings

Using page 06 as a guide

1. Check security of sub-frame mounting bolts/nuts.

(52) Check/report condition of steering unit/joints for security, backlash and gaiter condition

1. Check security of steering unit mounting and steering joints, using page 06 as a guide.
2. Check steering for backlash.
3. Check condition of steering gaiters.
4. Report any defects found.

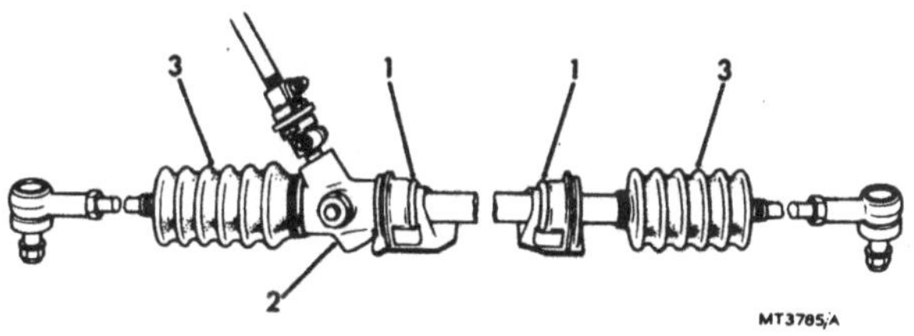

MT3785/A

(53) Adjust front hubs

See 60.25.13

(54)(55) Check/adjust/report front and rear wheel alignment with tracking equipment.

Front wheel alignment – See 57.65.01

Rear wheel alignment – See 64.25.17

(56) Inspect brake pads for wear and discs for condition

Front brakes

1. Jack up front of car and place safely on stands.
2. Remove front brake pads – See 70.40.02.

   CAUTION: Do NOT depress the brake pedal while pads are removed.

3. Report pad condition if the friction lining has been reduced to 0·125 inch (3 mm) or if there is not sufficient material to provide a thickness of 0·125 in (3 mm) at the completion of a further 3,000 miles (5,000 km) motoring.
4. Check brake discs for excessive scoring and report this if present.
5. Check brake discs for run out and report this if it exceeds 0·007 in (0·178 mm).
6. Refit front brake pads – See 70.40.02
7. Lower car off stands.

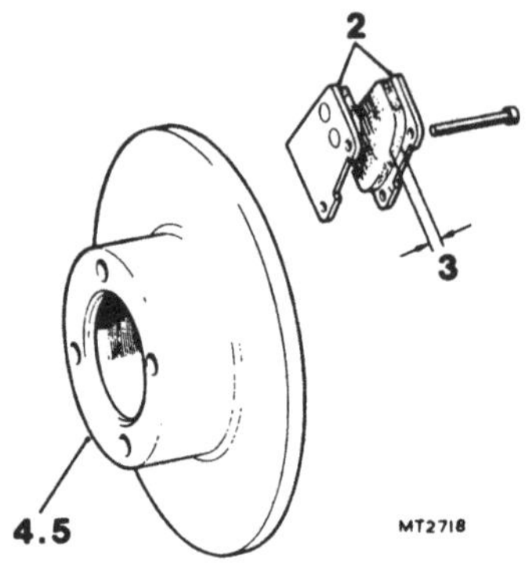

(57) Inspect and report brake linings for wear and drums for condition

1. Jack up car and place safely on stands.
2. Remove road wheel – See 74.20.01.
3. Remove brake drum – See 70.10.02 (front) or 70.10.03 (rear).
4. Check brake linings for wear and report if linings are excessively worn.
5. Check brake linings for damage and contamination by oil or grease and report if linings are damaged or contaminated.
6. Check brake drums for wear, scoring or other damage and report if drums are excessively worn, scored or damaged.
7. Remove dust, oil and grease from brake drum and backplate.
8. Refit brake drum – See 70.10.02 (front) or 70.10.03 (rear).
9. Refit road wheel – See 74.20.01.
10. Lower car off stands.

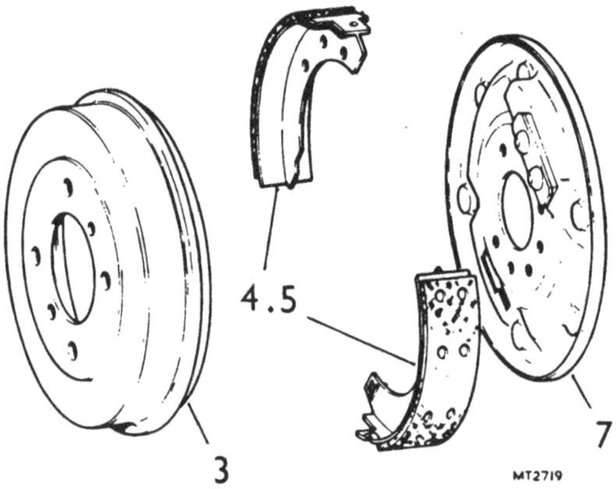

(58) **Check security of road wheel fastenings.**

Disc Wheels

For each wheel in turn:-

1. Check tightness of road wheel securing nuts, which if correct should be tightened to 38 to 48 lbf ft (5·2 to 6·6 kgf m).

Wire Wheels

For each wheel in turn

1. Visually check that adaptor is fitted on the correct side,
2. Remove road wheel – See 74.20.01
3. Check tightness of the adaptor securing nuts, which if correct should be tightened to 38 to 45 lbf ft (5·2 to 6·2 kgf m).
4. Replace road wheel – See 74.20.01 ensuring that the centre nut is correctly secured.

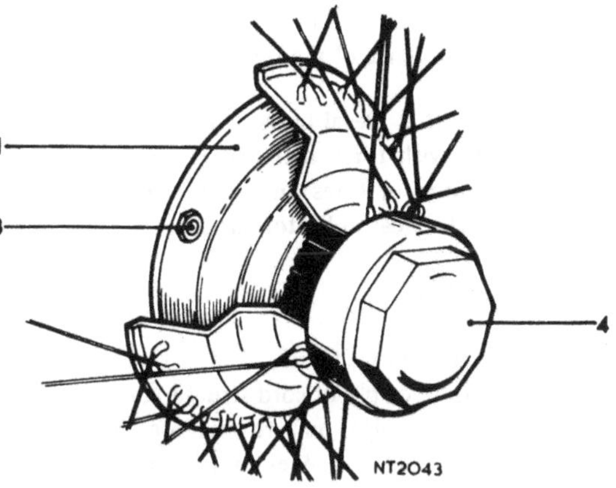

(59) **Check that tyres are in accordance with manufacturers specification**

For each road wheel and spare wheel:–

1. Check that tyres are in accordance with vehicle manufacturers recommendations for type and size and report any deviation.
2. Check for mixing of cross ply and radial ply tyres and report if both types are present on the vehicle (including spare wheel).

WARNING: It is illegal in the U.K. and highly dangerous to mix cross ply and radial ply tyres on the same axle or to fit radial ply tyres to the front wheels only.

(60) Check visually and report depth of tread, cuts in tyre fabric, exposure of ply or cord structure, lumps or bulges

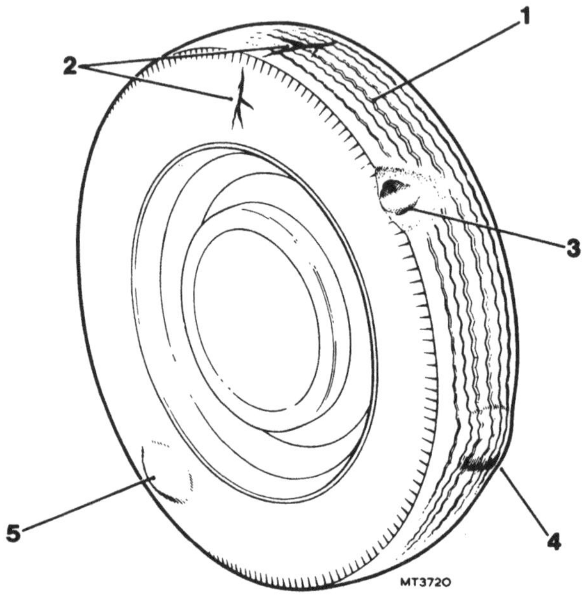

For each road wheel and spare wheel:–

1. Check tread depth, which if correct should show 1mm (0·039 in) of tread (excluding wear bars) over three quarters of the bredth for the entire circumference of the tyre.

WARNING: It is illegal in the U.K. to use a car of this type fitted with tyres that have a tread depth below this minimum or tyres on which the tread is worn level with the wear indicator bars.

Check for

2. Cuts in the tyre fabric,
3. Exposure of ply or cord structure,
4. Lumps or bulges on tyre circumference,
5. Lumps, bulges or other damage on tyre walls.

WARNING: It is illegal in the U.K. to use a car fitted with tyres in a damaged condition.

(61) Check/adjust tyre pressures (including spare wheel)

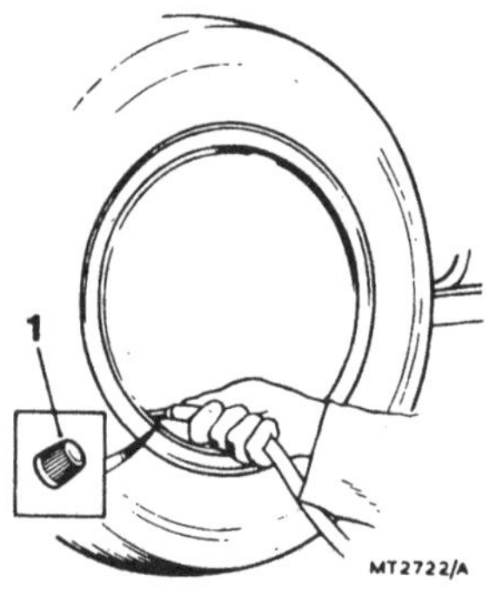

With all tyres at ambient temperature:–

1. Remove protective dust cap.
2. Using a tyre pressure gauge, tested for accuracy, check tyre rpessure.

   Recommended tyre pressures: 21 lb/in^2 Front
   26 lb/in^2 Rear

3. Adjust tyre pressure as necessary.
4. Replace dust cap or renew if missing.

WARNING: It is illegal in the U.K. to use a car with the tyres inflated to a pressure that is not suitable for the use to which the vehicle is put.

(62) Check/adjust/report headlamp alignment

See 86.40.18

(63) Check, if necessary replace windscreen wiper blades

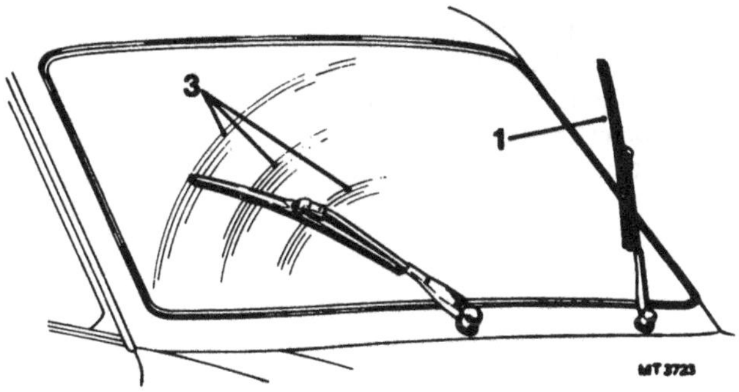

1. Examine each wiper blade in turn for damage.
2. With wiper blades in position and windscreen wet, operate wiper motor.
3. Check wiper blade operation for smearing and adequate removal of dirt.
4. Stop wiper motor.
5. If the checks in procedures 1 and 3 are not satisfactory, replace one or both wiper blades as necessary – See 84.15.05.

(64) Fuel tank filler cap – check seal for security

See 17.15.01

(65) Check brake pedal travel and handbrake operation, adjust if necessary.

1. With handbrake in 'off' position, check brake pedal for spongyness and excessive travel.
2. If brake pedal has spongy operation, bleed and adjust brakes - See 70.25.01.
3. If brake pedal travel is excessive, adjust brakes – See 70.25.03.
4. With foot clear of brake pedal, check handbrake for excessive travel.
5. If handbrake travel is excessive, adjust handbrake – See 70.35.10.

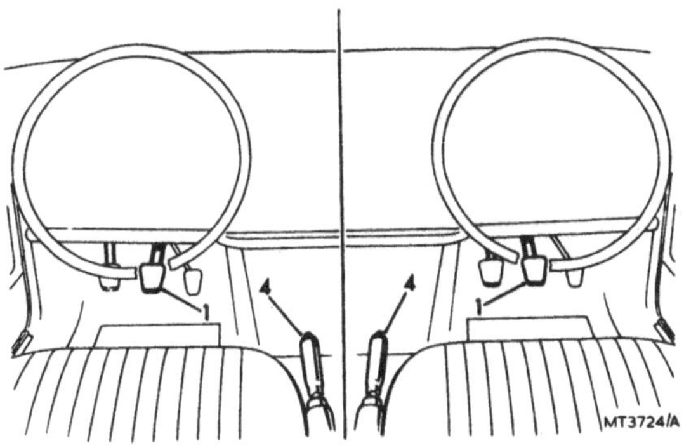

(66) Check/report brake pedal travel and handbrake operation.

1. With handbrake in 'off' position, check brake pedal for spongy operation and excessive travel.
2. Report brake pedal condition.
3. With foot clear of brake pedal, check handbrake for excessive travel.
4. Report handbrake operation.

(67) Check operation of window controls, locks and bonnet release.

Check operation of:-

1. Window raising and lowering controls,
2. Internal door locks,
3. External door locks,
4. Luggage compartment lock,
5. Bonnet release controls.
6. Report any defects found.

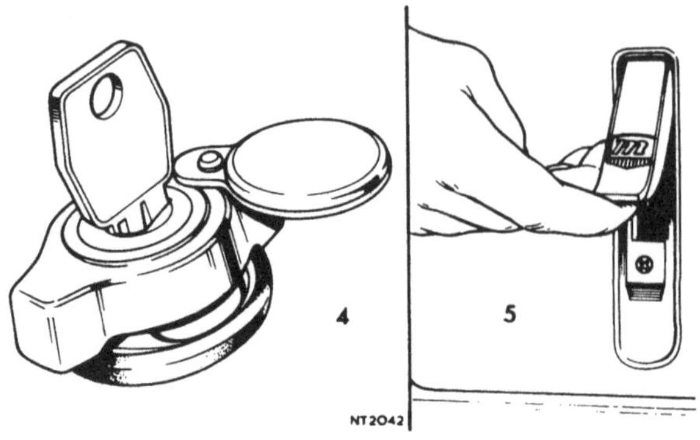

(68) Check function of all electrical systems and windscreen washer.

In sequence, check operation of:–

1. Side, tail and headlamps (including headlamp dip/main beam and 'flash' controls),
2. Instrument panel illumination.
3. Interior light,
4. Horn(s),
5. Auxiliary lights,

With ignition circuits energised, check operation of:

6. All warning lights (including 'hazard' warining lights if fitted),
7. Fuel level indicator,
8. Heater blower motor,
9. Windscreen washers,
10. Windscreen wipers,
11. Direction indicators,
12. Brake lights,
13. Reversing lights.
14. Start engine and note that oil pressure warning light has extinguished.

Check operation of:-

15. Charging system warning light in relation to engine speed,
16. Temperature indicator,
17. Radio (if fitted)
18. Switch off engine and return ignition switch to the auxiliary position, then recheck the function of any fitted accessories e.g. a radio, that are supplied with power from this switch position.
19. Report any defects found.

(69) Lubricate clutch and brake pedal pivots.

Using an oil can, lubricate
1. Clutch pedal pivot.
2. Brake pedal pivot.
3. Wipe away surplus oil to prevent staining the carpet.

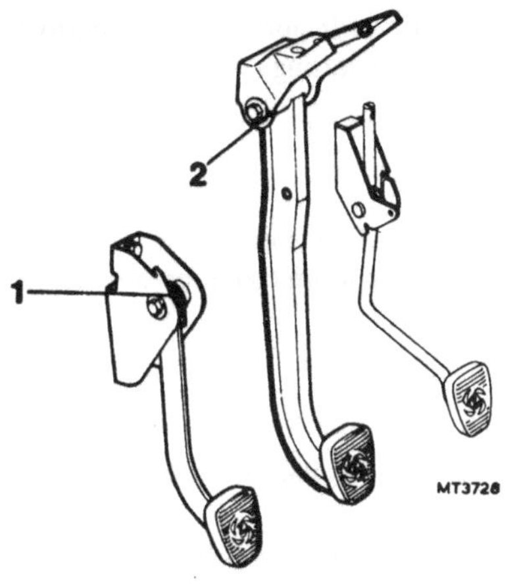

(70) Lubricate all locks, door hinges, strikers and bonnet release.

Using an oil can, lubricate
1. Door locks.
2. Door hinges.
3. Door strikers.
4. Luggage compartment lock.
5. Bonnet release mechanism.
6. Wipe away surplus oil.

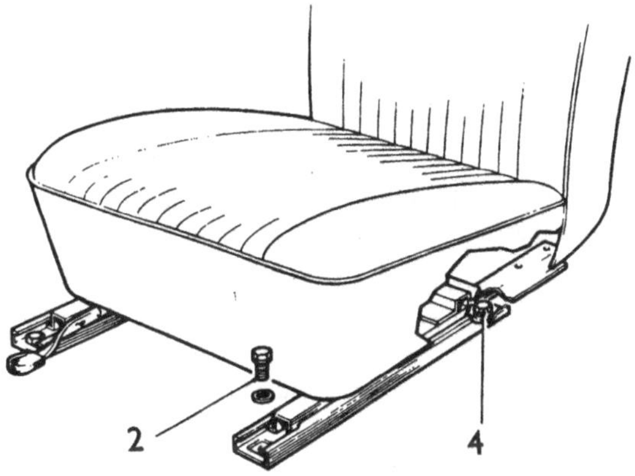

(71) Check/report condition and security of seats and seat belts.

1. Move driver's seat back to its fullest extent.
2. Check security of front bolts holding seat runner to floor.
3. Move driver's seat forward to its fullest extent.
4. Check security of rear bolts holding seat runner to floor.
5. With seat in middle position, check security of seat in runner.
6. Repeat the procedure in 1 to 5 for passenger seat.
7. Check seat tipping and lock mechanisms.
8. Check seat belts for wear and damage.
9. Check seat belt connections for wear and damage.
10. Check security of seat belt anchorage bolts, which if correct should be tightened to 24 to 32 lbf ft (3·3 to 4·4 kgf m).
11. Report any defects found.

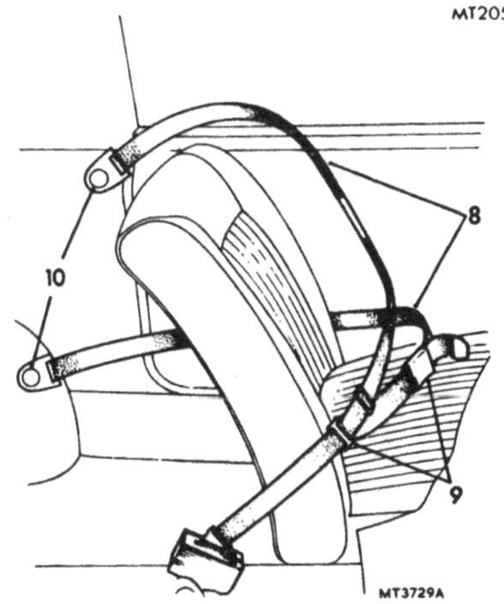

(72) Check/report rear view mirrors for looseness, cracks and crazing.

1. Check interior mirror for looseness, cracks and crazing.
2. Check external mirrors (when fitted) for looseness, cracks and crazing.
3. Report on condition of mirrors.

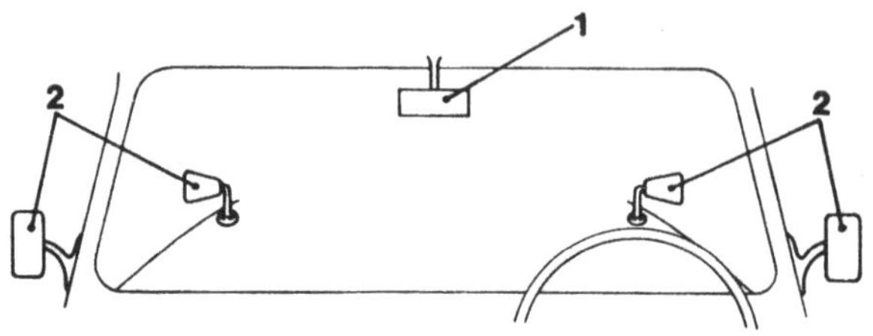

MT2812

(73) Road/Roller test and report additional work required

In addition to the general road test, pay particular attention to:–

1. The efficiency and function of the footbrake and handbrake.
2. The function of the steering mechanism.
3. The function of the speedometer.

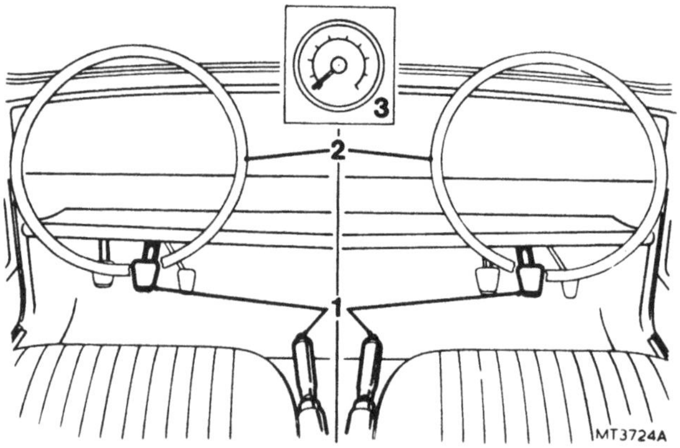

MT3724A

(74) Ensure cleanliness of controls, door handles, steering wheel etc.

1. Check steering wheel, gear lever, bonnet release control and fascia controls etc. for dirt and damage attributable to the service just completed.
2. Check door trims, locks and window controls for dirt and damage attributable to the service just completed.
3. Check seats, carpets and pedal rubbers for dirt and damage attributable to the service just completed.

ENGINE OPERATIONS

continued

continued

***All bracketed material ([ ]) in this section has been added by Robert Bentley, Inc.**

DISTRIBUTOR DRIVE SHAFT AND GEAR

– Remove and refit **12.10.22**

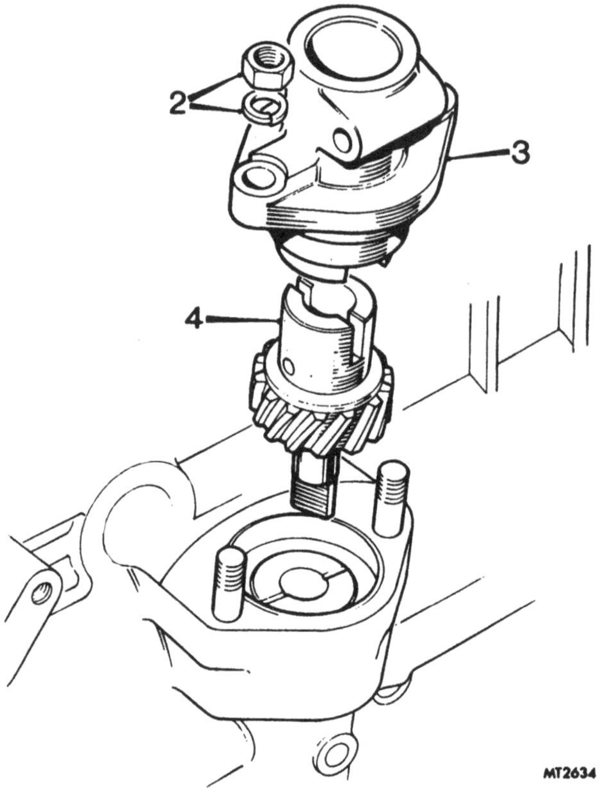

[NOTE: To prevent abrasive dirt from entering the engine, thoroughly clean the pedestal and the surrounding part of the cylinder block before you remove the pedestal.]

Removing

1. Remove the distributor 86.35.20.
2. Remove the two nuts and washers securing the pedestal to the cylinder block studs.
3. Remove the pedestal and gasket.
4. Lift out the drive shaft and gear.

Refitting

NOTE: An end-float of 0·005 in ±0·002 in (0·13 mm ±0·05 mm) must exist between the end of the distirbutor pedestal and the drive gear. The end-float is controlled by the selective use of gaskets on the pedestal/cylinder block interface. To measure the end-float adopt the following procedure.

5. Measure, note the thickness of, and fit a plain washer 0·5 in (12·7 mm) i.dia over the drive shaft below the gear.
6. Insert the shaft assembly and washer into position in the bush, ensuring that the oil pump drive is mated with the shaft.

NOTE: Use a screwdriver to turn the oil pump drive: try the drive gear in a different position in order to be sure that it is down correctly.

7. Fit the distributor pedestal into position without gasket.

8. Measure the gap between the pedestal and cylinder block (a) and subtract from (b) the thickness of the washer. The result is gear end-float (or load) which must be adjusted, by gaskets, to give a mean end-float of 0·005 in (0·13 mm).

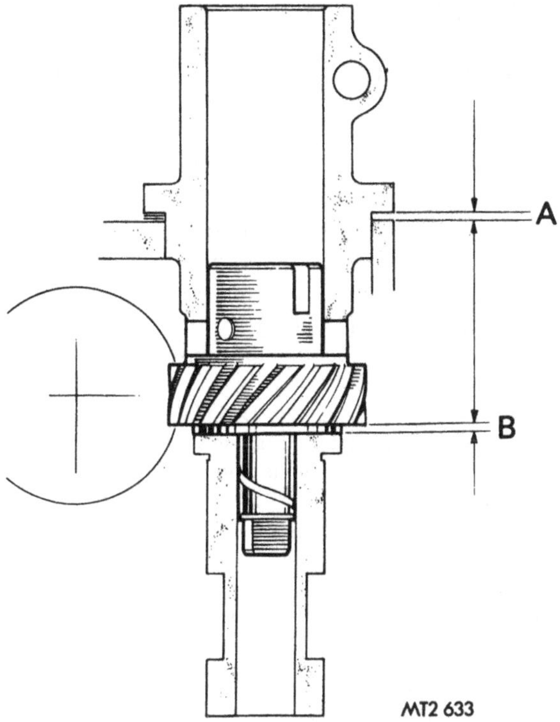

MT2 633

EXAMPLE 1

	in	mm
Thickness of washer *(b)*	0·100	2.54
Pedestal/cylinder block gap *(a)*	0·098	2·49
= gear end-float	+0·002	+0·05
add gasket of	0·003	0·08
for correct end-float of	0·005	0·13

continued

EXAMPLE 2

	in	mm
Thickness of washer *(b)*	0·100	2·45
Pedestal/cylinder block gap *(a)*	0·110	2·79
= Gear load of	–0·010	–0·25
add gasket of	0·015	0·38
for correct end-float of	0·005	0·13

9. Remove the pedestal, shaft assembly and washer.
10. Turn the crankshaft to bring No. 1 piston to T.D.C. on compression stroke.
11. Lower the drive gear into the bush, allowing it to turn as it meshes with the camshaft gear and ensuring that it engages with the oil pump drive dog.
12. The gear is correctly positioned when the offset slot is as shown in illustration.
13. Fit the pedestal and selected gasket.
14. Fit and tighten the nuts and washers.
15. Fit the distributor 86.35.20.

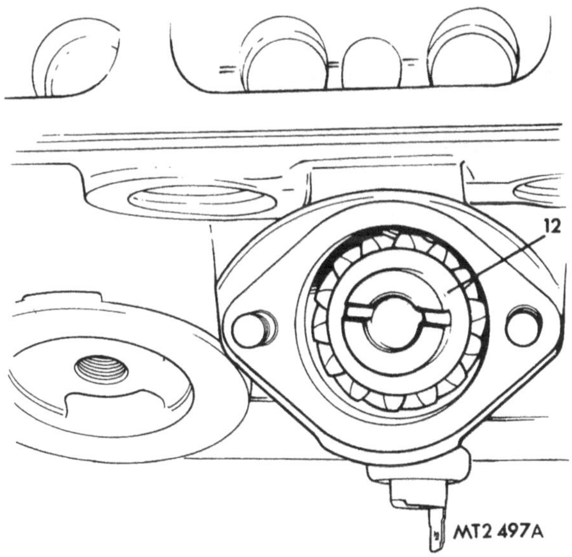

MT2 497A

[NOTE: To prevent undue wear when the engine is first started, lightly lubricate the gear with motor oil during installation.]

DATA

Drive gear end-float	0·003 to 0·007 in (0·08 to 0·18 mm)
Spindle diameter	0·4980 to 0·4985 in (12·65 to 12·67mm)
Bush bore	0·5005 to 0·5010 in (12·71 to 12·73 mm)

CAMSHAFT

– Remove and refit **12.13.01**

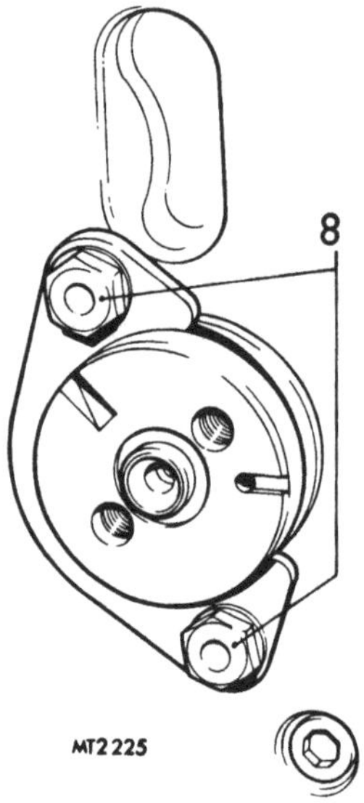

Removing

1. Isolate the battery.
2. Remove the radiator 26.40.01.
3. Remove the cylinder head 12.29.10.
4. Withdraw the cam followers, identifying for re-assembly. **[12.29.51]**
5. Remove the distributor drive shaft and gear. 12.10.22.
6. Remove the fuel pump. 19.45.08.
7. Remove the timing chain and sprockets 12.65.12.
8. Remove the two bolts and withdraw the camshaft location plate.
9. Withdraw the camshaft.

Refitting

10. Reverse instructions 1 to 9, and in addition:
    a. Check the camshaft end float and reduce if necessary by fitting a new camshaft plate. See instruction 55, 12.41.05.
    b. Time the valves. 12.65.08.

NOTE: Engines prior to No. 25000 used camshaft bearings.

[NOTE: Used cam followers should never be interchanged. The best way to identify them for reassembly is to wipe dry each follower as you remove it. Then use waterproof ink or a waterproof felt-tipped marker to write an identification number on the follower. During reassembly, lightly lubricate the cam followers with assembly lubricant (available from automotive supply stores) or with a thin coat of multipurpose grease. Doing so will prevent undue cam lobe and cam follower wear when the engine is first started.]

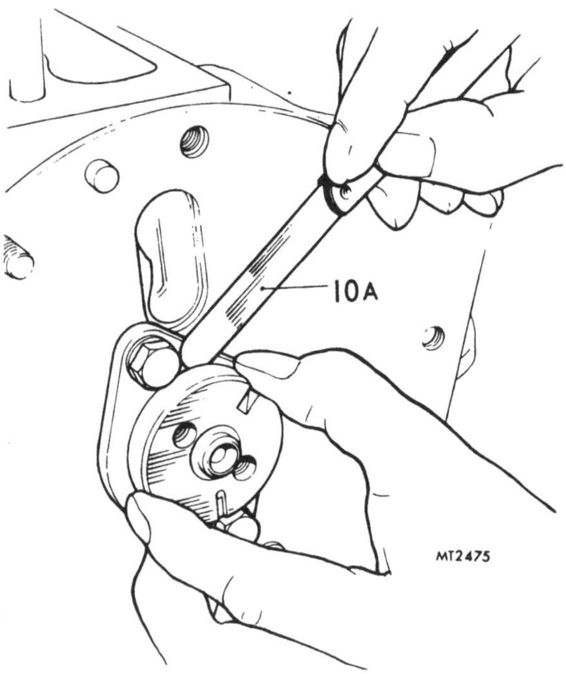

DATA

Journal diameter (Up to Engine No. 25000)	1·9649 to 1·9654 in (49·91 to 49·92 mm)
Journal diameter (from Engine No. 25001)	1·8402 to 1·8407 in (46·74 to 46·75 mm)
End-float	0·0042 to 0·0085 in (0·110 to 0·216 mm)
Bore in block	1·9680 to 1·9695 in (49·980 to 50·025 mm)

CONNECTING RODS AND PISTONS

– Remove and refit **12.17.01**

Removing

1. Isolate the battery.
2. Place the car on a ramp or over a pit.
3. Remove the cylinder head. 12.29.10
4. Remove the oil sump. 12.60.44.
5. Remove the oil pick-up strainer. 12.60.20. 2 to 4.
6. Turn the crankshaft to bring Nos. 1 and 4 connecting rod bolts to an accessible position.
7. Check the identifying marks on the connecting rods and bearing caps. Mark if necessary.
8. Remove the connecting rod bolts.
9. Withdraw the bearing caps and lower half bearing shells complete with the bolt bushes.
10. Push the connecting rods and pistons upwards and carefully withdraw, identifying for reassembly.
11. Mate up the top and bottom bearing shells and caps with their respective connecting rods and piston assemblies.
12. Repeat instructions 6 to 11 on Nos. 2 and 3 bearings.

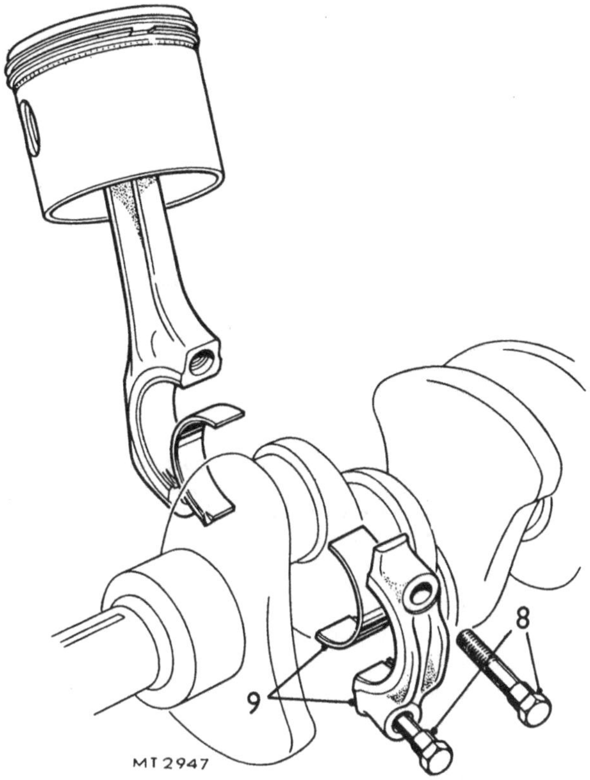

Refitting

13. Position Nos. 1 and 4 crankpins at B.D.C. and lubricate them with clean engine oil.
14. Smear the pistons and cylinder bores with clean engine oil.
15. Carefully insert the connecting rods and pistons into the bores, ensuring that the arrow (thus ▲) is pointing to the front of the engine.
16. Ensure that the open face of the big-end bearing is towards the non thrust side of the cylinder bore.
17. Stagger the piston ring gaps, avoiding a gap on the thrust side of the piston.
18. Using a piston ring compressor, gently push the pistons into the bore.
19. Fit the upper bearing shells to the connecting rod big-ends and pull the connecting rods onto the crankpins.
20. Check that the connecting rod bolt bushes are correctly positioned.
21. Fit the bearing shells to the bearing caps and fit to the connecting rods, securing with new bolts and tighten evenly (see 06-1 for correct torque).
22. Repeat instructions 13 to 21 on Nos. 2 and 3 cylinders.
23. Reverse instructions 1 to 5.
24. Replenish the sump with correct grade of engine oil.

[NOTE: Lubricate the pistons, the piston rings, and the cylinders with engine oil only. Other lubricants may interfere with the proper seating of new piston rings. If the engine is to be stored for some time following repairs — or if you are installing new bearing shells — it is advisable to use assembly lubricant (available from automotive supply stores) or a light coating of multipurpose grease on the crankpins rather than ordinary motor oil.]

PISTONS AND/OR RINGS – ENGINE SET

– Remove and refit	12.17.03
Pistons and/or rings – extra each	12.17.06

See 12.17.10.

CONNECTING RODS AND PISTONS

– Overhaul	12.17.10
Gudgeon pin bush – each – remove and refit	12.17.13

NOTE: Do not mix components during this operation

Service tools: 335, S336-4

[CAUTION: Every check and fitting operation described here is essential. If you lack the skills or the proper equipment, please turn such work over to your Authorized Dealer or to a qualified automotive repair shop.]

1. Remove the connecting rods and pistons 12.17.01.
2. Remove the circlips from the pistons.
3. Remove the gudgeon pin.
4. Remove the piston from the connecting rod.
5. Remove the piston rings with the expander tool and clean.
6. Clean the pistons, removing all carbon deposits, particularly from the piston ring grooves.
7. Remove the big-end bearing shells.
8. Examine the gudgeon pin and check for wear – see Data.
9. Check the piston dimensions to grade and bore – see Data.
10. Check the dimensions of piston ring grooves – see Data.
11. Check the piston ring gaps in the bores – see Data.

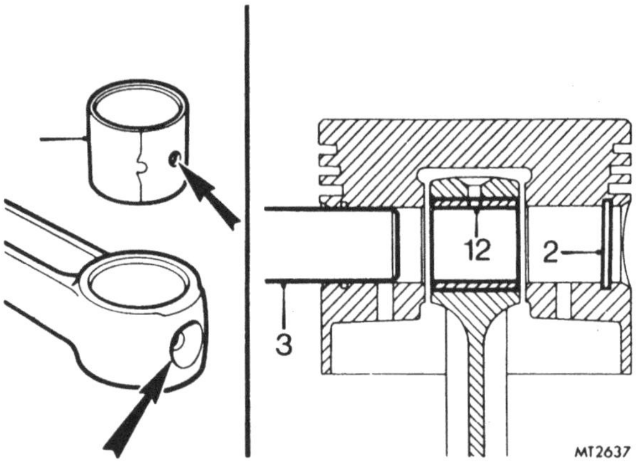

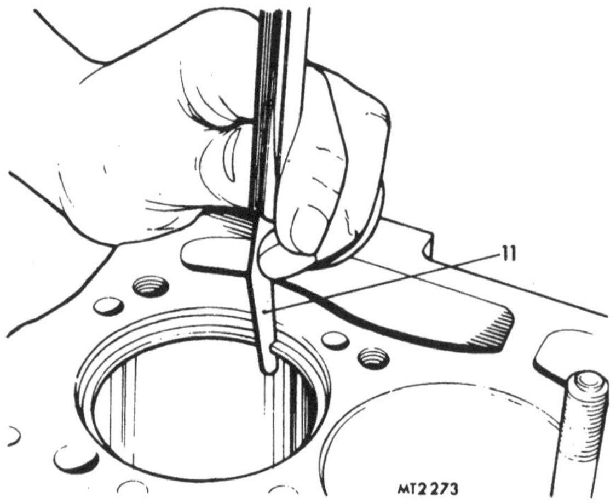

[NOTE: If wear has left ridges of metal around the tops of the cylinders (known as top cylinder ridges), you should not replace the piston rings unless these ridges are first removed with a special ridge reaming tool. If unremoved, the ridges may break the new top rings.]

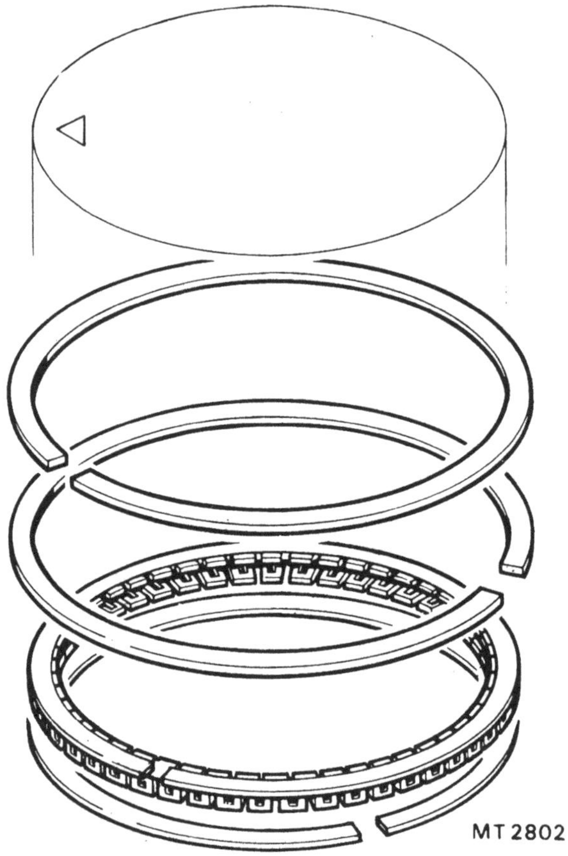

Triumph Spitfire Mk. IV Manual. Part No. 545254. Issue 2.

12. Check the small end bushes for wear, and if necessary renew using a suitable hand press to remove the old bush and fit new. Ensure that the oil hole in the bush corresponds with the connecting rod. Ream the new bush to size – see Data. When assembled dry, the gudgeon pin is a thumb push fit at 68°F room temperature. If the gudgeon pin passes through the bore under its own weight, it is too slack.
13. Check the connecting rods for bend and twist. Use tools 335, and arbor adaptor S336-4 – see Data. Rods that exceed the tolerances in both conditions should be renewed or re-aligned.

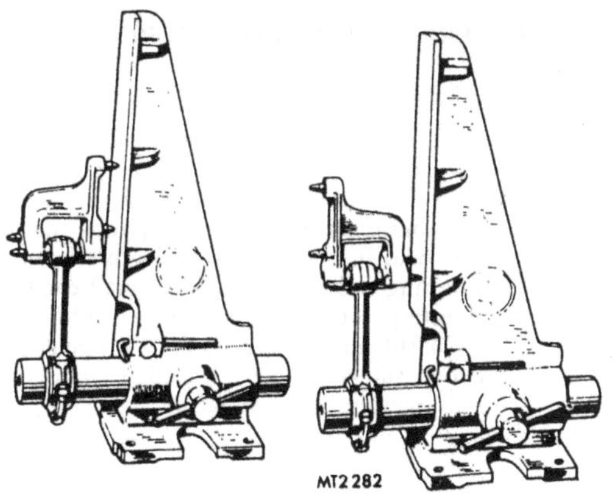

Refitting

14. Reverse instructions 2 to 5:
    - *a.* Fitting oversize pistons as necessary.
    - *b.* Fitting the piston rings with care and in the following order: (A) expander ring into bottom groove, ensure that ends are butting but not overlapping. (B) bottom rail of expander from bottom of piston, (C) top rail from top of piston, (D) scraper ring – step uppermost to second groove, (E) compression ring to top groove.
    - *c.* Separating the ring gaps equally on the non-thrust side of piston.
15. Refit the connecting rods and pistons 12.17.01.

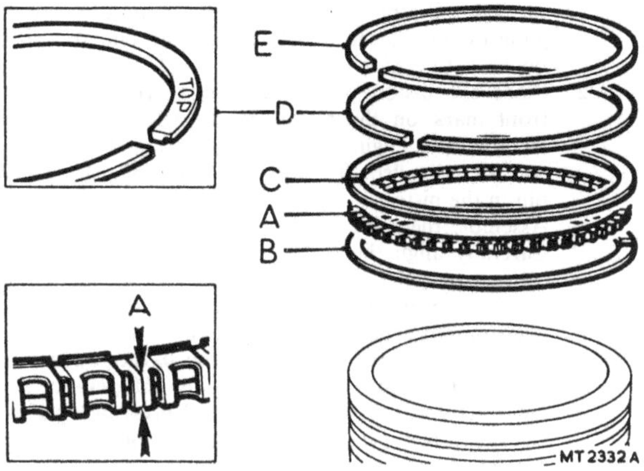

DATA

Unless otherwise stated data is applicable to all engines.

Pistons

Bore size: Grade F	2·899 to 2·900 in (73·64 to 73·66 mm)
Grade G	2·9010 to 2·9005 in (73·86 to 73·67 mm)
Piston top diameter: Grades F and G up to Eng. No. FH/FK25001	2·875 to 2·880 in (73·03 to 73·15 mm)
From Eng. No. FH/FK25001	2·871 to 2·874 in (72·93 to 73·01 mm)
From Eng. No FM 1	2·870 to 2·873 in (72·90 to 72·98 mm)
Piston Bottom diameter: Grade F Eng. Nos. Prefix FH/FK	2·8976 to 2·8981 in (73·59 to 73·61 mm)
Eng. Nos. Prefix FM	2·8984 to 2·8989 in (73·62 to 73·63 mm)
Grade G Eng. Nos. Prefix FH/FK	2·8982 to 2·8987 in (73·617 to 73·620 mm)
Eng. Nos. Prefix FM.	2·8990 to 2·8995 in (73·635 to 73·647 mm).
Pistons available	0·020 in (0·52 mm) oversize
Groove width: bottom	0·1578 to 0·1588 in (3·99 to 4·01 mm)
centre and top	0·064 to 0·065 in (1·625 to 1·650 mm)
Ring width: bottom (oil control) 3-part	0·1540 to 0·1560 in (3·90 to 3·96 mm)
centre and top (compression)	0·0620 to 0·0625 in (1·575 to 1·5787 mm)
Rings available	0·010, 0·020, 0·030 in (0·25, 0·52, 0·76 mm) oversize

Connecting rods

Small end bush (fitted) i.dia	0·8126 to 0·8129 in (20·64 to 20·65 mm)
Gudgeon pin diameter	0·8123 to 0·8125 in (20·63 to 20·64 mm)
Connecting rod bend not to exceed	0·0015 in (0·04 mm)
Connecting rod twist not to exceed	0·0045 in (0·114 mm)

Connecting rod and piston assemblies

Weight variation between heaviest and lightest assembly	Max. 4 drams.

INTERFERENCE FIT GUDGEON PINS

—Refit **[12.17.14]**

NOTE: The early Mk. III Spitfire engines have interference fit gudgeon pins. Pistons with such pins can be indentified by the absence of gudgeon pin circlips. There are no bushings in the connecting rods and the gudgeon pin is held in place solely by its interference fit in the connecting rod small end. The small end bore in such connecting rods must be between 0.8110 and 0.8115 in. (20.510 and 20.612 mm). A special tool, such as the one shown in the illustration, should be used to install interference fit gudgeon pins (Tool S334).

Refitting

1. Remove burrs, if present, from the gudgeon pin and the bore in the connecting rod. Then lubricate the gudgeon pin, the connecting rod bore and the piston bosses with engine oil.
2. Place the piston on the connecting rod so that the front mark on the piston crown will be toward the front of the engine when the piston and connecting rod are installed. Then partially insert the gudgeon pin in the piston.
3. Assemble the special tool as shown in the illustration. Carefully align the gudgeon pin with the bore in the connecting rod.
4. Using a torque wrench, tighten the nut (1) until the gudgeon pin contacts the limit stop.

CAUTION: If the torque required to draw in the gudgeon pin is less than 5 lbf. ft (0.7 kgf. m), the interference fit is incorrect and a new connecting rod must be obtained. Lubricate the special tool with engine oil only, otherwise the torque wrench reading may be inaccurate.

5. Remove the special tool. Then check that the piston rocks evenly on the gudgeon pin. If there are tight spots which cause the piston to bind, it is an indication that the pin has picked up metal from the connecting rod owing to misalignment during assembly. If such faults are left uncorrected, the piston bosses will wear rapidly.

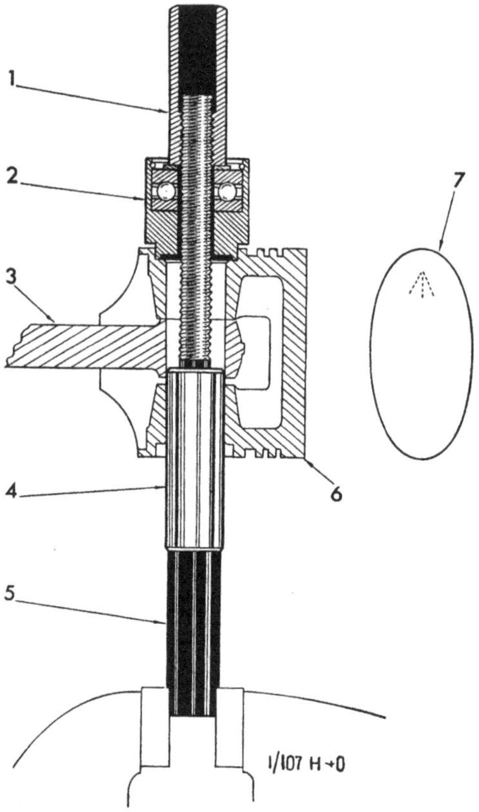

1. **Nut.**
2. **Thrust race**
3. **Connecting rod**
4. **Gudgeon pin**
5. **Pilot shaft**
6. **Piston**
7. **Front mark on piston crown**

Part No. 511243. page 1.133. Issue 1.

CONNECTING ROD BEARINGS – SET

– Remove and refit	12.17.16
Connecting rod bearing – one	12.17.17
Connecting rod bearing – extra each	12.17.18

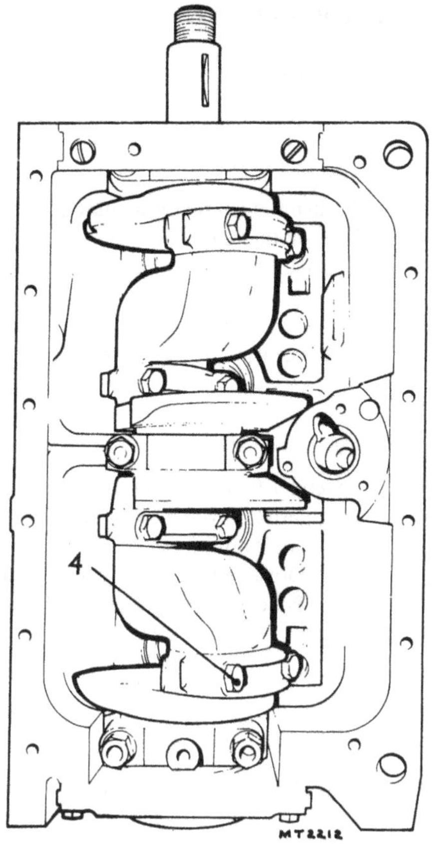

Removing

1. Remove the oil sump 12.60.44.
2. Remove the oil pick-up strainer. 12.60.20. 2 to 4.
3. Turn the crankshaft to bring the connecting rod bearing to be removed to an accessible position.
4. Remove the connecting rod bolts.
5. Remove the bearing cap complete with the shell bearing and connecting bolt bushes.
6. Push the connecting rod and piston upwards sufficiently to clear the crankpin to enable the upper shell bearing to be removed.

NOTE: Do not push the piston higher than T.D.C. or the top piston ring may be released necessitating removal of the cylinder head.

7. Remove the upper shell bearing.

[NOTE: If the engine is to be stored for some time following repairs, it is advisable to use assembly lubricant (available from automotive supply stores) or a light coating of multipurpose grease on crankpins rather than an ordinary engine oil. Many experienced mechanics choose to use assembly lubricant whenever new bearing shells are installed.]

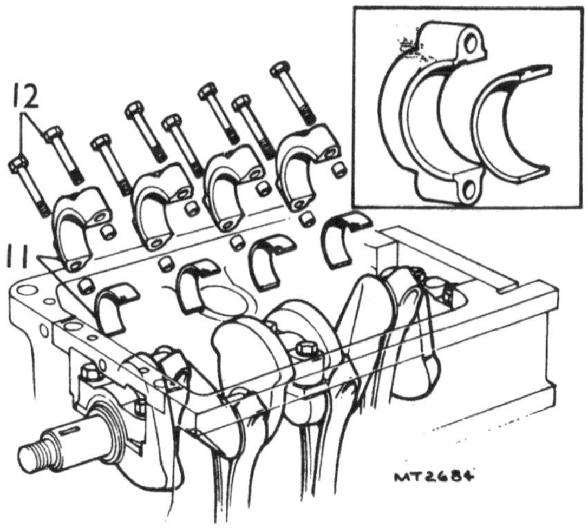

Refitting

8. Insert the shell bearing in the connection rod big-end.
9. Lubricate the crankpin with clean engine oil.
10. Pull the connecting rod onto the crankpin.
11. Fit the shell bearing to the bearing cap and fit to the connecting rod.

NOTE: When fitting the bearing shells, ensure that the lug on the bearing locates correctly in the recess in the connecting rod big-end and cap.

12. Fit and tighten the bolts (see 06-1 for correct torque).
13. Repeat the instructions on the remaining assemblies if necessary.
14. Reverse instructions 1 to 2.
15. Replenish the sump with the correct grade of engine oil.

NOTE: Mk. III Spitfire engines have crankpins with smaller diameters than those of later engines. However, the connecting rod side clearance for all engines should be between 0.0025 and 0.0086 in (0.063 and 0.218 mm).
If you check the rod bearing clearance with Plastigage® (available from automotive supply stores), the clearance should be between 0.0005 and 0.0020 in (0.012 and 0.050 mm).

CRANKSHAFT PULLEY

– Remove and refit **12.21.01**

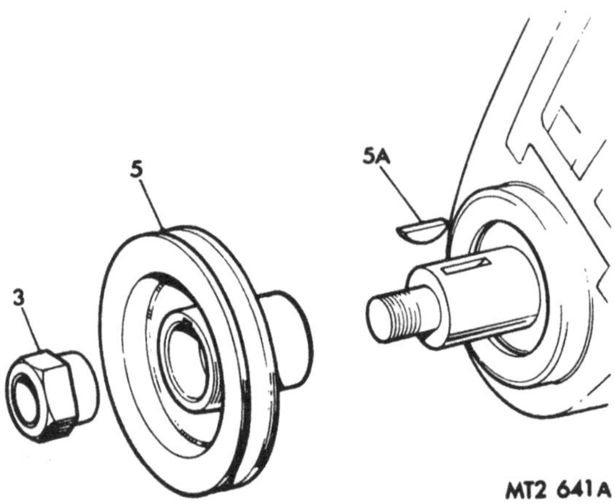

MT2 641A

Removing

1. Slacken the alternator mountings.
2. Remove the fan belt.
3. Working from below vehicle, remove pulley retaining nut.
4. Withdraw pulley, if necessary use extractor, bearing on the centre nut, loosely replaced.

Refitting

5. Reverse 1 to 4, ensuring:
   a. Drive key is square in the crankshaft keyway an· is not burred.
   b. Drive belt is correctly tensioned 86.10.05.
   c. Pulley retaining nut is tightened to 90 to 110 lbf. ft (12·4 to 15·2 kgf. m).

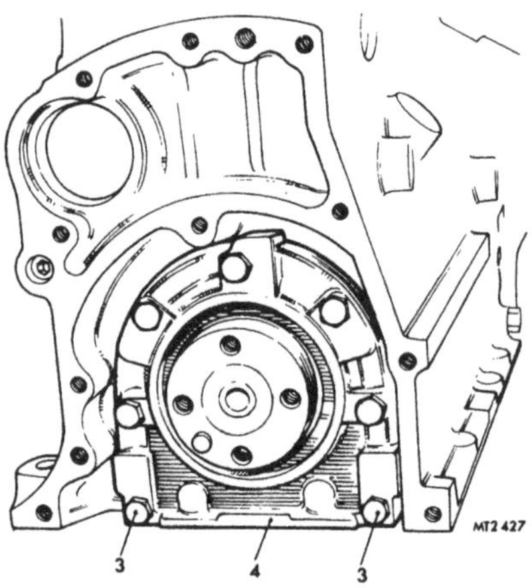

CRANKSHAFT REAR OIL SEAL

– Remove and refit **12.21.20**

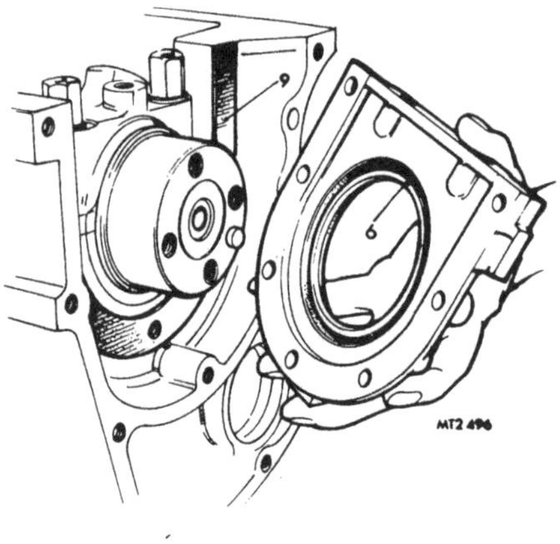

Removing

1. Remove the engine rear gearbox adaptor plate. 12.53.03.
2. Remove the two bolts securing the sump to the rear oil seal housing.
3. Remove the seven bolts holding the oil seal housing to the crankcase.
4. Withdraw the oil seal and housing complete, taking care not to tear the sump gasket.
5. Press out the the oil seal from the housing.

Refitting

6. Smear the outside diameter of the replacement oil seal with grease and press into housing, lip face to crankshaft.
7. Clean the crankcase joint face, removing all traces of old gasket and jointing compound.
8. Coat the crankcase face with sealing compound and smear the crankshaft with oil.
9. Place a new gasket in position on the crankcase joint face.
10. Carefully slide the seal housing over the crankshaft. Slide the oil seal housing into contact with the crankcase joint face.
11. Fit and tighten evenly the seal housing securing bolts to 16 to 20 lbf. ft (2·2 to 2·8 kgf. m) and ensure top bolt has a plain copper washer under it.

CRANKSHAFT END-FLOAT

– Check and adjust 12.21.26

Operation performed during Engine – strip and rebuild 12.41.05., Crankshaft – remove and refit 12.21.33., Front main bearing – remove and refit 12.21.41., and Centre and rear main bearing – remove and refit 12.21.42.

NOTE: Standard thrust washers are available in thicknesses from 0.091 to 0.093 in (2.31 to 2.36 mm). Oversize thrust washers are available in thicknesses from 0.096 to 0.098 in (2.44 to 2.49 mm).

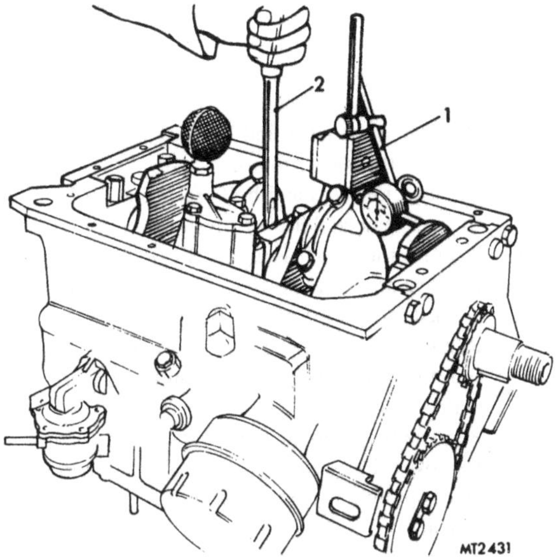

1. Secure a dial gauge to the crankcase so that the indicator rod rests in a loaded condition on a machined surface of a crankshaft throw.
2. Using a screwdriver between the centre main bearing cap and an adjacent crankshaft throw, force the crankshaft against the dial indicator rod.
3. Zero the dial indicator.
4. Force the crankshaft in the opposite direction and note the gauge reading, which should indicate an end-float of 0·004 to 0·008 in (0·1016 to 0·2032 mm).
5. If necessary, adjust the end-float by removing the rear main bearing cap and existing thrust washers and renewing with different washers of suitable thickness, ensuring:
   a. Thrust washers are fitted so that the thrust faces – identified by oil grooves – bear against the crankshaft faces.
   b. Locate in the register on both sides of the crankshaft bearing bore half.
6. Refit the main bearing cap complete with shell bearing and tighten the nuts to 50 to 65 lbf. ft (7 to 9 kgf. m).
7. Re-check the end-float.

NOTE: The end-float may be checked with a feeler gauge inserted between the thrust washer and crankshaft.

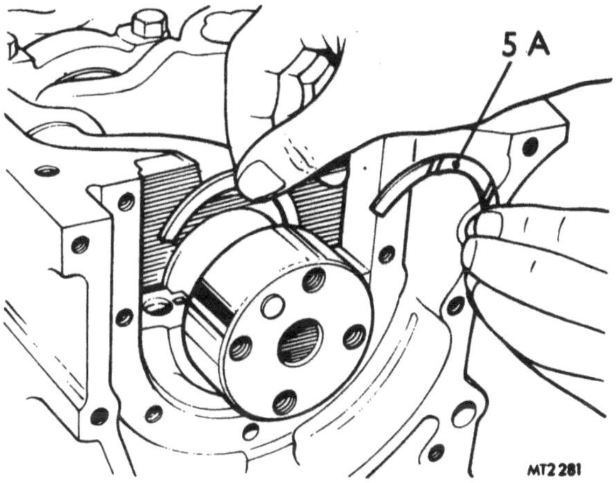

CRANKSHAFT

– Remove and refit **12.21.33**

Removing

1. Remove the engine and gearbox assembly 12.37.01.
2. Remove the starter motor.
3. Remove the nuts and bolts securing the bell housing to the engine and lift off the gearbox assembly.
4. Remove the clutch assembly 33.10.01.
5. Remove the four bolts securing the flywheel to the crankshaft and lift off the flywheel 12.53.07.
6. Remove the seven bolts securing the engine rear adaptor to the cylinder block and remove the adaptor plate. 12.53.03.
7. Remove the alternator 86.10.02.
8. Remove the water pump housing complete with water pump, thermostat housing and fan. 26.50.03 instructions 8 to 10.
9. Remove the crankshaft pulley.
10. Remove the timing chain cover 12.65.01. instructions 5 and 6.
11. Remove the oil thrower.
12. Remove the timing chain sprocket, remove key, and then remove the shims.
13. Remove the two bolts securing the camshaft location plate and remove the plate.
14. Remove the front mounting plate. 12.25.10. instructions 9 to 12.
15. Remove the dipstick.
16. Remove the sump.
17. Remove the oil strainer 12.60.20.
18. Remove the front sealing block.
19. Remove the crankshaft rear oil seal 12.21.20. instructions 3 and 4.
20. Disconnect the connecting rods from the crankshaft. 12.17.16 instructions 4 to 7. Check the identification marks for reassembly. DO NOT MIX.
21. Remove the front centre and rear main bearing caps and shell 12.21.39. instructions 6 to 8.
22. Remove the thrust washers from the rear main bearing.
23. Lift out the crankshaft.

[NOTE: If he engine is to be stored for some time following repairs, it is advisable to use assembly lubricant (available from automotive supply stores) or a light coating of multipurpose grease on the crankshaft journals and the crankpins rather than ordinary engine oil. Many experienced mechanics choose to use assembly lubricant whenever new bearing shells are installed.]

continued

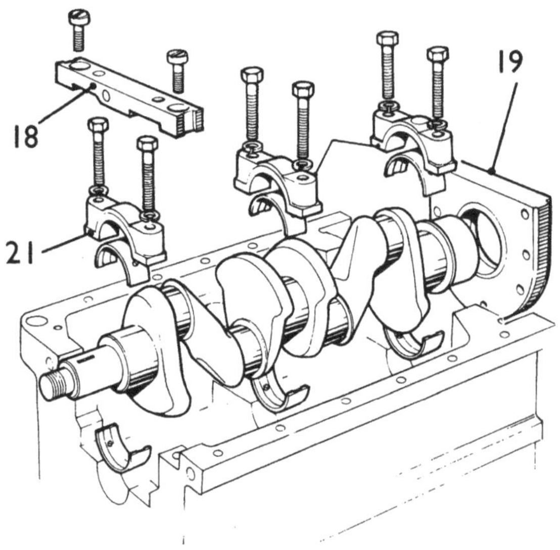

MT2 642

Refitting

Lubricate all parts before assembly.

[NOTE: There is little use in fitting new bearing shells if the crankshaft journals and crankpins are worn undersize, worn out-of-round or worn to a taper, Measure each journal and crankpin with a micrometer. Check both ends of each journal or crankpin, making two measurements at 90° to one another at each point. If there is a difference of more than 0.002 in (0.05 mm) between any of these measurements, the crankshaft should either be replaced or reconditioned to accept one of the undersize bearing sets listed under Data.]

24. Fit the uppper bearing shells to the crankcase and the thrust washers to the rear main bearing.
25. Fit the lower shells to the bearing caps.
26. Lower the crankshaft into the crankcase.
27. Fit the bearing caps and tighten the front and centre main bearing bolts and rear main nuts to 50 to 65 lbf. ft (7 to 9 kgf. m).
28. Check, and adjust if necessary, the crankshaft end-float 12.21.26.
29. Reconnect the connecting rods to the crankshaft, 12.17.16 instructions 8 to 12 and fit the front sealing block, 12.21.39 instructions 14 to 17.
30. Fit the rear main oil seal. 12.21.20. instructions 6 to 11.
31. Fit the oil pump strainer 12.60.20.
32. Fit the sump, renewing the gasket. Tighten the retaining bolts evenly to 16 to 20 lbf. ft (2·2 to 2·8 kgf. m).
33. Fit the flywheel. 12.53.07. instructions 5 to 8, ensuring that the crankshaft spigot bush is in position.
34. Fit the clutch assembly 33.10.01.
35. Fit the front mounting plate and gasket, ensuring that:
    a. All traces of old gasket and jointing compound are removed from the cylinder block joint face.
    b. The cylinder block face of the new gasket is smeared with jointing compound.
    c, The mounting plate is located correctly over the dowels.
    d. The retaining bolts and screws are tightened evenly.
36. Fit the timing chain and sprockets 12.65.12. instructions 6 to 15.
37. Check the valve timing 12.65.08.
38. Fit the oil thrower, ensuring that the dished perphery faces the timing cover.
39. Fit the timing cover 12.65.01.
40. Fit the crankshaft pulley, tightening the retaining nut to 150 lbf. ft (20·7 kgf. m)
41. Fit the water pump housing complete with water pump, fan and thermostat housing. Ensure that the new gasket is fitted with jointing compound.
42. Fit the alternator and drive belt 86.10.02.
43. Fit the gearbox.
44. Fit the starter motor.
45. Fit the engine and gearbox assembly to the car. 12.37.01.
46. Fit the dipstick and check the oil sump level.

DATA

Crankshaft end-float tolerance	0·004 to 0·008 (0·10 to 0·20 mm)
Thrust washer oversizes	0·005 in (0·13 mm)
Main bearing journals diameter (3)	2·3115 to 2·3120 in (58·713 to 58·725 mm)
[Mk. III main bearing journals diameter (3)	**2·000 to 2·0010 in (50·81 to 50·83 mm)]**
Crankpins diameter (4)	1·8750 to 1·8755 in (47·625 to 47·638 mm)
[Mk. III crankpins diameter (4)	**1·6250 to 1·6255 in (41·27 to 41·28 mm)]**
Maximum run-out of centre journal (with front and rear supported)	0·003 in (0·076 mm)
Maximum out-of-balance of shaft (with key and dowel fitted)	0·3 oz. in (3·36 g. cm).
Crankshaft end-float tolerance	0·004 to 0·008 in (0·10 to 0·20 mm)
Main bearings and big-end bearings are available in the following undesizes	0·010, 0·020, 0·030 in (0·25, 0·51, 0·76 mm)

[NOTE: Standard thrust washers are available in thicknesses from 0.091 to 0.093 in (2.31 to 2.36 mm). Oversize thrust washers are available in thicknesses from 0.096 to 0.098 in (2.44 to 2.49 mm).]

MAIN BEARING – SET

– Remove and refit	12.21.39
Front main bearing 1 to 6, and 9 to 18	12.21.41
Centre and rear main bearing 1, 2, 7, 8, 9 to 13 and 18	12.21.42

Removing

1. Remove the sump 12.60.44.
2. Remove the oil pump strainer.
3. Remove the front mounting plate gasket. 12.25.10.
4. Remove the two screws securing the front sealing block to the crankcase.
5. Withdraw the sealing block.
6. Remove the front main bearing cap bolts and withdraw the cap complete with lower shell. Remove the upper shell and check that the cap is marked for reassembly.
7. Remove the centre main bearing bolts and withdraw the cap and lower shell. Check identification marks for reassembly and withdraw the upper shell.
8. Remove the rear main bearing cap bolts and withdraw the cap and lower shell. Check that the cap is marked for reassembly, and remove the upper shell.

[NOTE: Mk. III Spitfire engines have smaller-diameter main bearing journals than later engines. However, bearing clearances are the same for all engines. If you check the main bearing clearance with Plastigage® (available from automotive supply stores), the clearance should be between 0.001 and 0.002 in (0.025 and 0.050 mm).]

continued

MT2 647

[NOTE: If the engine is to be stored for some time following repairs, it is advisable to use assembly lubricant (available from automotive supply stores) or a light coating of multipurpose grease on the crankshaft journals and the crankpins rather than ordinary engine oil. Many experienced mechanics choose to use assembly lubricant whenever new bearing shells are installed.]

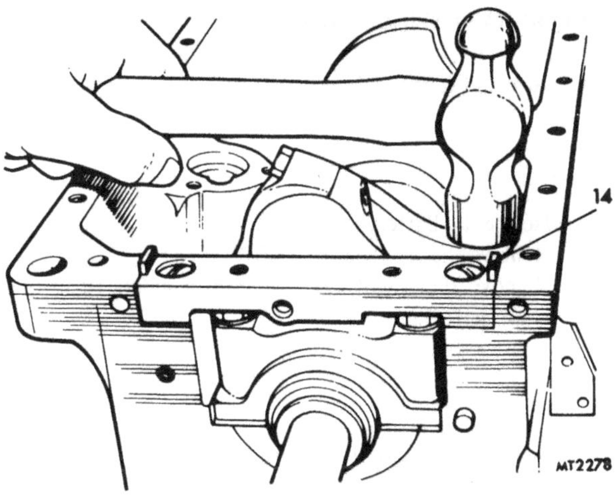

Refitting

9. Fit the shell bearings to the front, centre and rear main bearing caps, ensuring that the tags locate in the recesses.
10. Lubricate and feed the upper shells – tag end last – between the crankshaft journals and bearing bores, ensuring that the tags locate in the resesses.
11. Lubricate the lower shells and refit the caps, tightening the front, centre and rear main bearing cap bolts to 50 to 65 lbf. ft (7 to 9 kgf. m).
    Ensure that the identification marks coincide.
12. Check and adjust if necessary, the crankshaft end float 12.21.26.
13. Smear the front sealing block gaskets with jointing compound and fit to the crankcase, loosely tightening the screws.
14. Drive wedges into the slots.
15. Align the face of the sealing block with the crankcase and tighten the retaining screws.
16. Trim the protruding ends of the wedges flush with the crankcase – do not under-cut.
17. Reverse instructions 1 to 3, ensuring that:
    a. All gaskets are renewed.
    b. The sump is replenished with the correct grade of oil to the top mark on the dipstick.

[NOTE: There is little use in fitting new bearing shells if the crankshaft journals and crankpins are worn undersize, worn out-of-round or worn to a taper. Measure each journal and crankpin with a micrometer. Check both ends of each journal or crankpin, making two measurements at 90° to one another at each point. If there is a difference of more than 0.002 in (0.05 mm) between any of these measurements, the crankshaft should either be replaced or reconditioned to accept one of the undersize bearing sets listed under Data 12.21.33.]

CRANKSHAFT SPIGOT BUSH

– Remove and refit 12.21.45

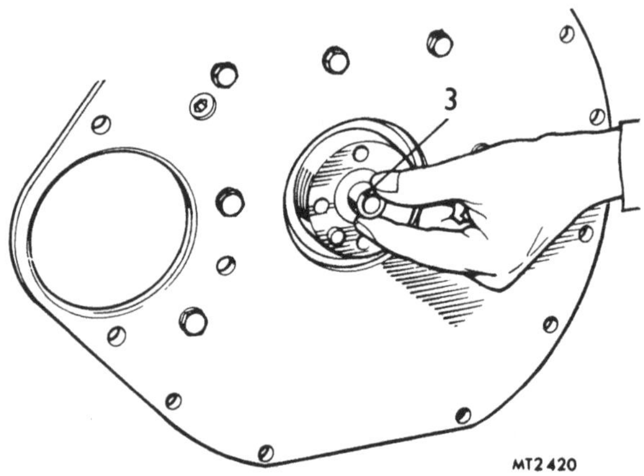

Removing

1. Isolate the battery.
2. Remove the flywheel 12.53.07.
3. Remove the spigot bush which is a sliding fit in the crankshaft flange.

Refitting

4. Smear the spigot bush with zinc oxide grease and refit.
5. Reverse instructions 1 and 2.

CYLINDER PRESSURES

– Check 12.25.01

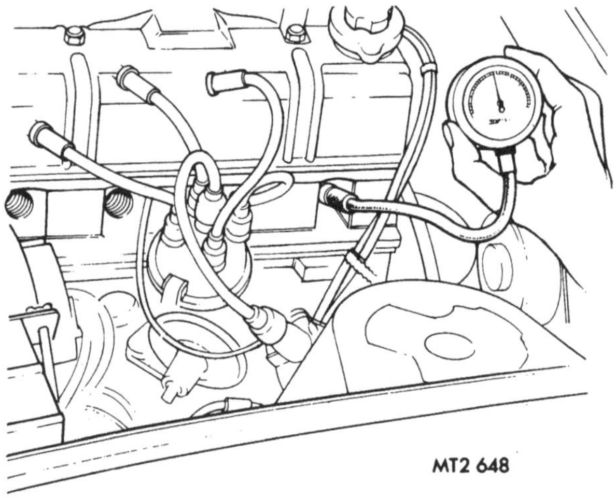

1. Run the engine until normal temperature is attained as indicated by the temperature gauge.
2. Remove the spark plugs.
3. Fit a compression gauge to No. 1 cylinder.
4. Turn the engine over with the starter motor with the throttle wide open.
5. Note and record the gauge reading.
6. Repeat on the remaining cylinders.

NOTE: All cylinders should have pressures within 10 lb/in^2 (0·70 kg/cm^2) Differential.

CYLINDER BLOCK DRAIN TAP

– Remove and refit **12.25.07**

Removing

1. Drain the cooling system 26.10.01.
2. Remove the drain tap and adaptor assembly.

Refitting

3. Reverse instructions 1 to 2 and check for leaks.

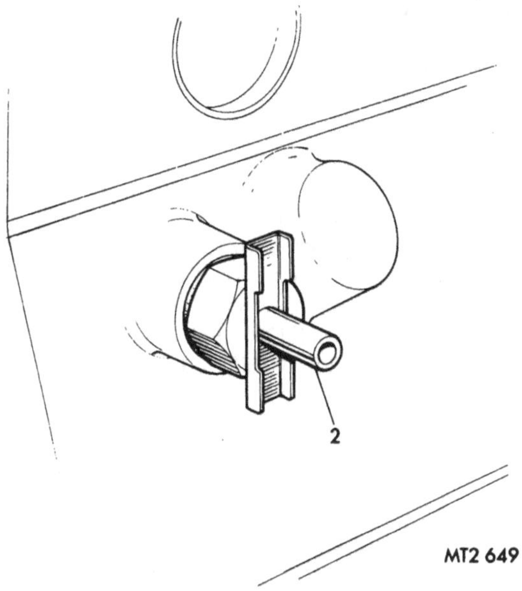

CYLINDER BLOCK FRONT MOUNTING PLATE GASKET

– Remove and refit **12.25.10**

Removing

1. Isolate the battery.
2. **Remove the [generator or the] alternator 86.10.02.**
3. Remove the water pump housing 26.50.02, 1 to 9.
4. Remove the timing chain cover 12.65.01, 3 to 6.
5. Remove the timing chain and sprockets. 12.65.12, 2 to 5.
6. Remove the two bolts securing the camshaft locating plate and remove the plate.
7. Support the engine under the sump.
8. Remove the two nuts and bolts (two each side), securing the L.H. and R.H. engine mounting bolts to the front suspension turrets.
9. Remove the four bolts securing the mounting plate to the cylinder block.
10. Remove the two screws securing the mounting plate to the sealing block.
11. Remove the front mounting plate complete with mounting rubbers.
12. Remove the front mounting plate gasket.

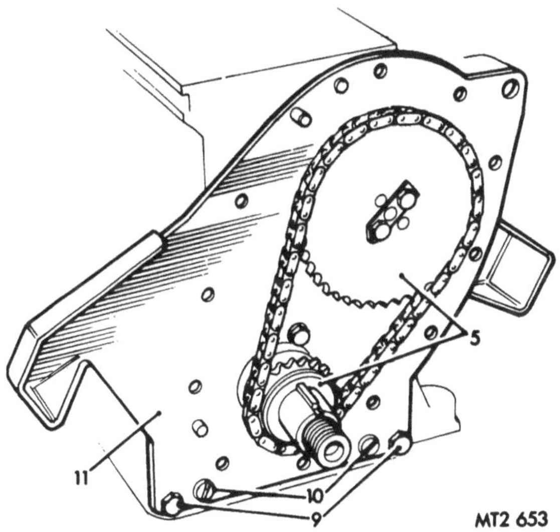

Refitting

13. Remove all traces of old jointing compound from the cylinder block face.
14. Coat the cylinder block face of the new gasket with Wellseal jointing compound.
15. Fit the new gasket.
16. Reverse instructions 1 to 11, ensuring:
    a. The camshaft sprocket is refitted in its original position and that the timing marks line up.
    b. A new water-pump housing gasket is fitted, using sealing compound.
    c. The mounting plate locates correctly over the dowels.

CYLINDER BLOCK

– Rebore 12.25.23

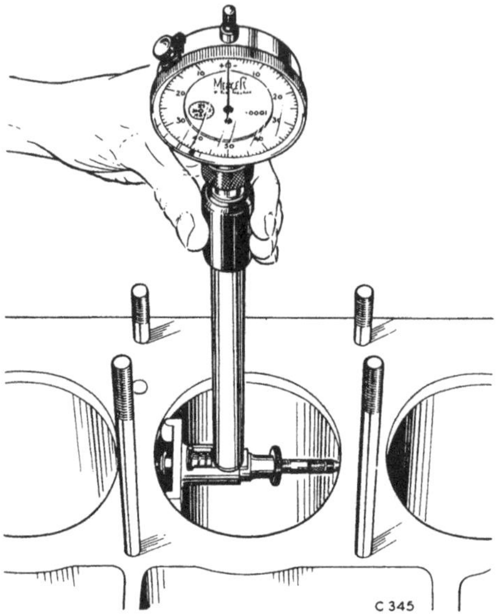

1. Strip engine. 12.41.05.
2. Measure bores for taper, ovality and maximum wear.
3. Rebore to dimensions in Data.

NOTE: Maximum rebore size is +0·020 in. Cylinders that cannot be rebored within this limit may be sleeved to restore them to the original size as follows:

a. **[1500 and] 1300 engine bore out cylinders to 3·0302 to 3·0272 in dia.**

4. Rebuild engine 12.41.05.
   a. Fit new pistons to rebored dimensions, or
   b. Fit standard graded pistons to relined bores.
   c. Check alignment of connecting rods. 12.17.10. instruction 13.

DATA

Original size bore: Grade F	2·8995 to 2·9000 in (73·64 to 73·66 mm).
Grade G	2·9001 to 2·9005 in (73·66 to 73·77 mm).
Maximum rebore size	+0·020 in (+0·51 mm).

CYLINDER HEAD

– Remove and refit 12.29.10

Which includes:
Cylinder head gasket 12.29.01

Removing

1. Isolate the battery.
2. Drain the cooling system 26.10.01.
3. Remove the inlet and exhaust manifold assembly. 30.15.01.
4. Disconnect the plug leads and remove the distributor cap.
5. Remove the rocker shaft. 12.29.54.
6. Remove the push rods, suitably identifying them for reassembly in the corresponding cam followers.

[NOTE: A convenient way to keep the pushrods in order is to push them through a piece of cardboard in which you have punched eight holes. If necessary, you can number the holes to identify the location of each pushrod in the engine. If the old cylinder head gasket has been leaking, check the cylinder head and the cylinder block for warping before you reassemble the engine. To check, lay a straightedge diagonally across the mating surfaces of both the head and the block. Using feeler gauges of various thicknesses, measure the clearance, if any, between the surface, and the straightedge. Then repeat the check with the straightedge at the opposite diagonal. If the clearance at either diagonal exceeds 0.004 in (0.10 mm), the part should be milled or ground by a competent machinist in order to obtain a true surface.]

7. Remove the water pump housing. 26.50.03. 1 to 7.
8. **Remove the [generator or the] alternator. 86.10.07.**

[CAUTION: To avoid distortion, tighten the cylinder head nuts to 38 to 45 lbf. ft (5.2 to 6.2 kgf.m) in the sequence given in 12.29.27.]

9. Slacken and remove the 10 nuts securing the cylinder head to the block reversing the tightening sequence. 12.29.27.
10. Lift off the cylinder head.
11. **Remove the cylinder head gasket and discard.**

Refitting

12. Reverse instructions 1 to 11, ensuring:
    - *a.* Cylinder head and cylinder block faces are clean.
    - *b.* New cylinder head gasket is fitted ensuring that the top side is uppermost (marked).
    - *c.* Cylinder head nuts are tightened to the correct torque figure and in the correct sequence.
    - *d.* All gaskets are renewed.
13. Adjust the valve clearances 12.29.48.

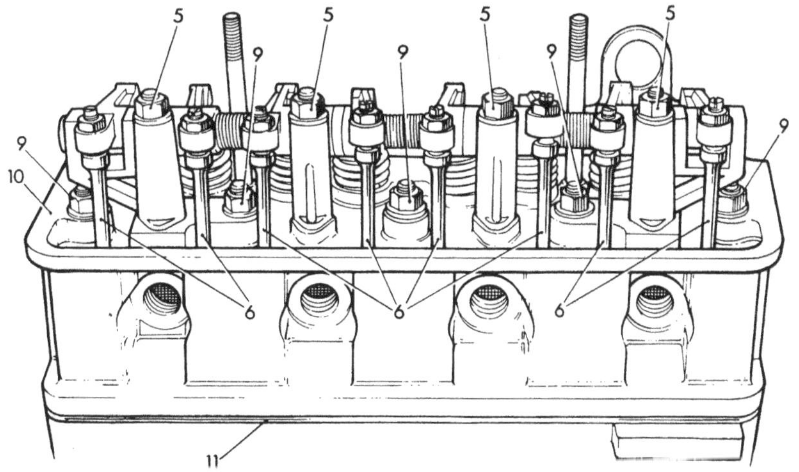

CYLINDER HEAD

– Overhaul **12.29.18**

Which includes:

Valves – inlet and exhaust – remove and refit	**12.29.62**
Valves – inlet – remove and refit	**12.29.63**
Valves – exhaust – remove and refit	**12.29.64**
Valve guide – inlet – remove and refit	**12.29.70**
Valve guide – exhaust – remove and refit	**12.29.71**
Inlet valve seat – remove and refit	**12.29.76**
Exhaust valve seat – remove and refit	**12.29.77**

Service tools: 18G, 106, 6118B, 60A, S60A-2A or S60A-2, S60A-6.

[CAUTION: Please read the instructions thoroughly before you attempt cylinder head repairs. If you lack the necessary skills or tools we suggest you leave such work to an Authorized Dealer or other qualified shop. Before taking a removed cylinder head in for repair, thoroughly clean its exterior but do not undertake any disassembly.]

1. Remove the cylinder head. 12.29.10.
2. Remove the spark plugs.
3. Using the valve spring compressor 18G 106 or 6118B, remove the inlet exhaust valve spring collars or split cotters.
4. Remove the inlet and exhaust valves, springs and identify for reassembly.
5. Check the inlet and exhaust valve guides for wear, and renew if necessary. Using Service tool S60A-6, assemble the replacement valve guide in the tool, with the chamfered end uppermost (leading). Position the tool on the combustion chamber face and pull the replacement guide in and the old one out. Ensure that the guide protrusion above the cylinder head top face is correct – see Data.
6. Examine the valve seats for pitting and wear. Reface where necessary, removing the minimum of material.
7. Fit valve seat inserts where the seats cannot be restored by refacing. Machine the cylinder head to the dimensions given (see Data).

NOTE: When fitting a pair of valve inserts, it will be necessary to bore the head and fit the first before boring the head and fitted inserts for the second.

8. Press in the valve insert and peen over the cylinder head casting material insert to secure.
9. Cut seats in the inserts to the dimensions given below.
10. Examine the valves and renew any with bent or worn stems (see Data) and heads where the thickness is reduced to 0·0312 in (0·8 mm). Reface where necessary.
11. Test the valve springs for fatigue to the dimensions in Data. Renew as necessary.

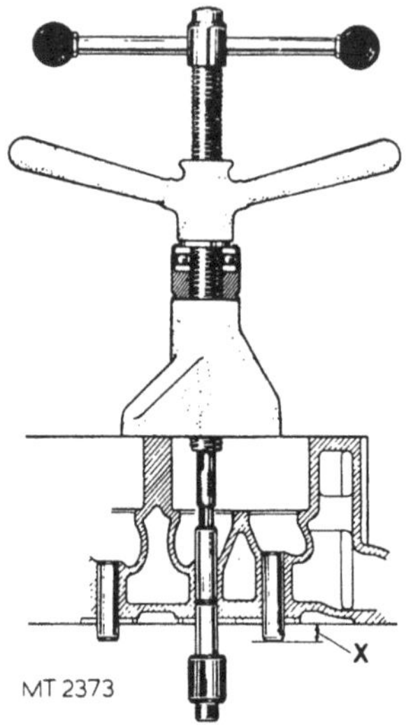

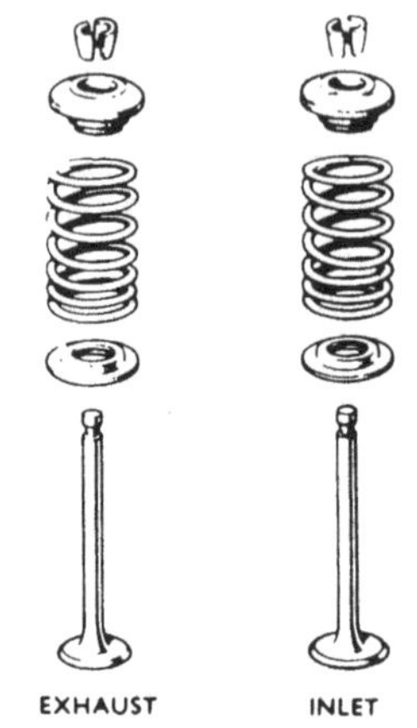

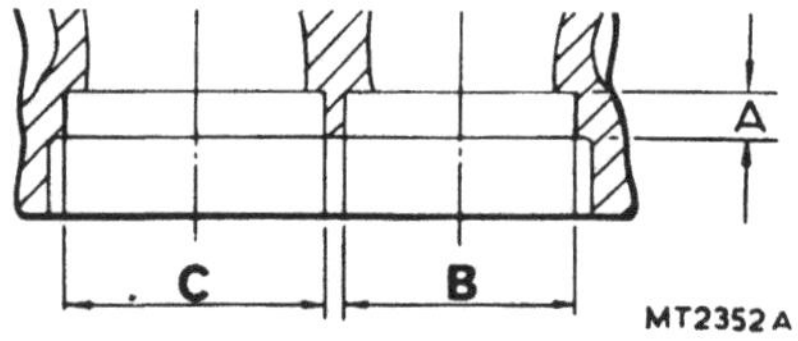

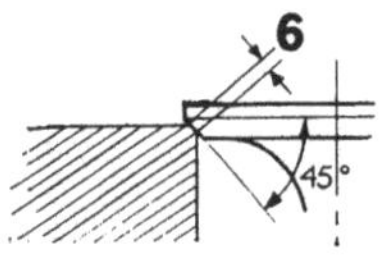

6 = 0·060" (1·52 mm)

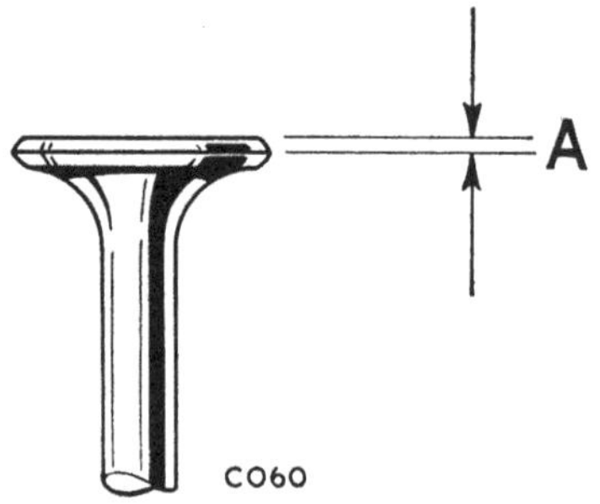

[NOTE: After refacing, dimension **A** must not be less than 1/32 in (0.80 mm).**]**

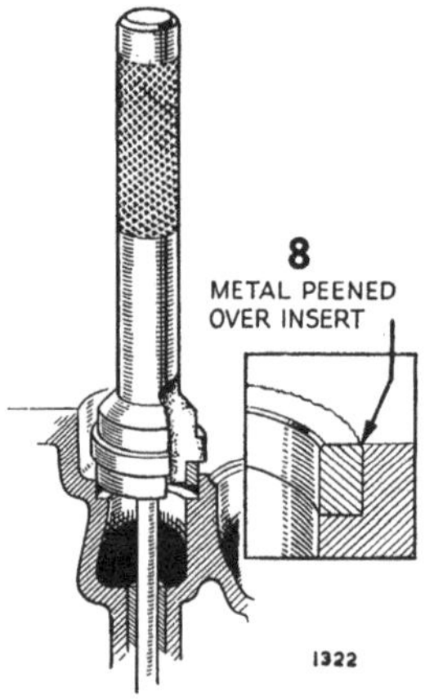

Triumph Spitfire Mk IV Manual. Part No. 545254. Issue 2
Part No. 511243, pages 0.123 - 5, Issue 1.

Reassembling

12. Lap-in all valves, using coarse followed by fine carborundum paste, until a gas-tight seal is obtained. Remove all traces of paste and maintain identification of each valve.

13. Lubricate the valve stems with clean engine oil and assemble to the cylinder head. Ensure that the valve spring retaining collars or split collars are correctly positioned.

14. Refit the cylinder head, 12.29.10, fitting a new cylinder head gasket.

DATA

Unless otherwise stated data is applicable to all engines

Valves	Inlet		Exhaust
	Up to Eng. No. FH/FK 25001	From Eng. No. FH/FK 25001 and FM1	
Head diameter	1·304 to 1·308 in (33·12 to 33·22 mm)	1·437 to 1·443 (36·50 to 36·65 mm)	1·168 to 1·172 in (29·66 to 29·76 mm)
Stem diameter	0·3107 to 0·3112 in (7·87 to 7·90 mm)	0·3107 to 0·3112 in (7·87 to 7·90 mm)	0·3100 to 0·3105 in (7·874 to 7·887 mm)
Valve stem to guide clearance	0·0008 to 0·0023 in (0·02 to 0·06 mm)	0·0008 to 0·0023 in (0·02 to 0·06 mm)	0·0015 to 0·0030 in (0·03 to 0·07 mm)

Seat face angle 90° plus 30′ minus 0′

Valve guides

Length	2·0625 in (52·387 mm)
Bore	0·312 to 0·313 in (7·92 to 7·95 mm)
Outside diameter	0·501 to 0·502 in (12·72 to 12·75 mm)
Valve guide height above cylinder head	0·749 to 0·751 in (19·025 to 19·075 mm)

Valve springs – inlet and exhaust valves

	Up to Eng. No. FH/FK 25001	From Eng. No. FH 25001	From Eng. No. FK 25001 and Eng. No. FM 1
		Inner	Outer
Internal diameter	0·795 in (20·19 mm)	0·578 in (14·68 mm)	0·795 in (20·19 mm)
Working coils	3½	4	3¾
Load at length	1·074 in=105 to 115 lbf (27·28 mm=47·6 to 52·1 kgf)	0·657 in=40 to 43 lbf (16·68 mm=18·1 to 19·5 kgf)	0·989 in=123 to 133 lbf (25·1 mm=55·8 to 60·03 kgf)
Free length (approx.)	1·59 in (40·38 mm)	1·14 in (28·95 mm)	1·52 in (38·6 mm)
Solid length (max.)	0·96 in (24·38 mm)	0·552 in (14·48 mm)	0·875 in (22·2 mm)
Rate	235 lbf/in (4196 kgf/m)	85·6 lbf/in (1526 kgf/m)	240 lbf/in (4286 kgf/m)

Valve seat inserts	Inlet		Exhaust	
	Up to	From	Up to	From
	Eng. No. FH/FK 25001 and FM1		Eng. No. FH/FK 25001 and FM1	
Bore into cylinder head dimension 'A'	0·216 to 0·219 in (5·48 to 5·56 mm)	0·250 to 0·255 in (6·35 to 6·48 mm)	0·248 to 0·250 in (6·30 to 6·35 mm)	0·250 to 0·255 in (6·35 to 6·48 mm)
Bore into cylinder head dimension 'B' (Exhaust)			1·249 to 1·250 in (31·72 to 31·75 mm)	
Bore into cylinder head dimension 'C' (Inlet)	1·375 to 1.376 in (34·92 to 34·95 mm)	1·5625 to 1·5635 in (39·68 to 39·71 mm)		

Machined fitted inserts to 89° inclusive angle. Valve seats are to be concentric with guide bore to within 0·002 in (0·0508 mm) total indicator reading.

NOTE: When fitting valve inserts to units up to Engine No. FH 25001, FK 25001 and FM 1 it will be found that a maximum interference of 0·013 in (0·33 mm) can exist between the inlet valve and the exhaust valve insert positions. It will therefore be necessary to fit one insert first then bore into it when fitting the second insert.

DECARBONIZE, REFACE ALL VALVES AND SEATS, GRIND-IN VALVES, TUNE ENGINE

12.29.21

[NOTE: **In removing carbon from the pistons, you can easily avoid scoring the piston crown by using a wooden or plastic scraper. To prevent distortion, tighten the cylinder head nuts to 38 to 45 lbf. ft (5.2 to 6.2 kgf. m) in the sequence given in 12.29.27.]**

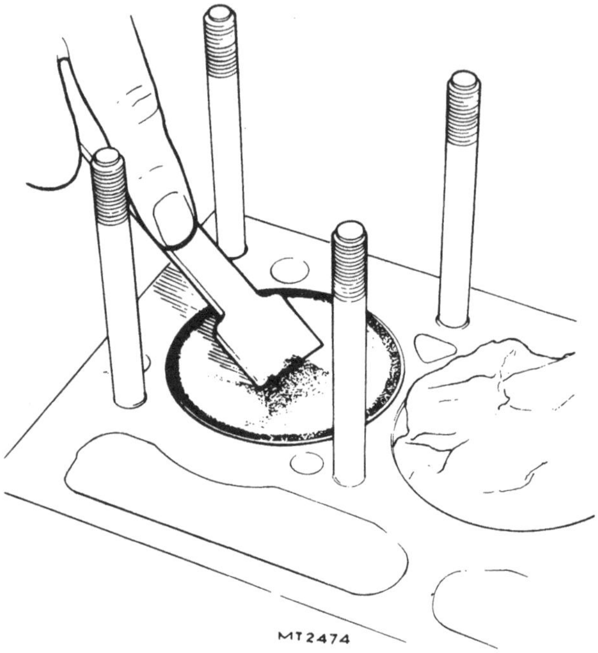

Dismantling

1. Remove the cylinder head 12.29.10.
2. Remove the inlet and exhaust valves 12.29.62.
3. Remove the carbon deposits from combustion chambers and ports.
4. Clean the face of the cylinder head, removing all traces of carbon and high spots.
5. Clean out the water-ways.
6. **Reface all seats, removing the minimum of material 12.29.18.**
7. **Clean carbon from all valves and re-face 12.29.18.**
8. Lap-in the valves, using coarse followed by fine grinding paste until a gas-tight joint is obtained. Do not mix the valves.
9. Turn the crankshaft until Nos. 1 and 4 pistons are at T.D.C.
10. Fill Nos. 2 and 3 cylinders with clean non-fluffy rag to prevent carbon falling into the bores, and cover the cam follower apertures.
11. Carefully, without scoring the piston crown, remove carbon deposits leaving a band of carbon round the periphery of the piston crown. Avoid carbon particles falling into the cylinder block water-ways.
12. Repeat instructions 9 to 11 on Nos. 2 and 3 cylinders.
13. Clean the cylinder block face, removing all traces of carbon and high spots.

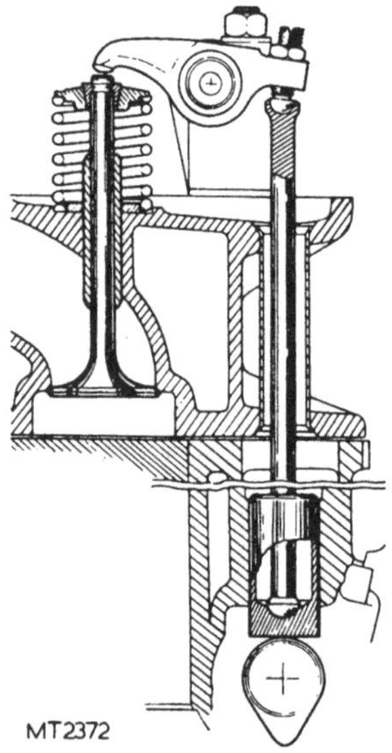

Reassembling

14. Reverse instructions 1 and 2.
15. Tune the engine 12.49.02.

CYLINDER HEAD – RENEW CASTING

– Dismantle and reassemble 12.29.22

Dismantling

1. Remove the cylinder head 12.29.10.
2. Remove the inlet and exhaust valves 12.29.62.
3. Remove the spark plugs.

Reassembling to new casting

4. Fit the rocker shaft pedestal studs.
5. Fit the rocker cover studs.
6. Fit the inlet and exhaust manifold studs.
7. Lap-in all valves, removing all traces of compound.
8. Reverse instructions 1 and 2.
9. Tune the engine. 12.49.02.

CYLINDER HEAD NUTS

– Tighten 12.29.27

1. Remove the rocker cover 12.29.42.
2. Remove the rocker shaft 12.29.54.
3. To avoid distortion, tighten the cylinder head nuts to 38 to 45 lbf. ft (5·2 to 6·2 kgf. m) in the following sequence A, B, C, D, E, F, G, H, I, J.
4. Refit the rocker shaft 12.29.54.
5. Refit the rocker cover. 12.29.42.

ROCKER COVER

– Remove and refit 12.29.42

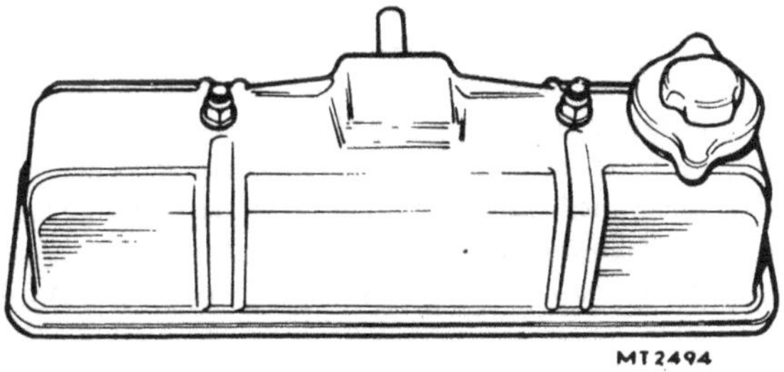

Removing

1. Disconnect the engine breather hose from the carburetter.
2. Remove the two nuts complete with plain and fibre washers securing the rocker cover to the cylinder head.
3. Lift off the rocker cover complete with gasket.

Refitting

4. Reverse instructions 1 to 3 ensuring that the rocker cover retaining nuts are tightened evenly and not beyond 1 to 2 lbf. ft (0·14 to 0·3 kgf. m) to avoid distortion of cover. Renew the gasket.

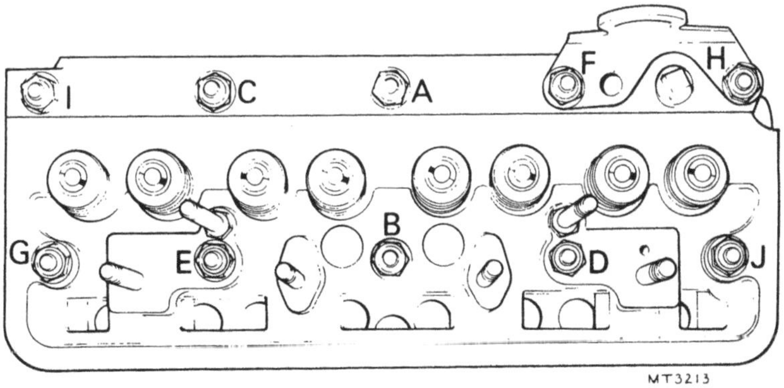

VALVE CLEARANCE

– Check and adjust 12.29.48

1. Isolate the battery.
2. Remove the rocker cover. 12.29.42.
3. Remove the spark plugs.
4. Counting from the front of the engin, turn the crankshaft until Nos. 8 and 6 valves are open, i.e. the valve springs fully compressed.
5. Using a 0·010 in (0·25 mm) feeler gauge, check the gap between the rocker pad and valve tip of Nos. 1 and 3 valves.
6. If adjustment is required, insert a screwdriver blade in the slot in the adjustment pin and slacken the locknut. Turn the adjustment pin clockwise to decrease and anti-clockwise to increase the gap.
7. Check and adjust the remaining valve clearances in the following sequence:
   Adjust Nos. 5 and 2 valves with Nos. 4 and 7 valves open.
   Adjust Nos. 8 and 6 valves with Nos. 1 and 3 valves open.
   Adjust Nos. 4 and 7 valves with Nos. 5 and 2 valves open.

Refitting

8. Reverse instructins 1 to 3.

ROCKER SHAFT

– Remove and refit 12.29.54

Removing

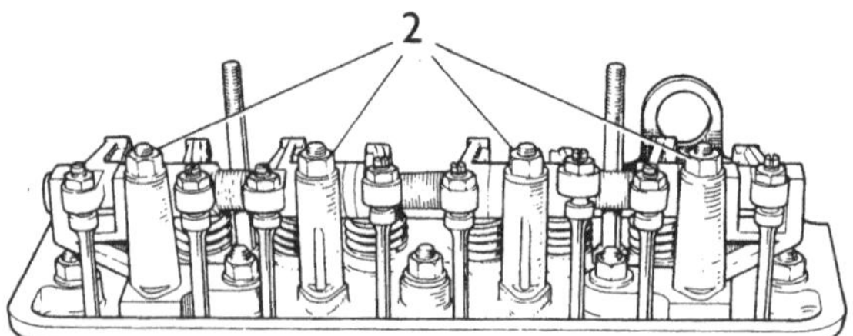

1. Remove the rocker cover 12.29.42.
2. Remove the four nuts complete with washers securing the rocker shaft pedestals to the cylinder head.
3. Lift off the rocker shaft.

Refitting

4. Reverse instructions 1 to 3, ensuring that:
   a. The rocker adjustment screws are located correctly in the push-rod cups.
   b. The pedestal nuts are tightened evenly to 26 to 32 lbf. ft (3.6 to 4.4 kgf. m)
5. Adjust the valve clearances 12.29.48.

ROCKER SHAFT ASSEMBLY

– Overhaul 12.29.55

1. Remove the rocker shaft 12.29.42.
2. Withdraw the cotter pin from the front end of the rocker shaft.
3. Slide off the rockers, pedestals, springs and spacers from the front end of the shaft, noting the order for reassembly.
4. Remove the screw locating the rear pedestal to the shaft.
5. Withdraw the rear pedestal and rocker.

Reassembling

6. Reverse instructions 1 to 5, observing:
   a. Renew all worn components. The grinding of worn rocker pads is not recommended.
   b. Ensure that oil-ways in rockers and shafts are clear.
   c. Ensure that the rear pedestal locating screw engages properly in the rocker shaft.
7. Adjust the valve clearances. 12.29.48.

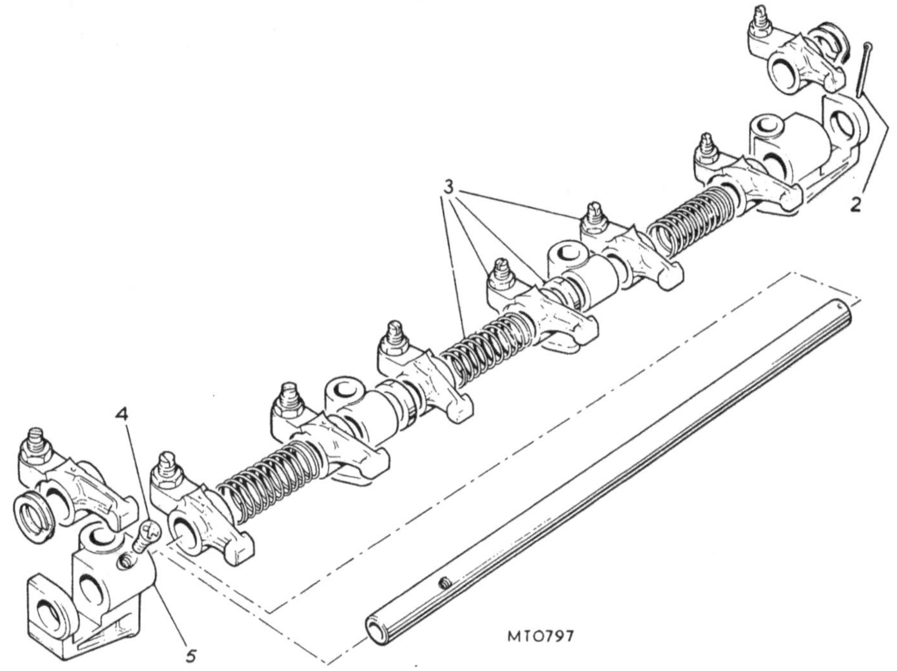

ROCKER ADJUSTING SCREWS – SET

– Remove and refit 12.29.56

Removing

1. Remove the rocker shaft assembly 12.29.54.
2. Remove the rocker adjusting screw locknuts.
3. Remove the rocker adjusting screws.

Refitting

4. Reverse instructions 1 to 3.
5. Adjust the valve clearances 12.29.48.

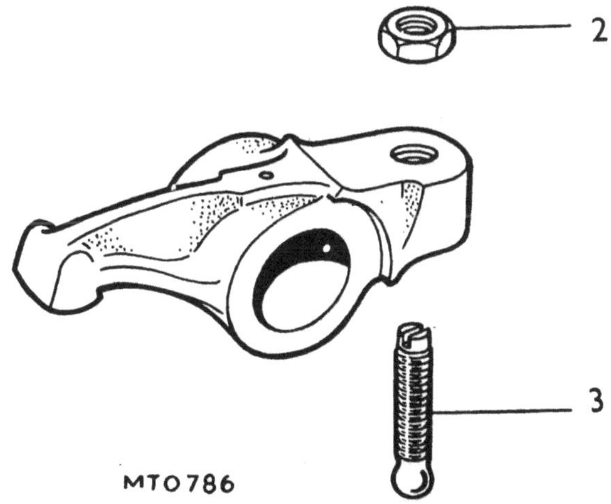

CAM FOLLOWERS – SET

– Remove and refit 12.29.57

[NOTE: Used cam followers should never be interchanged. The best way to identify them for reassembly is to wipe dry each follower as you remove it. Then use waterproof ink or a waterproof felt-tipped marker to write an identification number on the follower. During reassembly, many experienced mechanics prefer to lightly lubricate the cam followers with assembly lubricant (available from automotive supply stores) or with a thin coat of multipurpose grease.]

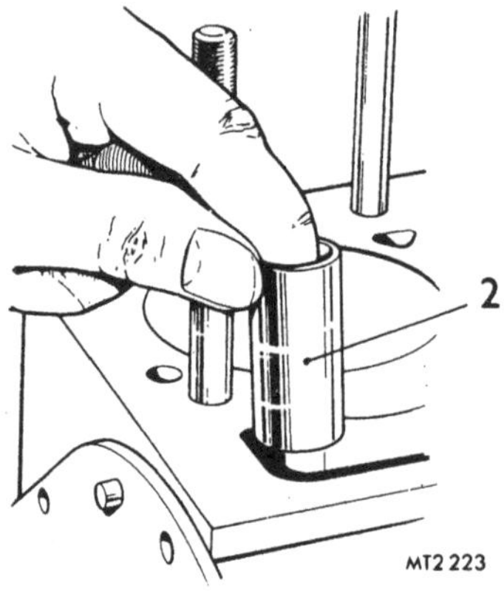

Removing

1. Remove the cylinder head 12.29.10.
2. Lift out the eight cam followers and identify for reassembly.

Refitting

3. Reverse instructions 1 and 2, ensuring:
   a. Worn or pitted cam followers are renewed.
   b. Lubricated with clean engine oil before resetting.
   c. Each follower is free to rotate and slide in its locating bore.

PUSH RODS – SET

– Remove and refit 12.29.59

[NOTE: Used pushrods should never be interchanged. A convenient way to keep pushrods in order is to push them through a piece of cardboard in which you have punched eight holes. If necessary, you can number the holes to identify the location of each pushrod in the engine.]

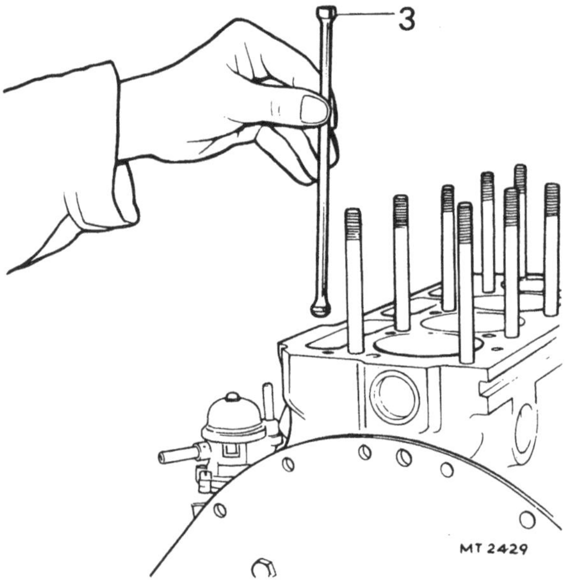

Removing

1. Remove the rocker cover 12.29.42.
2. Remove the rocker shaft assembly 12.29.54.
3. Lift out the push-rods, identifying them for reassembly.

Refitting

4. Reverse instructions 1 to 3, ensuring:
   a. Push rods with worn or pitted cup and ball ends and bent shafts are renewed.
   b. Ball and cup ends are lubricated.
   c. Cup ends are located correctly in the rocker adjusting screw ball and ball ends are seated properly in the cam followers.
5. Adjust the valve clearances 12.29.48.

[NOTE: All other valve work is described in 12.29.18.]

ENGINE AND GEARBOX ASSEMBLY

– Remove and refit **12.37.01**

Removing

1. Remove bonnet. 76.16.01.
2. Disconnect and remove battery. 86.15.01.
3. Drain coolant. 26.10.01.
4. Drain sump oil. See 'Maintenance'.
5. Disconnect top and bottom hoses at engine, plus hose at expansion bottle. 26.30.01. 26.30.07.
6. Remove radiator – four bolts each side. 26.40.01.
7. Disconnect two heater hoses at engine.
8. Disconnect throttle cable at linkage 19.20.06.
9. Disconnect choke cable from rocker cover and carburetter.
10. Remove three nuts and washers securing exhaust front pipe to manifold. 30.10.09.
11. Disconnect at the distributor, the high and low tension leads.
12. Disconnect starter cable. 86.60.01.
13. Disconnect multi-socket connector from the alternator.
14. Disconnect oil pressure switch wire.
15. Disconnect water temperature transmitter wire.
16. Disconnect battery earth lead from alternator mounting bracket.
17. Disconnect fuel feed pipe at fuel pump (plug end to prevent fuel siphoning or dirt entering the system).
18. Remove gearbox tunnel cover assembly 76.25.07.
19. Disconnect propeller shaft at gearbox (raise one rear wheel to turn shaft) 47.15.02.
20. Remove top cover.

[NOTE: After you have removed the top cover from the gearbox, fit a temporary cardboard cover to prevent the entry of foreign matter. Alternatively, you can remove the shift lever from the extension, leaving the extension and top cover in place on the gearbox.]

21. Remove two nuts, securing rear mounting to gearbox support bracket. 12.45.08.
22. Remove nut and bolt securing exhaust pipe clip to gearbox.
23. Disconnect speedometer drive cable 88.30.06.
24. Remove bolt and nut securing clutch slave cylinder to gearbox – withdraw and move cylinder into engine compartment 33.35.01.
25. Attach lifting equipment to engine lifting eyes.
26. Remove two nuts and bolts securing each front mounting to the front suspension turrets. 12.45.04.
27. Lift engine to raise sump above chassis, pull forward, then raise and manoeuvre engine and gearbox clear of vehicle.

Refitting

28. Reverse instructions 1 to 27, ensuring:
    - *a.* Sump filled with correct grade of engine oil to high mark on dipstick.
    - *b.* No pipes or wires are trapped between engine and frame.
    - *c.* Cooling system is filled, checked for leaks and then topped up, after the engine has been run.
    - *d.* All mountings are secure and tightened to specific torque.
    - *e.* Gearbox oil level checked.

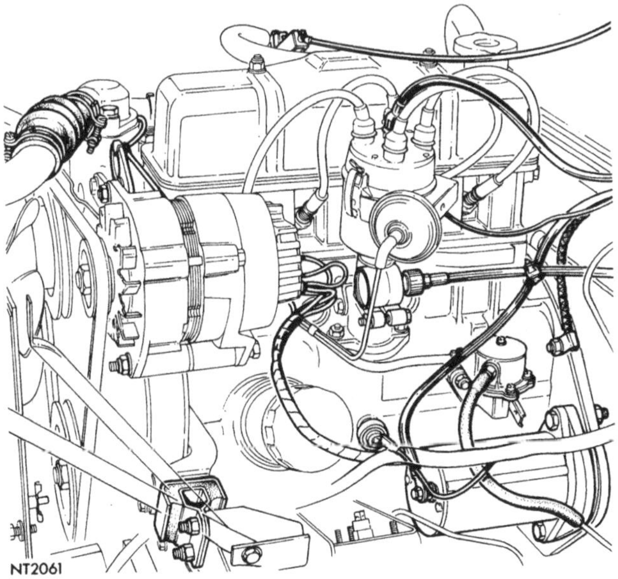
NT2061

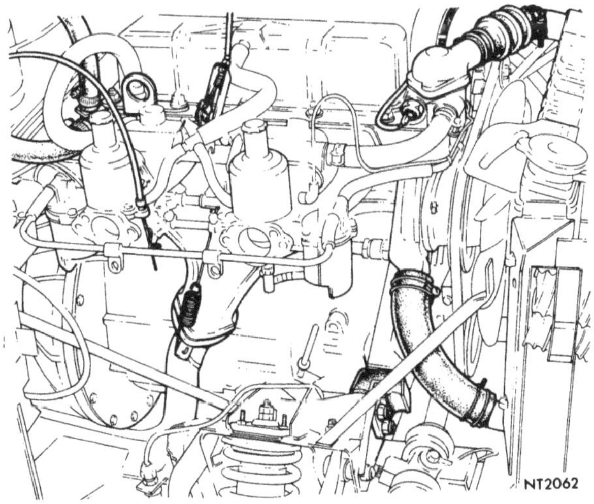
NT2062

ENGINE AND GEARBOX ASSEMBLY

– Remove, change ancillary equipment and refit 12.37.03.

Removing

1. Remove the engine and gearbox assembly 12.37.01.
2. Remove the water pump housing complete with thermostat housing, water pump, fan, pulley and belt. 26.50.03.
3. Remove the alternator and brackets.
4. **Remove the [generator or the] alternator and brackets.**
5. **Remove the air cleaner.[See 19.10.01 for data on single-carburetter engines.]**
6. Remove the inlet and exhaust manifolds.
7. Remove the dipstick.
8. Remove the distributor. 86.35.20.
9. Remove the spark plugs.
10. Remove the oil filter, 12.60.01.
11. Remove the starter motor.
12. Remove the fuel pump.
13. Remove the clutch housing complete with gearbox.
14. Remove the clutch unit.
15. Remove the engine mounting rubbers.
16. Remove the oil filler cap.

Refitting

17. Reverse instructions 1 to 16, ensuring:
    a. New gaskets and seals are fitted where applicable.
    b. Drive belt tension is adjusted 86.10.05.
    c. Distributor contact gap is adjusted 86.35.14.
    d. Check oil sump level, refill or top-up with recommended grade to high mark on dipstick.
    e. Refit the cooling system. 26.10.01.

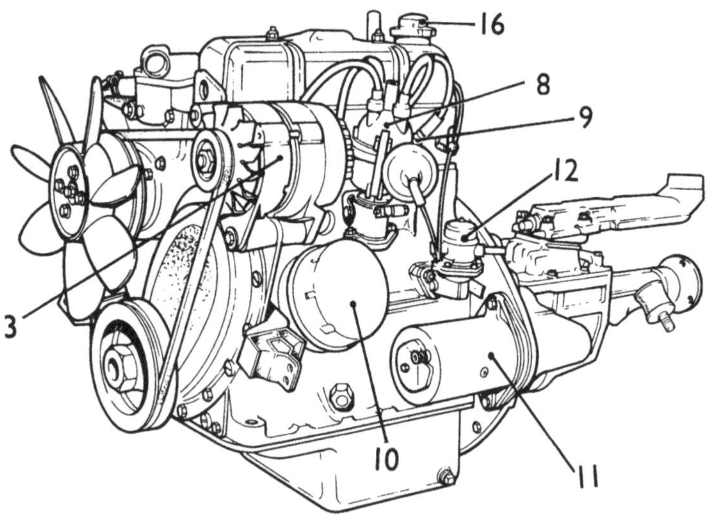

NT2063

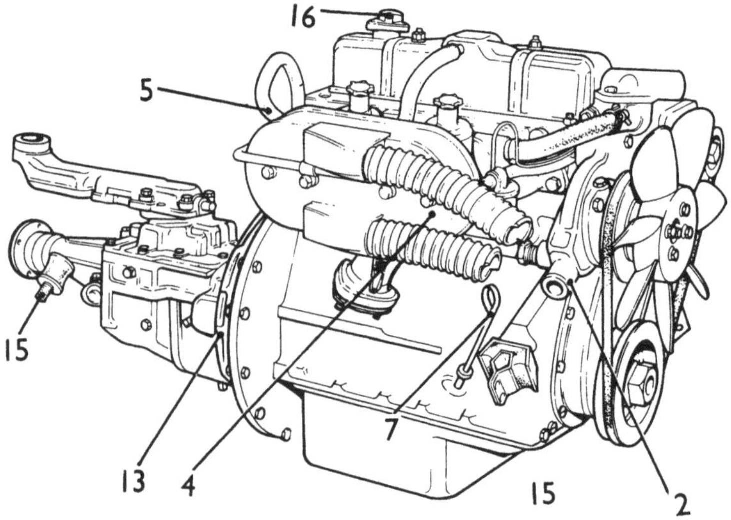

NT2064

ENGINE ASSEMBLY

– Strip and rebuild 12.41.05

[CAUTION: There are many parts of the engine which must not be intermixed or which require special techniques to remove. If you are not an experienced Spitfire mechanic, please read carefully the other parts of this Manual where each specific removal task is described.]

Stripping

1. Remove the engine and gearbox assembly and remove ancillary equipment 12.37.03.
2. Remove the rocker cover.
3. Remove the rocker shaft. 12.29.54. instructions 2 and 3.
4. Remove the cylinder head. 12.29.10 instructions 9 to 11, ensuring that the nuts are released in correct sequence 12.29.27.
5. Remove the valves and springs. 12.29.62. instructions 3 and 4, and identify for reassembly.
6. Remove the push rods and identify for reassembly.
7. Withdraw the cam followers and identify for reassembly. 12.29.57.
8. Remove the crankshaft pulley.
9. Remove the timing chain cover. 12.65.01 instructions 5 and 6.
10. Remove the timing chain and sprockets 12.65.12 instructions 2 and 5, ignoring reference to moving the crankshaft and camshaft.
11. Remove the distributor pedestal and drive gear. 12.10.22, instructions 2 to 4.
12. Remove the two bolts securing the camshaft locating plate to the cylinder block. Remove the plate and withdraw the camshaft.
13. Remove the cylinder block front mounting plate 12.25.10, instructions 9 to 12.
14. Remove the flywheel 12.53.07, instructions 3 and 4.
15. Remove the spigot bush from the crankshaft.
16. Remove the engine rear adaptor plate 12.53.03, instructions 2 and 3.
17. Turn the cylinder block over so that the sump is uppermost.
18. Remove the bolts securing the sump to the crankcase and lift off the sump complete with gasget.

19. Remove the oil pump strainer.
20. Remove the oil pump 12.60.26, instructions 2 and 3.
21. Remove the oil pressure relief valve 12.60.56.
22. Remove the crankshaft rear oil seal 12.21.20, instructions 3 to 5.
23. Remove the two screws securing the front sealing block to the crankcase and lift out the sealing block.
24. Remove the eight big-end bolts.
25. Withdraw the connecting rod bearing caps complete with lower bearing shells DO NOT MIX. Check identification marks for reassembly.
26. Push the connecting rods and pistons down the cylinder bores and remove the upper bearing shells.
27. Turn the cylinder block on its side and with care withdraw the pistons and connecting rod assemblies.
28. Marry-up the connecting rods and piston assemblies with their respective bearing caps and shells.
29. Remove the front and centre main bearing cap bolts.
30. Remove the rear main bearing cap bolts.
31. Withdraw the main bearing caps, complete with lower shells. Check identification marks for reassembly. DO NOT MIX.
32. Lift out the crankshaft.
33. Remove the upper bearing shells and thrust washers from the rear main bearing.

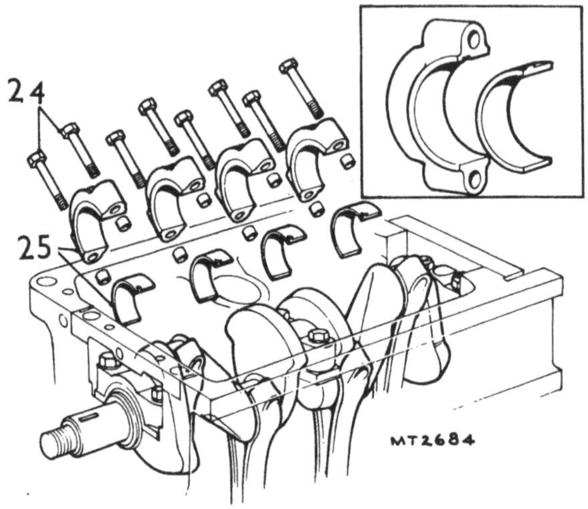

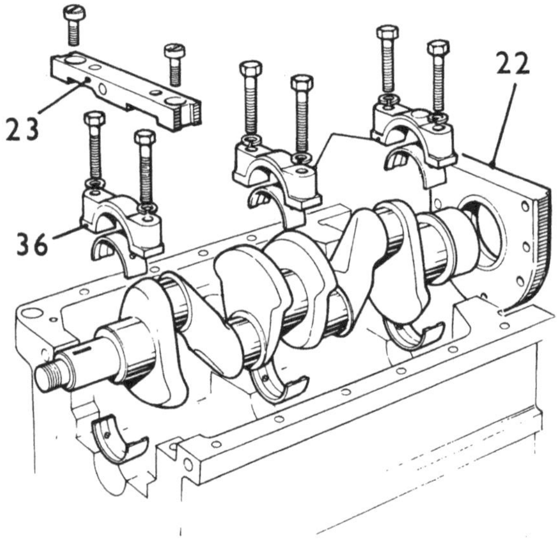

MT2 642

Rebuilding

Overhaul all sub-assemblies.
Clean and degrease all components before assembly. Lubricate all parts with clean engine oil before assembly, unless otherwise stated.

34. Fit the main bearing shells to the crankcase, ensuring that the tags locate in the recesses.
35. Check that the crankshaft oilways are clear, and lower the crankshaft into the crankcase.
36. Fit the bearing shells to the main bearing caps, ensuring that the tags locate in the recesses.
37. Fit the centre and front main bearing caps and tighten to 50 to 65 lbf. ft (7 to 9 kgf. m).
38. Fit the thrust washers to the rear main bearing, ensuring that the thrust faces, identified by oil grooves, bear against the crankshaft faces and locate in the registers on both sides of the crankcase bearing bore half.

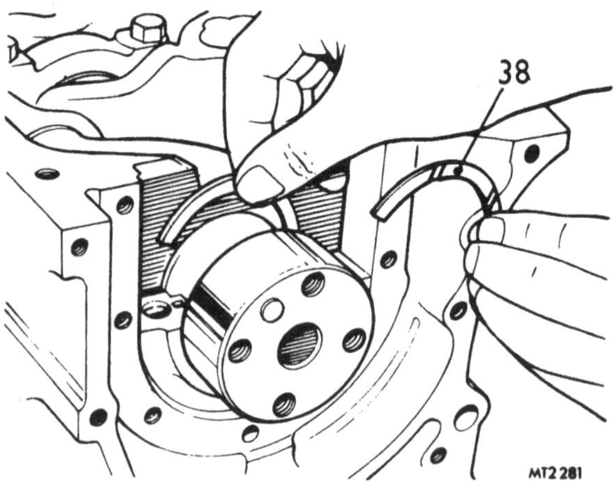

39. Check the crankshaft end-float. 12.21.26.
40. Fit the rear main bearing cap and tighten the bolts to 50 to 65 lbf. ft (7 to 9 kgf. m).
41. Fit the piston and connecting rod assemblies to the cylinder block bores. 12.17.01, instructions 14 to 21.
42. Fit the crankshaft rear oil seal. 12.21.20, instructions 6 to 11.
43. Fit the front sealing block 12.21.39, instructions 14 to 17.
44. Fit the oil pump, ensuring that the spring washers are located under the bolt heads, tighten the bolts evenly.
45. Fit the oil strainer to the pump inlet.
46. Fit the sump with a new gasket and evenly tighten the retaining bolts.
47. Fit the oil pressure relief valve 12.60.56.
48. Turn the cylinder block over, and secure in an up-right position.
49. Fit the engine adaptor plate, ensuring correct location over the studs and dowels and that the arrow is at the top. Evenly tighten the retaining bolts.
50. Fit the crankshaft spigot bush, which should be a loose fit in the crankshaft.
51. Fit the flywheel. 12.53.07, instructions 5 to 8.
52. Coat the cylinder block face of a new front mounting plate gasket with jointing compound and place it in position on the joint face.
53. Fit the mounting plate in position over the dowels. Secure it in position with three bolts into the cylinder block and three screws into the front sealing block.
54. Assemble the camshaft into the cylinder block, secure it in position with the camshaft location plate, and tighten the retaining bolts.
55. Check the camshaft end-float, which should be within 0·004 to 0·008 in (0·10 to 0·20 mm). Pull the camshaft out against the location plate and insert a feeler gauge between the groove and location plate. Reduce end-float by fitting a new location plate.

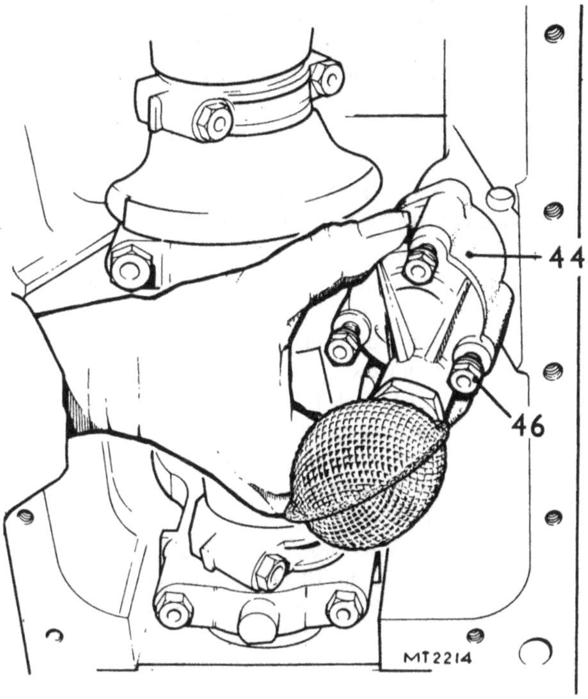

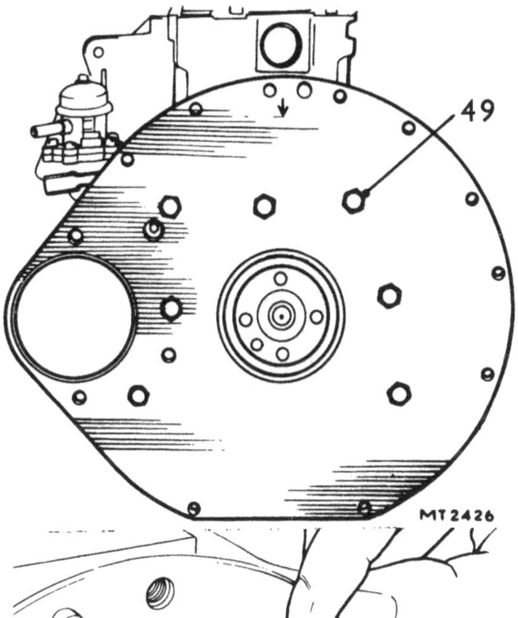

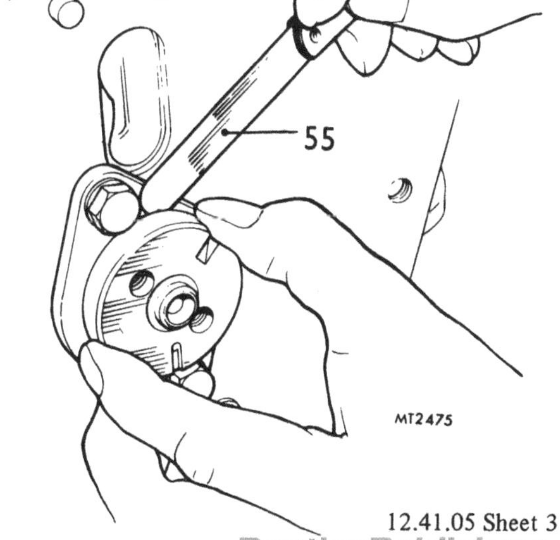

57. Reassemble the valves and springs to the cylinder head. 12.29.18, instructions 5 to 13.
58. Fit the cam followers to the cylinder block. 12.29.57, instruction 3.
59. Rotate the crankshaft until No. 1 piston is at T.D.C. by using a dial gauge on the piston crown. **DO NOT MOVE THE CRANKSHAFT AGAIN UNTIL THE VALVES HAVE BEEN TIMED.**
60. Fit the cylinder head assembly with a new gasket. Tighten the retaining nuts in correct sequence and torque loading. 12.29.27.
61. Fit the push-rods, ensuring that the ball ends locate in the cam followers.
62. Fit the rocker shaft assembly, ensuring that the rocker adjusting screw balls locate in the push-rods cups.
63. Tighten the rocker shaft pedestal nuts evenly to 26 to 32 lbf. ft (3·6 to 4·4 kgf. m).
64. Adjust the valve clearances of Nos. 7 and 8 valves only to 0·050 in (0·397 m) to give a working clearance. 12.29.48, instructions 6 and 7.

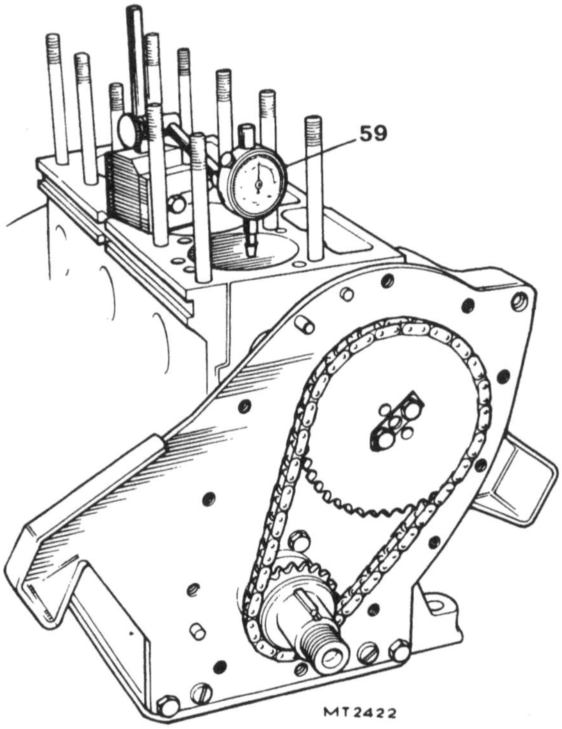

65. Turn the camshaft until Nos. 7 and 8 valves are on the 'rock', i.e. inlet valve about to open and exhaust valve about to close. This may be determined by oscillating the camshaft while measuring the rocker clearances of these two valves with feeler gauges. When the two clearances are the same, the valves and camshaft are in their correct relationship with the crankshaft.

66. Encircle both sprockets with the timing chain and fit to crankshaft and camshaft respectively, keeping the chain taut on the drive side.

NOTE: The camshaft timing sprocket is provided with four holes which are equally spaced but offset from a tooth centre. Half-tooth adjustment is obtained by rotating the sprocket 90 degrees from its original position. A quarter tooth adjustment is possible by turning the sprocket 'back to front . By rotating the sprocket 90 degrees in this position three-quarters of a tooth variation is available.

67. If new sprockets have been fitted, make a punch mark (A) on the crankshaft and camshaft sprockets on a line (C) scribed through the diameters of the two sprockets. Make also a punch mark (B) on the end of the camshaft through one of the sprocket holes and a corresponding punch mark on he sprocket.

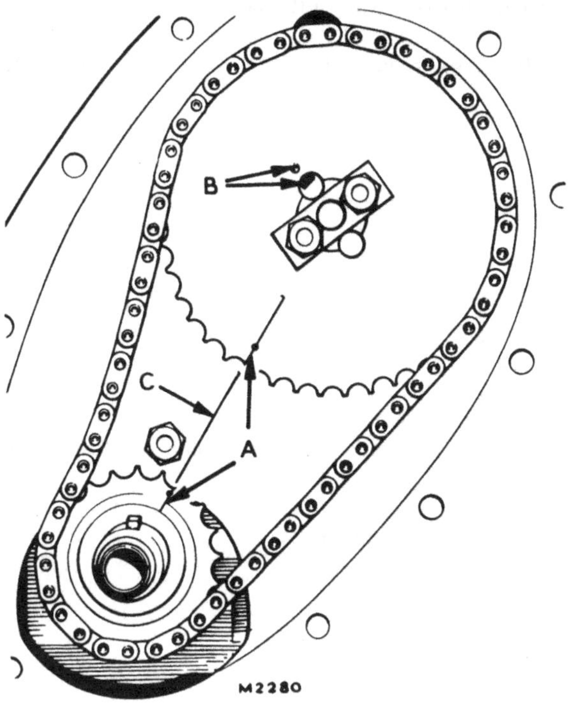

68. Check the sprockets for alignment 12.65.12.
69. Fit a new camshaft sprocket lock plate and tighten the bolts and bend over the tabs.
70. Fit the oil thrower to the crankshaft, ensuring that the dished periphery faces the timing cover.
71. Fit the timing chain cover complete with a new gasket. 12.65.01.
72. Fit the crankshaft pulley and tighten the retaining nut to 90 to 110 lbf. ft (12·4 to 15·2 kgf. m).
73. Adjust all valve clearances. 12.29.48, instructions 4 to 7.
74. Fit the distributor drive shaft and gear. 12.10.22.
75. Fit the rocker cover.
76. Fit the ancillary equipment. 12.37.03.
77. Fit the clutch assembly.
78. Fit the gearbox assembly.
79. Fit the engine and gearbox assembly to the car. 12.37.01.
80. Fill the oil sump with engine oil of the correct grade to the top mark on the dipstick, and replace the dipstick.
81. Fill the cooling system 26.10.01, instructions 5 to 10.
82. Time the ignition 86.35.16.
83. Tune and adjust the carburetters. 19.15.02.

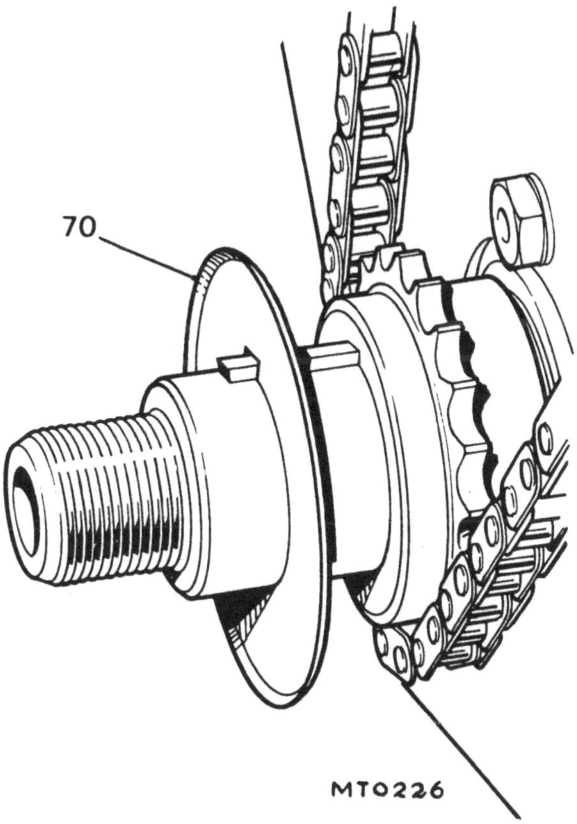

ENGINE MOUNTINGS – FRONT

– L.H. – remove and refit 12.45.01

– R.H. – remove and refit 12.45.03

– Set – remove and refit 12.45.04

Removing

1. Employ a jack or hoist to take the weight of the engine.
2. Remove the nuts and bolts securing the mounting to frame (two each side).
3. Remove the nut securing the mounting to the engine mounting plate.
4. Raise the engine slightly.
5. Remove the mounting(s).

Refitting

6. Reverse instructions 1 to 5.

ENGINE MOUNTINGS REAR L.H.

– LH – Remove and refit 12.45.07
– RH – Remove and refit 12.45.09
– Set – Remove and refit 12.45.10

Removing

1. Remove the front floor carpets 76.49.02.
2. Remove the gearbox cover carpet 76.49.01.
3. Remove the gearbox tunnel cover 76.25.07.
4. Place jack under gearbox.
5. Remove nuts.
6. Remove washers.
7. Jack up gearbox until the mounting rubbers are clear of the support bracket.
8. Unscrew the mounting rubber.

Refitting

9. Screw the new mounting rubber in place.
10. Lower the gearbox and ensure the mounting locates in the support bracket.
11. Replace the nuts and washers.
12. Refit the gearbox tunnel cover 76.25.07.
13. Refit gearbox cover carpet 76.49.01.
14. Refit the front floor carpets 76.49.02.

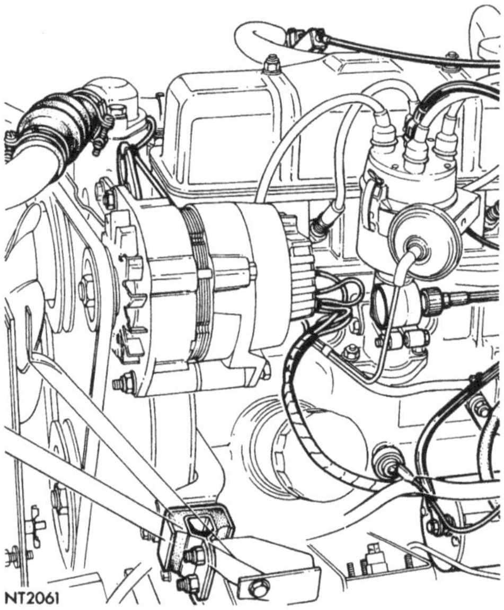
NT2061

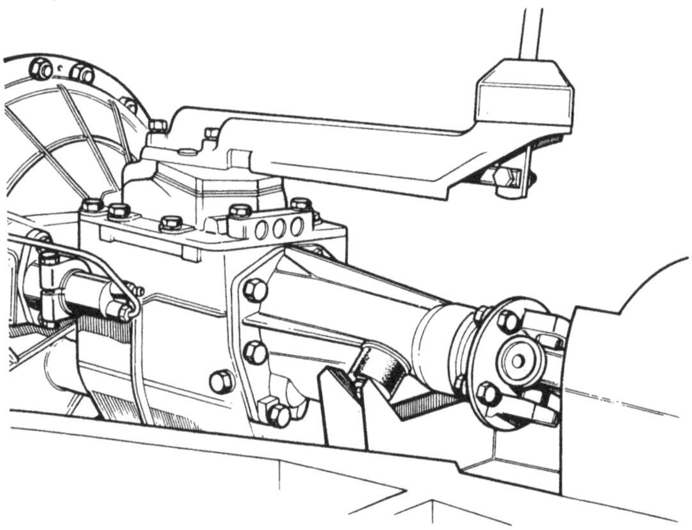

ENGINE TUNE

– Check and adjust distributor points, spark plugs, ignition timing, tune carburetters, clean fuel pump filter and road test. 12.49.02

1. Check the condition and gap of distributor points, renew or adjust as required. 86.35.13.
2. Remove the spark plugs and check their condition, renew if necessary. Clean and adjust the gaps. 86.35.01.
3. Check the ignition timing, and adjust if necessary. 86.35.16.
4. Tune the carburetters 19.15.02.
5. Clean the fuel pump filter. 19.45.02.

ENGINE REAR GEARBOX ADAPTOR PLATE

– Remove and refit 12.53.03

Removing

1. Remove the flywheel 12.53.07.
2. Remove the seven bolts securing the adaptor plate to the cylinder block.
3. Remove the adaptor plate.

Refitting

4. Reverse instructions 1 to 3, ensuring:
   a. The adaptor plate is located correctly over the dowel at the top of the cylinder block.
   b. The adaptor plate retaining bolts are tightened evenly to 16 to 20 lbf. ft (2·2 to 2·8 kgf. m).

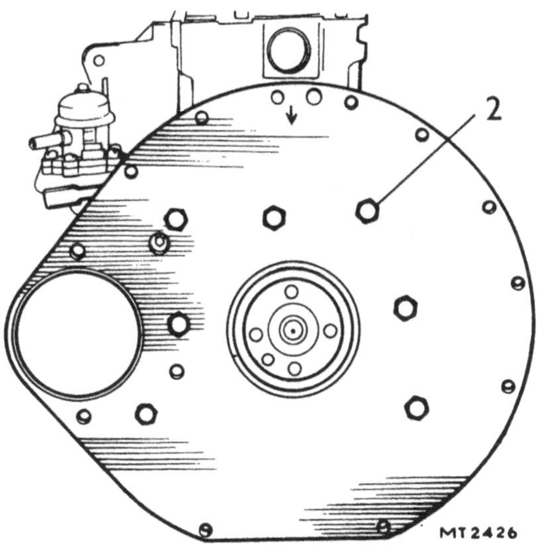

FLYWHEEL

– Remove and refit **12.53.07**

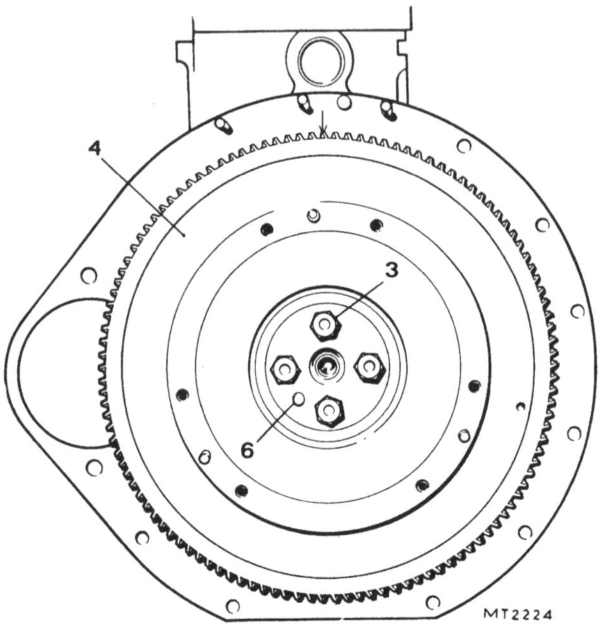

Removing

1. Remove the gearbox 37.20.01.
2. Remove the clutch assembly 33.10.01.
3. Remove the four bolts securing the flywheel to the crankshaft.
4. Lift off the flywheel.

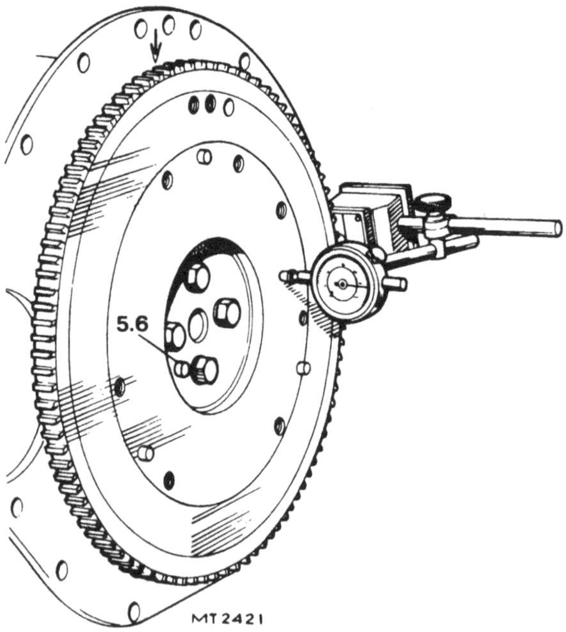

Refitting

5. Clean the flywheel mating face with the crankshaft and crankshaft flange. Check the dowel for damage, and that the crankshaft spigot bush is in position.
6. Fit the flywheel to the crankshaft, locating over the dowel in the crankshaft.
7. Tighten the flywheel retaining bolts evenly to 38 to 45 lbf. ft. (5·2 to 6·2 kgf. m).
8. Using a dial indicator gauge, check the flywheel for run-out, not to exceed 0·002 in (0·051 mm) at 3·0 in (76·2 mm) radius from spigot centre. Check concentricity, not to exceed 0·004 in (0·100 mm).
9. Refit the clutch assembly 33.10.01.
10. Refit the gearbox 37.20.01.

STARTER RING GEAR

– Remove and refit 12.53.19

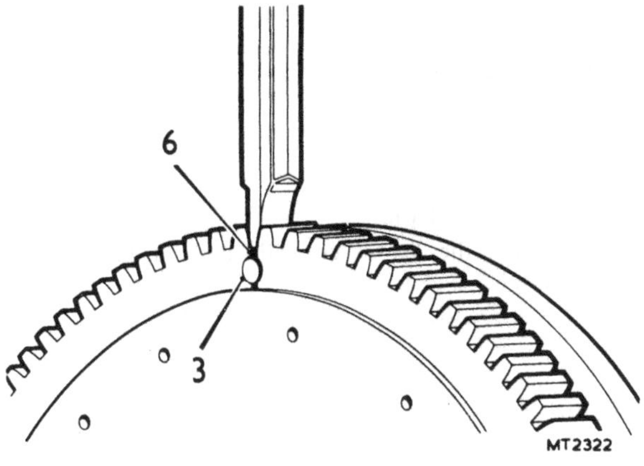

Removing

1. Isolate the battery.
2. Remove the flywheel. 12.53.07.
3. Drill a hole ¼ in (6·35 mm) diameter at the point of intersection of a scribe line between any two teeth and a scribe line midway between the root diameter and inside diameter of the ring gear.
4. Hold the flywheel assembly in a soft jaw vice.
5. Place a cloth of heavy material over the ring gear for protection against flying fragments.

WARNING: **Ensure adequate protection, particularly for the eyes, to prevent injury from flying fragments when the ring gear is split.**

6. Place a cold chisel immediately above the centre line of the drilled hole and strike sharply to split the ring gear.

Refitting

7. Heat the replacement starter ring gear uniformly to a maximum of 200°C.
8. Place the flywheel on a flat surface, clutch face side uppermost, and clean the ring gear locating register.
9. Locate the ring gear and hold in position until it contracts sufficiently to grip the flywheel.
10. Allow the ring gear to cool gradually to avoid distortion. A maximum gap of 0·025 in (0·635 mm) is tolerable between the ring gear and the flywheel face in any one area of 6 in (15 cm) around the circumference.
11. Refit the flywheel, 12.53.07, and check that ring gear eccentricity does not exceed 0·010 in (0·254 mm).

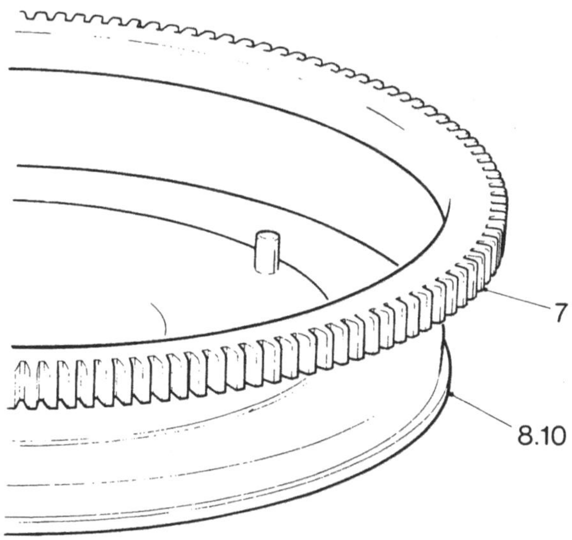

OIL FILTER

– Remove and refit **12.60.01**

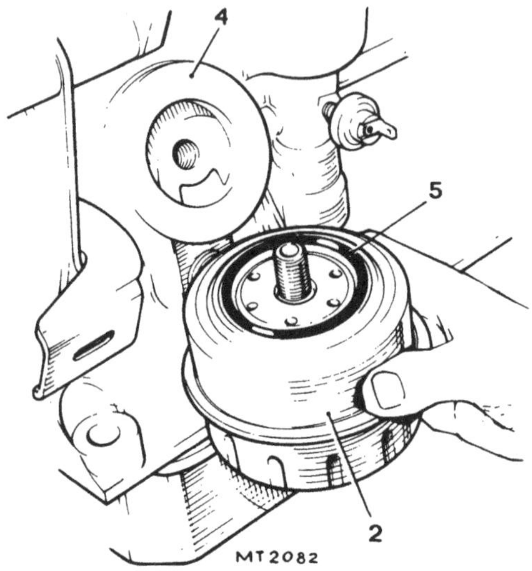

Removing

1. Isolate the battery.
2. Grasp the oil filter with both hands and turn anti-clockwise.
3. Remove the filter.

Refitting

4. Clean the cylinder block face.
5. Smear the 'O' ring with grease to prevent 'picking up'.
6. Fit the filter, turning clockwise, and tighten.
7. Start the engine and check for oil leaks between the cylinder block and filter.

OIL PICK-UP STRAINER

– Remove, clean and refit **12.60.20**

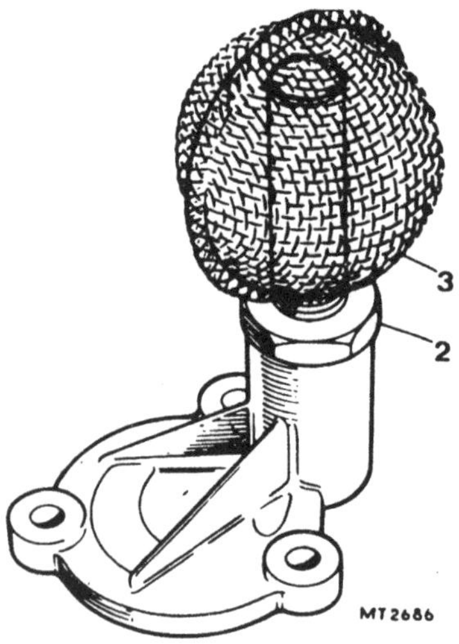

Removing

1. Remove the sump 12.60.44.
2. Release the oil strainer locknut.
3. Unscrew the oil strainer and remove it.
4. Wash the oil strainer in petrol and dry it before refitting.

Refitting

5. Reverse instructions 1 to 3.

OIL PUMP

– Remove and refit 12.60.26

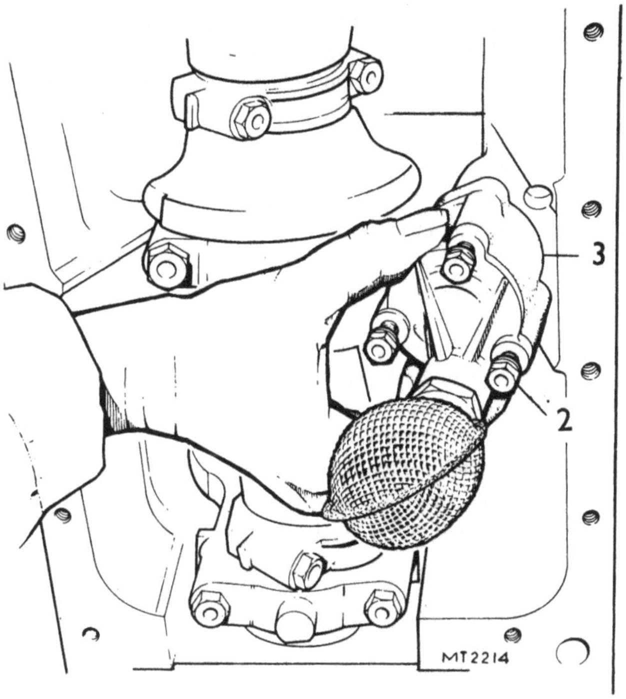

Removing

1. Remove the oil sump 12.60.44.
2. Slacken and remove the three bolts complete with washers securing the oil pump to the crankcase.
3. Remove the oil pump complete with cover-plate.

Refitting

4. Reverse instructions 1 to 3, ensuring:
   - *a.* The oil pump drive shaft engages correctly into the drive gear shaft.
   - *b.* The securing bolts are evenly tightened.
   - *c.* Absolute cleanliness is observed.

OIL PUMP

– Overhaul 12.60.32

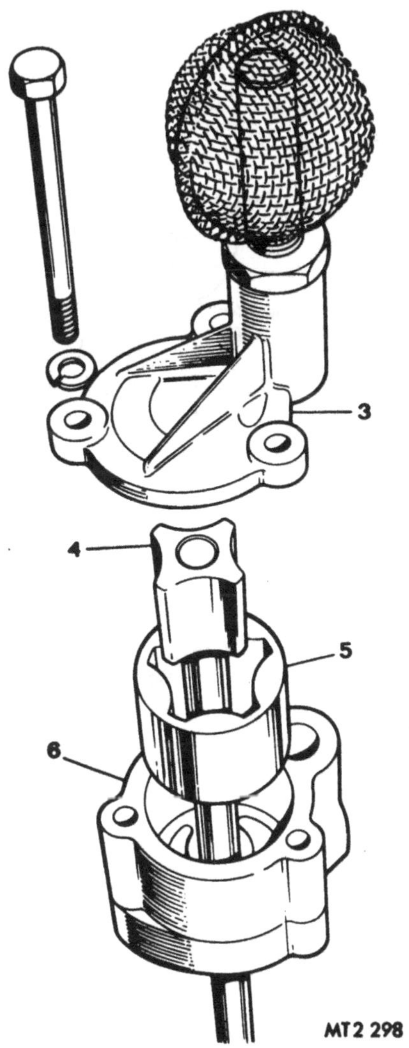

1. Isolate the battery.
2. Remove the oil pump from the crankcase. 12.60.26.
3. Remove the cover-plate complete with strainer.
4. Remove the inner rotor and shaft assembly.
5. Remove the outer rotor.
6. Clean oil from the body and rotors, and reassemble.

7. Place the oil pump in a vice and using a feeler gauge:
   a. With a straight-edge across the pump body face, check the clearance between the rotors and straight-edge. This clearance must not exceed 0·004 in (0·1 mm).
   b. Check the clearance between the inner and outer rotors. This must not exceed 0·010 in (0·25 mm).
   c. Check the clearance between the outer rotor and body. This must not exceed 0·008 in (0·2 mm).
8. Check the cover-plate for scoring, and test on a surface plate for distortion. Renew if necessary.
9. Check the pump spindle bearing surface in the pump body for excessive wear.
10. Reassemble the pump, fitting any new parts necessary to satisfy the above quoted tolerances. Renew the complete pump if a satisfactory condition cannot be achieved.

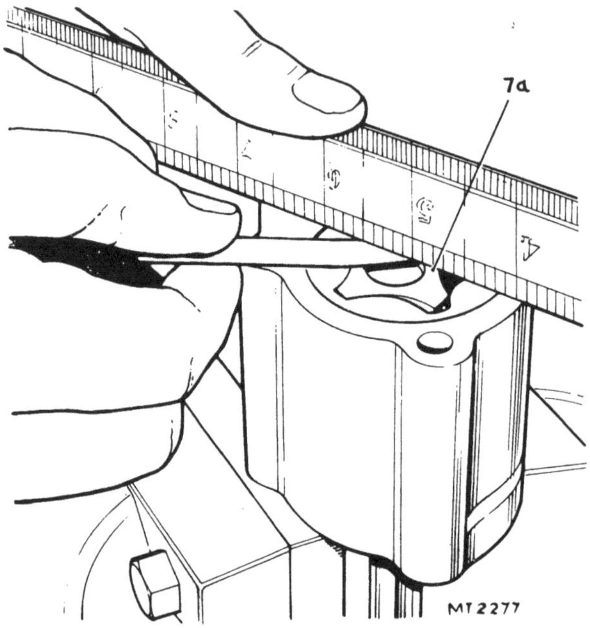

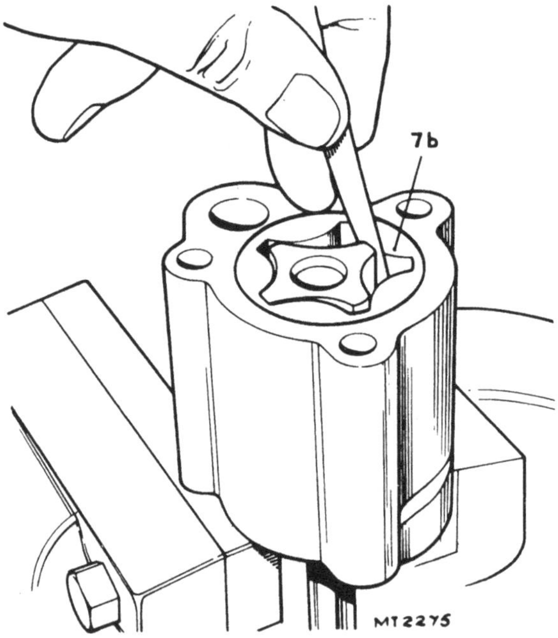

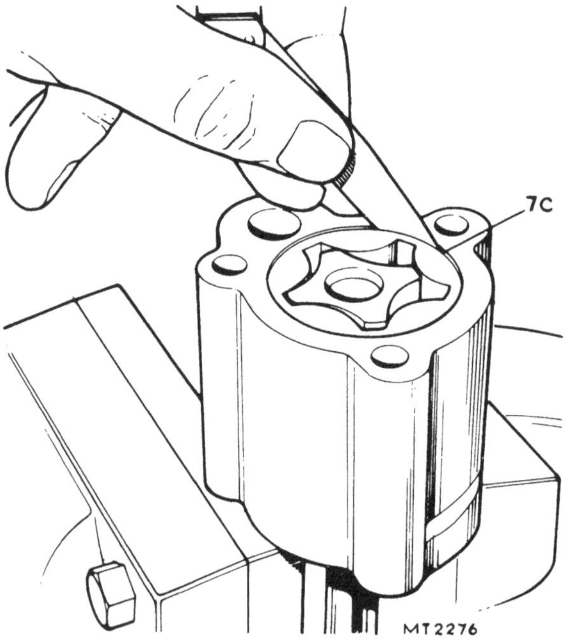

OIL SUMP

– Remove and refit **12.60.44**

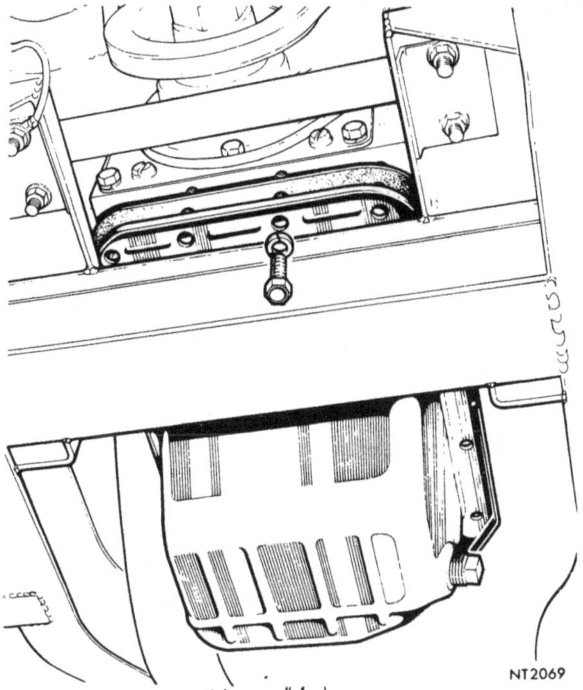

Removing

1. Drain the sump oil.
2. Remove the 16 bolts securing the sump to the crankcase.
3. Raise the engine sufficiently to lower the sump, turn it and withdraw.

Refitting

4. Reverse instructions 1 to 3 noting:
   a. Clean the crankcase and sump faces.
   b. Fit a new gasket
   c. The longer bolts are fitted at the rear of the sump.
   d. Tighten the retaining bolts to 16 to 20 lbf. ft (2·2 to 2·8 kgf. m).

OIL PRESSURE RELIEF VALVE

– Remove and refit **12.60.56**

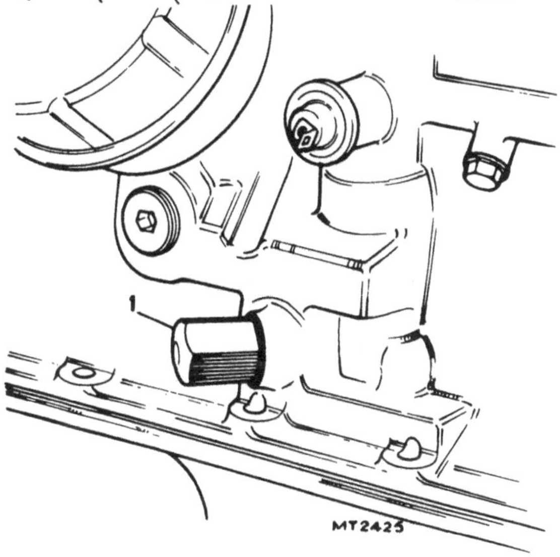

Removing

1. Unscrew the relief valve body from the cylinder block.
2. Remove the washer.
3. Take out the plunger.
4. Remove the spring.

Refitting

5. Reverse instructions 1 to 4 ensuring:
   a. Cleanliness.
   b. A new spring is fitted if free length is not in accordance with Data.

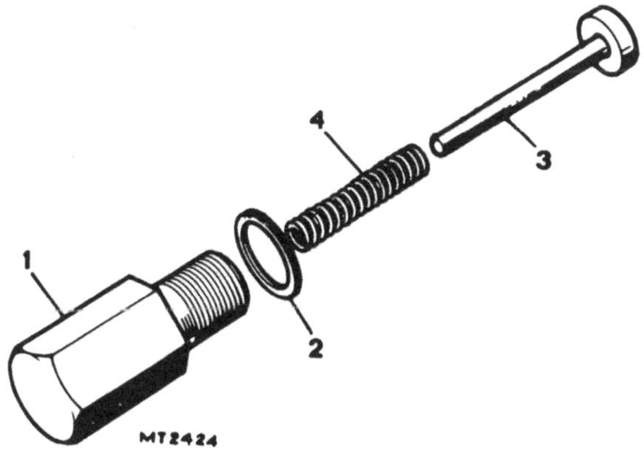

DATA

Oil pressure relief valve spring free length . . . . . 1·53 in (38·8 mm).

TIMING CHAIN COVER

– Remove and refit **12.65.01**

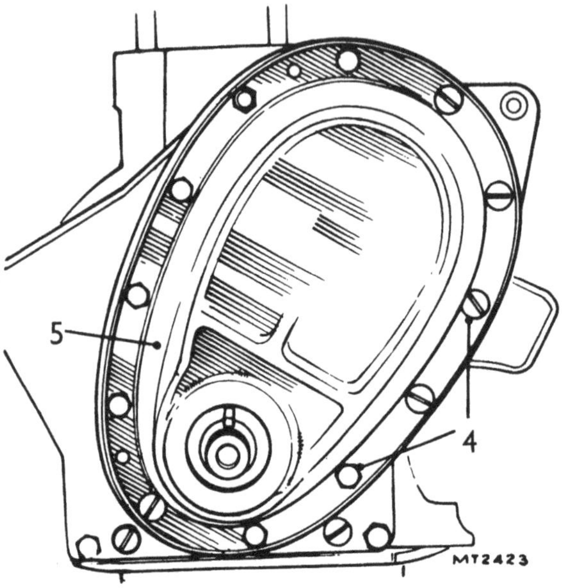

Removing

1. Remove the fan belt.
2. Remove the fan blades 26.25.06.
3. Remove the crankshaft pulley. 12.21.01.
4. Remove the five screws, six bolts and one nut securing the timing chain cover to the cylinder block.
5. Remove the timing chain cover complete with gasket, taking care not to damage the oil seal.

Refitting

6. Reverse instructions 1 to 4, noting:
   - *a.* Fit a new cover gasket.
   - *b.* To facilitate refitment of the cover, compress the chain tensioner with a suitable bent length of rod taking care not to damage the gasket when withdrawing the rod.
   - *c.* The cover gasket locates on dowels.

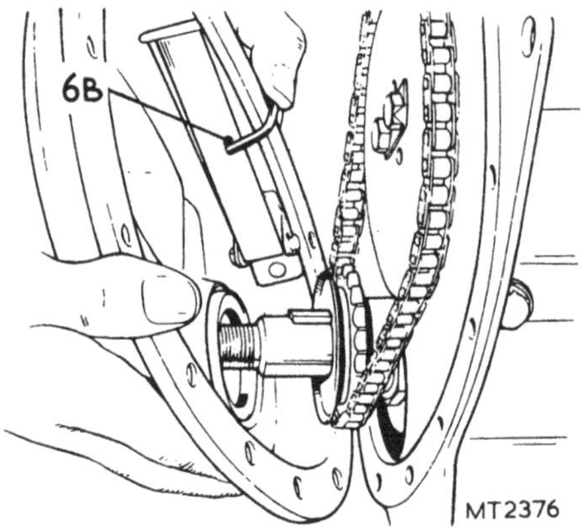

TIMING CHAIN COVER OIL SEAL

– Remove and refit **12.65.05**

Removing

1. Remove the timing chain cover 12.65.01.
2. Carefully remove the oil seal, avoiding damage to the timing cover.

Refitting

3. Reverse instructions 1 and 2, ensuring that the new oil seal is:
   - *a.* Smeared with oil before fitting.
   - *b.* Fitted correctly, i.e. cavity face towards the engine.
   - *c.* Pressed home squarely.

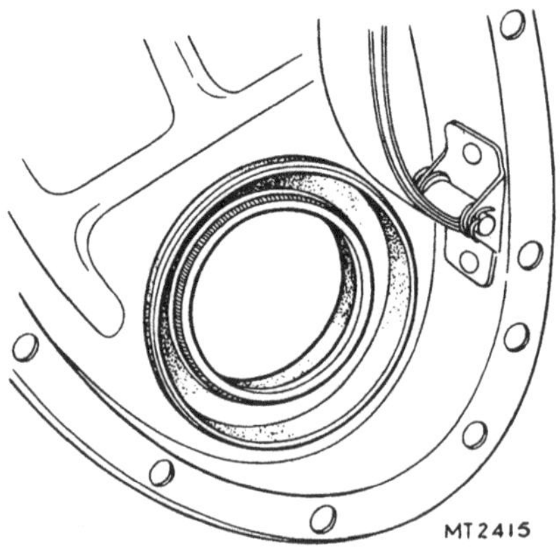

VALVE TIMING

– Check **12.65.08**

1. Remove the rocker cover 12.29.42.
2. Adjust the rocker clearances of Nos. 7 and 8 valves to 0·050 in (1·27 mm) 12.29.48, instructions 6 and 7, to give a working clearance.
3. Turn the crankshaft until No. 1 piston is at T.D.C. on compression stroke, indicated by the mark on the crankshaft pulley coinciding with the pointer on the timing cover.
4. Check that Nos. 1 and 2 valves are fully closed by inserting a feeler gauge between the valve tip and rocker pad to ascertain clearance.
5. Using two feeler gauges of the same thickness, check that the rocker clearances on Nos. 7 and 8 valves are the same. Oscillate the crankshaft to achieve this condition, but ensure that when the rocker clearances are the same the conditions in instructions 3 and 4 are maintained within a few degrees.

NOTE: The actual valve clearance does not matter providing the clearances are the same.

6. Should the valve timing prove to be incorrect, retiming will be necessary, 12.41.05, instructions 59 to 67.
7. Re-adjust Nos. 7 and 8 valves to 0·010 in (0·25 mm) 12.29.48, instructions 6 and 7.
8. Refit the rocker cover 12.29.42.

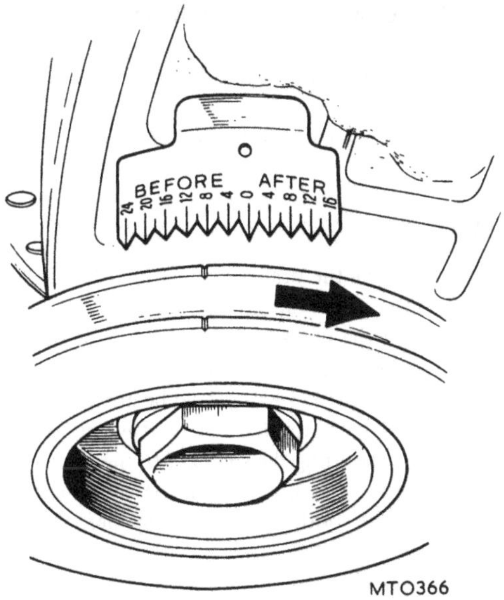

MTO366

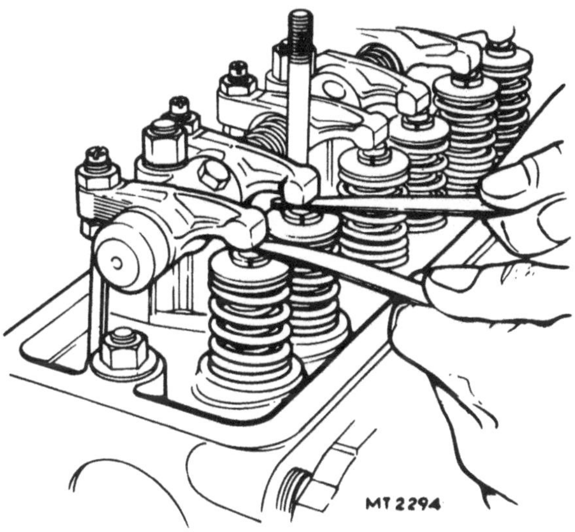

MT 2294

DATA

Valve timing	English	American
inlet opens	25° B.T.D.C.	18° B.T.D.C.
closes	65° A.B.D.C.	58° A.B.D.C.
Exhaust opens	65° B.B.D.C.	58° B.B.D.C.
closes	25° A.T.D.C.	18° A.T.D.C.

TIMING CHAIN AND SPROCKETS

– Remove and refit 12.65.12

Timing chain 1 to 5, 12, and 15 to 17 12.65.14

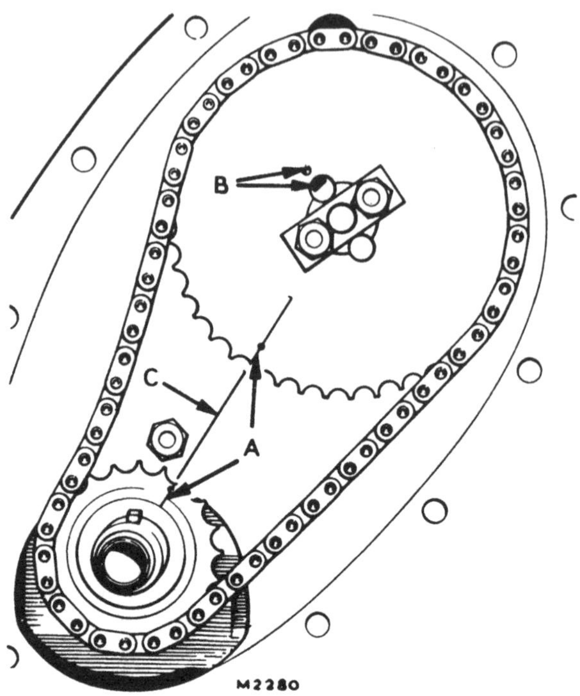

Removing

1. Remove the timing chain cover 12.65.01.
2. Remove the oil thrower.
3. Turn the crankshaft until the mark 'A' is in line with scribe mark 'C' and marks 'B' correspond. The crankshaft keyway should be at 12 o'clock.
4. Bend back the lock plate tabs and remove the two bolts securing the sprocket to the camshaft.
5. Taking care not to turn the crankshaft or camshaft, remove both sprockets together with the timing chain.

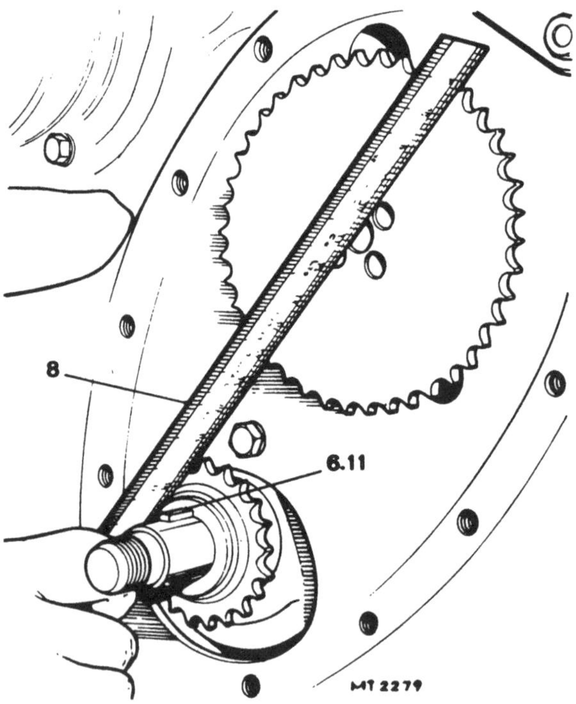

Refitting

6. Remove the crankshaft drive key.
7. Temporarily refit both sprockets.
8. Check the alignment of the sprockets by placing a straight edge across the teeth of both sprockets.

9. Correct any misalignment by fitting selective shims behind the crankshaft sprocket.
10. Remove the sprockets.
11. Refit the drive key.
12. Encircle the sprockets with the timing chain and refit ensuring that the marks on the sprockets and engine plate line up.
13. Refit the camshaft sprockets securing bolts (temporarily).

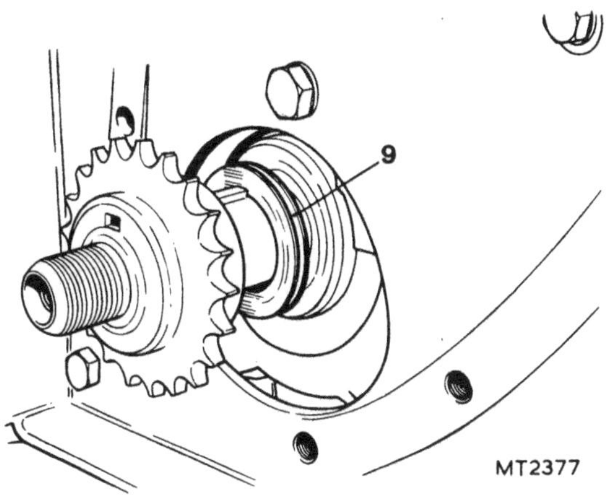

MT2377

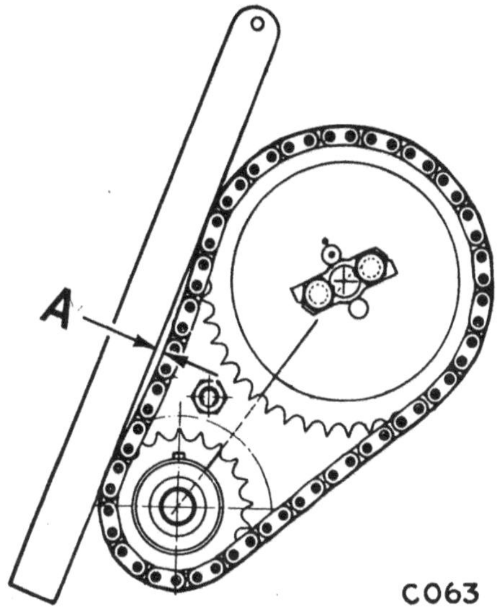

C063

14. Check timing chain wear by placing a straight edge along the slack run of chain. If movement at the mid-point 'A' exceeds 0·4 in (10mm), renew the chain.
15. Fit a new camshaft sprocket lock plate, refit the bolts and bend the tabs.
16. Refit the oil thrower with the dished periphery towards the timing chain cover. 12.41.05
17. Refit the timing chain cover 12.65.01

NOTE: If new sprockets are fitted see instructions 67 and 68, 12.41.05.

TIMING CHAIN TENSIONER

– Remove and refit 12.65.28

Removing

1. Remove the timing chain cover. 12.65.01
2. Prise open the tensioner blades.
3. Slide the tensioner off the anchor pin.

Refitting

4. Reverse instructions 1 to 3, ensuring that the tensioner is fitted correctly i.e. convex surface towards the timing chain.

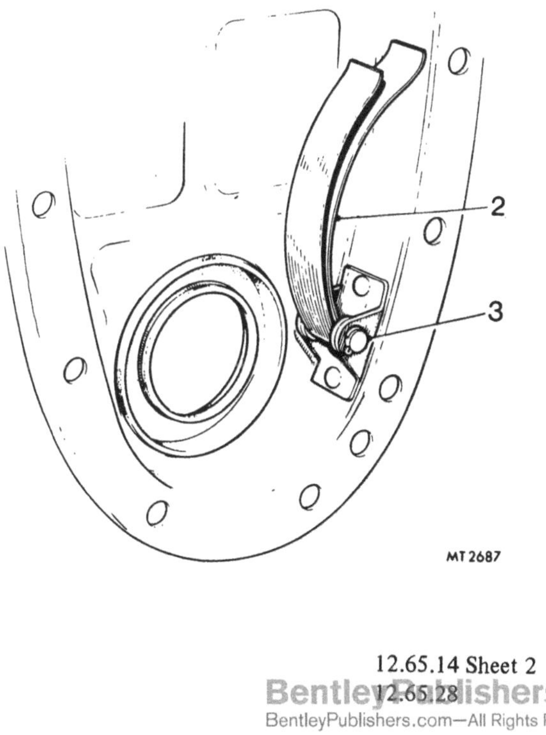

MT2687

EMISSION CONTROL OPERATIONS

***All bracketed material ([]) in this section has been added by Robert Bentley, Inc.**

EVAPORATIVE LOSS CONTROL SYSTEM

–Leak testing 17.15.01

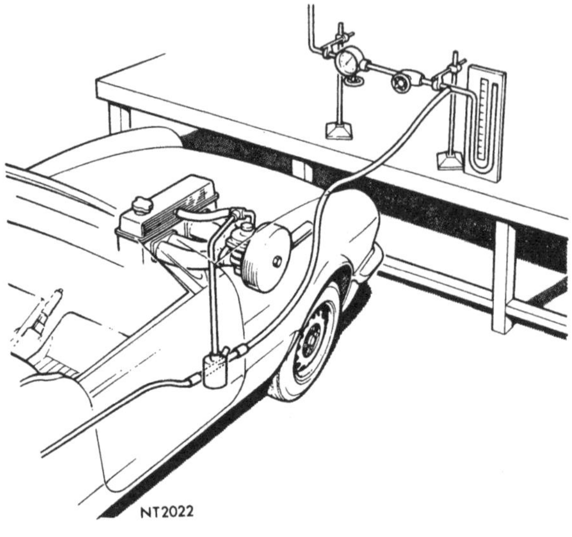

Test apparatus: Water manometer, pressure regulating valve, pressure sealing valve and pressurized air supply. An alternative to the manometer is an accurate pressure gauge reading 0 to 30 in of water with which a pressure relief valve or weak rubber connection will be necessary to protect against over-pressurizing the system.

WARNING: **During the test, pressure will be applied to the fuel tank breather system. The pressure, albeit very low, may displace pipe connections or cause an emission of fuel vapour. It is, therefore, very dangerous to allow smoking or naked lights in the test area, or exceed the pressure quoted.**

Test procedure

1. At the carbon canister, detach the tube from the separator tank and insert the pressurizing tube of the test apparatus.
2. Apply 20 in of water pressure to the system and close the sealing valve. DO NOT EXCEED 25 INCHES OF WATER PRESSURE.
3. After two minutes, check the pressure in the system. If this has fallen more than 2 in of water, investigation and rectification is necessary.

CRANKCASE BREATHER VALVE

— Cleaning **[17.15.03]**

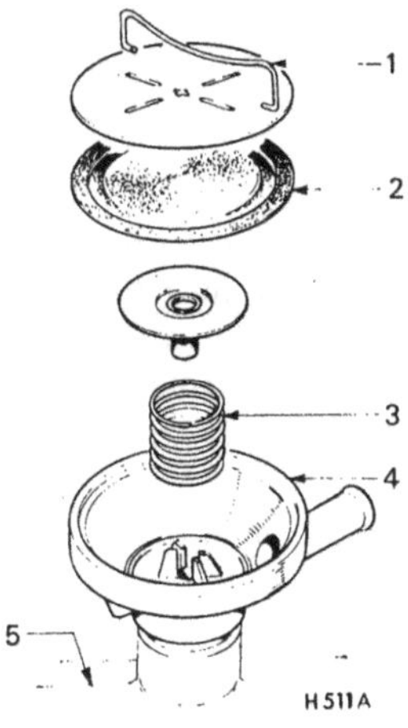

Removing

1. Disengage the clip.
2. Lift out the cover and the diaphragm.
3. Lift out the valve pin and the spring.
4. If necessary, unscrew the body from the inlet manifold.

Servicing and installing

5. Clean the threads in the inlet manifold.
6. Clean all valve components in methylated spirits and check them to see that they are in serviceable condition.
7. Reverse instructions 1 to 4.

ADSORPTION CANISTER AIR FILTER

–Remove and refit 17.15.07

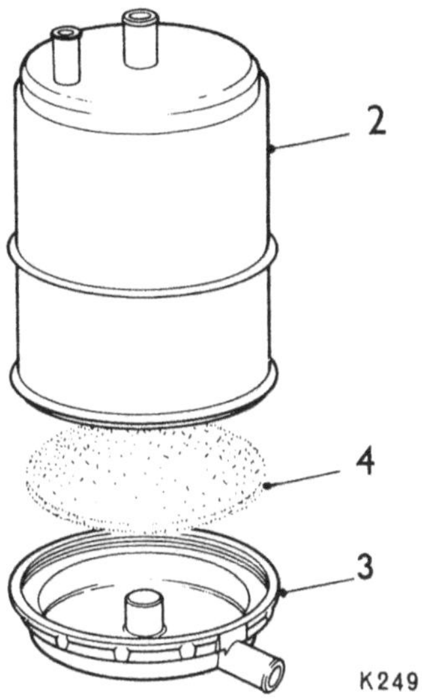

Removing

1. Slacken nut on retaining clip bolt.
2. Lift canister clear of mounting bracket.
3. Unscrew base of canister.
4. Remove filter.

Refitting

5. Reverse instructions 1 to 4.

[**NOTE:** Always replace the filter.]
[**WARNING: Because of the possible presence of fuel fumes, avoid naked lights or actions likely to cause sparks.]**

ADSORPTION CANISTER

–Remove and refit 17.15.13

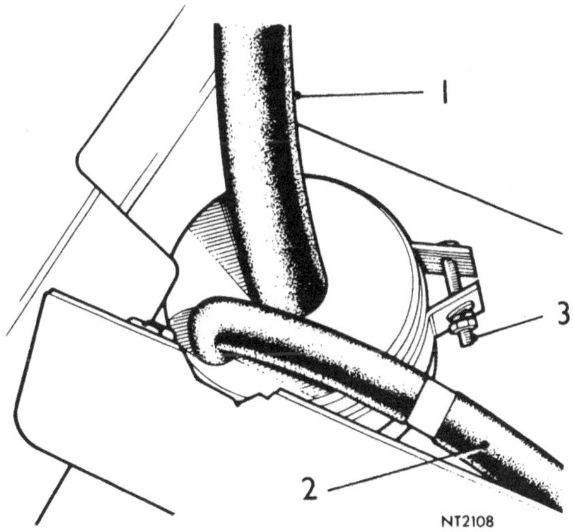

Removing

1. Disconnect the canister purge pipe.
2. Disconnect the pipe to tank.

[**NOTE:** There are two pipes on 1973 and 1974 vehicles.]
[**WARNING: Because of the possible presence of fuel fumes, avoid naked lights or actions likely to cause sparks.]**

3. Slacken the screw on the canister retaining clip.
4. Lift out canister.

Refitting

5. Reverse instructions 1 to 4 ensuring that the restrictor is in position in the canister nozzle.

Triumph Spitfire Mk IV Manual. Part No. 545254. Issue 1.
Part No. 511243. page 1.336. Issue 1.

SEPARATOR TANK

–Remove and refit **17.15.19**

Removing

1. From inside the driving compartment remove the six screws securing the rear squab trim pad.
2. Slacken the clamp bracket screw.
3. Disconnect and remove the separator tank.

Refitting

4. Reverse instructions 1 to 3.

CONTROL PIPE – CANISTER TO TANK

–Remove and refit **17.15.24**

Purge pipe – Canister to crankcase breather **17.15.36**

1. All piping in the evaporative control system is push fitted, access to the tank end of the control pipe is via the luggage compartment trim pad. When refitting pipes ensure that any restictor is replaced into line.

CARBURETTER EMISSION PACK – RED

–Remove and refit **17.20.07**

1. Fit the gaskets and sealing washers of this pack in accordance with the overhaul procedure detailed in 19.15.17.

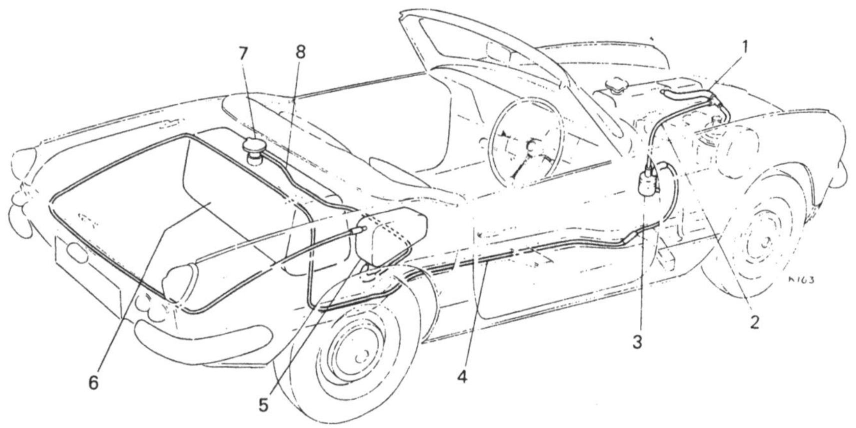

1 Pipe — Rocker cover to carburettor 17.15.36
2 Pipe — Canister purge
3 Activated carbon canister 17.15.13
4 Pipe — overflow tank to canister 17.15.24
5 Overflow tank 17.15.19
6 Main fuel tank
7 Sealed filler cap
8 Pipe — Filler tube to overflow tank

Mk. III (from Jan. 1, 1970)

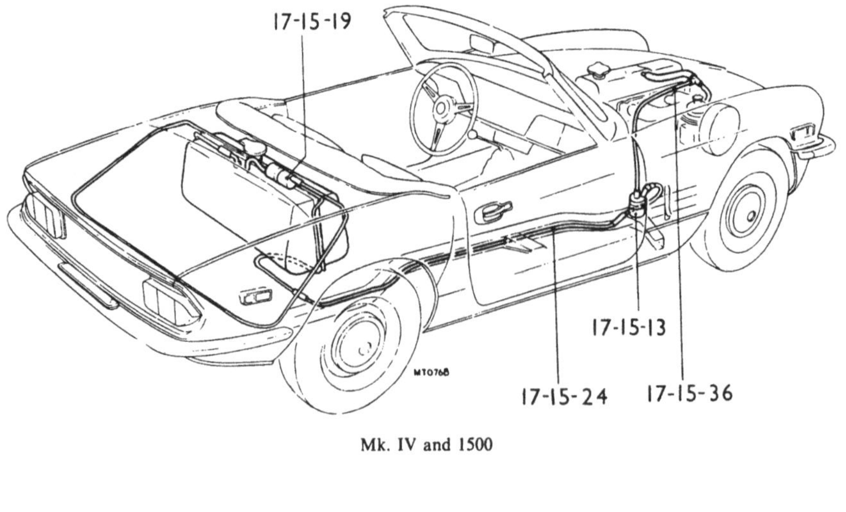

Mk. IV and 1500

C O LEVEL AT IDLE

– Check 17.35.01

Special tools: Approved infra-red gas analyser.

1. Attain normal engine running temperature.
2. Set the idle speed to 800-850 rev/min 19.15.01 s.c.
3. Check ignition timing – reset if necessary 86.35.16.
4. Re-check idle speed – adjust if necessary.
5. Insert the exhaust gas analyser probe as far as possible into the exhaust pipe.
6. Check the CO reading. (Compare with emission control lable).
7. *a)* Adjust mixture if necessary 19.15.01 s.c.
   b) Check idle speed – adjust if necessary.
8. Withdraw the analyser probe.
9. Switch off the ignition.

NOTE: Do not allow the engine to idle for longer than 3 minutes without a 'clear out' burst of 1 minute at 2,000 rev/min.

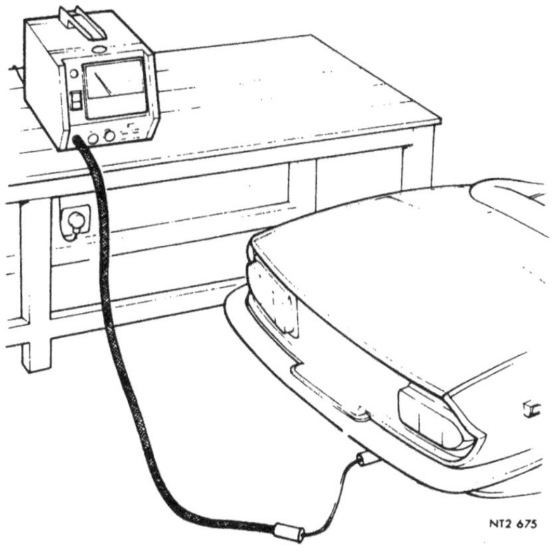

THERMOSTATIC SWITCH (FITTED TO LATER CARS ONLY)

—Remove and refit [17.20.31]

Removing

1. Drain the coolant.
2. Disconnect the two vacuum pipes at the switch.
3. Slacken the two hose clips.
4. Remove the switch complete with mounting tube.
5. Secure the tube in a vice and unscrew the switch.

Refitting

6. Reverse instructions 1 to 5, apply sealing compound to the thread of the switch.

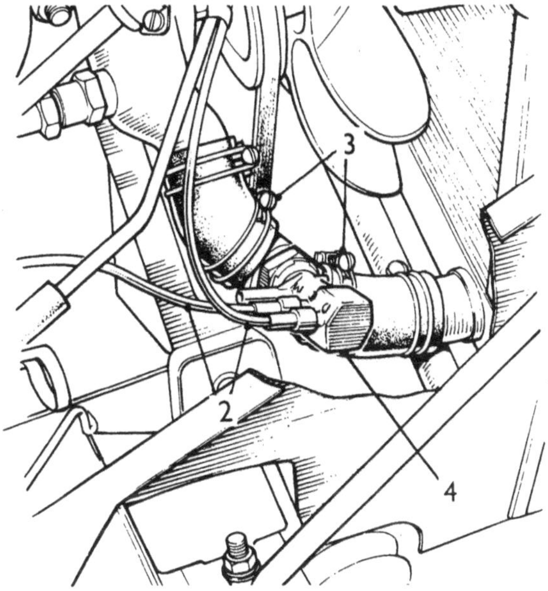

RUNNING-ON CONTROL VALVE (FROM 1973 ONLY)

—Remove and refit [17.40.01]

Removing

1. Remove the carbon canister 17.15.07.
2. Disconnect the two electrical leads to the solenoid at the top of the valve.
3. Disconnect the vacuum signal pipe.
4. Remove the securing bolts and lift off the valve complete with mounting bracket.

Refitting.

5. Reverse instructions 1 to 4.

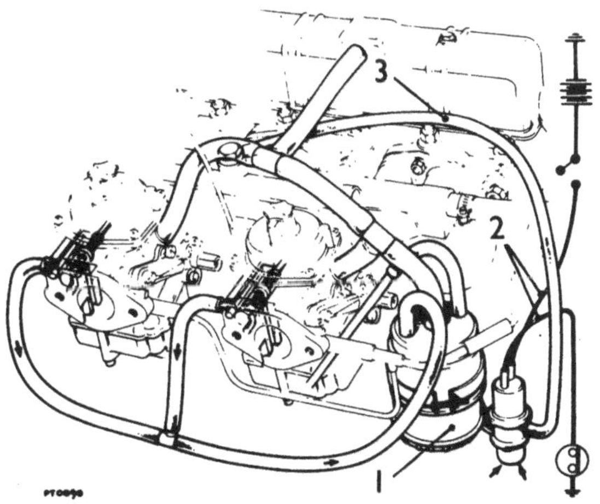

FUEL SYSTEM OPERATIONS (TWIN CARBURETTER)

***All bracketed material ([]) in this section has been added by Robert Bentley, Inc.**

AIR CLEANER

– Remove and refit 19.10.02

Removing

1. Disconnect the two intake hoses.
2. Remove the four bolts and washers.
3. Remove the fuel pipe and withdraw the air cleaner complete with two gaskets.

Refitting

4. Reverse instructions 1 to 3, ensuring that new gaskets are fitted.

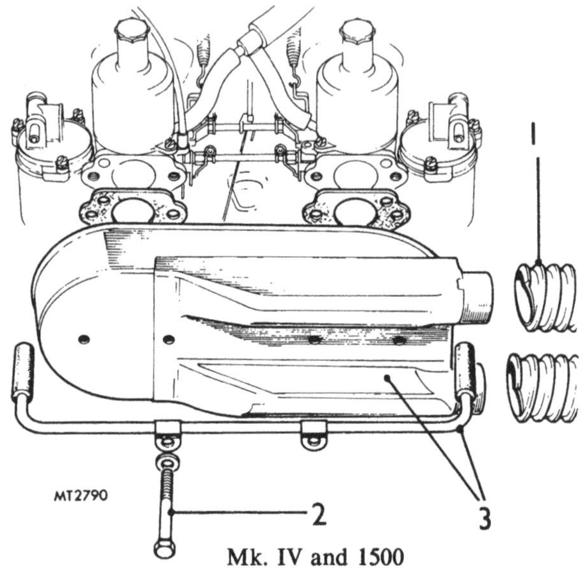

Mk. IV and 1500

[NOTE: On early Mk. III Spitfires individual air cleaners are fitted as shown in the illustration. On such engines, the fuel pipe may be lifted up in order to remove the air cleaners.]

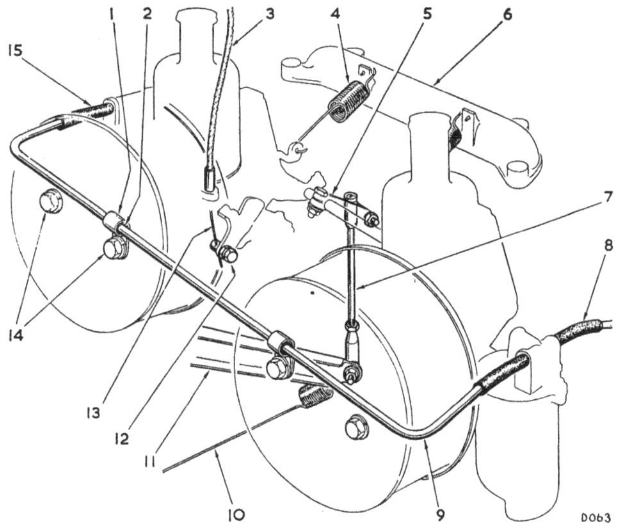

Mk. III

AIR CLEANER – RENEW ELEMENTS

– Remove and refit 19.10.09

Removing

1. Remove the air cleaner assembly 19.10.02
2. Remove the bolt securing the cover-plate to the container.
3. Separate the cover-plate from the container.
4. Remove and discard the paper elements.

Refitting

5. Reverse instructions 1 to 4, ensuring:
   a. New elements are fitted.
   b. Gaskets and cover-plate seal are renewed if originals are unsatisfactory.

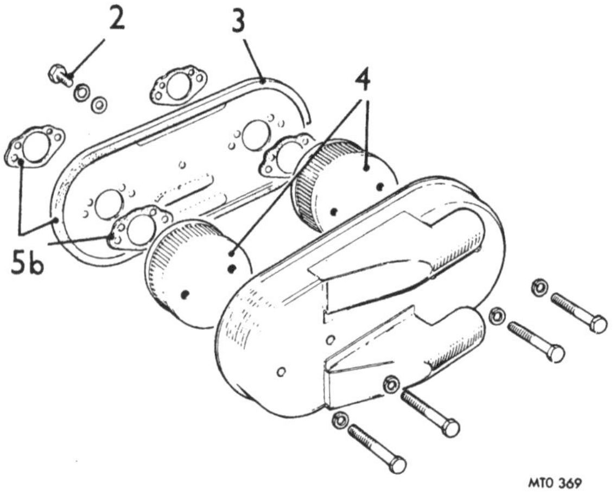

Triumph Spitfire Mk IV Manual. Part No. 545254. Issue 1.
Part No. 511243. page 1.309. Issue 1.

CARBURETTERS

– Tune and adjust **19.15.02**

Idling – check and adjust 1 to 11 and 16

Mixture – check and adjust 1 to 20

[NOTE: Much of the twin carburetor information given on this and following pages does not apply to cars sold in the U.S.A. When servicing a U.S.A. Spitfire, use only those tuning procedures which are identified as applicable to U.S. Mk. III models.]

1. Check each piston damper, and top-up if necessary with a recommended engine oil until the level is ½in (13 mm) above the top of the hollow piston rod.
2. Check the throttle action for signs of sticking, and ensure that the mixture control cable has 1/16 in (1·6 mm) free movement.
3. Remove the air cleaner. 19.10.02.
4. Connect up a suitable tachometer.
5. Start and run the engine until it reaches normal running temperature, as indicated by the temperature gauge. Continue running the engine at 2,500 rev/min for approximately one minute before commencing tuning. Repeat this operation as often as necessary.
6. Stop the engine and raise each carburetter piston with the respective lifting pins. Release the pin and check that each piston falls freely on to the carburetter bridge with a soft metallic click. Should either piston fail to fall freely, refer to 19.15.18.
7. Start the engine and check the idling speed with the tachometer, which should be 750-800 rev/min.
   a. If the reading is not correct, continue with instructions 8 to 11 and 15 and 16.
   b. If the idle speed is correct but is not consistent with smooth running, continue with instructions 8 to 21.
8. Slacken both throttle interconnection clamp nuts.
9. Slacken both jet control interconnection clamp nuts.
10. Slacken the trunnion bolts securing the mixture control inner cable.
11. With the engine running, check the carburetter balance with a meter or by comparing the intake hiss in each carburetter by holding one end of a length of small bore rubber tubing to each carburetter intake in turn and the other end to the ear. Turn the throttle adjusting screw on each carburetter until the intensity of the hiss is equal in both intakes with the correct idling speed of 750-800 rev/min as indicated by the tachometer.
    a. If with the carburetters correctly balanced the idling is still erratic, continue with instructions 12 to 20.
    b. If the idle speed is correct and consistent with smooth running, follow instructions 15 to 20.

continued

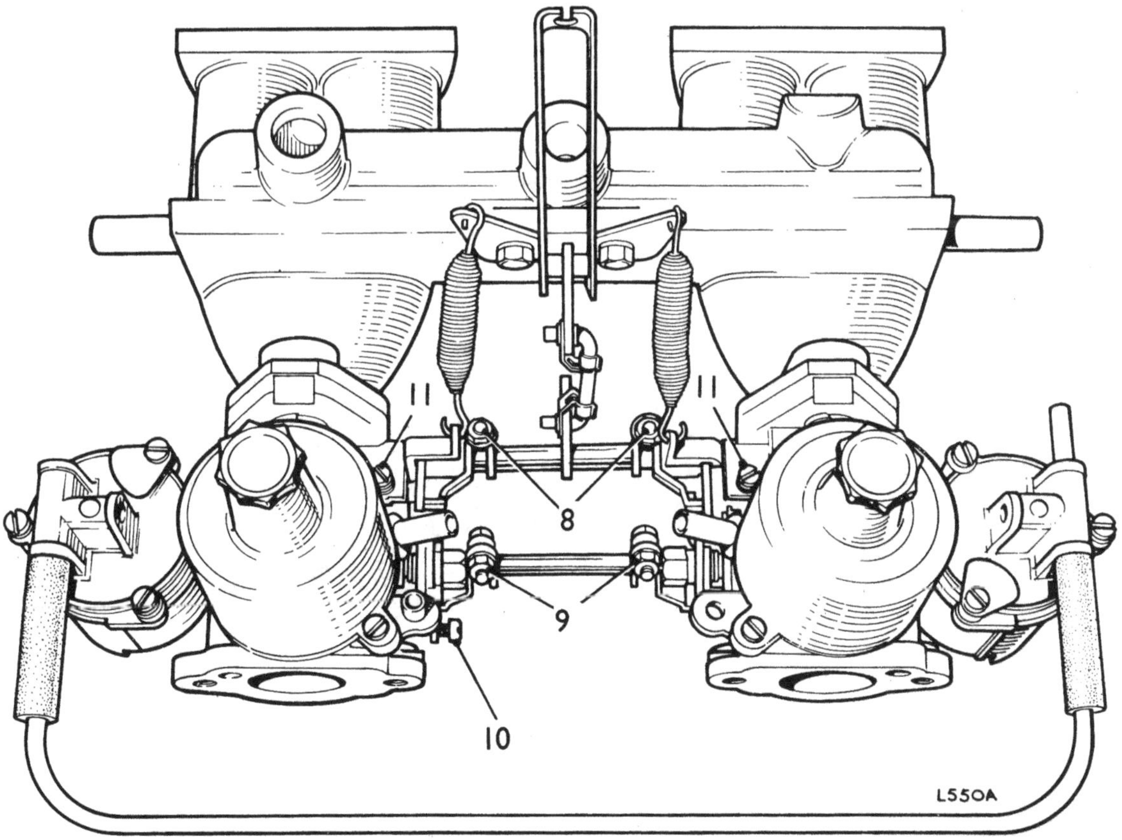

12. Tighten the jet adjusting nuts until the jet is just level with the bridge and then turn each one down two complete turns.
13. Run the engine as in instruction 5, turn the jet adjusting nut on both carburetters one flat at a time up or down until the fastest engine speed consistent with smooth running is obtained. Turn each nut down one flat at a time until the engine speed starts to decrease, and then turn each nut up one flat.
14. Re-check the idling speed, 750-800 rev/min, and adjust by turning each throttle adjusting screw an equal amount; see instruction 11.
15. Set the throttle interconnection clamping levers until the lever pins rest on the lower arm of the forks.
16. Insert feeler gauges to the value of 0·030 to 0·035 in (0·76 to 0·89 mm) between the heel of the fulcrum plate and the machined surface of the manifold and tighten the clamping nuts while holding the lever pins on the lower arm of the fork.
Following this instruction a clearance should exist between the lever pins and the bottom arm of the forks. The actual clearance is not important providing it is the same on both forks.
17. Position the choke control interconnecting rod with approximately 1/32 in (0·8 mm) end clearance, and tighten the clamp nuts.
18. Pull out the mixture control approximately ½ in (13 mm) until the linkage is just about to move the jet. Start the engine and adjust the fast idle screws to give an engine speed of 1,100 to 1,200 rev/min.
19. Refit the air cleaner. 19.10.02.
20. Disconnect the tachometer.

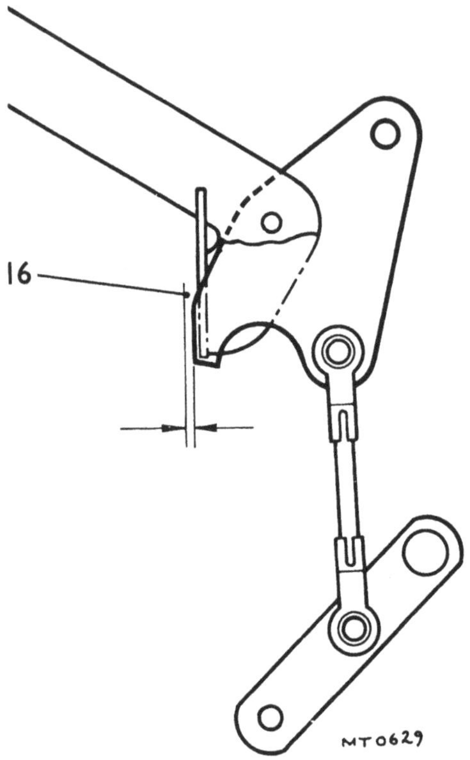

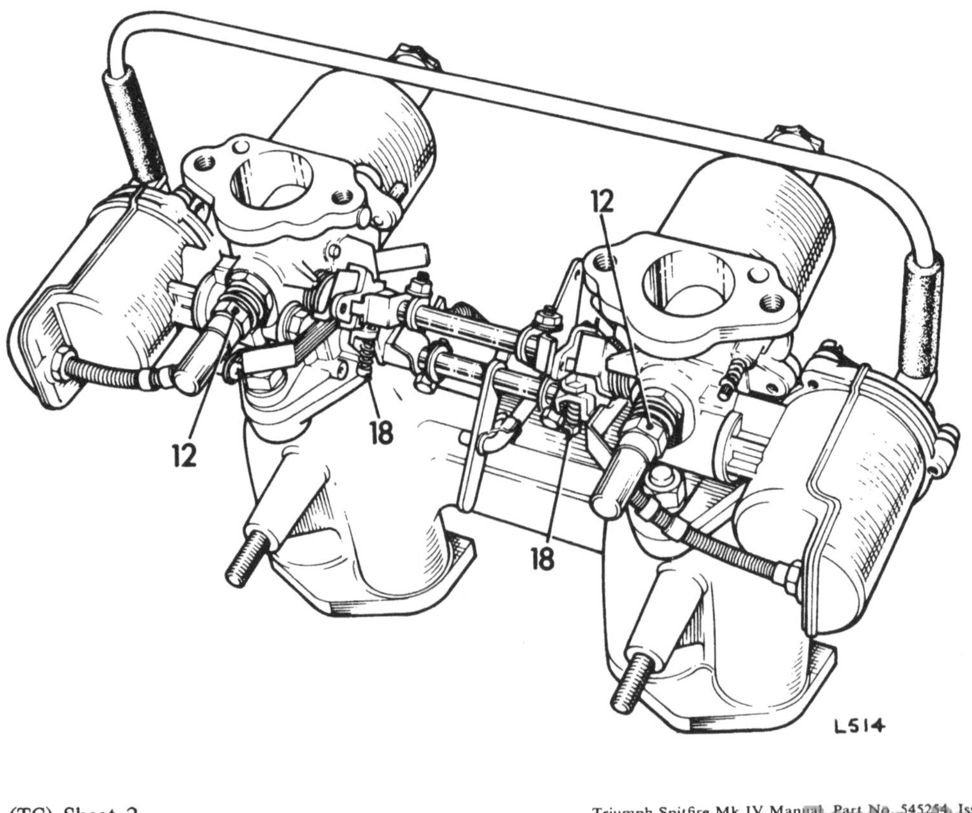

CARBURETORS

—Tune and adjust **[19.15.02 US]**

NOTE: The following information applies specifically to the emission control twin carburetors fitted to Mk. III Spitfires manufactured for the U.S. prior to January, 1970. Twin carburetor tuning and adjusting procedures which appear elsewhere in this manual should not be applied to the Mk. III models imported into the United States. Such inappropriate procedures are labeled as not applicable to cars sold in the U.S.

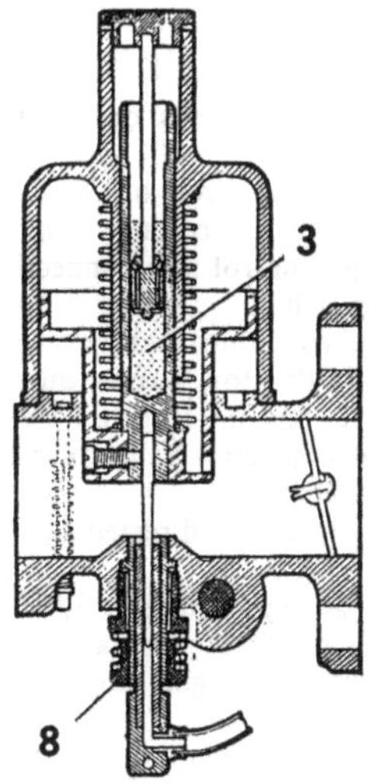

Basic tuning

1. Connect up a suitable tachometer.
2. Remove the air cleaners and gaskets.
3. Check each piston damper, and top-up if necessary with a recommended engine oil until the level is ½ in. (13 mm) above the top of the hollow piston rod.
4. Run the engine at approximately 1,000 rev/min until five minutes after normal operating temperature is reached. Then run the engine an additional one-half minute at 2,500 rev/min.
5. Commence tuning at once and repeat the half minute at 2,500 rev/min every three minutes until the tuning is completed.
6. Using a balancing meter such as the one shown in the illustration, ascertain whether the carburetors are in balance.
7. If the carburetors are out of balance, refer to "Complete tuning" operations 11 and 12.
8. If, after balancing, a satisfactory idle cannot be achieved, adjust the mixture by turning the jet adjuster nut on each carburetor, by the same amount within the limits of the restrictors, until you obtain the maximum speed consistent with smooth running.
9. Re-check the idling speed and adjust, if necessary, by altering both idling screws by the same amount. Re-check the balance.
10. With the carburetors in balance and the engine idling smoothly at 800 to 850 rev/min, check to see that the actuating pins of the interconnecting clamping levers are set 0.015 in (0.38 mm) from the lower edge of the fork and that there is 1/32 in (0.80 mm) end play between the interconnecting clamping levers and the throttle nuts. Adjust as necessary.
11. Check that when the mixture control is operated both jets commence to move simultaneously. Check that the carburetors are in balance at 1,500 rev/min. Ensure that there is 1/16 in (1.60 mm) free movement of the choke control wire before it starts to actuate the jet levers. If any of these points require attention refer to "Final adjustment."

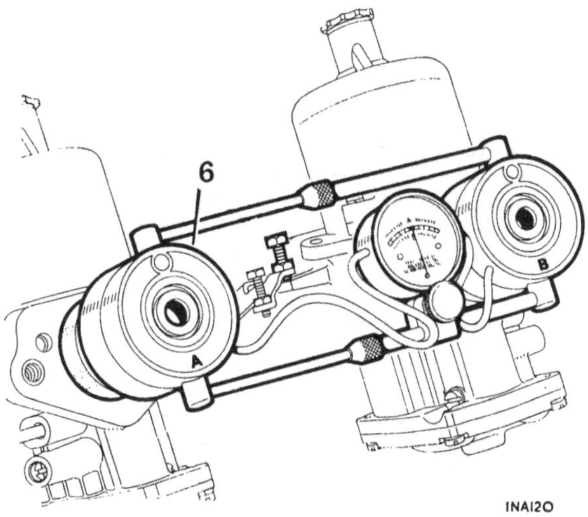

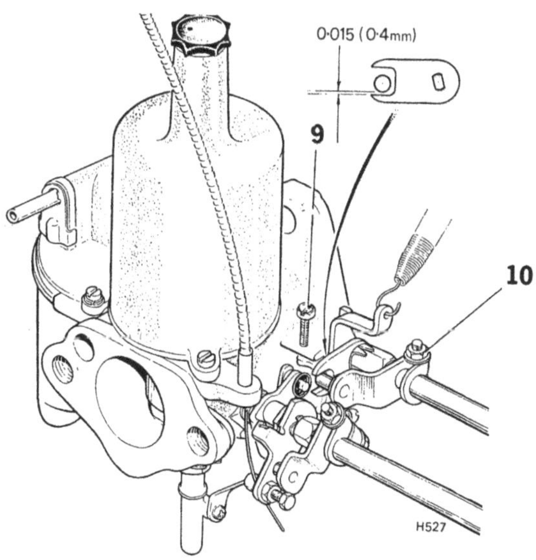

Part No. 511243. pages 1.329-30. Issue 1.
Part No. AKD 4957 (6th ed.) p. 4-17.

Complete tuning

Preparation

1. Slacken both clamping bolts on the throttle spindle interconnections.
2. Unscrew the fast idling screw on each carburetor until both screws are well clear of their cams.
3. Disconnect the jet control interconnection by slackening the clamping bolts.
4. Disconnect the choke control cable.
5. Unscrew the throttle stop screws until they are just clear of their stops and the throttles are closed.
6. Set each throttle stop screw half a turn open.
7. Top up the dampers.
8. Connect a tachometer as directed by the instrument manufacturer.
9. Run the engine at approximately 1,000 rev/min until five minutes after normal operating temperature is reached. Then run the engine an additional one-half minute at 2,500 rev/min.
10. Commence tuning at once and repeat the half minute at 2,500 rev/min every three minutes until the tuning is complete.

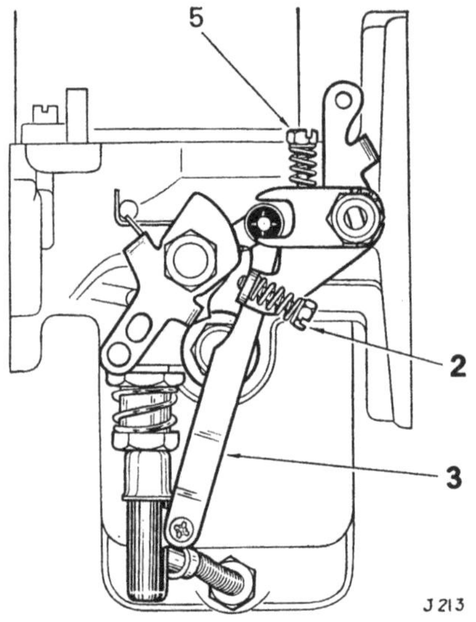

Balancing

11. Adjust each throttle stop screw by the same amount to attain an idling speed of 800 to 850 rev/min.
12. Using a balancing meter as illustrated earlier, measure the air intake of each carburetor and balance the carburetors as described in "Basic tuning". Maintain the idling speed at 800 to 850 rev/min by adjusting throttle stop screws by equal amounts.

NOTE: An inability to balance the carburetors may be due to vacuum leaks in the intake system (i.e., brake servo, inlet manifold, etc.). If no such cause can be found, refer to 19.18.18 (TC).

Mixture setting

13. Turn each jet adjusting nut by the same amount — turning them upward to weaken the mixture, downward to enrich the mixture. To ensure that the piston and needle move to their proper relationship with the jet, gently tap the neck of the suction chamber with a non-metallic tool such as a screwdriver handle after each re-adjustment of the jet adjusting nut.
14. Check that the idling speed and carburetor balance remain correct.
15. Using an exhaust gas analyser (either a CO meter or an air/fuel ratio meter), check the percentage CO reading or air/fuel ratio to see that it is within the specified limits (3.5 to 4.5% CO; air/fuel ratio approximately 13:1). If the reading falls outside these limits adjust both adjusting nuts by the minimum amount necessary to bring the reading just within the limits 17.35.01.

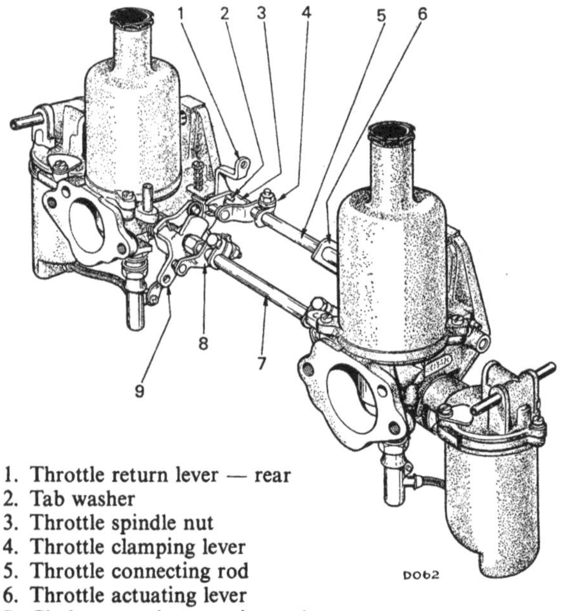

1. Throttle return lever — rear
2. Tab washer
3. Throttle spindle nut
4. Throttle clamping lever
5. Throttle connecting rod
6. Throttle actuating lever
7. Choke control connecting rod
8. Choke clamping lever
9. Choke actuating lever

Carburettor linkages

Part No. 511243. pages 1.330-331. Issue 1.

16. Hold the jet adjusting nut on each carburetor to prevent it from turning, then rotate the adjustment restrictor (a sheet metal strap) around the nut until the vertical tag contacts the carburetor body on the left-hand side when viewed from the air cleaner flange. In this position, bend down the small tag on the adjustment restrictor so that the restrictor locks to the nut and will follow its movement.

NOTE: Normally, the restrictor need only be adjusted when a new carburetor is installed or after carburetor dismantling or reassembling.

Final adjustments

17. Set the throttle interconnection clamping levers so that the actuating pins are 0.015 in (0.35 mm) away from the lower edge of the fork. See illustration under "Basic tuning". Ensure that there is 1/32 in (0.80 mm) end play between the interconnecting clamping levers and the throttle nuts.
18. With both jet levers pressed down to their lowest points, set the jet interconnection lever clamping bolts so that both jets commence to move simultaneously.
19. Run the engine at 1,500 rev/min and check the carburetor balance.
20. Reconnect the mixture control cable allowing 1/16 in (1.60 mm) free movement before it starts to actuate the jet levers.
21. Pull out the choke knob on the dash panel as far as possible without actually moving the jets. Then adjust the fast idling screws to give an engine speed of 1,100 rev/min when hot.

CARBURETTERS

– Remove and refit 19.15.11

[except early Mk. III vehicles for U.S.A.]

Removing

1. Pull off the breather pipes from the carburetters.
2. Disconnect the choke cable at the trunnion.
3. Disconnect the link rod between the cable linkage and throttle interconnection shaft lever.
4. Disconnect the two throttle return springs.
5. Disconnect the vacuum pipe.
6. Disconnect the main fuel feed pipe.
7. Remove the four carburetter flange to manifold nuts (two per carburetter) complete with washers.
8. Remove the carburetters complete with air cleaner assembly and gaskets.
9. Remove the four bolts securing the air cleaner assembly to the carburetter intake flanges. 19.10.02.
10. Remove the air cleaner assembly complete with gaskets.

Refitting

11. Reverse instructions 1 to 10, ensuring that all gaskets are renewed.

[—Remove and refit (early Mk. III vehicles for U.S.A.)

Removing

1. Remove the air cleaners.
2. Disconnect the fuel inlet pipe, choke cable, the accelerator cable and return springs.
3. Remove the four nuts and spring washers securing the carburetors to the inlet manifold.
4. Lift off the carburetors as a pair complete with interconnecting linkages.

Refitting

5. Reverse instructions 1 to 4, ensuring that all gaskets are renewed.]

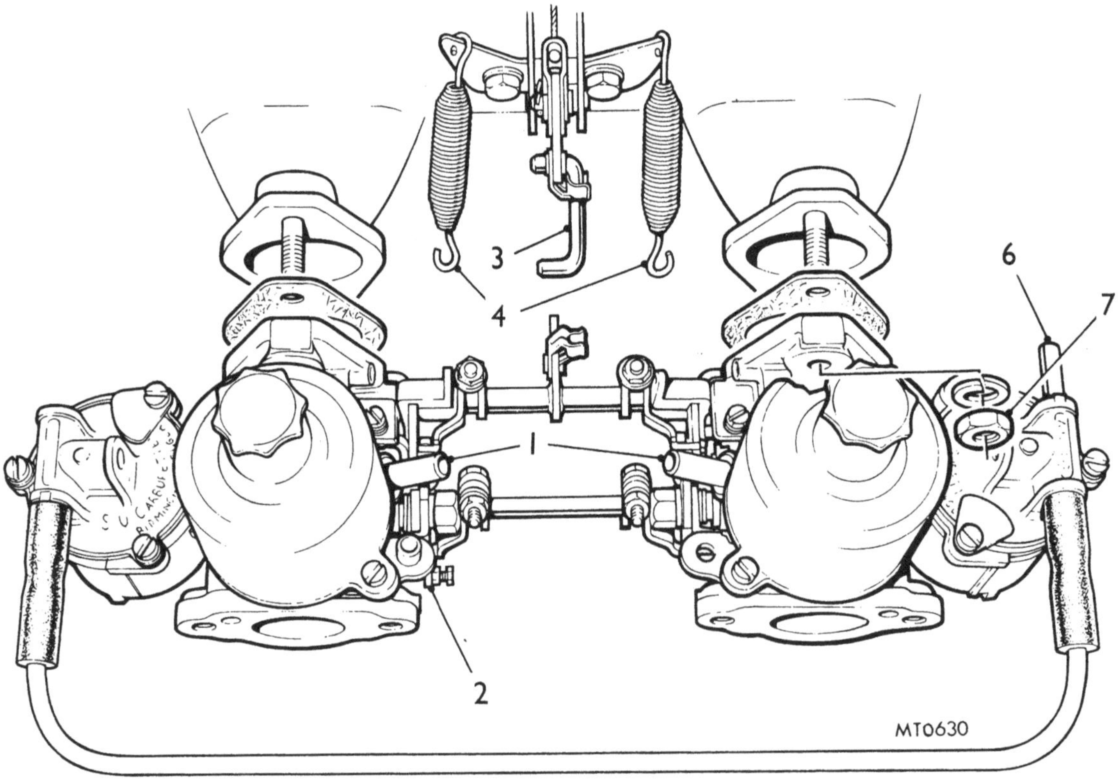

CARBURETTERS

– Overhaul and adjust 19.15.18

[CAUTION: Please read the instructions thoroughly before you attempt carburetor repairs. If you lack the necessary skills or tools we suggest you leave such work to an Authorized Dealer or other qualified shop. Before taking a removed carburetor in for repair, thoroughly clean its exterior but do not undertake any disassembly.]

Dismantling

1. Remove the carburetters 19.15.11.
2. Thoroughly clean the outside of the carburetters.
3. Mark the relative position of the suction chamber and the carburetter body.
4. Remove the damper and its washer.
5. Unscrew the suction chamber securing screws and lift off the chamber.
6. Remove the piston spring.
7. Carefully lift out the piston assembly and empty the oil from the piston rod.
8. Remove the guide locking screw and withdraw the needle assembly, taking care not to bend the needle Withdraw the needle from the guide and remove the spring from the needle.
9. Push the piston lifting pin upwards, detach its securing circlip and withdraw the pin and spring downwards.
10. Release the pick-up lever return spring from its retaining lug.
11. Support the plastic moulded base of the jet and remove the screw retaining the jet pick-up link and link bracket (when fitted).
12. Unscrew the flexible jet tube sleeve nut from the float-chamber and withdraw the jet assembly. Note the gland, washer and ferrule at the end of the jet tube.
13. Remove the jet adjusting nut and spring.
14. Unscrew the jet locking nut and detach the nut and jet bearings ; withdraw the bearing from the nut.

continued

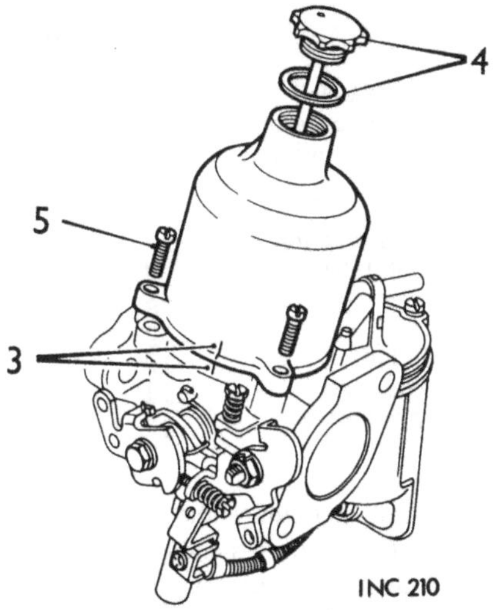

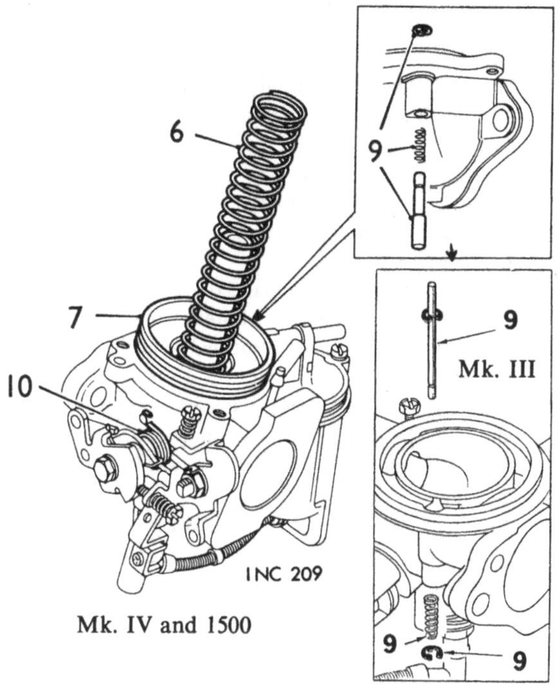

Mk. IV and 1500

Early Mk. III models for U.S.

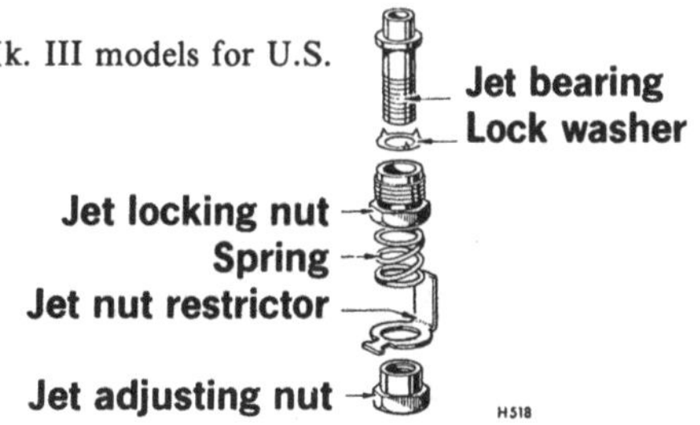

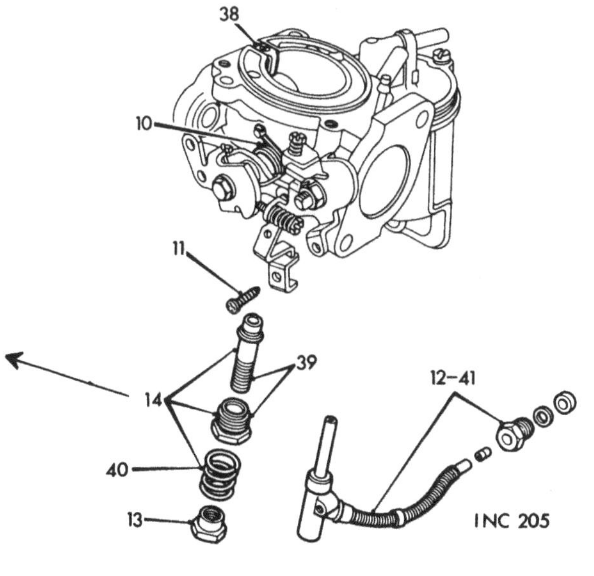

Mk. IV and 1500

Early Mk. III Model

1. Lever pivot bolt
2. Double spring washer
3. Pick-up lever spring
4. Link retainer
5. Pick-up lever
6. Lever return spring
7. Pivot bolt tube
8. Cam lever
9. Washer
10. Pick-up link
11. Ferrule spacer
12. Lever pivot screw

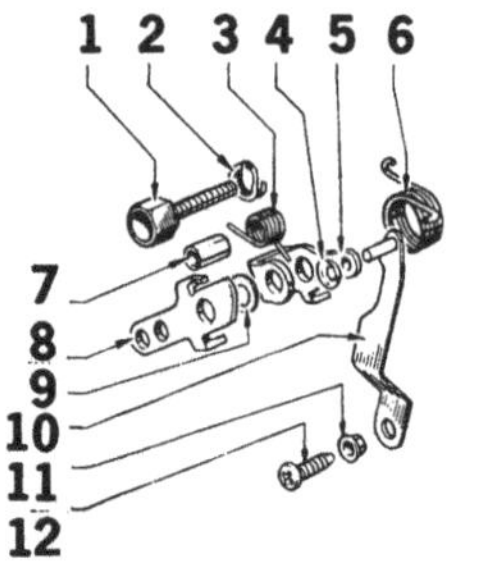

Mk. IV and 1500

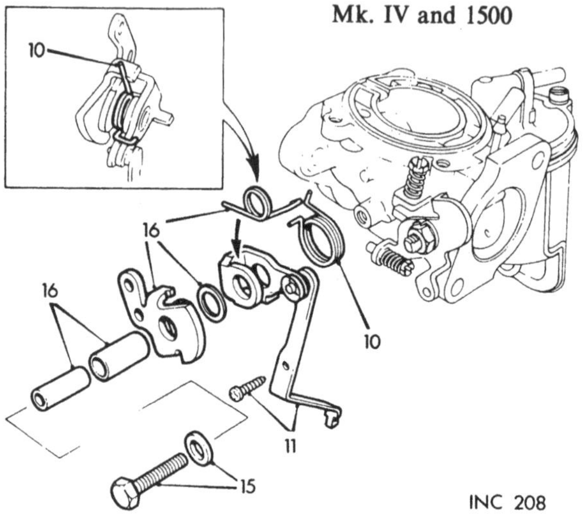

15. Unscrew and remove the lever pivot bolt and spacer.
16. Detach the lever assembly and return springs noting the pivot bolt tubes, skid washer and the location of the cam and pick-up lever springs.
17. Remove the float-chamber securing bolt and the chamber.
18. Mark the float-chamber lid location.
19. Remove the lid securing screws and detach the lid with its joint washer and float.
20. Hold the float hinge pin at its serrated end and withdraw the pin and float.
21. Extract the float needle from its seating and unscrew the seating from the lid.
22. Close the throttle and mark the relative position of the throttle disc and the carburetter flange.

CAUTION: Do not mark the throttle disc in the vicinity of the limit valve.

23. Unscrew the disc retaining screws, open the throttle and ease the disc from its slot in the throttle spindle. Store the disc in a safe place until required for assembly.
24. Tap back the tabs of the lock washer securing the spindle nut· remove the nut and detach the lever arm, washer and throttle spindle· note the location of the lever arm in relation to the spindle and carburetter body.

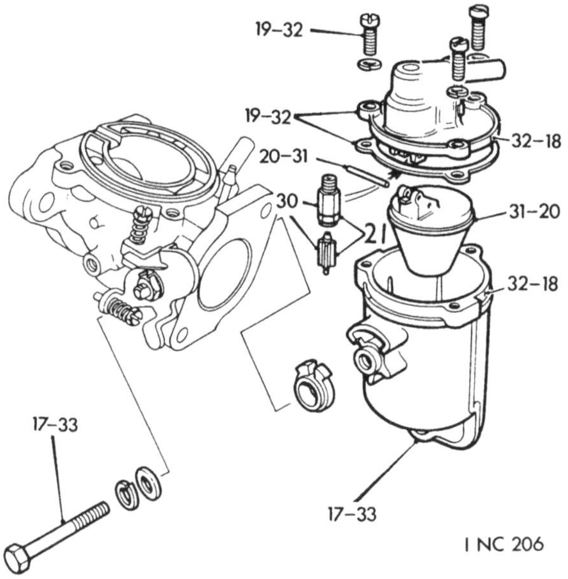

Inspecting

25. Examine the components as follows:
    - *a.* Check the throttle spindle in the body for excessive play, and renew if necessary.
    - *b.* Examine the float needle for wear, i.e. small ridges or grooves in the seat of the needle, and ensure that the spring-loaded plunger on the opposite end operates freely. Replace the needle and seating if necessary.
    - *c.* Inspect all other components for wear and damage· renew unserviceable components.

continued

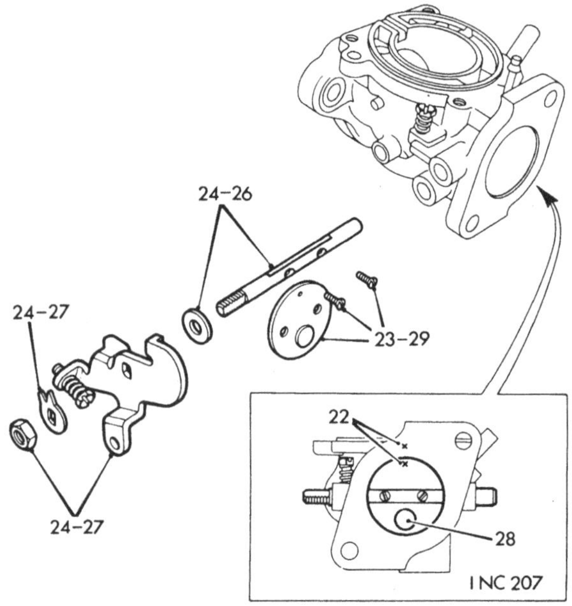

Triumph Spitfire Mk IV Manual. Part No. 545254. Issue 2
Part No. 511243. page 1.333. Issue 2.

Reassembling

26. Refit the spindle to the body, with the countersunk holes in the spindle facing outwards.
27. Assemble the spacing washer, lever, lock washer and securing nut, ensure that the idling stop on the lever is against the idling screw abutment on the body in the closed throttle position. Tighten the spindle nut and lock with the tab washer.
28. Insert the throttle disc into the spindle slot note the markings for reassembling, i.e. the limit valve positioned at the bottom of the disc with the head of the valve towards the engine. Manoeuvre the disc in the spindle until the throttle can be closed, snap the throttle open and closed to centralize it in the bore of the carburetter.
29. Fit new disc retaining screws but do not fully tighten; check that the disc closes fully and adjust its position as necessary. Tighten the screws fully and spread their split ends just enough to prevent them turning.
30. Screw the seating into the float-chamber; do not overtighten. Insert the needle coned-end first into the seating.
31. Refit the float to the chamber lid and insert the hinge pin.
32. Refit the float-chamber lid with a new joint washer, noting the assembly markings, tighten the securing screws evenly.
33. Refit the float-chamber to the body and tighten the retaining bolt.
34. Refit the piston lifting pin, spring and circlip.
35. Clean fuel deposits off the suction chamber and piston with fuel or methylated spirit and wipe dry.

CAUTION: Do not use abrasives.

continued

Mk. IV and 1500

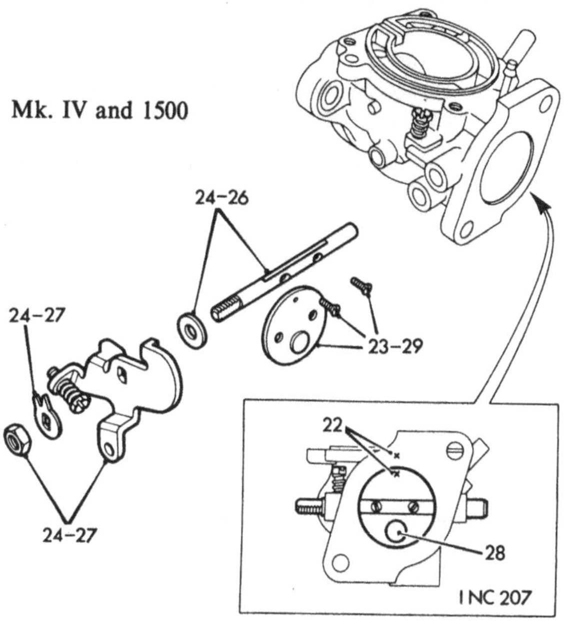

Early Mk. III model

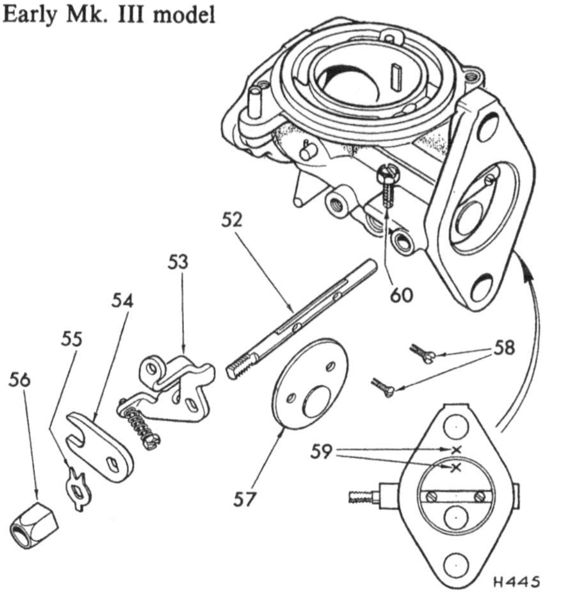

52 Throttle spindle
53 Throttle lever
54 Fork lever
55 Tab washer
56 Lever retaining nut
57 Throttle disc assembly
58 Throttle disc screws
59 Marks for re-assembly
60 Throttle adjusting screw

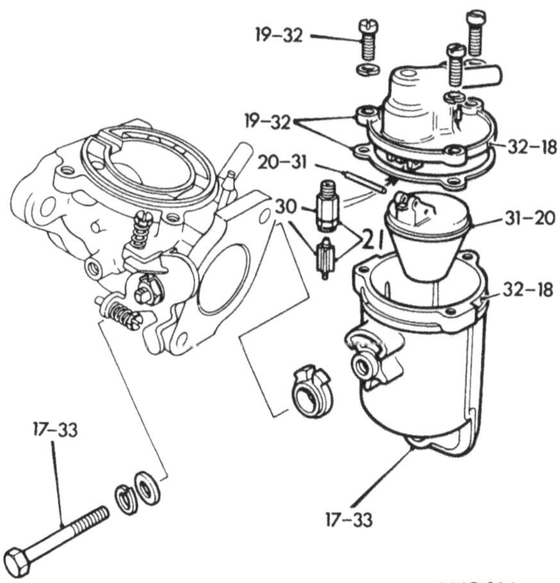

I NC 206

36. Check the operation of the suction chamber and piston (without the spring fitted) as follows:
   a. Refit the damper and washer to the suction chamber, temporarily plug the piston transfer holes with rubber plugs or Plasticine and insert the piston fully into the suction chamber.
   b. Secure a large flat washer to one of the fixing holes with a screw and nut so that it overlaps the bore.
   c. With the assembly upside-down, hold the piston and check the time taken for the suction chamber to fall the full extent of its travel. The time taken should be five to seven seconds, if this time is exceeded, check the piston and chamber for cleanliness and mechanical damage. Renew the assembly if the time taken is still not within these limits.

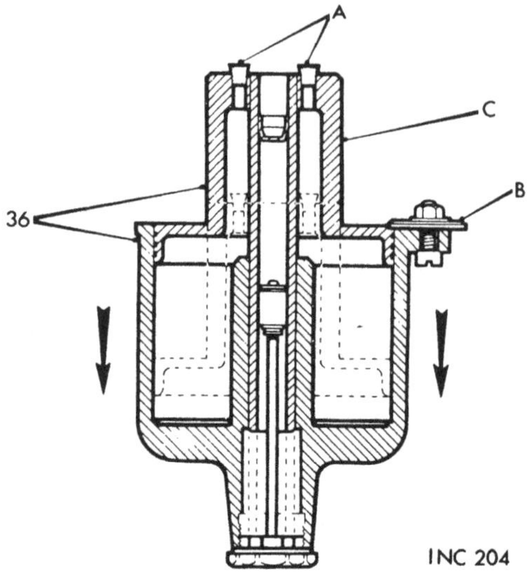

INC 204

[**NOTE:** The piston used in early Mk. III Spitfire carburetors, which has only one piston transfer hole, should fall the full extent of its travel in three to five seconds. In making the test described here, there should be no oil in or on any of the parts.]

37. Fit the spring and guide to the needle and insert the assembly into the piston, ensuring:
   a. The lower edge of the guide is flush with the face of the piston.
   b. The guide is positioned so that the etched locating mark on its lower face is adjacent to, and in line with, the centre line between the two piston transfer holes.
   c. A new guide locking screw is fitted.

NOTE: Alternative needle guides may be fitted which have a flat machined on the guide and must be positioned so that the guide locking screw tightens down on to the flat. If the guide is incorrectly positioned, the locking screw will not tighten down on the flat and will remain proud of the piston resulting in damage to the piston bore.

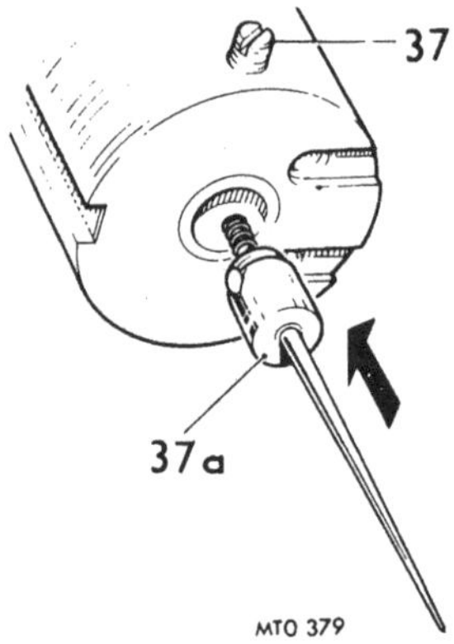

MTO 379

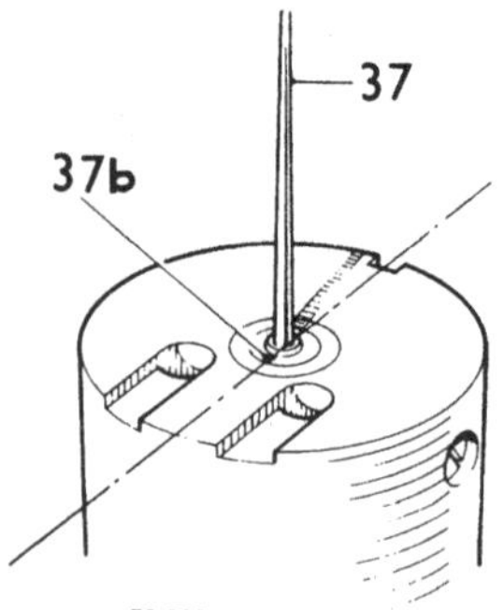

MTO 380

continued

38. Check the piston key in the body for security.
39. Refit the jet bearing, fit and tighten the jet locking nut.
40. Refit the spring and jet adjustment nut, screw the nut up as far as possible.
41. Insert the jet into the bearing, fit the brass sleeve nut, washer and gland to the end of the flexible tube (if removed). The tube must project a minimum of 3/16 in (4·8 mm) beyond the gland. Tighten the brass sleeve nut until the gland is compressed; overtightening can cause leakage.

[NOTE: In assembling the jet bearing — especially for the early Mk. III model for the U.S. — please refer to the illustration given in the dismantling procedure.]

42. Refit the piston, spring and suction chamber to the body (noting the assembly marks) and tighten the securing screws evenly.
43. Reverse the instructions 15 to 14.
44. Hold up the choke lever to relieve pressure on the jet pick-up link, refit the link bracket (when fitted). Support the end of the moulded jet and tighten the securing screw.
45. Screw the jet adjusting nut down two complete turns (12 flats) to provide the initial setting.
46. Refit the carburetters 19.15.11.
47. Tune the carburetters 19.15.02.

FLOAT-CHAMBER NEEDLE AND SEAT

– Remove and refit 19.15.24

Removing

1. Disconnect the fuel hose from the float-chamber.
2. Mark the lid and float-chamber for assembly.
3. Remove the lid securing screws and detach the lid.
4. Hold the float hinge pin at its serrated end and withdraw the pin and float.
5. Extract the float needle from its seating.

Inspecting

6. Examine the float needle for wear, i.e. small ridges or grooves in the seat of the needle, also check that the spring-loaded plunger on the opposite end operates freely. Renew the needle and seating if the needle is worn.

[NOTE: On early Mk. III Spitfires the float has a metal pivot. If necessary, the metal pivot should be carefully bent to obtain the clearance indicated in the illustration.]

Refitting

7. Clean any sediment from the float-chamber, and fit a new joint washer if required.
8. Reverse instructions 1 to 5.

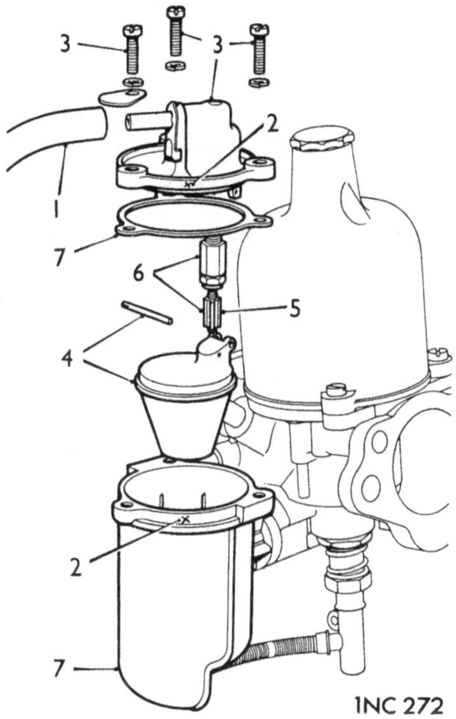

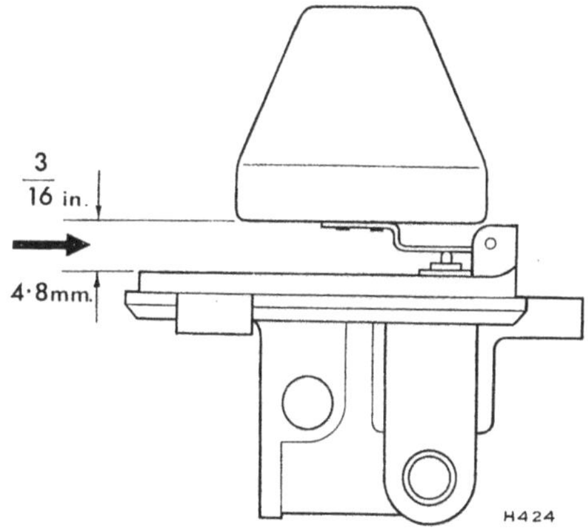

Early Mk. III float height

PISTON AND SUCTION CHAMBER

– Remove, clean and refit **19.15.30**

Removing

1. Mark the relative position of the suction chamber and the carburetter body.
2. Remove the damper and its washer.
3. Unscrew the suction chamber securing screws and lift off the chamber.
4. Remove the piston spring.
5. Carefully lift out the piston assembly and empty the oil from the piston rod.
6. Clean fuel deposits off the suction chamber and piston with fuel or methylated spirit and wipe dry.

CAUTION: **Do not use abrasives.**

7. Check the operation of the suction chamber and piston (without the spring fitted) as follows:
   - *a.* Refit the damper and washer to the suction chamber, temporarily plug the piston transfer holes with rubber plugs or Plasticine and insert the piston fully into the suction chamber.
   - *b.* Secure a large flat washer to one of the fixing holes with a screw and nut so that it overlaps the bore.
   - *c.* With the assembly upside-down, hold the piston and check the time taken for the suction chamber to fall the full extent of its travel. The time taken should be five to seven seconds, if this time is exceeded, check the piston and chamber for cleanliness and mechanical damage. Renew the assembly if the time taken is still not within these limits.

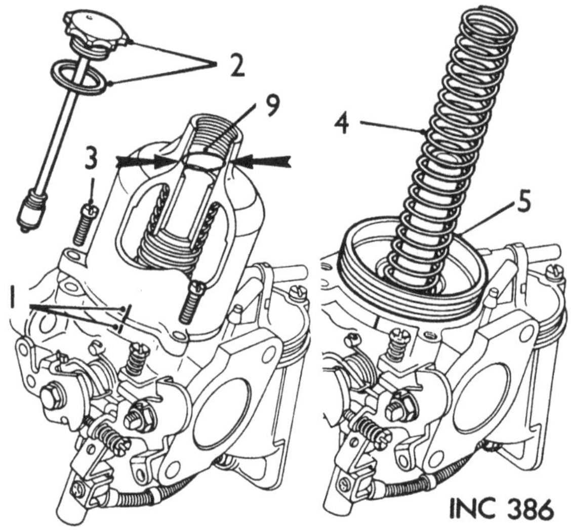

INC 386

Refitting

8. Refit the piston, spring and suction chamber to the carburetter (noting the assembly marks) and tighten the screws evenly.
9. Top up each piston damper with a recommended engine oil until the level is ½ in (13 mm) above the top of the hollow piston rod.
10. Refit each piston damper with its washer.

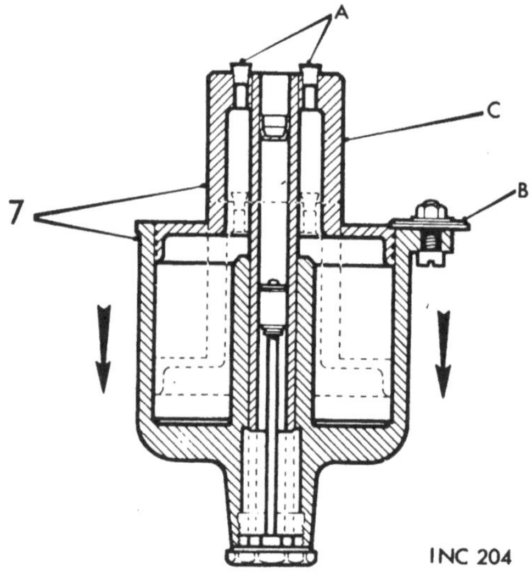

INC 204

THROTTLE PEDAL [except early Mk. III]
– Remove and refit 19.20.01

Removing

1. Remove the cable retaining clip.
2. Lift out the exposed inner throttle cable from the fork end of the pedal.
3. Remove the two bolts complete with plain and spring washers securing the pedal bracket to the bulkhead.
4. Remove the bracket.
5. Remove the nut, bolt and plain washer securing the pedal to the bracket.
6. Remove the pedal and return spring.

THROTTLE CABLE [except early Mk. III]
– Remove and refit 19.20.06

Removing

1. Remove the cable retaining clip from the throttle pedal.
2. Withdraw the exposed end of the inner cable from the fork end of the throttle pedal.
3. Pull the outer cable complete with inner cable from the bulkhead.
4. Remove the cotter pin from the linkage clevis pin.
5. Withdraw the clevis pin complete with washer.
6. Slacken the cable adjuster locknut.
7. Screw the lower adjuster nut off the end of the cable.
8. Remove the inner and outer cable complete through the slot in the adjuster bracket.

Refitting

9. Reverse instructions 1 to 8, ensuring:
   a. The cable clip on the pedal is correctly fitted so that the cable will not slip out of the fork.
   b. The cable is not kinked or subjected to sharp bends.
   c. Cable tension is adjusted so that the heel of the fulcrum plate is just touching the machined face of the manifold.

Refitting

7. Reverse instructions 1 to 6, ensuring:
   a. The return spring is positioned so that the pedal returns positively.
   b. The cable clip is correctly positioned so that the inner cable is gripped and cannot slip out.

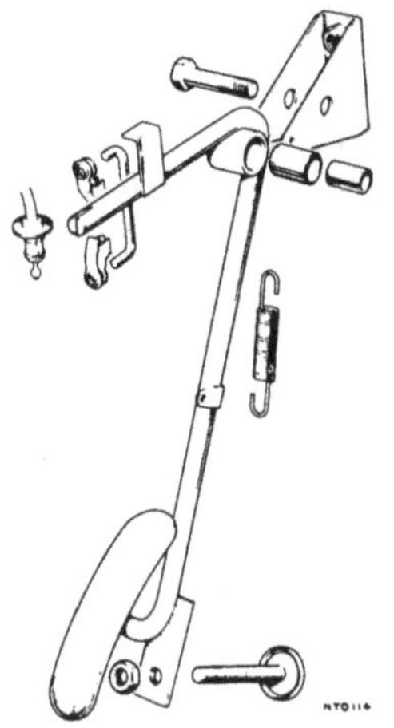

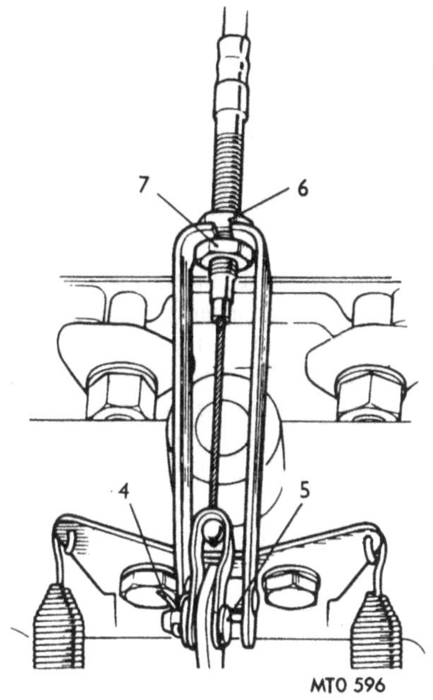

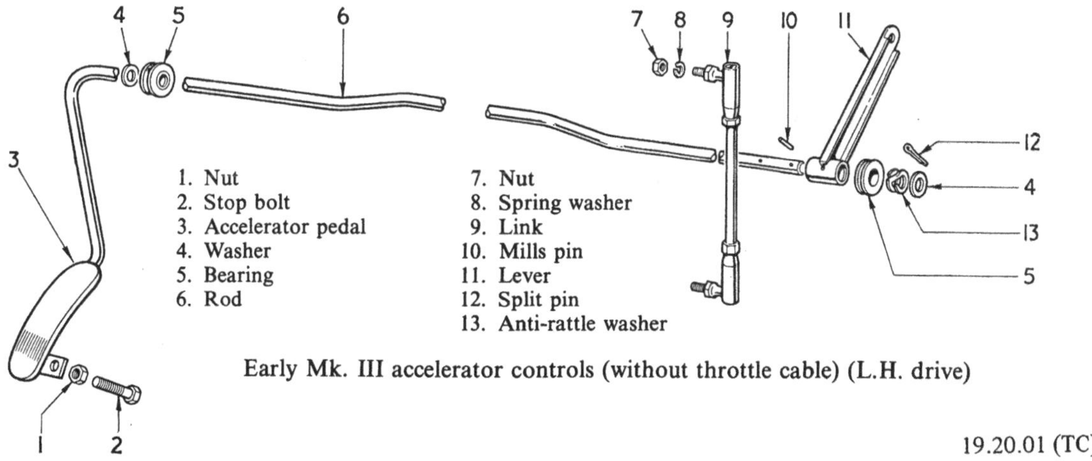

1. Nut
2. Stop bolt
3. Accelerator pedal
4. Washer
5. Bearing
6. Rod
7. Nut
8. Spring washer
9. Link
10. Mills pin
11. Lever
12. Split pin
13. Anti-rattle washer

Early Mk. III accelerator controls (without throttle cable) (L.H. drive)

THROTTLE LINKAGE [except early Mk. III]

– Remove and refit 19.20.07

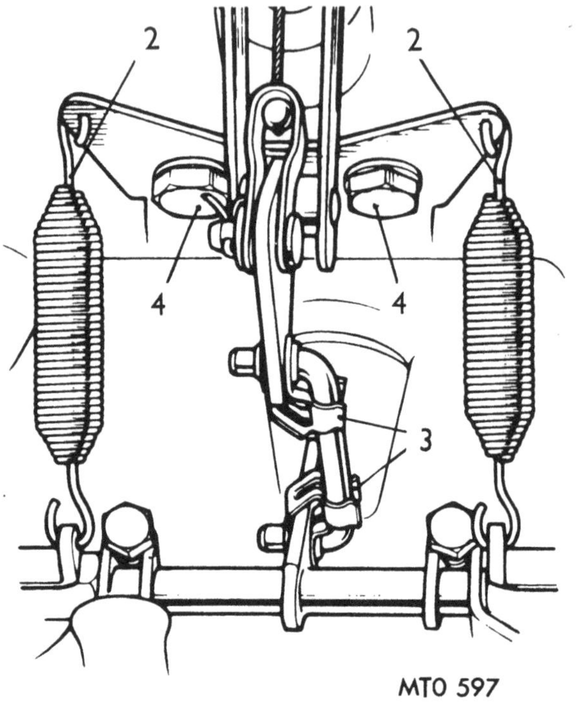

MTO 597

Removing

1. Disconnect the throttle cable from the linkage. 19.20.06, instructions 4 to 8.
2. Disconnect the two return springs from the bracket attached to the manifold.
3. Release the clips securing the link rod between the throttle interconnection lever and fulcrum plate and remove the rod.
4. Remove the two bolts securing the linkage bracket to the manifold and remove the bracket complete.

Refitting

5. Reverse instructions 1 to 4.

MIXTURE CONTROL CABLE

– Remove and refit 19.20.13

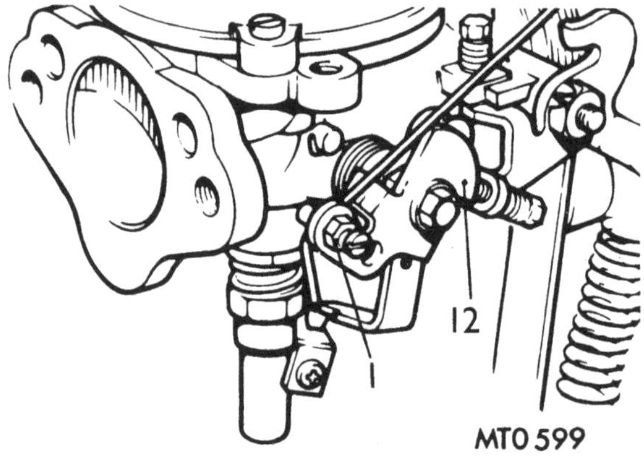

MTO 599

Removing

1. Disconnect the inner cable from the carburetter.
2. Separate the inner cable from the outer by pulling the control knob on the control panel.
3. Carefully unscrew and remove the ferrule securing the outer cable to the control panel.
4. Remove the outer cable from the control panel.
5. Drop the outer cable below the control panel and pull the cable through the bulkhead grommets into the interior of the car, and remove.

Refitting

6. Feed the control panel end of the outer cable below the panel and through the bulkhead grommets.
7. Secure the outer cable to the control panel and with the ferrule and locknut.
8. Feed the inner cable into the outer cable from the interior of the car.
9. Connect the inner cable to the carburetter.
10. Allow 1/16 in (1·6 mm) free movement of cable before the linkage is about to move the cam.

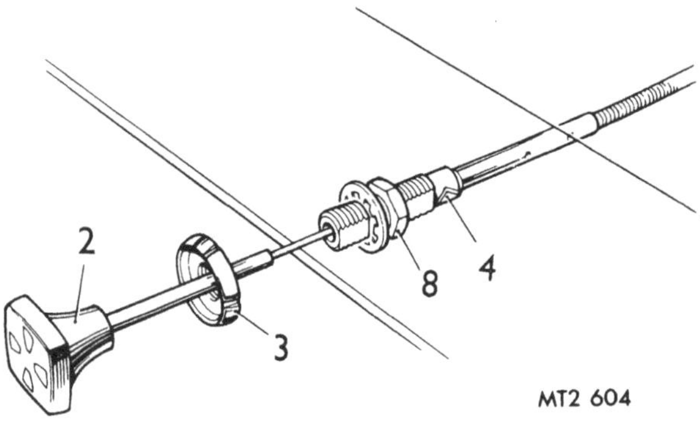

MT2 604

FUEL MAIN FILTER (specified markets only)

– Remove and refit 19.25.02

Removing

1. Open the luggage compartment lid.

[NOTE: On some earlier Mk. IV models, and on most Mk. III Spitfires, the filter (where specified) is positioned near the fuel pump in the engine compartment and is secured by a bolt on the bracket. When replacing the filter ensure that it is inserted into the piping with the side marked "IN" correctly located.]

2. Remove three top screws securing the tank trim pad to the body.
3. Ease forward the pad.
4. Disconnect the fuel feed pipes.

NOTE: Plug the pipe ends to prevent loss of fuel and ingress of dirt.

5. Withdraw filter.

Refitting

6. Reverse instructions 1 to 5, ensuring the filter is inserted into piping with the side marked 'IN' uppermost.

NT2009

FUEL PIPE –TANK END

– Remove and refit 19.40.02

Removing

1. Disconnect the pipe at the tank end and blank off the connector to prevent syphoning of fuel.
2. Disconnect the pipe at the pump end.
3. Pull the pipe away from the clips.

Refitting

4. Reverse instructions 1 to 3, ensuring that the pipe does not chafe or foul the body or other components.

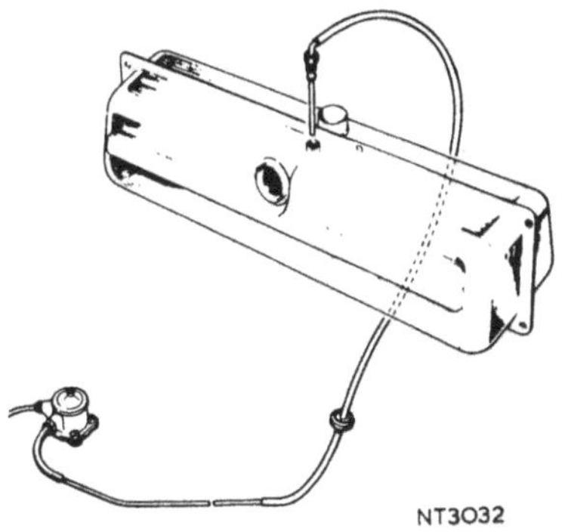
NT3032

FUEL PIPE – ENGINE END

– Remove and refit 19.40.04

Removing

1. Unscrew the union from the fuel pump.
2. Remove the bolt securing the pipe retaining clip from the thermostat housing.
3. Disconnect the pipe from the rubber connector at the carburetter end.
4. Detach the clip securing the vacuum pipe to the fuel pipe.
5. Remove the fuel pipe.

Refitting

6. Reverse instructions 1 to 5.

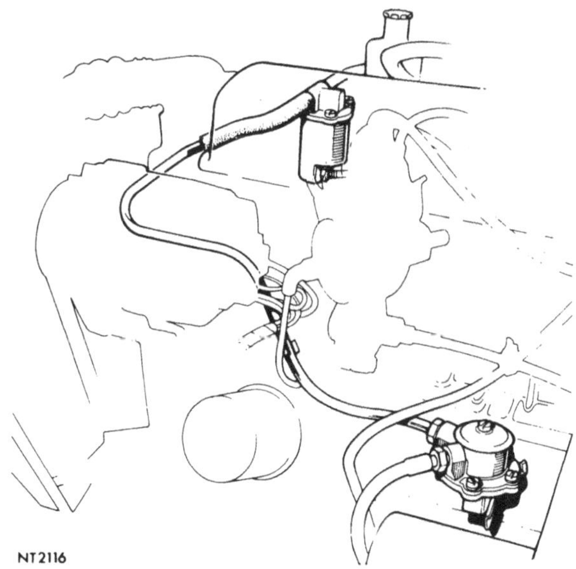
NT2116

Triumph Spitfire Mk IV Manual. Part No. 545254. Issue 1.

HOSE – FILLER TO TANK

–Remove and refit **19.40.19**

Removing

1. Remove the rear squab trim pad – six screws.
2. Slacken the two clips securing the hose to the tank and filler cap.
3. Remove the filler cap.
4. Detach the hose.

Refitting

5. Reverse instructions 1 to 4.

FUEL PUMP

– Test on vehicle **19.45.01**

1. Connect a pressure gauge into the pump to carburetter fuel line.
2. Start the engine, and observe the pressure which should be 2·5 to 3·5 to lbf/in^2 (0·029 to 0·040 kgf m.)
3. Remove the pressure gauge.

NOTE: Where pressure is high, it may be reduced by fitging extra paper washers between the pump and cylinder block. Where pressure is too low, overhaul the pump, 19.45.15, or renew, 19.45.08.

FUEL PUMP

– Clean filter **19.45.05**

1. Remove the screw in the top of the pump.
2. Lift off the dome cover.
3. Lift off the cover sealing ring.
4. Lift out the filter.
5. Use clean lint-free cloth and soak up the fuel in the pump.
6. Clean sediment from the pump and blow out the filter.
7. Examine the dome seal, renew if necessary.
8. Fit the filter and dome, tighten the retaining screw.
9. Reverse instructions 1 to 4.

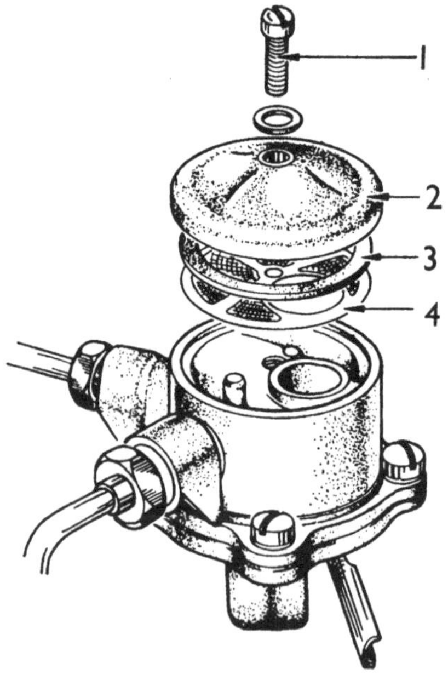

H 002

FUEL PUMP

– Remove and refit 19.45.08

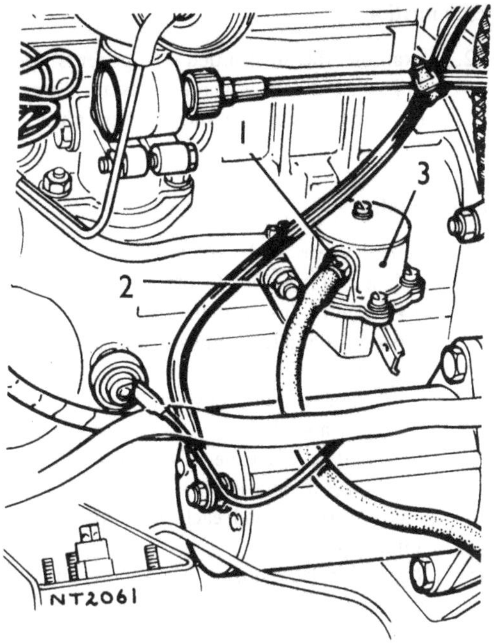

Removing

1. Unscrew the pipe unions from the pump, plug the pipes and pump to prevent dirt entering the system.
2. Remove one plain and one special nut securing the pump to the cylinder block.
3. Remove the pump and gasket.

Refitting

4. Reverse instructions 1 to 3.

FUEL PUMP

– Overhaul 19.45.15

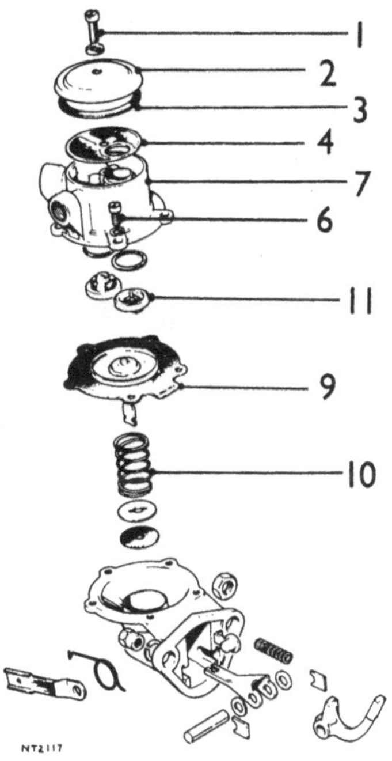

[**NOTE:** This operation does not apply to the alternative pump installed beginning with the 1973 models. Though filter cleaning is possible, overhaul of this alternative fuel pump is not recommended.]

Dismantling

1. Remove screw complete with plain washer securing dome cover to upper body.
2. Remove dome cover.
3. Remove sealing ring.
4. Lift out filter.
5. Mark for reassembly the relationship between upper and lower body of pump.
6. Remove five screws complete with spring washers securing upper body to lower body.
7. Lift off upper part of body.
8. Mark for reassembly the relationship of diaphragm to lower body.
9. Turn diaphragm clockwise or anti-clockwise until it can be released from lower body.
10. Remove diaphragm spring from diaphragm assembly.
11. Using a screwdriver, prise out the valves, and remove the gaskets.

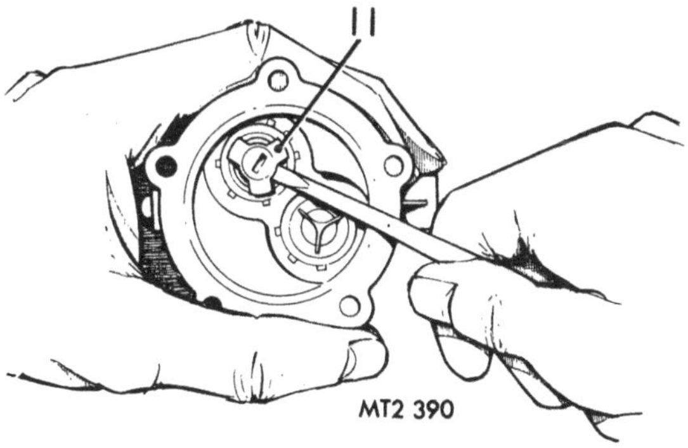

Examination

12. Examine the dismantled components for wear and damage and renew where necessary. Since the valves are damaged during removal, these must be renewed.

Reassembling

13. Renew valves and gaskets. Press valves into casting with a suitable tool (a piece of steel tubing 9/16 in (14·28 mm) inside diameter and ¾ in (19·05 mm) outside diameter. Ensure valves are positioned correctly, i.e.
    a. Inlet to pump valve, press in so raised side faces downwards.
    b. Outlet to engine valve, press in so concave side faces downwards.
14. Stake casting round each valve in six places with a suitable punch.
15. Reverse instructions 1 to 10, renewing all seals and washers.

FUEL TANK

– Remove and refit **19.55.01**

Removing

1. Isolate the battery and extinguish all naked lights.
2. Siphon the fuel out of the tank.
3. Disconnect the two Lucar connectors, noting positions for reconnection.
4. Remove the rear squab trim pad – six screws.
5. Slacken the clip securing the filler hose to the tank.
6. Disconnect the separator tank from the piping (U.S.A. models only).
7. Unscrew the nut securing the breather pipe to the tank.
8. Support the tank and remove the five bolts and plain washers.

Refitting

9. Reverse instructions 1 to 8.

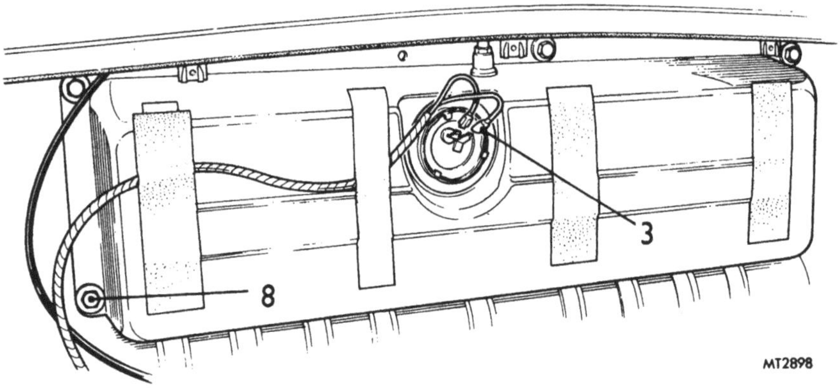

FUEL SYSTEM OPERATIONS (SINGLE CARBURETTER)

*All operations marked thus are detailed in the twin carburetter section.

EMISSION CARBURETTER – FAULT FINDING

NOTE: Before undertaking extensive carburetter servicing it is recommended that other engine factors and components such as cylinder compressions, valve clearance, distributor, sparking plugs, air intake temperature control system, etc., are checked for correctness of operation.

SYMPTOM	CAUSE	ACTION
1. Poor idle quality	*a.* Air leakage on induction manifold joints.	Remake joints as necessary. Check idle carbon monoxide level with CO meter.
	b. Throttles not synchronized.	Re-balance carburetters and re-set linkage.
	c. Air valve or valves sticking in piston guide-rods.	Clean air valve rods and guides and reassemble. Check piston free movement by hand; unit should move freely and return to carburetter bridge with an audible click.
	d. Partially or fully obstructed float-chamber or diaphragm ventilation holes.	Check that gasket(s) are not causing obstruction or piping obstructed.
	e. Incorrect fuel level caused by maladjusted float assemblies or worn or dirty needle valve.	Reset float heights and clean or replace needle valves worn.
	f. Metering needle incorrectly fitted or wrong type of needle fitted.	Ensure shoulder of needle is flush with face of air valve and that needle bias is correct.
	g. Diaphragm incorrectly located or damaged.	Check location with air valve cover removed, piston depression holes should be in line with and face towards the throttle spindle. Renew diaphragm with correct type if damage is in evidence.
	h. Leakage from ignition advance or retard pipe connections.	Re-make connections and re-check ignition settings.
	i. Temperature compensator faulty.	With engine and carburetter cold, check that compensator cone is seated, and free to move off seat. If any doubt exists, replace unit with new assembly.
	j. After considerable service leakage may occur at throttle spindle or secondary throttle spindles	Replace spindle seals or spindles as required.

continued

SYMPTOM	CAUSE	ACTION
2. Hesitation or 'flat spot' *a, b, c, d, e, f, g* and *h* plus:	Piston damper inoperative	Check damper oil level and top up with specified oil; re-check damper operation by raising piston by hand, whereupon resistance should be felt.
	Air valve spring missing or wrong part fitted	Check correct grade of spring and refit as required.
	Ignition timing incorrect	Check and reset as required.
	Throttle linkage operation incorect	Check operation of linkage between carburetters and operation of secondary throttle links; reset or replace parts as required.
3. Heavy fuel consumption 1 and 2 plus:	Leakage from the fuel connections, float-chamber joints or sealing plug 'O' rings	Replace gaskets and 'O' rings as required.
4. Lack of engine braking	Faulty by-pass valve	Replace by-pass valve with new unit.
	Sticking throttles	Check throttle operation and reset as required.
	Ignition retard inoperative	Check ignition setting at idle and ensure correct functioning of retard system.
5. Lack of engine power	Damaged diaphragm	Inspect, and replace if incorrectly fitted or damaged.
	Low fuel flow	Check discharge from fuel pump. Inspect needle valve seating.

NOTE: To ensure compliance with exhaust emission legislative requirements the following items MUST NOT be changed or modified in any way:

The fuel jet assembly; the air valve; the depression cover; the position of the fuel metering needle.

The following items must not be adjusted in service but should be replaced completely by factory-set units.

The temperature compensator; the air valve return spring; the by-pass unit; the starter assembly.

AIR CLEANER

– Remove and refit	19.10.01
– Renew element	19.10.08

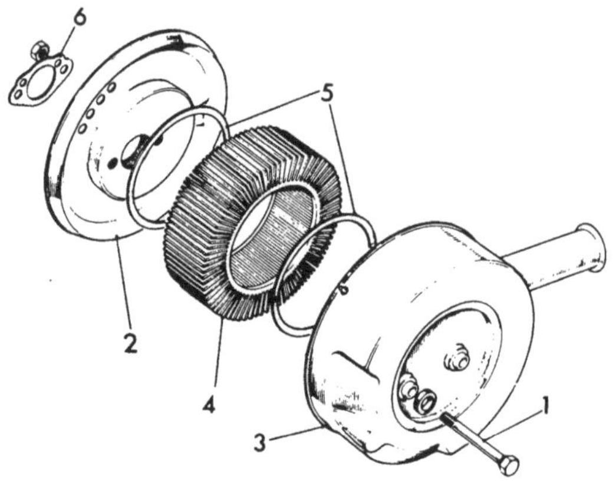

MTO 563

Removing

1. Remove two bolts, plain and rubber washers, securing the air cleaner to the carburetter.
2. Lift off the air cleaner.
3. Separate the components.
4. Remove the gasket from the carburetter.

Refitting

5. Clean or renew the element if required and fit new gasket.
6. Reverse instructions 1 to 3. Do not overtighten air cleaner securing bolts.

CARBURETTER

– Tune and adjust **19.15.02**

Service Tool: S353

NOTE: The following operation assumes all adjustments to be incorrectly set.
Omit directions not relavent.

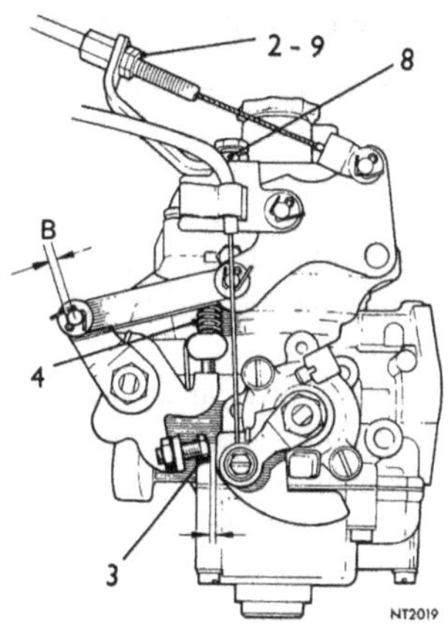

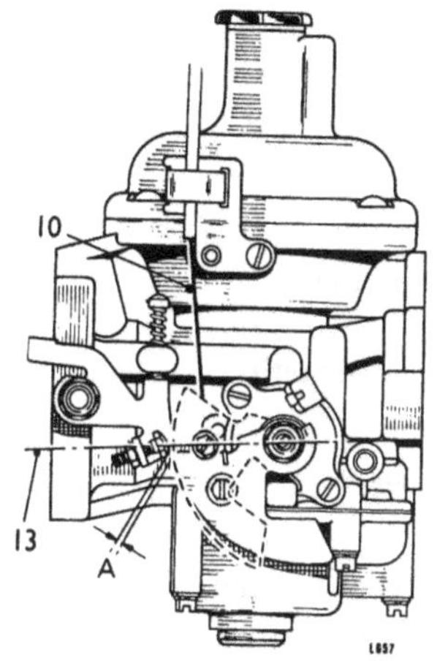

Mk. III without throttle cable

Setting control linkage and idling speed.

1. Remove the air cleaner (two bolts).
2. Slacken the throttle cable abutment nuts right back.
3. **[Where applicable,] slacken the throttle cable abutment nuts right back.**
4. Turn the idling screw until the throttle is just closed, then turn screw 1½ turns clockwise to provide a datum setting.
5. Ensure that the mixture control knob is pushed fully 'IN' on the fascia and that the cam is fully back against its stop.
6. Start the engine and attain normal running temperature.
7. Set the engine speed to 800 to 850 rev/min by turning the idling screw, stop engine.
8. Slacken the locknut and set the linkage adjusting screw until the clevis pin is moved to the engine side of the slots in the linkage straps, tighten locknut

NOTE: By moving the throttle spindle bracket independantly it is possible to ascertain the position of the clevis pin in the slot 'B'. Free play is necessary in the linkage to enable a fast idle setting without interfering with the closed position of the throttle.

9. **[Where applicable,] tighten the throttle cable abutment nuts to give a cable condition of no tightness no play.**

Fast idle adjustment

10. Ensure that the mixture control cable is correctly adjusted with no slackness or tightness.
11. Adjust the gap 'A' between the fast idle screw and the cam to 0·020 to 0·025 in
12. Start engine.
13. Pull the mixture control knob until the cam is turned to a position where the cam pivot, cable clamp screw and fast idle screw are in alignment. Adjust the fast idle speed if necessary to attain an engine speed of 1100 to 1300 rev/min, tighten the locknut.
14. Push the mixture control knob fully 'IN' at fascia.
15. Carry out C.O. level check.

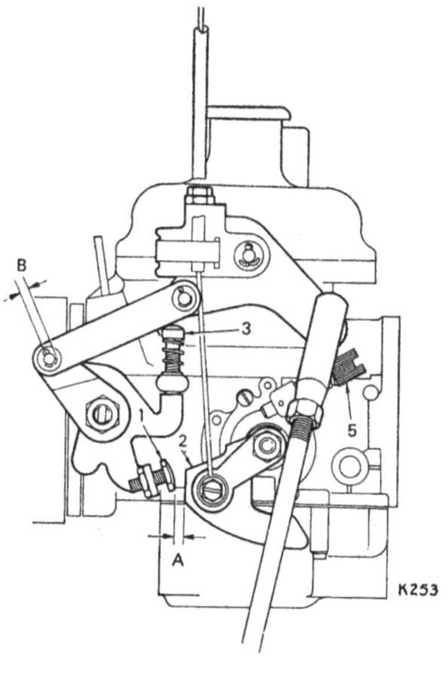

Where necessary adjust mixture as follows:

16. Remove carburetter damper.
17. Slowly insert tool S353 into the dashpot until the outer tool engages in the air valve and the inner tool engages the hexagon in the needle adjuster plug.
18. Hold the tool firmly and turn the inner tool clockwise to enrich the mixture anti-clockwise to weaken.
19. Check as in 15 and repeat as necessary.
20. Top up the dashpot (see maintenance item 7), replace the damper.
21. Fit the air cleaner.
22. Check and if necessary adjust idling speed to 800 to 850 rev/min.
23. Switch off engine.

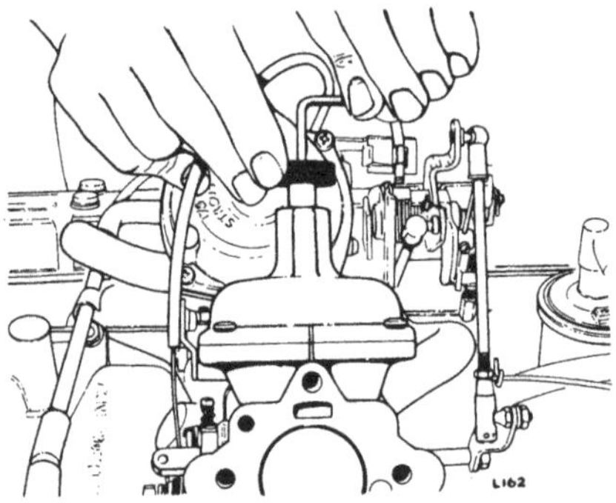

L162

CARBURETTER

– Remove and refit **19.15.09**

R emoving

1. Remove the air cleaner (two bolts).
2. Disconnect the fuel feed pipe.
3. Disconnect the pipe to rocker cover.
4. Release the throttle return spring.
5. **[On Mk. III models, remove the nut that holds the throttle operating link, then disconnect the link; on later models], remove the split pin, clevis pin and washers, slacken the abutment nuts and disconnect the throttle cable.**
6. Release the mixture control cable by slackening the clamp screw and prising off the abutment clip.
7. Disconnect the ignition vacuum pipe.
8. Remove the two nuts securing the carburetter.
9. Lift off the carburetter.

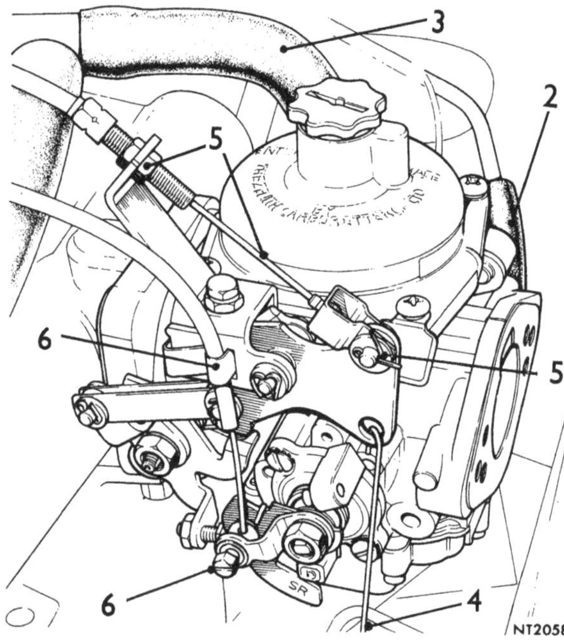

NT2058

Refitting

10. Renew carburetter and air cleaner gaskets.
11. Reverse instructions 1 to 9 and check settings.

1.50 C.D.S.E. carburetter as fitted to Mk. III Spitfire

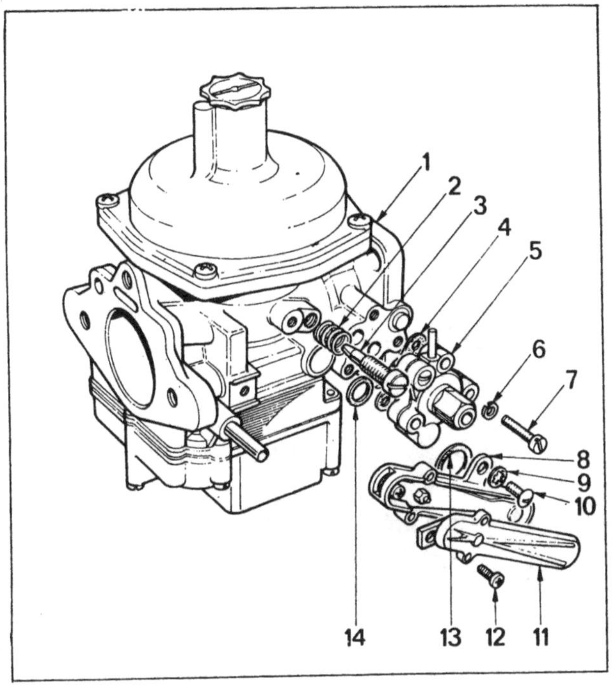

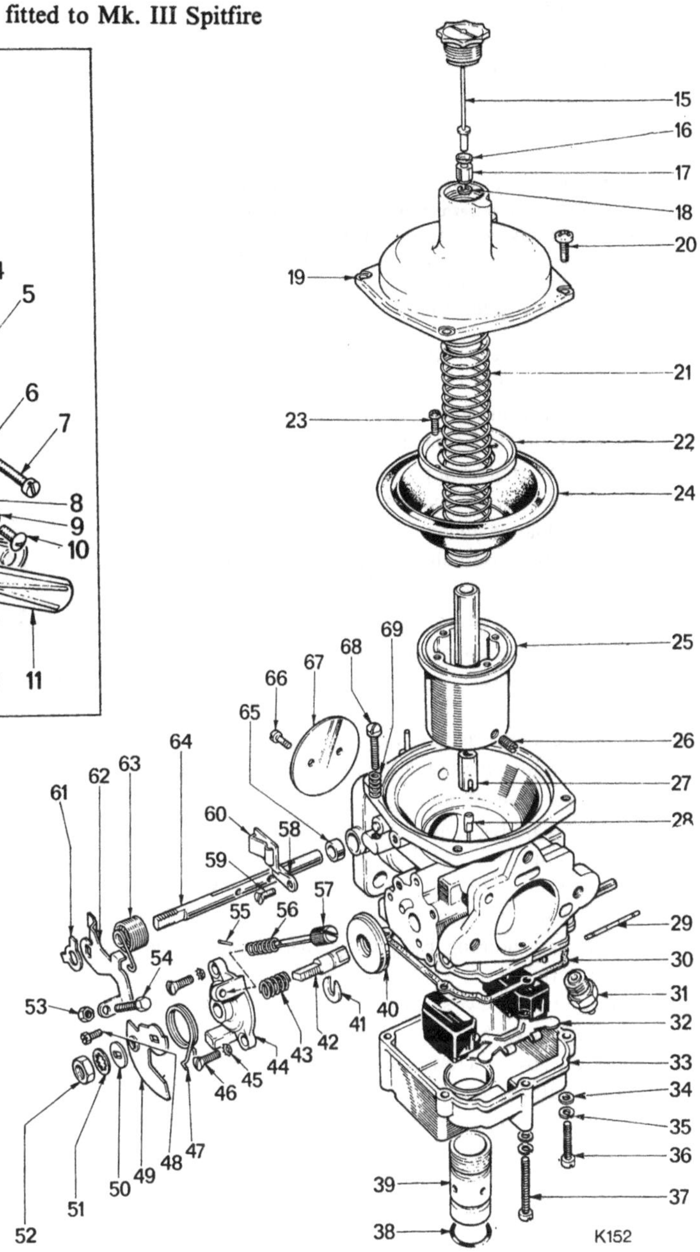

1 Carburettor
2 Spring—idle trimming screw
3 Idle trimming screw
4 Gasket—by pass valve
5 By pass valve
6 Lockwasher under (7)
7 Screw—securing (5)
8 Temperature compensator unit
9 Lockwasher under (10)
10 Screw—securing (8)
11 Cover—temperature compensator
12 Screw—securing (11)
13 Seal—on compensator body
14 Seal—inside carburettor
15 Damper rod } Damper Assembly
16 Washer } Damper Assembly
17 Distance sleeve } Damper Assembly
18 Circlip
19 Cover—air valve
20 Screws—securing (19)
21 Spring—air valve return
22 Ring—diaphragm attachment
23 Screw—securing (22) (24)
24 Diaphragm
25 Air valve
26 Screw—securing (28)
27 Needle holder assembly
28 Needle assembly
29 Float pivot pin
30 Gasket—float chamber
31 Needle valve
32 Float assembly
33 Float chamber cover
34 Washer
35 Spring washer
36 Screw—securing
37 Screw—securing
38 Rubber "O" ring
39 Plug
40 Valve plate
41 Retainer
42 Spindle
43 Spring
44 Starter box cover
45 Shakeproof washer
46 Screw
47 Return spring
48 Screw cable attachment
49 Lever
50 Washer
51 Shakeproof washer
52 Nut
53 Locknut—securing
54 Screw—fast idle
55 Choke stop pin
56 Stop spring
57 Choke stop
58 Cable abutment bracket
59 Screw—securing
60 Spring clip
61 Lockwasher—retaining
62 Lever—throttle
63 Spring—throttle return
64 Throttle spindle
65 Seal—throttle spindle
66 Screw—securing
67 Throttle disc
68 Idle adjusting screw
69 Spring—idle adjusting screw

Part No. 511243. page 1.340-41. Issue 1.

CARBURETTER

– Overhaul and adjust **19.15.17**

Service tool S353

[CAUTION: Please read the instructions thoroughly before you attempt carburetor repairs. If you lack the necessary skills or tools we suggest you leave such work to an Authorized Dealer or other qualified shop. Before taking a removed carburetor in for repair, thoroughly clean its exterior but do not undertake any disassembly.]

1. Remove carburetter. 19.15.09
2. Remove damper.
3. Remove bottom plug.
4. Drain carburetter of oil and fuel.
5. Remove 'O' ring from plug.
6. Remove six screws securing float-chamber to body.
7. Remove float-chamber.
8. Remove float assembly by gently prising spindle from clip each end.
9. Remove needle valve.
10. Remove four screws securing top cover to body.
11. Remove top cover.
12. Remove spring.
13. Remove air valve assembly.
14. Remove four screws securing diaphragm and retaining ring to air valve assembly.
15. Remove diaphragm and retaining ring.
16. Slacken grubscrew in side of air valve.
17. Insert tool S353 in stem of air valve, turn anticlockwise approximately two turns, withdraw needle and housing by pulling firmly and straight with the fingers.
18. Remove two screws securing starter box to body.
19. Remove starter box.
20. Remove two screws securing the temperature compensator to body.
21. Remove the temperature compensator and two rubber washers of different diameters.
22. Remove three (slotted) screws securing the by-pass valve to body.
23. Remove the by-pass valve and gasket.
24. Remove two screws securing butterfly to spindle.
25. Turn spindle, remove butterfly.
26. Release spindle return spring.
27. Withdraw spindle and spring.
28. Remove spindle seals from body by hooking out with small screwdriver.
29. Wash all components in clean fuel, allow to drain dry or use compressed air. Place all components on a clean surface. Discard all seals and gaskets.
30. Examine the condition of all components for wear, paying special attention to needle and seat, air valve and diaphragm which should be renewed unless in exceptionally good condition.
31. Use clean compressed air to blow through all ports, needle valve and starter box.

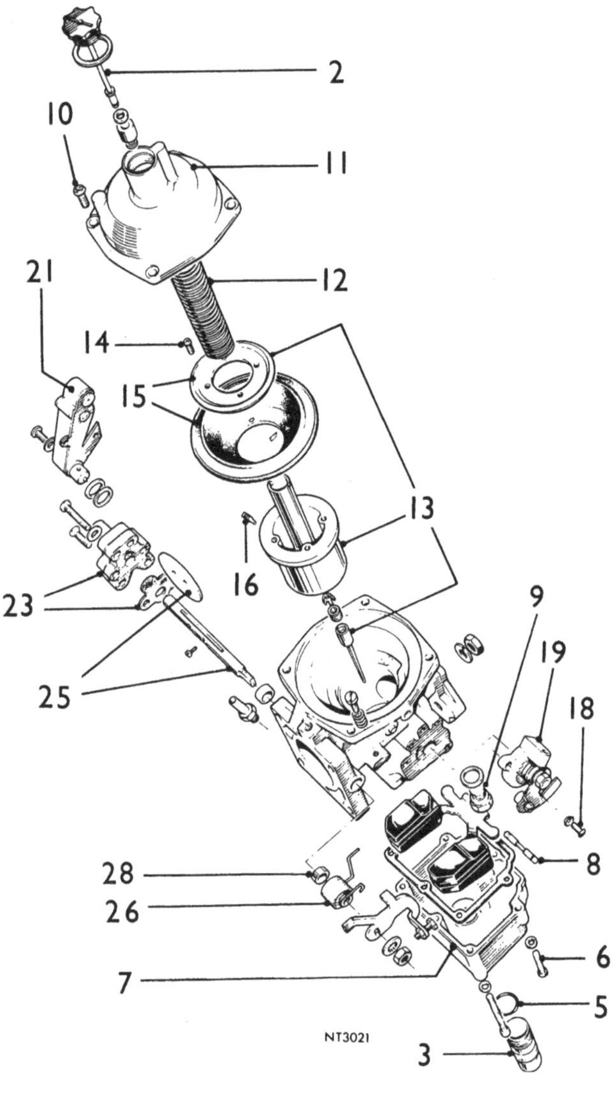

continued

32. Fit spindle seals to body, tapping gently into position, with metal casing of seals flush with body of carburetter.
33. Insert spindle, loading and locating spindle return spring whilst so doing.
34. Insert butterfly with two protruding spots, outboard and below spindle, tighten screws.

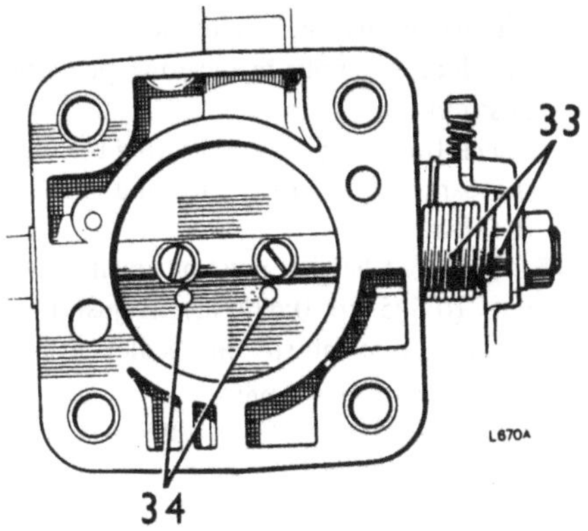

35. Fit starter box, tighten screws.
36. Fit by-pass valve and gasket, tighten screws.
37. Fit temperature compensator, tighten screws.
38. Insert needle housing assembly into the bottom of the air valve.
39. Fit tool S353, turning clockwise to engage threads of needle valve assembly with adjusting screw; continue turning until slot in needle housing is aligned with grub screw.
40. Tighten grub screw.

NOTE: The grub screw does not tighten on the needle housing but locates into the slot. This ensures that, during adjustment, the needle will remain in its operating position, i.e. biased, by a spring in the needle housing, towards the air cleaner side of the carburetter.

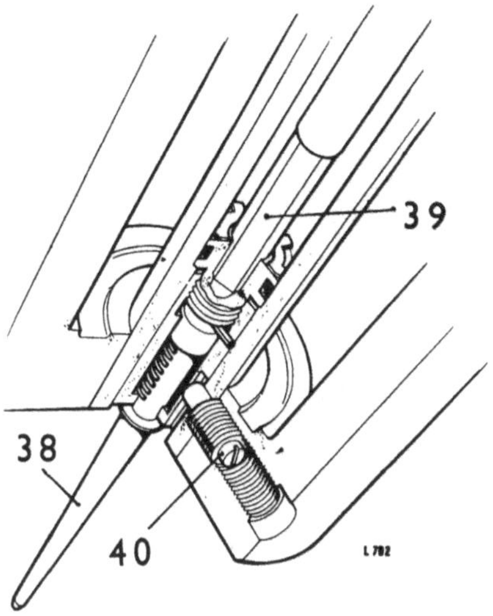

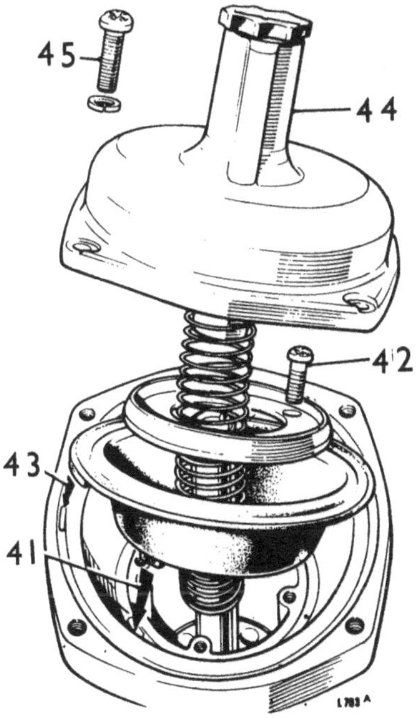

41. Fit diaphragm, locating inner tag into recess in air valve.
42. Fit diaphragm retaining ring; secure with four screws.
43. Fit air valve assembly, locating outer tag and rim of diaphragm in complementary recesses in carburetter body.
44. Fit carburetter top cover with bulge on housing neck towards air intake.
45. Fit and evenly tighten top cover screws.

continued

46. Fit needle valve and sealing washers; tighten.
47. Fit float assembly by levering pivot pin gently into position.
48. Check float height by measuring the distance between the carburetter gasket face and the highest point of the floats.

NOTE: The float heights must be equal and set to 16 to 17 mm (0·625 to 0·672 in). Adjust by bending tabs ensuring that tab sits on needle valve at right angles.

49. Fit float-chamber gasket.
50. Fit float-chamber, secure with six screws.
51. Fit 'O' ring to bottom plug.
52. Fit bottom plug.

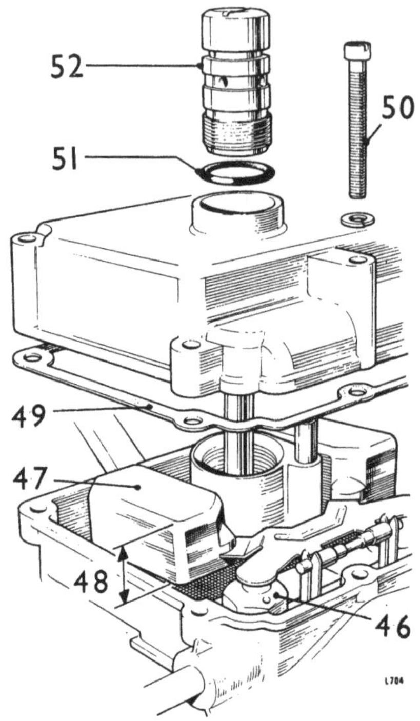

53. Fit carburetter. 17.15.09.
54. Fill carburetter damper dashpot with seasonal grade of engine oil until, using the damper as a dipstick, the threaded plug is 6 mm (0·25 in) above the dashpot when resistance is felt.
55. Fit damper.
56. Tune carburetter. 19.15.01.

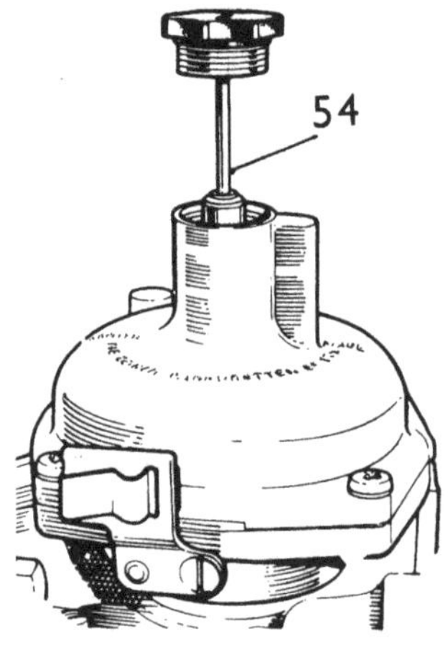

FLOAT-CHAMBER NEEDLE VALVE

–Remove and refit 19.15.24

Removing

1. Remove carburetter. 19.15.09.
2. Remove six screws securing float-chamber to body.
3. Remove float chamber.
4. Remove gasket.
5. Remove float assembly by gently prising spindle from clip each end.
6. Remove needle valve.

Refitting

7. Reverse instructions 1 to 6–renew gasket.

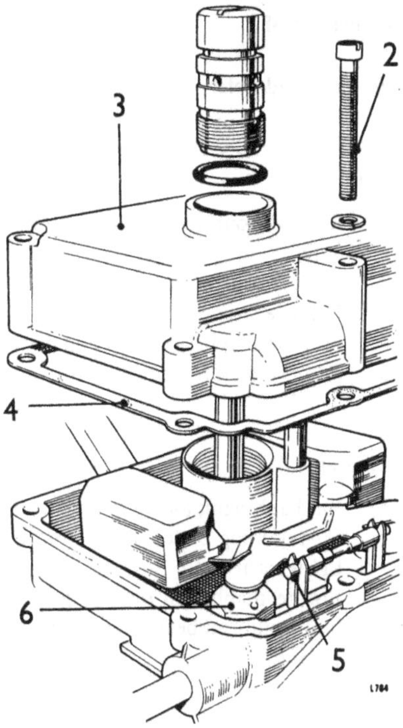

L784

FLOAT-CHAMBER LEVELS

–Check and adjust 19.15.32

1. Remove carburetter. 19.15.09.
2. Remove six screws securing float-chamber to body.
3. Remove float chamber.
4. Remove gasket.
5. Check distance between gasket face on carburetter body to highest point of each float.

NOTE: The height of each float must be equal and correct to datum setting which is 16 to 17 mm (0·625 to 0·672 in).

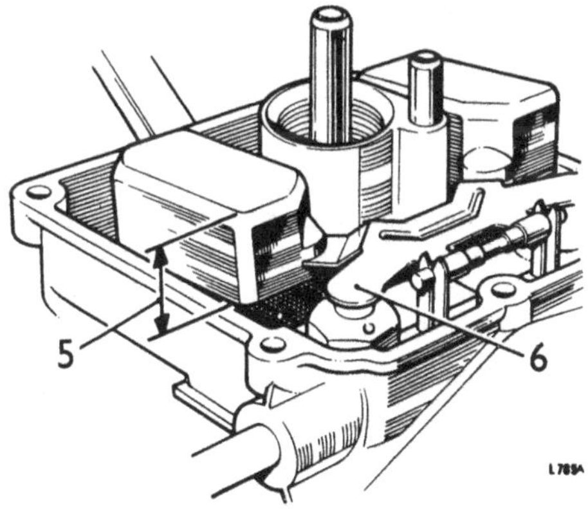

L785A

To adjust

6. Bend tab over to obtain correct setting ensuring that the tab sits on needle valve at right angles.
7. Fit new gasket, reverse instructions 1 to 3.

DIAPHRAGM

–Remove and refit **19.15.35**

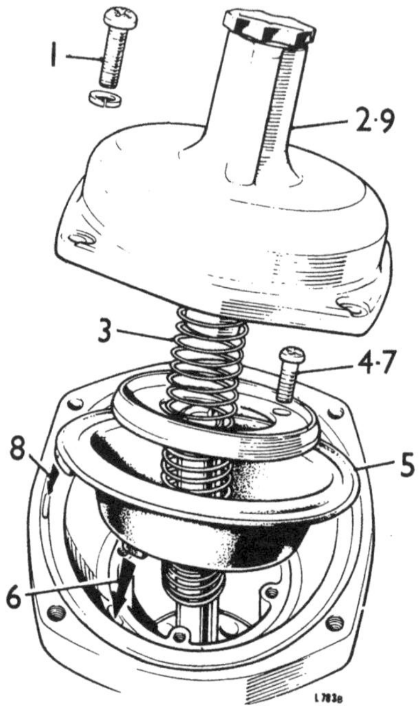

Removing

1. Remove four screws securing top cover to carburetter body.
2. Lift off top cover.
3. Remove spring.
4. Remove diaphragm retaining plate (four screws).
5. Remove diaphragm.

Refitting

6. Fit diaphragm, locating inner tag in air valve recess.
7. Fit retaining plate, ensure correct diaphragm seating, tighten screws.
8. Locate diaphragm outer tag in recess in carburetter body, and rim bead in annular groove.
9. Fit top cover, evenly tighten securing screws.

THROTTLE LINKAGE

Remove and refit **19.20.07**

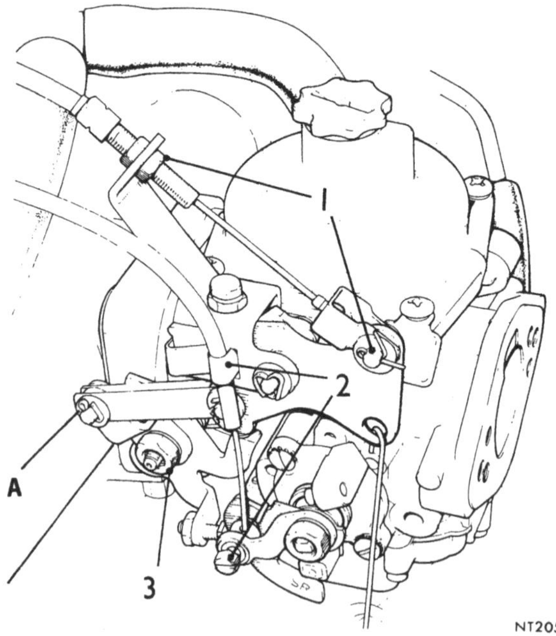

NT2058

Removing

1. **[On Mk. III models, remove the nut that holds the throttle operating link, then disconnect the link; on later models], remove the split pin, clevis pin and washers, slacken the abutments and disconnect the throttle cable.**
2. Release the mixture control cable by slackening the clamp screw and prising off the abutment clip.
3. Remove the nut on throttle spindle.
4. De-tab and remove the two screws securing the upper abutment bracket to carburetter.
5. Lift off linkage.

Refitting

6. Ensure that all linkage joints actuate freely and are not worn, lightly oil and reverse instructions 1 to 5.

NOTE: Free play is built into the linkage at point 'A' and should not be confused with wear nor should any attempt be made to adjust it out.

COOLING OPERATIONS

COOLANT

– Drain and refill 26.10.01

Draining

1. Set the heater control to 'HOT' position.
2. Remove the radiator filler cap. **CAUTION**: If the engine is hot, exercise care. Turn the cap slowly and allow any pressure in the system to release completely before removing the cap.
3. Open the drain tap at the rear right hand rear of the cylinder block.
4. Open the radiator drain tap (if fitted). If the radiator is not fitted with a drain tap release the bottom hose at the radiator.

Refitting

5. Close the drain tap on the cylinder block.
6. Close the radiator drain tap or refit the bottom hose to the radiator.
7. Fill the cooling system with clean soft water.
8. Fit the radiator filler cap and run the engine at a fast idling speed for approximately 1 to 2 minutes.
9. Remove the radiator cap and top up as necessary.
10. Replace the radiator filler cap and inspect the cooling system for leaks.
11. Top up the overflow reservoir as necessary.

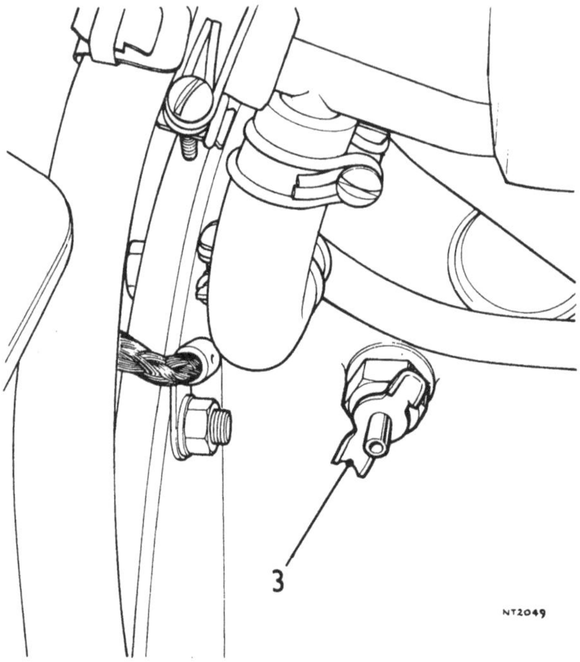

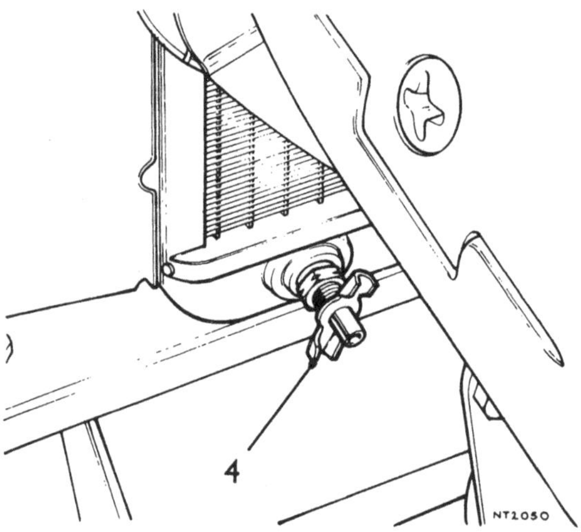

COOLING SYSTEM

– Pressure test 26.10.07

Radiator

1. Rinse radiator filler cap in clean water.
2. Whilst still wet fit the cap to the pressure tester.
3. Pump up pressure until the gauge pointer ceases to rise. Reject the filler cap if it will not register and maintain a pressure of 1 lb in^2 (0·06 kg/cm^2) below the figure stamped on the filler cap for a period of at least ten seconds without additional pumping.

Radiator

4. Warm the engine.
5. Remove the radiator filler cap.
6. Top up the cooling system of required.
7. Fit the pressure tester to the radiator.
8. Pump up to 13 p.s.i. (0·78 kg/cm^2).
9. Check that pressure will hold for approximately ten seconds without further pumping.
10. Check for leaks whilst system is pressurised.
11. If pressure falls within ten seconds and no leaks are visible, internal leakage is indicated.

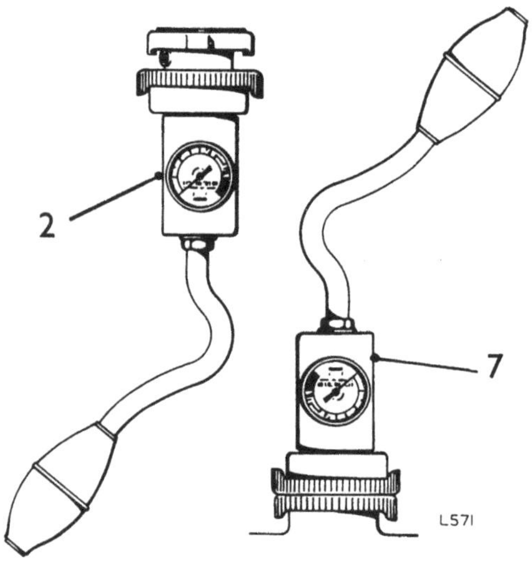

Triumph Spitfire Mk IV Manual. Part No. 545254. Issue 1.

FAN BELT

– Remove and refit 26.20.07

[NOTE: Mk. III Spitfires equipped with generators have the same fan belt adjustment specification as the later models with alternators.]

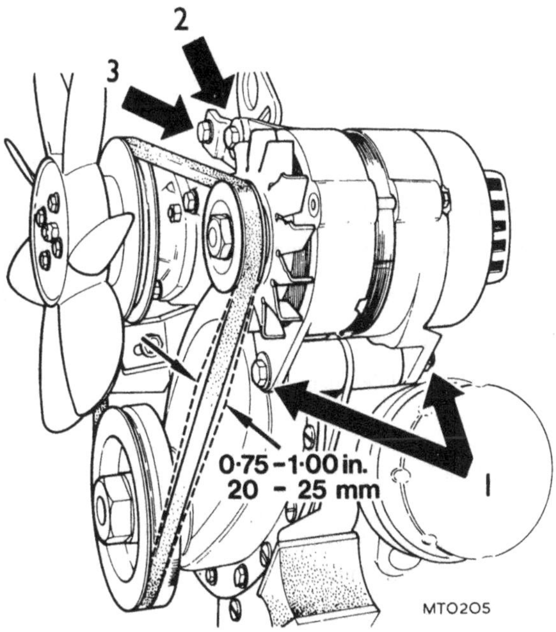

Removing

1. Slacken the pivot bolt and nut on the underside of the alternator.
2. Slacken the bolt securing the alternator to the adjusting plate.
3. Slacken the bolt securing the adjusting plate to the water pump.
4. Pivot the alternator towards the engine to release tension from the fan belt.
5. Release the fan belt from the alternator, water pump and engine crankshaft pulleys.

Refitting

6. Locate the fan belt on the engine crankshaft, water pump and alternator pulleys.
7. Adjust fan belt tension and tighten bolts 1, 2, and 3.

FAN

– Remove and refit 26.25.06

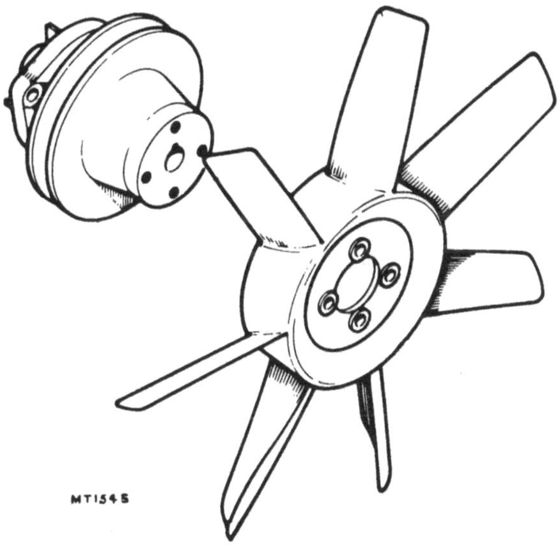

Remove

1. Drain the cooling system and remove the radiator 26.40.01
2. Remove the four bolts securing the fan to the water pump pulley and withdraw the fan.

Refitting

3. Reverse instructions 1 and 2.

HOSES

– Remove and refit

Cylinder head to heater valve	26.30.39
Heater to water pump pipe	26.30.18
Heater valve to heater	26.30.40
Inlet manifold to water pump pipe	26.30.17
Radiator bottom hose	26.30.07
Radiator top hose	26.30.01
Thermostat to inlet manifold	26.30.46

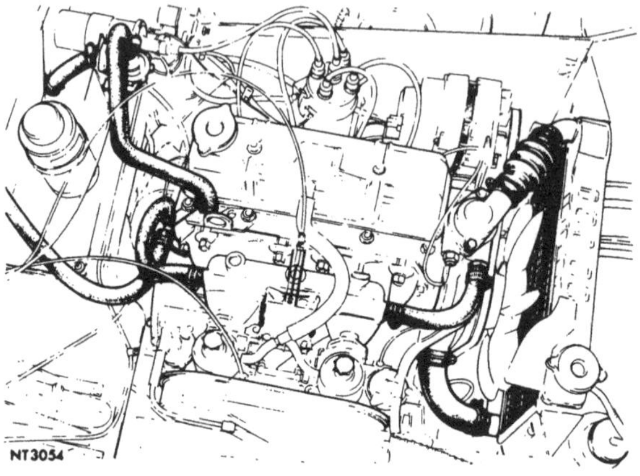

Removing

1. Drain the cooling system.
2. Slacken the hose clips.
3. Remove the hose.

Refitting

4. Reverse instructions 1 to 3.

RADIATOR

– Remove and refit 26.40.01

Removing

1. Remove the radiator filler cap and drain cooling system.
2. Disconnect radiator bottom hose. (If not already disconnected to drain system. Models without radiator drain taps only).
3. Disconnect radiator top hose.
4. Disconnect overflow bottle plastic hose from radiator.
5. Remove four bolts and nuts (2 either side) securing radiator to mounting bracket.
6. Withdraw the radiator.

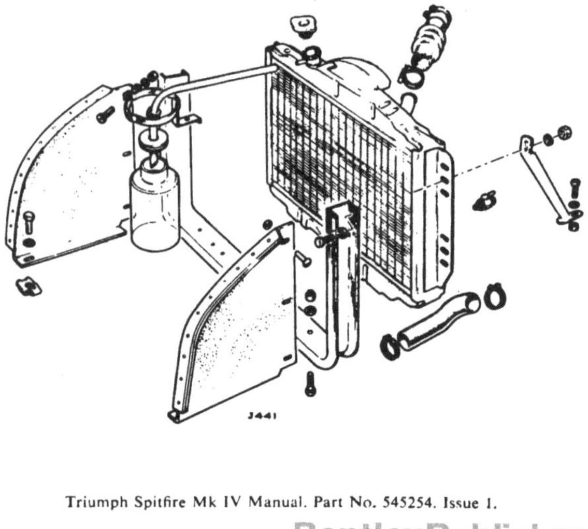

Refitting

7. Reverse instructions 2 to 6.
8. Fill the cooling system.

THERMOSTAT

– Remove and refit **26.45.01**

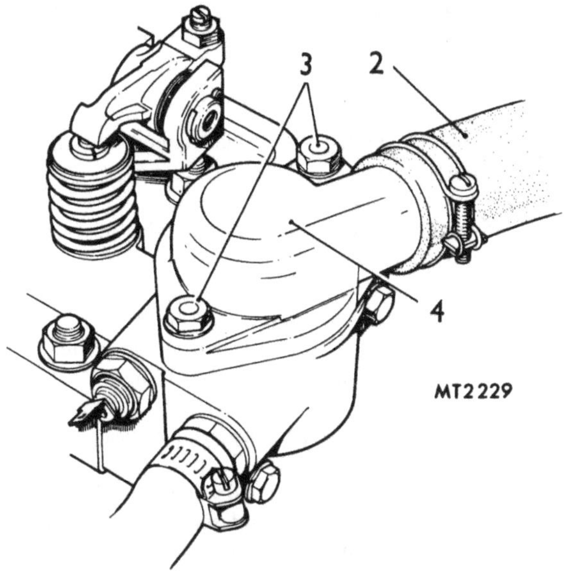

Removing

1. Drain the cooling system.
2. Remove the radiator top hose.
3. Remove the two bolts and spring washers securing the thermostat elbow to the thermostat and water pump housing.
4. Lift off the thermostat elbow.
5. Lift out the thermostat.

Refitting

6. Clean the mating faces of the thermostat elbow and the water pump housing.
7. Place the thermostat in the water pump housing.
8. Fit a new gasket and the thermostat elbow. Fit and tighten the two securing bolts and spring washers.
9. Fit the radiator top hose.
10. Top up the cooling system.

THERMOSTAT

– Test **26.45.09**

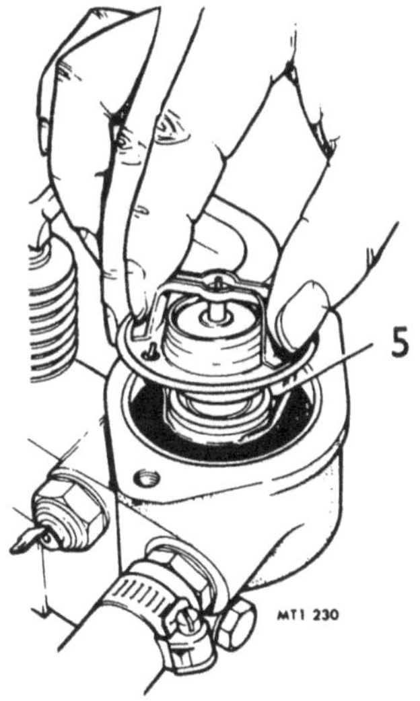

[NOTE: At room temperature, the thermostat must be fully closed. Otherwise the engine will not warm up sufficiently and the heater will deliver inadequate heat in cold weather. In replacing the thermostat, select the correct temperature range as specified in the General Specification Data for your particular year Spitfire.]

1. Determine the opening temperature of the thermostat stamped on the flange or base.
2. Immerse the thermostat in water heated to the opening temperature of the thermostat. Renew the thermostat if it fails to open.

THERMOSTAT AND WATER PUMP HOUSING

– Remove and refit **26.45.10**

Removing

1. Drain the cooling system and remove the radiator. 26.40.01.
2. Remove the fan belt.
3. Disconnect the induction manifold hose at the thermostat outlet.
4. Disconnect the petrol feed hose to the carburetter.
5. Disconnect the temperature transmitter.
6. Disconnect the return water pipe union at the rear of the thermostat and water pump housing.
7. Remove the three bolts securing the thermostat and water pump housing to the cylinder head. Note the length and position of each bolt.
8. Withdraw the thermostat and water pump housing.
9. Remove the three nuts and spring washers securing the water pump to the housing and withdraw the water pump and fan.
10. Remove the temperature transmitter.

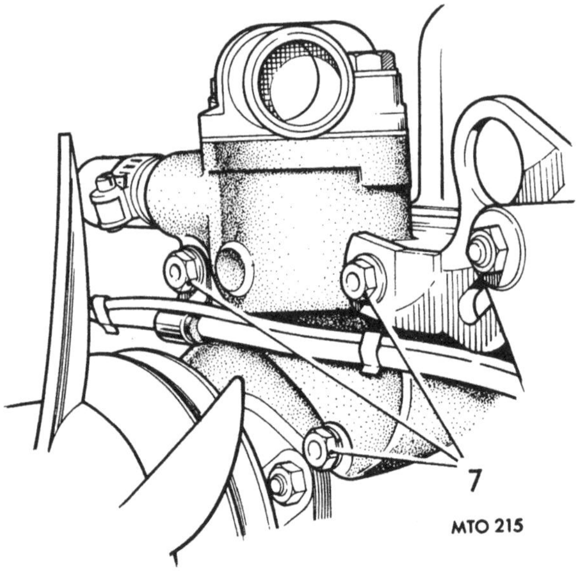

MTO 215

Refitting

11. Reverse instructions 1 to 10. Use new gaskets when fitting the water pump to the housing and the housing to the cylinder head.
12. Fill the cooling system.

WATER PUMP **26.50.01**

– Remove and refit

Removing

1. Remove the radiator cap and drain the cooling system.
2. Remove the radiator 26.40.01.
3. Remove the fan belt.
4. Remove the three nuts and spring washers securing the water pump flange to the thermostat and water pump housing.
5. Withdraw the water pump.
6. Remove the fan.

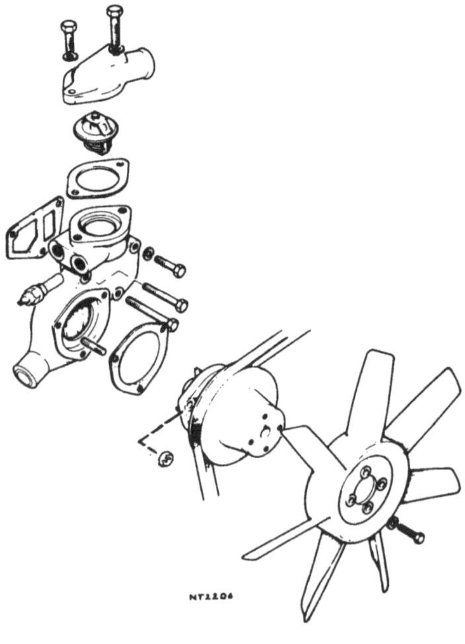

Refitting

7. Ensure that the mating faces of the water pump flanges and the thermostat and water pump housing are clean and free of burrs.
8. Using a new gasket fit the water pump and evenly tighten the three securing nuts.
9. Fit the fan.
10. Reverse instructions 1 to 3.
11. Fill the cooling system.

MANIFOLD AND EXHAUST SYSTEM OPERATIONS

*All bracketed material ([ ]) in this section has been added by Robert Bentley, Inc.

EXHAUST SYSTEM COMPLETE

– Remove and refit 30.10.01

[NOTE: The Spitfire Mk. III exhaust system is different from that of later models and is shown in the accompanying illustration. Similar clamps and supports are used on all models.]

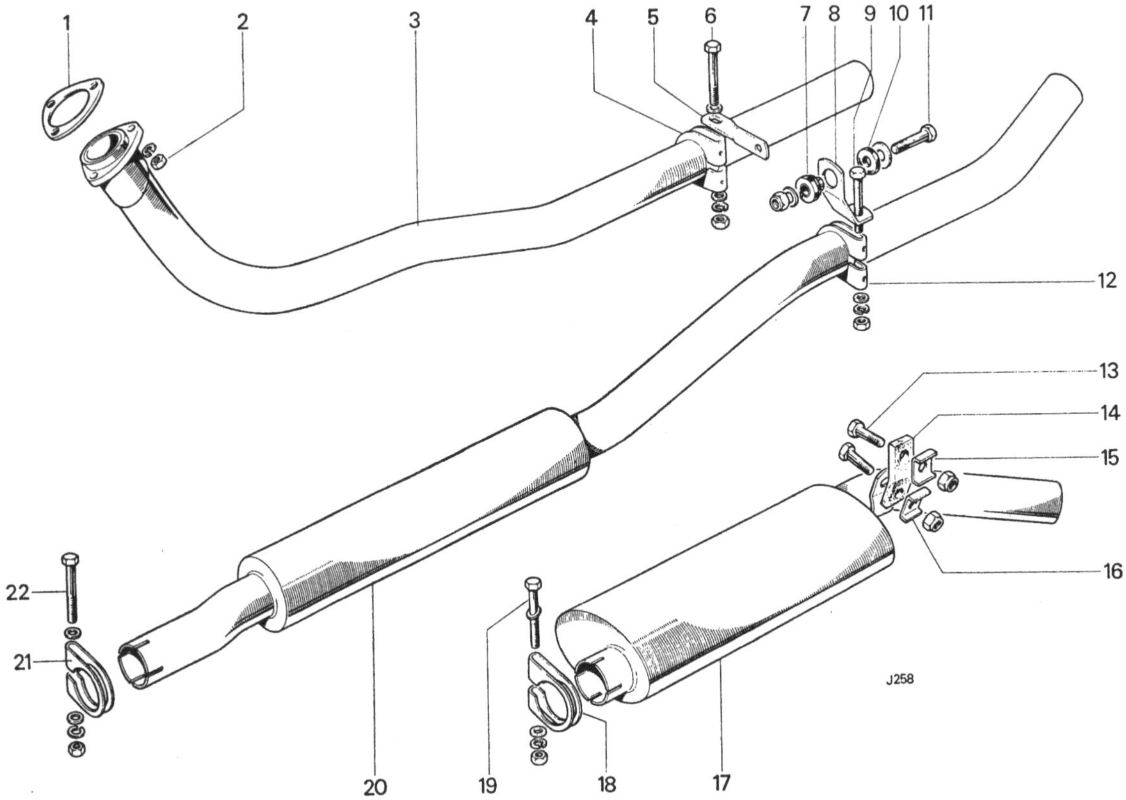

Spitfire Mk. III Exhaust Details

1	Gasket front pipe to manifold	12	Pipe clip
2	Nut	13	Bolt
3	Front exhaust pipe	14	Flexible strip
4	Clip	15	Clamp plate
5	Bracket	16	Clamp plate
6	Bolt	17	Rear silencer and tail pipe
7	Rubber grommet	18	Clip
8	Mounting bracket	19	Bolt
9	Bolt	20	Front silencer
10	Grommet washer	21	Clip
11	Bolt	22	Bolt

Part No. 511243. page 1.409. Issue 1.

Removing

1. Remove front pipe 30.10.09.
2. Remove silencer assembly 30.10.22.
3. Remove exhaust manifold 35.15.01.

Refitting

4. Refit exhaust manifold 35.15.01.
5. Refit silencer assembly 30.10.22.
6. Refit front pipe 30.10.09.

FRONT PIPE

– Remove and refit **30.10.09**

Removing

1. Remove nut and bolt holding the front pipe to front support bracket.
2. Remove nut and bolt holding front pipe to rear support bracket.
3. Release clip holding front pipe to silencer.
4. Remove three nuts holding front pipe to manifold.
5. Remove front pipe.

Refitting

6. Slide front pipe into silencer.
7. Place new front pipe gasket in position.
8. Align front pipe over manifold studs and replace and tighten the nuts.
9. Replace nut and bolt through front pipe front support bracket and tighten.
10. Tighten clamp around front pipe and silencer.

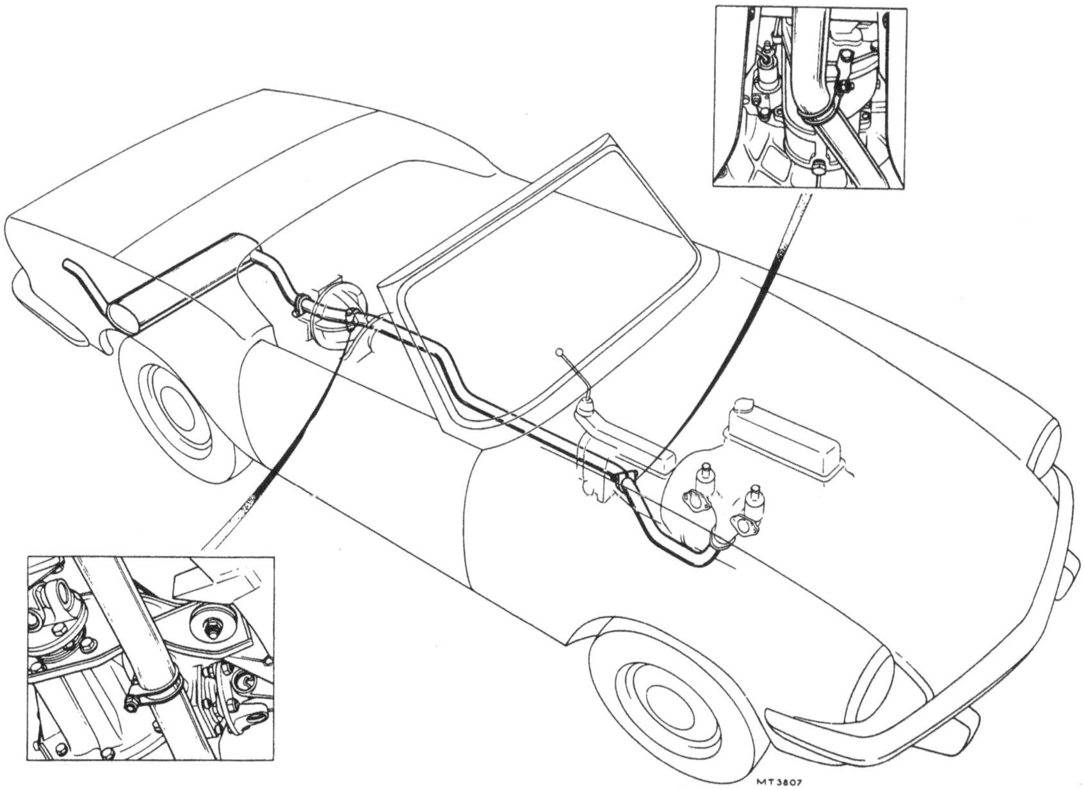

TAIL PIPE OR SILENCER [except Mk. III]

– Remove and refit 30.10.22

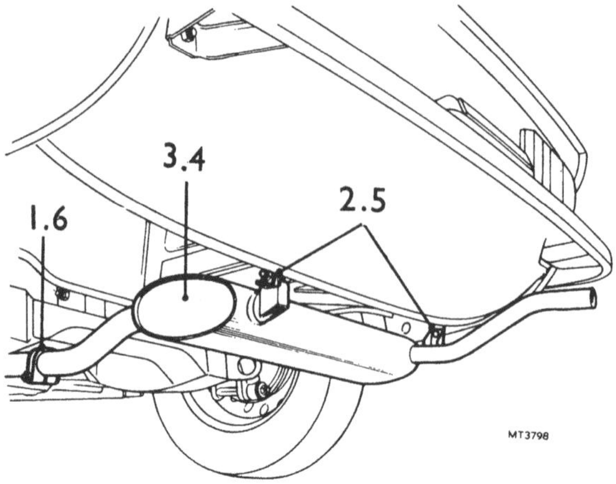

Removing

1. Release clip holding silencer to front pipe.
2. Remove two nuts and bolts holding silencer to the rubber straps.
3. Remove the silencer.

Refitting

4. Align and slide the silencer over the front pipe.
5. Replace the tail pipe mounting rubber straps.
6. Tighten the clip around the tail pipe and front pipe.

DOWN PIPE FLANGE PACKING

– Remove and refit 30.10.26

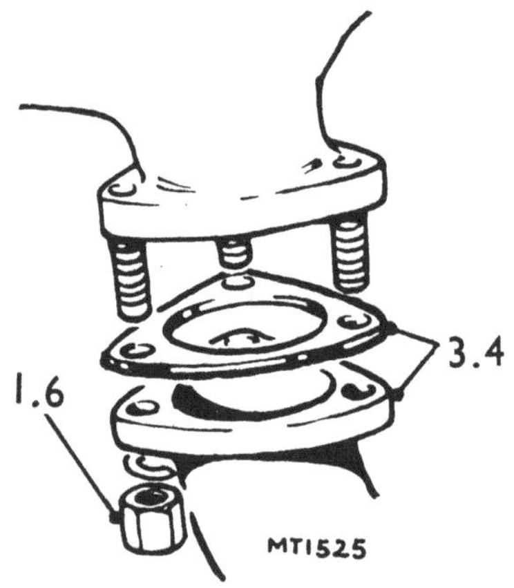

Removing

1. Remove three nuts holding the front pipe to manifold.
2. Remove nut and bolt holding front pipe to support bracket.
3. Pull down front pipe and remove flange packing.

Refitting

4. Pull front pipe down and replace packing.
5. Refit front pipe and tighten manifold stud nuts.
6. Refit nut and bolt in front pipe mounting bracket.

EXHAUST MANIFOLD

– Remove and refit **30.15.01**

Removing

1. Remove carburetter and inlet manifold as in operation 19.15.11.
2. Remove three nuts on down-pipe flange and pull clear.
3. Remove down-pipe flange packing.
4. Remove two nuts holding exhaust manifold to cylinder head.
5. Pull manifold clear of studs and remove from engine.
6.. Remove induction and exhaust manifold gasket.

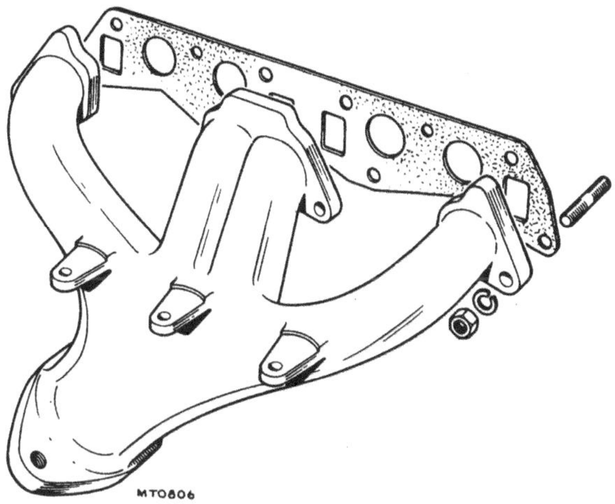

Refitting

7. Renew and refit the manifold gasket up to cylinder head.
8. Refit the manifold onto the head studs.
9. Tighten nuts holding manifold to head.
10. Renew down-pipe flange packing and fit over studs.
11. Replace and tighten three nuts holding front pipe in position.
12. Refit carburetter and inlet manifold as in operation 19.15.11.

CLUTCH OPERATIONS

CLUTCH ASSEMBLY

– Remove and refit 33.10.01

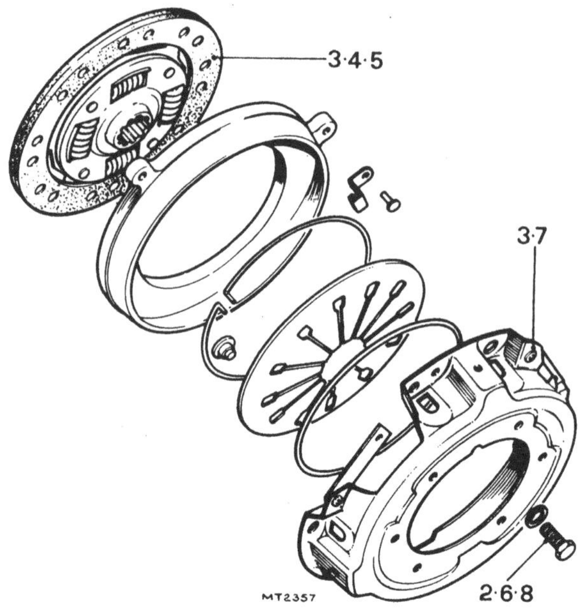

Removing

1. Remove the gearbox 37.20.01.
2. Working from inside the car, remove six bolts holding clutch to flywheel.
3. Remove the clutch driven plate and pressure plate.

Refitting

4. With the longer boss of the splined hub towards the gearbox, offer up the driven plate to the flywheel.
5. Centralise the driven plate with the bush at the rear of the crankshaft.
6. Insert the six bolts holding the clutch to the flywheel.
7. Ensure that the pressure plate locates correctly on the dowels on the flywheel.
8. Tighten the six bolts holding the pressure plate to the flywheel.
9. Refit the gearbox 37.20.01.

HYDRAULIC SYSTEM

– Bleed **33.15.01**

1. Ensure that the reservoir is topped up to ¼in from the top.
2. Remove the gearbox cover 76.25.07.
3. Wipe the bleed nipple clean.
4. Attach a piece of small bore tubing to the nipple and let it hang in a container partially filled with hydraulic fluid.
5. Unscrew the bleed nipple one complete turn.

NOTE: During bleeding, care should be taken to ensure that the reservoir does not become empty, resulting in air being drawn into the system. Ensure that fluid used is clean or air-free.

6. Depress the clutch pedal fully and let it return without assistance.
7. Repeat this operation with a slight pause between each depression of the pedal.
8. Observe the fluid being discharged from the pipe.
9. When it is air free, hold the pedal in the depressed position.
10. Securely tighten the bleed screw.
11. Remove the pipe from the slave cylinder.
12. Refit the gearbox cover 76.25.07.

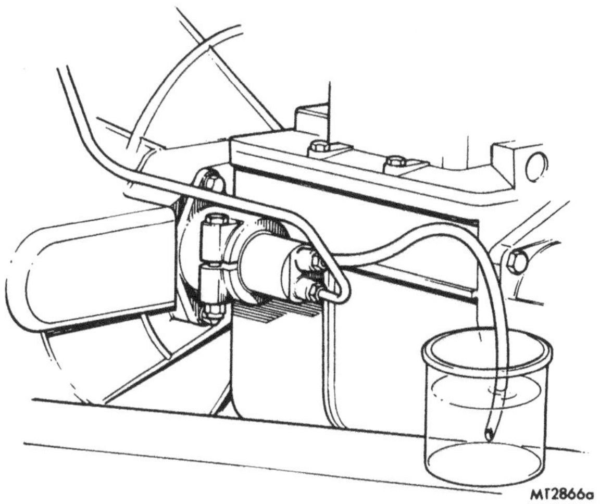

FLUID PIPE

– Remove and refit 33.15.09

Removing

1. Drain the hydraulic system.
2. Unscrew the pipe from slave cylinder.
3. Unscrew pipe from master cylinder.
4. Remove pipe from car.

Refitting

5. Align new pipe in approximate position in car.
6. Screw pipe into the master cylinder.
7. Screw pipe into the slave cylinder.
8. Top up master cylinder with new brake fluid.
9. Bleed hydraulic system 33.15.01.

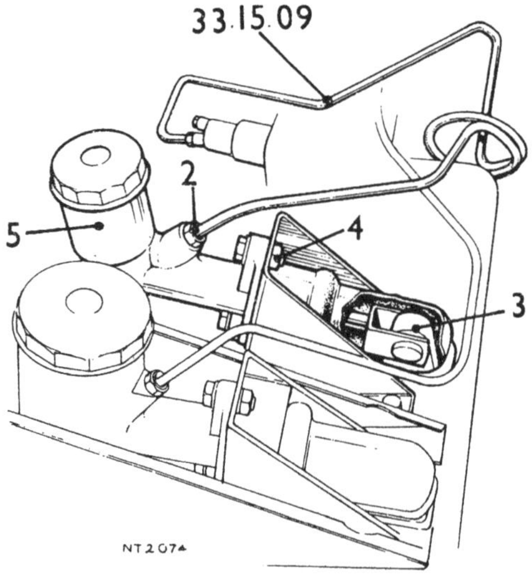

MASTER CYLINDER

– Remove and refit 33.20.01

Removing

1. Drain the fluid system.
2. Disconnect the fluid pipe at the master cylinder, plug open ends.
3. Lift the rubber boot, remove split pin and cotter pin from top of pedal.
4. Remove two bolts securing master cylinder to bracket.
5. Remove master cylinder from car.

Refitting

6. Hold master cylinder in place in engine compartment.
7. Renew two bolts securing master cylinder to bracket and fit.
8. Refit cotter pin, renew split pin, refit boot.
9. Connect the fluid pipe to the master cylinder.
10. Bleed the system 33.15.01.

MASTER CYLINDER

– Overhaul 33.20.07

[CAUTION: Wash hydraulic cylinder components in brake fluid only. Never use gasoline or other petroleum-based solvents since they are damaging to rubber parts in the system.]

1. Drain the fluid reservoir.
2. Remove the master cylinder from the car 33.20.01.
3. Slide the rubber along the push rod.
4. Remove the circlip from the end of the master cylinder and withdraw the push rod and washer.
5. Withdraw the piston spring and seal assembly from the master cylinder. Withdrawal may be facilitated by applying a compressed air line to the fluid outlet union.
6. Straighten the prong of the spring thimble and remove the thimble and spring from the piston.
7. Release the valve stem from the key hole slot in the thimble.
8. Slide the valve seal spacer along the valve stem.
9. Remove the valve seal from the valve stem and fit a new seal.
10. Assemble the spacer, spring and thimble to the valve stem.
11. Remove the seal from the piston and fit a new seal (seal lip towards the spring).
12. Engage the spring thimble on the piston and carefully depress the thimble prong.
13. Lubricate the bore of the master cylinder with clean brake fluid and insert the seal assembly spring and piston.
14. Fit a new rubber to the push rod.
15. Fit the push rod and washer to the master cylinder and secure with the clip.
16. Slide the rubber into position on the master cylinder.
17. Refit the master cylinder to the car 33.20.01.
18. Bleed the cylinder.

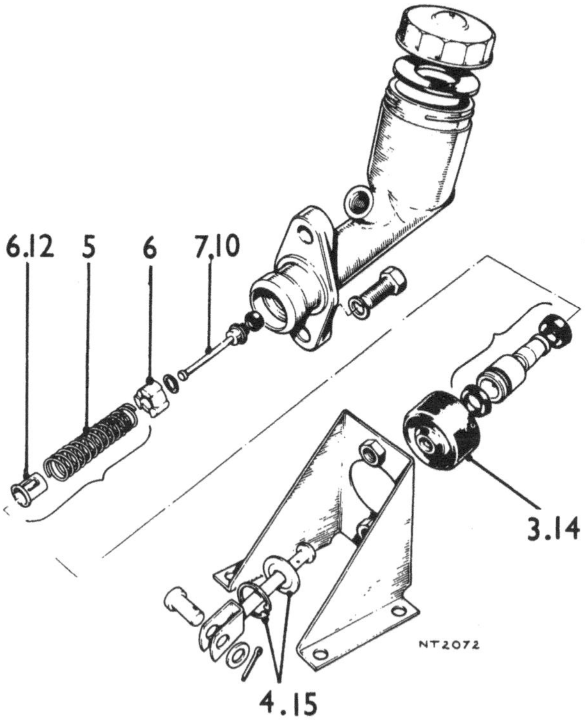

RELEASE BEARING ASSEMBLY

– Remove and refit 33.25.12

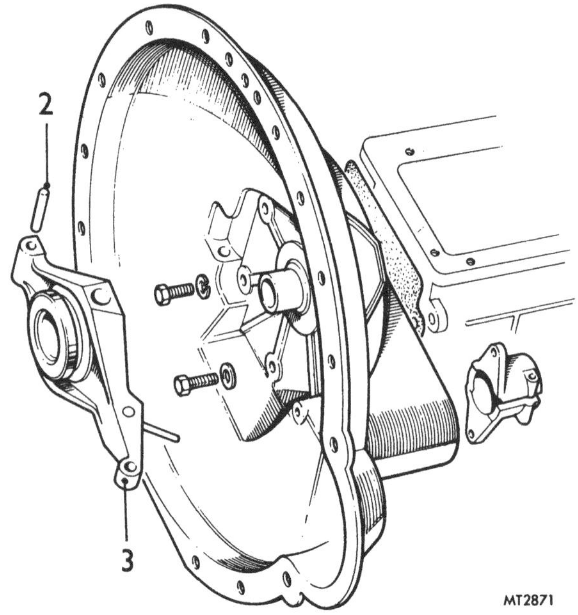

Removing

1. Remove the gearbox 37.20.01.
2. Drive the pin from the clutch housing.
3. Remove the operating lever from the housing.
4. Press out mills pin and release bearing sleeve by extracting plugs.
5. Withdraw the bearing from the sleeve.

Refitting

6. Reverse procedure 1 to 5.

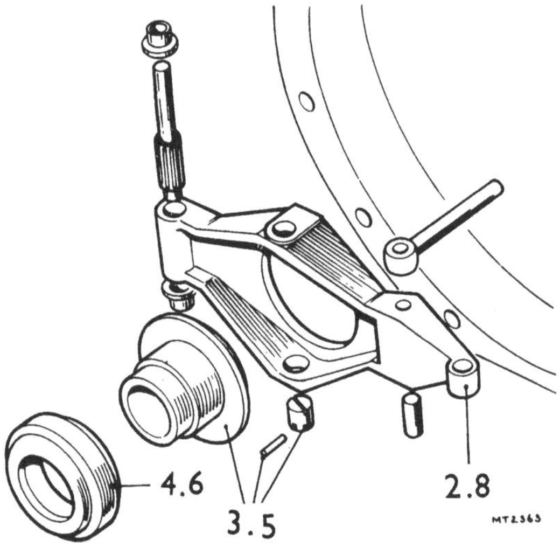

BEARING ASSEMBLY

– Overhaul 33.25.17

1. Remove the gearbox 37.20.01.
2. Remove operating lever assembly.
3. Remove bearing sleeve.
4. Remove bearing.
5. Renew bearing sleeve.
6. Renew bearing.
7. Grease bearing and sleeve.
8. Replace and refit operating lever.
9. Refit gearbox 37.20.01.

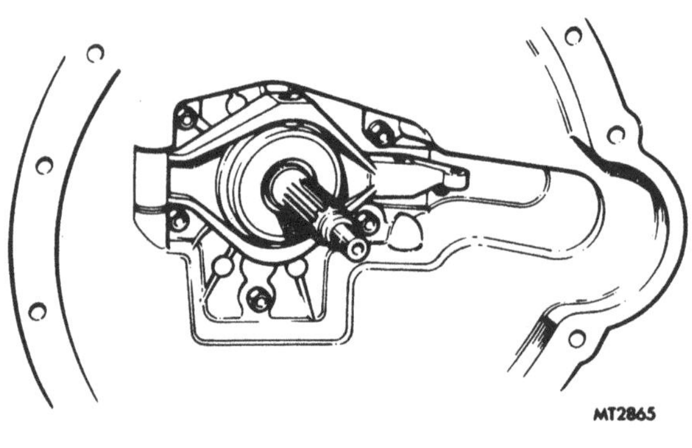

CLUTCH PEDAL

– Remove and refit **33.30.02**

Removing

1. Remove split pin, washer and clevis pin.
2. Remove return spring.
3. Remove eight bolts holding clutch pedal bracket to body.
4. Remove bracket and pedal from car.
5. Remove one circlip from the pivot pin.
6. Push pivot pin clear of pedal.
7. Push bush out of pedal.
8. Remove pedal from bracket.

Refitting

9. Reposition pedal in bracket.
10. Push bush through pedal.
11. Push lubricated pivot pin through bush.
12. Replace circlip and ensure it locates properly in the groove.
13. Reposition bracket in the car.
14. Refit eight screws in position and tighten.
15. Refit clevis pin and washer.
16. Renew split pin and fit.
17. Refit return spring.

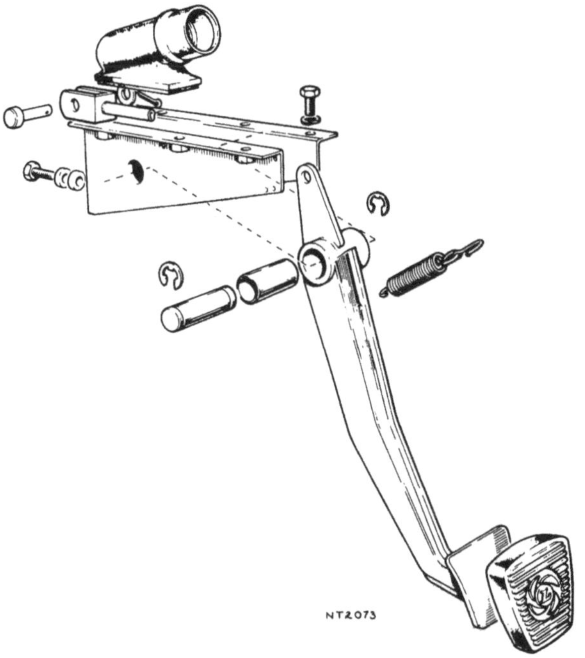

CLUTCH PEDAL RETURN SPRING

– Remove and refit **33.30.03**

Removing

1. Lever spring from hole in pedal.
2. Remove spring from bracket.

Refitting

3. Fit new return spring to bracket.
4. Spread spring and position it through hole in pedal.

CLUTCH PEDAL

– Overhaul **33.30.09**

1. Remove clutch pedal 33.30.02.
2. Remove bush from pedal.
3. Remove pedal rubbers.
4. Fit new bushes and lubricate.
5. Renew pedal rubber and fit.
6. Refit clutch pedal 33.30.02.

SLAVE CYLINDER

– Remove and refit 33.35.01

Removing

1. Drain the clutch system.
2. Disconnect hydraulic pipe and push clear of slave cylinder.
3. Remove locating bolt, nut and washer.
4. Pull slave cylinder clear of housing.

Refitting

5. Centralise push rod in housing.
6. Push slave cylinder into housing.
7. Line up groove with hole and place bolt in through hole and tighten nut.
8. Line up hydraulic pipe and tighten nut up in slave cylinder.
9. Bleed the system 35.15.01.

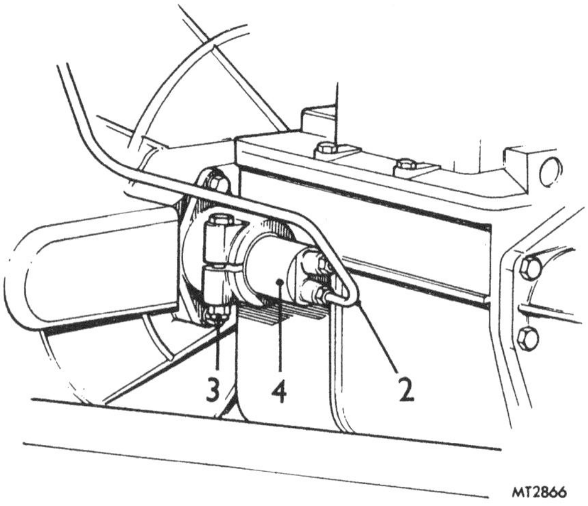

SLAVE CYLINDER

– Overhaul 33.35.07

[CAUTION: Wash hydraulic cylinder components in brake fluid only. Never use gasoline or other petroleum-based solvents since they are damaging to rubber parts in the system.]

1. Remove slave cylinder from car 33.35.01.
2. Remove dust cover retainer.
3. Remove dust cover.
4. Remove circlip.
5. Remove plunger.
6. Remove the seal.
7. Remove the spring.
8. Fit new seal, spring circlip, dust cover and retainer and lubricate.
9. Refit slave cylinder to car 33.35.01.

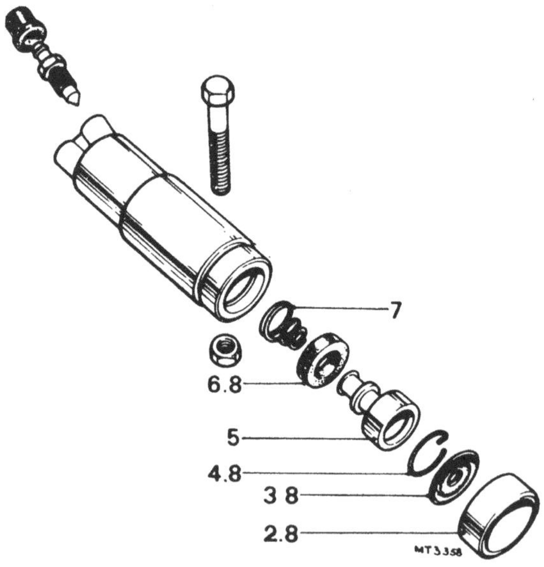

Triumph Spitfire Mk IV Manual. Part No. 545254. Issue 1.

SYNCHROMESH GEARBOX OPERATIONS

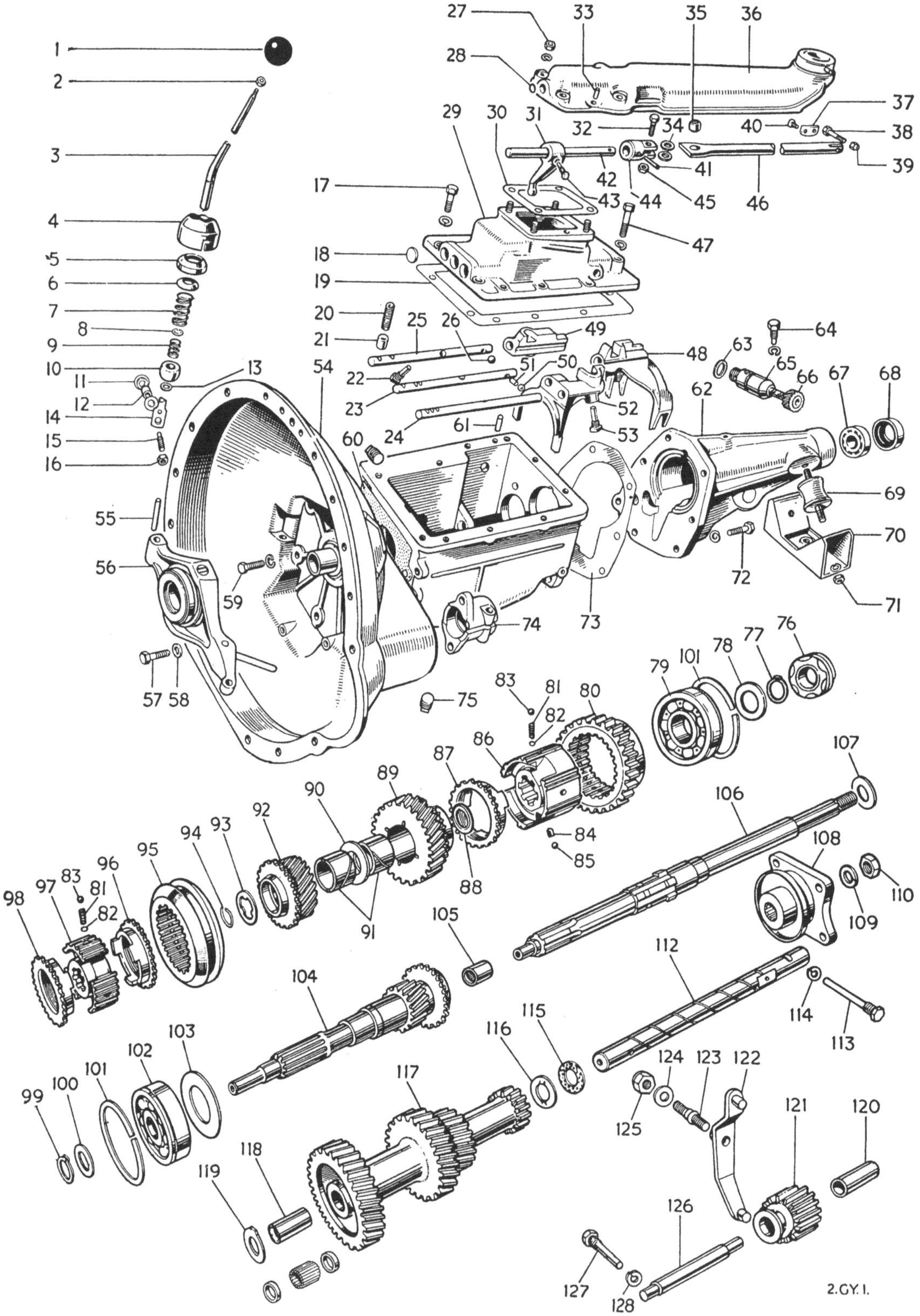

Exploded Arrangement of Mk. III Gearbox Details

Part No. 511243, page 2.204, Issue 1.

[NOTE: The Spitfire Mk. III transmission, shown in this exploded view, does not have a synchronized 1st gear. In order to accomodate the sychronized 1st gear found on the Mk. IV and 1500 models, the transmission case had to be made slightly longer. The all-synchro transmission is shown on the following page.

Though many parts of the late all-synchro gearbox are not interchangeable with corresponding parts in the Mk. III gearbox, there are only detail differences between the two kinds of gearbox. Most of the illustrations on the following pages show only the late all-synchro gearbox. Nevertheless, the repair procedures are the same for the early gearbox unless otherwise stated.]

Key to Exploded Arrangement of Mk. III Gearbox Details

1 Knob
2 Locknut
3 Gear change lever
4 Cover
5 Shield
6 Plate
7 Spring
8 Circlip
9 Spring
10 Nylon sphere
11 Stepped nylon washer
12 Bush
13 Washer
14 Lever end
15 Reverse stop pin
16 Locknut
17 Bolt
18 Welch plug
19 Gasket
20 Spring
21 Plunger
22 Taper locking pin
23 1st/2nd selector shaft
24 3rd/top selector shaft
25 Reverse selector shaft
26 Interlock ball
27 Nut
28 Rubber "O" ring
29 Top cover
30 Gasket
31 Selector ball-end
32 Bolt
33 Dowel
34 Washer
35 Bonded rubber bush
36 Gear change extension
37 Reverse stop
38 Bolt
39 Nyloc nut
40 Screw
41 Mills pin
42 Remote control shaft (front)
43 Taper locking pin
44 Fork
45 Nut
46 Remote control shaft (rear)
47 Bolt
48 1st/2nd selector fork
49 Reverse selector
50 Interlock ball
51 Interlock plunger
52 Top/3rd selector fork
53 Taper locking pin
54 Clutch housing
55 Pin
56 Clutch release mechanism
57 Wedgelock bolt
58 Plain washer
59 Bolt
60 Gasket
61 Dowel
62 Rear extension
63 Rubber "O" ring
64 Peg bolt
65 Speedo drive gear housing
66 Speedo drive gear
67 Extension ball race
68 Oil seal
69 Gearbox mounting rubber
70 Mounting bracket
71 Nut
72 Bolt
73 Gasket
74 Clutch slave cylinder bracket
75 Sump plug
76 Speedo driving gear
77 Circlip
78 Distance washer
79 Ball race
80 1st speed gear
81 Spring
82 Shim
83 Synchromesh ball
84 Plunger
85 Ball
86 2nd speed synchro hub
87 2nd speed synchro cup
88 Thrust washer
89 2nd speed mainshaft gear
90 Thrust washer
91 Bushes
92 3rd speed mainshaft gear
93 Thrust washer
94 Circlip
95 3rd/top synchro sleeve
96 3rd speed synchro cup
97 3rd/top inner synchro hub
98 Top synchro cup
99 Circlip
100 Distance washer
101 Circlip
102 Ball race
103 Oil deflector
104 Input shaft
105 Torrington needle roller bearing
106 Mainshaft
107 Distance washer
108 Driving flange
109 Spring washer
110 Nut
112 Countershaft
113 Peg bolt
114 Spring Washer
115 Rear fixed thrust washer
116 Rear rotating thrust washer
117 Countershaft gear cluster
118 Countershaft bush
119 Front fixed thrust washer
120 Reverse gear bush
121 Reverse gear
122 Reverse gear actuator
123 Actuator pivot
124 Plain washer
125 Nyloc nut
126 Reverse gear shaft
127 Reverse shaft retaining bolt.
128 Spring washer

[CAUTION: Please read the instructions thoroughly before you attempt transmission repairs. If you lack the necessary skills or tools we suggest you leave such work to an Authorized Dealer or other qualified shop. Before taking a removed transmission in for repair, thoroughly clean its exterior but do not undertake any disassembly.]

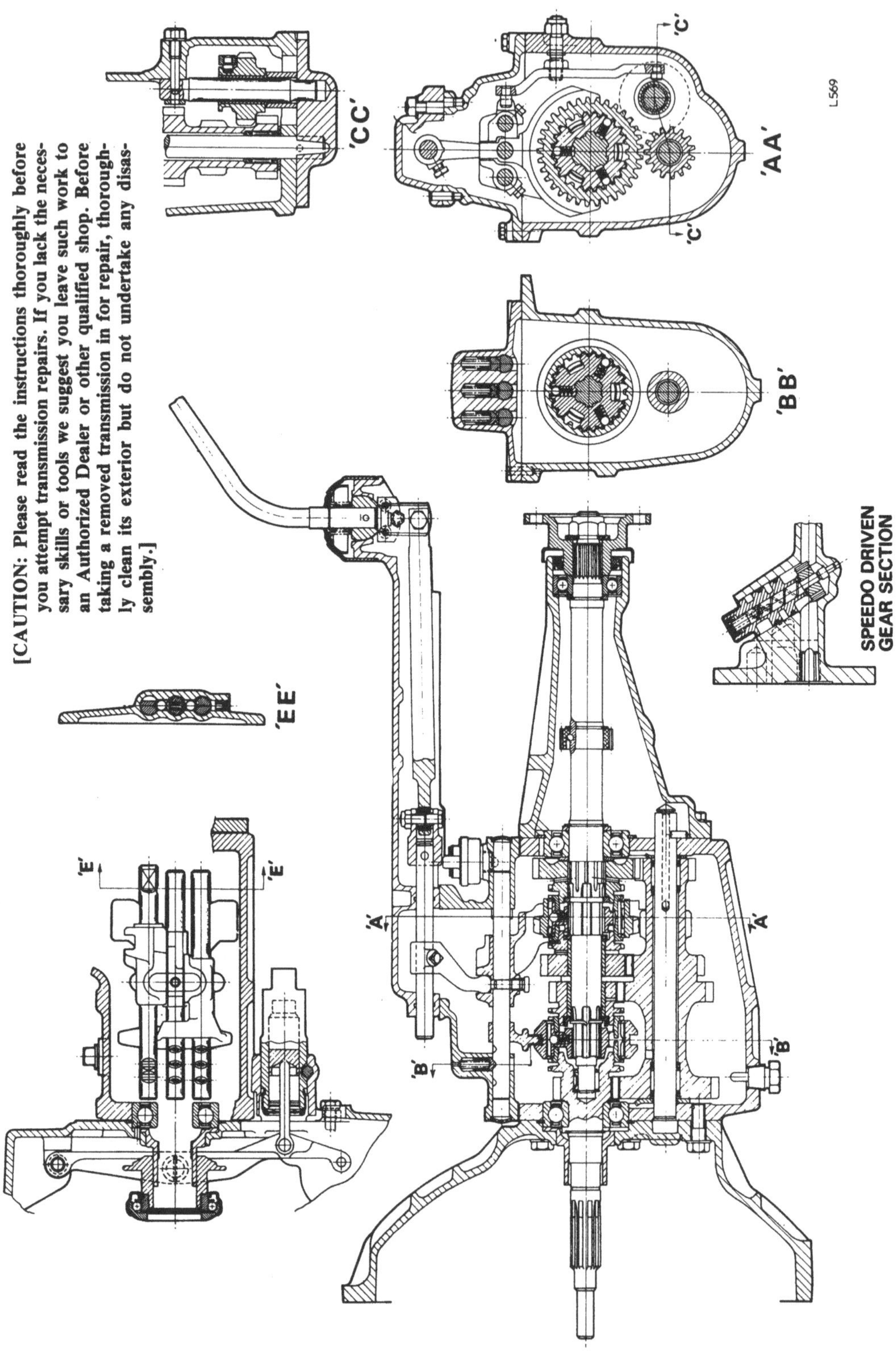

Section through gearbox assembly

DRIVE FLANGE

– Remove and refit **37.10.01**

Removing

1. Drive the vehicle onto a ramp and raise the ramp.
2. Disconnect the propeller shaft from the gearbox drive flange.
3. Select first gear.
4. **Remove split pin [where applicable.]**
5. Unscrew the flange nut and remove the washer.
6. Carefully remove the flange.

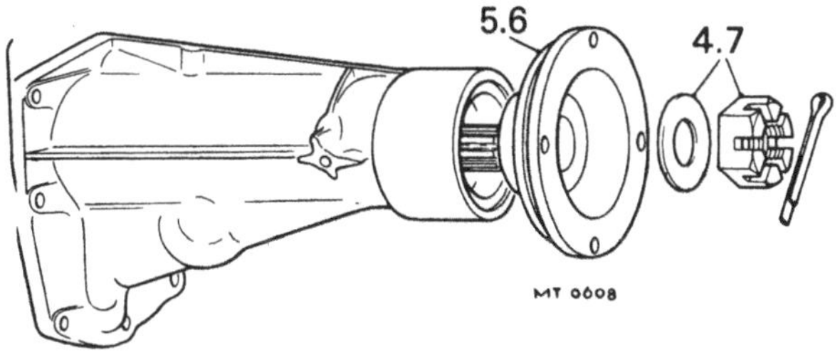

Refitting

7. Locate the flange in position.
8. Fit and tighten the nut and washer.
9. **Refit the split pin [where applicable.]**
10. Attach the propeller shaft to the drive flange.
11. Lower the ramp.

REAR EXTENSION

– Remove and refit **37.12.01**

Removing

1. Drive the vehicle onto a ramp. Raise the ramp and drain the gearbox oil.
2. Remove the propeller shaft assembly.
3. **Remove split pin [where applicable], nut and washer.**
4. Remove the drive flange.
5. Support the engine under the sump, using a ramp jack (interpose a piece of wood between the jack head and sump to spread the load).
6. Disconnect the exhuast support bracket from the front pipe.
7. Remove the gearbox mounting assembly.
8. Disconnect the speedometer cable from the rear extension.
9. Remove the bolts and spring washers.
10. Withdraw the rear extension assembly joint washer and mainshaft washer.

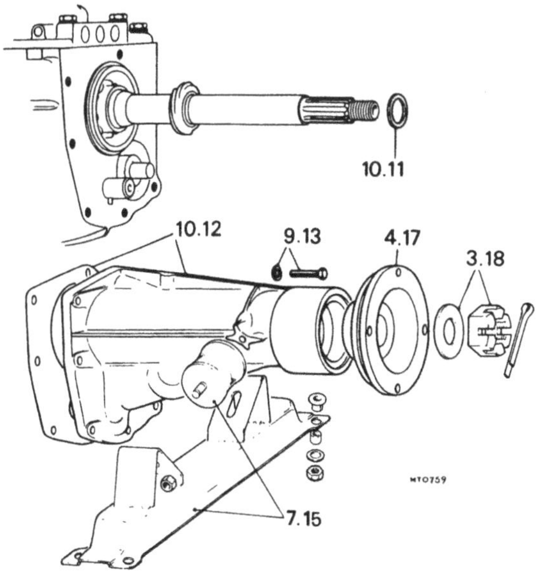

Refitting

11. Locate the washer on the end of the mainshaft.
12. Refit the rear extension and a new joint washer.
13. Fit and tighten the bolts and spring washers.
14. Reconnect the speedometer drive cable.
15. Refit the gearbox mounting assembly and remove the ramp jack.
16. Attach the exhaust front pipe to the support bracket.
17. Refit the drive flange.
18. **Fit washer and nut, tighten, fit split pin [where applicable.]**
19. Refit the propeller shaft assembly.
20. Refill the gearbox with oil.

REAR EXTENSION

– Overhaul **37.12.04**

1. Remove the rear extension 37.12.01 or 37.20.04.

Dismantling

2. Unscrew and remove the peg bolt and washer.
3. Withdraw the speedometer driven gear and housing assembly.
4. Extract the seal.
5. Drive out the bearing.

Reassembling

6. Press the bearing into the extension.
7. Press a new seal into the housing.
8. Refit the speedometer driven gear and housing assembly.
9. Fit and tighten the peg bolt and washer.
10. Refit the rear extension 37.12.01 or 37.20.04.

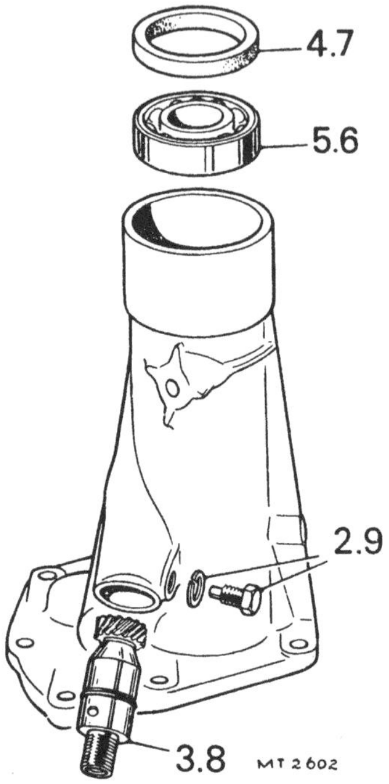

CLUTCH HOUSING

– Remove and refit **37.12.07**

1. Remove the gearbox assembly 37.20.01.

Removing

2. Remove the clutch release mechanism 33.35.12. operation 2.
3. Unscrew and remove the five bolts.
4. Remove the clutch housing and joint washer.

Refitting

5. Replace the clutch housing and a new joint washer.
6. Fit and tighten the five bolts· ensure that the copper washer and bolt with plain shank is fitted at the bottom.
7. Refit the clutch release mechanism 33.35.12 operation 2.
8. Refit the gearbox assembly 37.20.01.

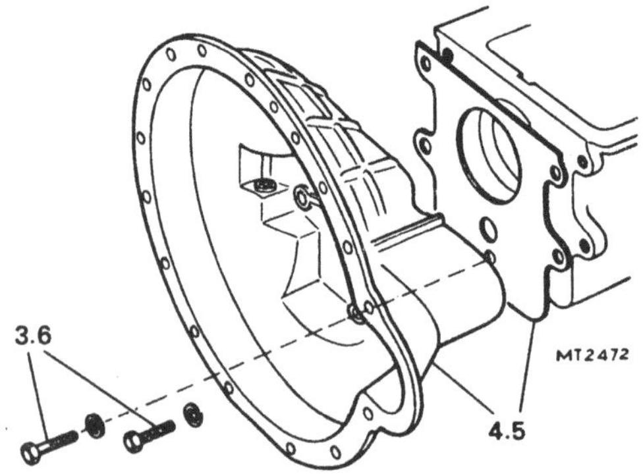

TOP COVER EXTENSION

– Remove and refit 37.12.10

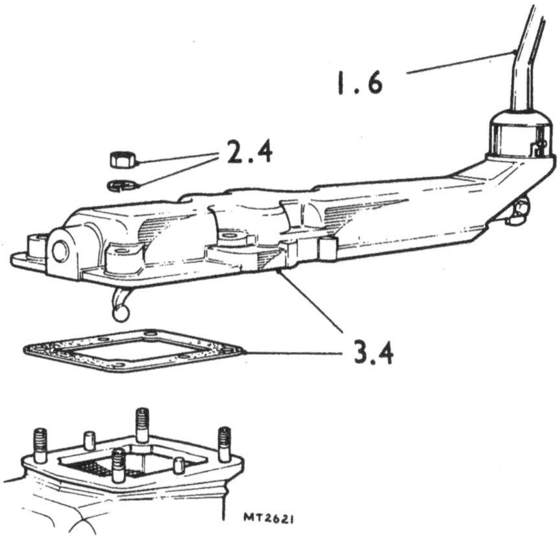

Removing

1. Remove the gear change lever 37.16.04.
2. Working inside the car, unscrew the nuts and washers securing the extension to the top cover.
3. Lift off the extension and remove the joint washer.

Refitting

4. Fit the extension and a new joint washer on to the top cover.
5. Fit and tighten the nuts and washers.
6. Refit the gearchange lever 37.16.04.

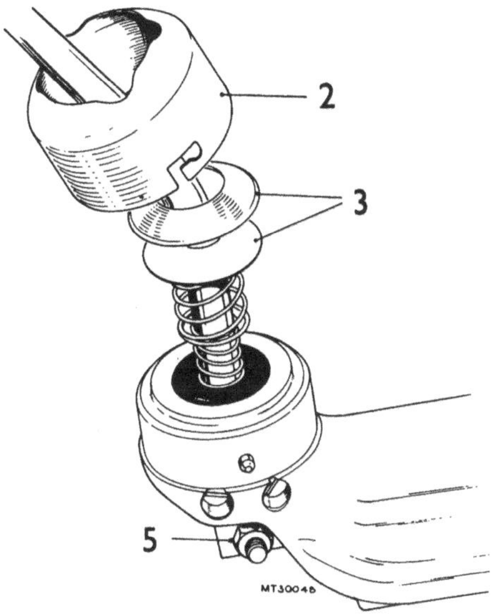

TOP COVER EXTENSION

– Overhaul 37.12.13

Dismantling

1. Remove the top cover extension 37.12.10.
2. Remove the cap.
3. Lift off the steel and nylon cups.
4. Remove the spring.
5. Release the lever from the operating shaft.
6. Withdraw the lever.
7. Remove the circlip.
8. Take off the spring.
9. Remove the nylon sphere.
10. Unscrew the stop-bolt and locknut.
11. Drill out the two rivets.
12. Remove the reverse stop plate.
13. Unscrew the taper locking pin.
14. Withdraw the shaft assembly.
15. Remove the actuator.
16. Remove the 'O' rings from the extension housing bores.
17. Unscrew and remove the nut.
18. Withdraw the bolt.
19. Separate the shafts and remove the washers.
20. Press out the bush.
21. Drive out the Mills pin.
22. Remove the fork end.

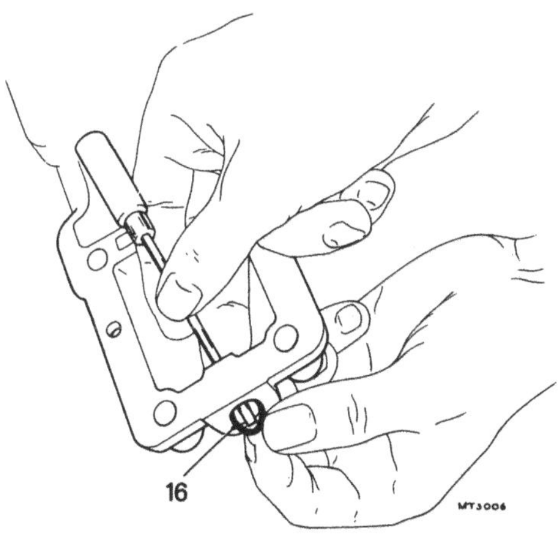

continued

Reassembling

23. Fit the fork end onto the actuator shaft.
24. Refit the Mills pin.
25. Press the bush into the primary shaft.
26. Fit washers on both sides of the bush.
27. Assemble the shafts and fit the bolt from the underside.
28. Fit and tighten the nut to a torque loading of 8 to 10 lbf ft (1·1 to 1·4 kgf m).
29. Fit the 'O' ring to the extension housing bores.
30. Place the reverse stop plate in position.
31. Secure the stop plate with pop rivets.
32. Fit the actuator shaft through the rear bore of the extension housing.
33. Locate the actuator onto the shaft.
34. Push the shaft into position and align the taper pin holes in the actuator and shaft. Fit the taper locking pin.
35. Loosely fit the stop-bolt and locknut to the lever.
36. Fit the nylon sphere to the lever.
37. Fit the reverse baulk spring.
38. Secure the circlip.
39. Fit the lever assembly into the extension.
40. Secure the lever to the shaft.
41. Fit the spring.
42. Refit the nylon and steel cups.
43. Refit the cap.
44. Refit the top cover extension and adjust the reverse stop bolt 37.12.13 and 37.16.01.

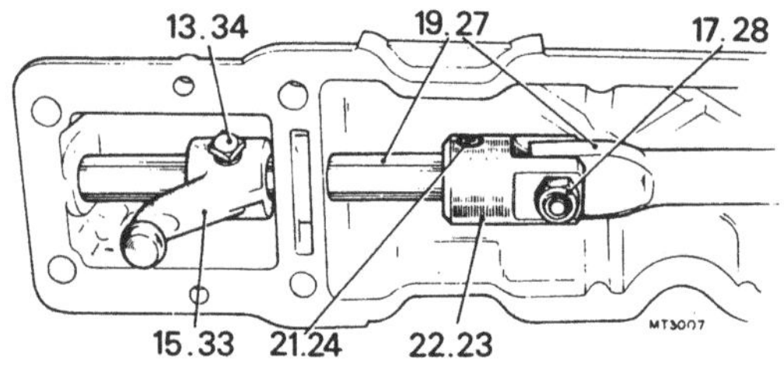

TOP COVER

– Remove and refit **37.12.16**

Removing

1. Select first gear.
2. Remove the top cover extension 37.12.13.
3. Working from inside the car, unscrew and remove the bolts and washers.
4. Remove the top cover and joint washer.

Refitting

5. Ensure that the first gear position is selected.
6. Fit the top cover and a new joint washer.
7. Fit and tighten the bolts and washers.
8. Refit the top cover extension 37.12.13.
9. Check that each gear position may be selected.

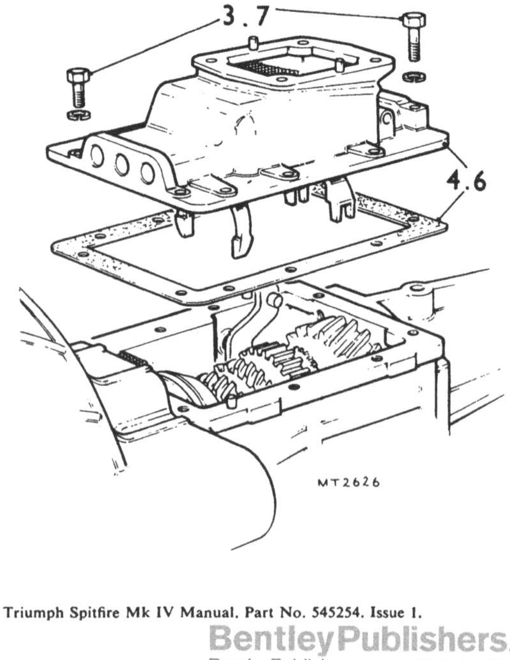

TOP COVER

– Overhaul 37.12.19

1. Remove the top cover 37.12.16.

Dismantling

2. Using a 0·125 in (3·0 mm) dia pin punch, drive out the selector shaft welch plugs.
3. Unscrew the taper locking pins.
4. Drive out the 1st/2nd selector shaft towards the rear of the top cover.
5. Remove the 1st/2nd selector fork.
6. Remove the sleeve.
7. Remove the detent plunger and spring.
8. Remove the interlock plunger from the shaft.
9. Shake out the interlock balls.
10. Drive out the 3rd/top selector shaft towards the rear of the top cover.
11. Remove the detent plunger and spring.
12. Remove the 3rd/top selector fork.
13. Drive out the reverse selector fork.
14. Remove the detent plunger and spring.
15. Remove the reverse actuator.

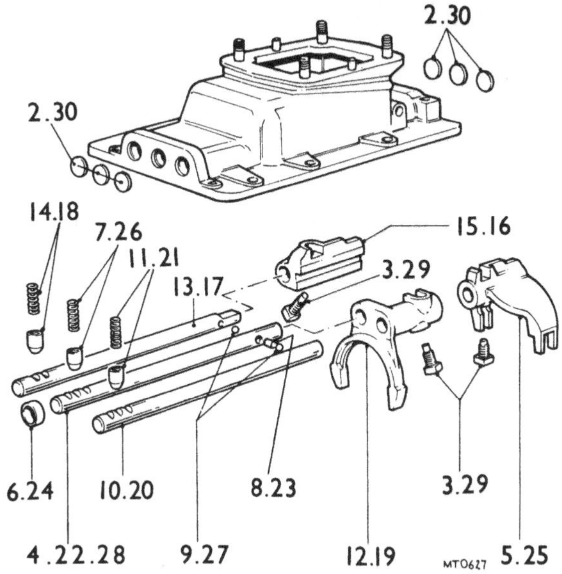

Reassembling

16. Invert the top cover and place the reverse actuator inside.
17. Slide the reverse selector shaft through the rear end of the top cover and through the reverse actuator.
18. Fit the spring and detent plunger into the top cover and push the selector shaft into the neutral position (2nd detent).
19. Place the 3rd/top selector fork into the top cover.
20. Slide the 3rd/top selector shaft through the rear end of the top cover and through the 3rd/top fork.
21. Fit the spring and detent plunger into the top cover and push the shaft into the neutral position (middle detent).
22. Slide the 1st/2nd selector shaft through the rear end of the top cover.
23. Fit the interlock plunger into the shaft.
24. Fit the sleeve over the shaft.
25. Fit the 1st/2nd selector fork in position and slide the shaft through both selector forks.
26. Fit the spring and detent plunger into the top cover and push the shaft through until it just protrudes through the front end of the cover.
27. Locate the interlock balls in the top cover.
28. Push the 1st/2nd selector shaft rearwards until it locates in the neutral (middle detent) position.
29. Fit and tighten the three taper locking pins.
30. Apply sealing compound to the edges of six new welch plugs and fit them into the top cover.
31. Refit the top cover 37.12.16.

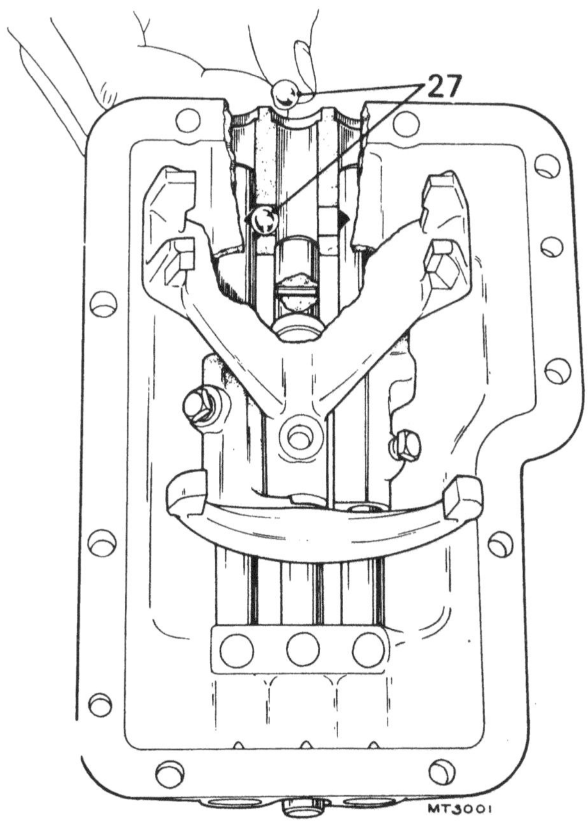

GEARCHANGE LEVER

– Check and adjust **37.16.01**

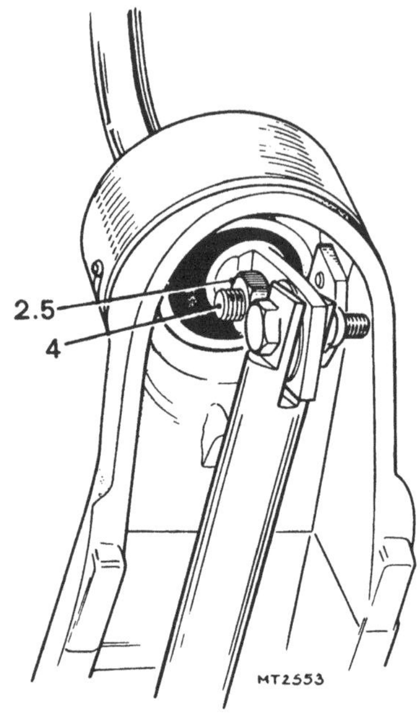

1. Drive the vehicle onto a ramp and raise the ramp.
2. Slacken the locknut.
3. Place the gear lever in the neutral position of the 1st/2nd gate.
4. Adjust the stop bolt to prevent inadvertent selection of the reverse gate without depressing the lever. Check that the reverse gate can be selected by depressing the lever.

NOTE: Failure to obtain the required adjustment will necessitate the renewal of the reverse stop bolt and reverse stop plate.

5. Tighten the locknut.

GEARCHANGE LEVER

– Remove and refit **37.16.04**

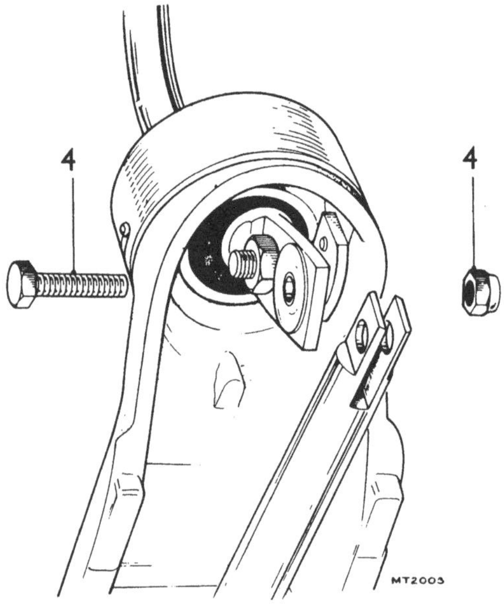

1. Drive the vehicle onto a ramp.
2. Remove the gear lever grommet 37.16.05.

Removing

3. Remove the fabric washer (if fitted).
4. Unscrew the nut and bolt and remove the bottle washer.
5. Remove the bushes, washers, pinch sleeve.
6. Release the cap.
7. Remove the steel and nylon cups.
8. Remove the spring.
9. Withdraw the gear change lever.

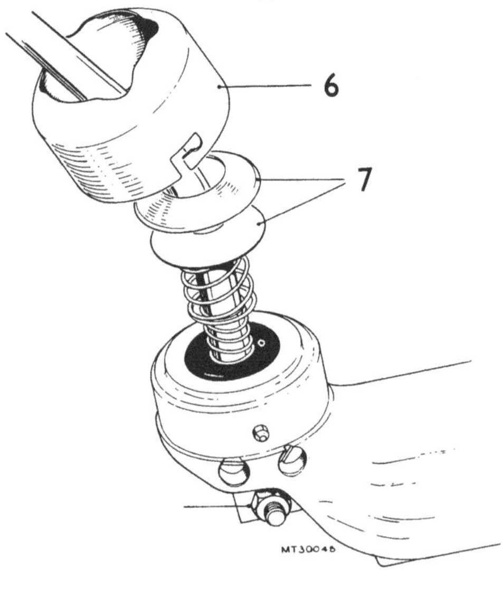

Refitting

10. Fit the gear change lever into the extension.
11. Replace the spring.
12. Replace the nylon and steel cups.
13. Refit the cap.
14. Refit the bushes, washers and pinch sleeve.
15. Assemble the lever, to the selector shaft and fit the bolt and washer.
16. Fit and tighten the nut to a torque loading of 6 to 8 lbf. ft (0·8 to 1·1 kgf. m).
17. Adjust the reverse stop bolt 37.16.01.
18. Refit the fabric washer.
19. Refit the gear lever grommet 37.16.05.

Triumph Spitfire Mk IV Manual. Part No. 545254. Issue 1.

GEARLEVER DRAUGHT EXCLUDER

– Remove and refit **37.16.05**

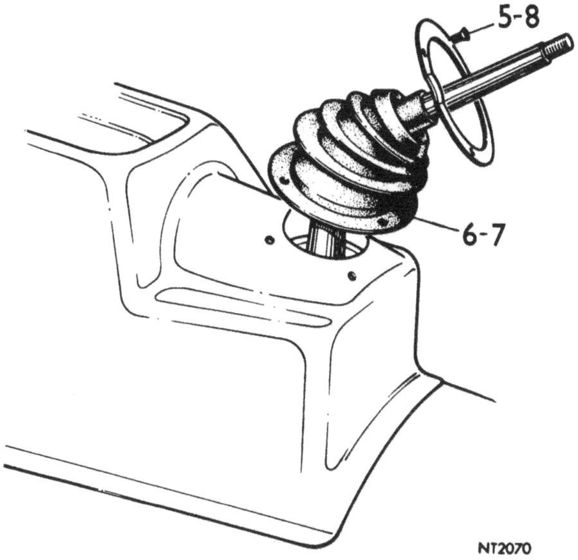

Removing

1. Remove gearlever knob.
2. Remove gearbox trim pads.
3. Remove facia support bracket 76.46.09.
4. Remove gearbox cover carpet 76.49.01.
5. Remove four screws holding draught excluder to gearbox cover.
6. Remove draught excluder.

Refitting

7. Replace draught excluder and refit.
8. Reposition metal ring and refit four screws.
9. Refit gearbox cover carpet 76.49.01.
10. Refit facia support bracket 76.46.09.
11. Refit gearbox trim pads.
12. Refit gearlever knob.

GEARCHANGE LEVER

– Overhaul **37.16.10**

Dismantling

1. Remove the gear- change lever 37.16.04.
2. Take out the pinch sleeve.
3. Remove the bushes and washers.
4. Slacken and remove the locknut and unscrew the stop bolt.
5. Remove the snap-ring.
6. Take off the spring.
7. Remove the nylon sphere.

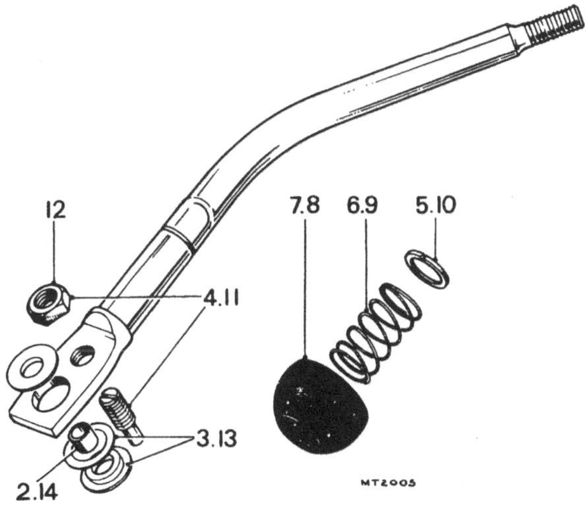

Reassembling

8. Fit a new nylon sphere to the lever.
9. Fit a new spring.
10. Fit a new snap-ring.
11. Screw a new reverse stop bolt into the lever.
12. Refit the locknut.
13. Assemble new bushes and the washers to the lever.
14. Refit the pinch sleeve.
15. Refit the gear-change lever assembly 37.16.04.

GEARBOX ASSEMBLY

– Remove and refit **37.20.01**

1. Drive the vehicle onto a ramp and disconnect the battery.
2. Take out the gearbox cover 76.25.07.

Removing

3. Disconnect the propeller shaft from the gearbox drive flange.
4. Disconnect the speedometer cable from the gearbox.
5. Take out the pinch-bolt and withdraw the clutch, slave cylinder.
6. Raise the ramp, drain the gearbox oil and support the sump with a ramp jack.
7. Disconnect the exhaust pipe from the support bracket.
8. Remove the gearbox mounting nuts and washers.
9. Take out the bell housing bolts accessible from below.
10. Lower the ramp and open the bonnet.
11. Take out the starter motor attachment bolts.
12. Working inside the vehicle, remove the top cover extension nuts.
13. Detach the gearbox mounting assembly.
14. Remove the remaining bell housing bolts and nuts.
15. Carefully withdraw the gearbox assembly and remove it from the vehicle.

Refitting

16. Check that the clutch driven plate is centralized.
17. Fit the gearbox in position, ensuring that the constant pinion shaft is not allowed to hang on the clutch plate splines.
18. Fit the dowel bolt and nut.
19. Fit and tighten the nuts and washers.
20. Secure the upper bell housing bolts and nuts.
21. Refit the gearbox mounting assembly.
22. Fit and tighten the bolts.
23. Refit the clutch slave cylinder.
24. Fit and tighten the pinch-bolt and nut.
25. Refit the speedometer drive cable.
26. Refit the propeller shaft to the gearbox drive flange.
27. Working under the bonnet, refit the starter motor.
28. Raise the ramp.
29. Tighten the dowel bolt and nut.
30. Fit and tighten the remaining bell housing bolts.
31. Lower and remove the ramp jack.
32. Fit and tighten the gearbox mounting nut.
33. Refit the exhaust front pipe to the support bracket.
34. Lower the ramp.
35. Refill the gearbox with oil.
36. Refit the gearbox cover.
37. Refit the parcel shelf.
38. Connect the battery and close the bonnet.

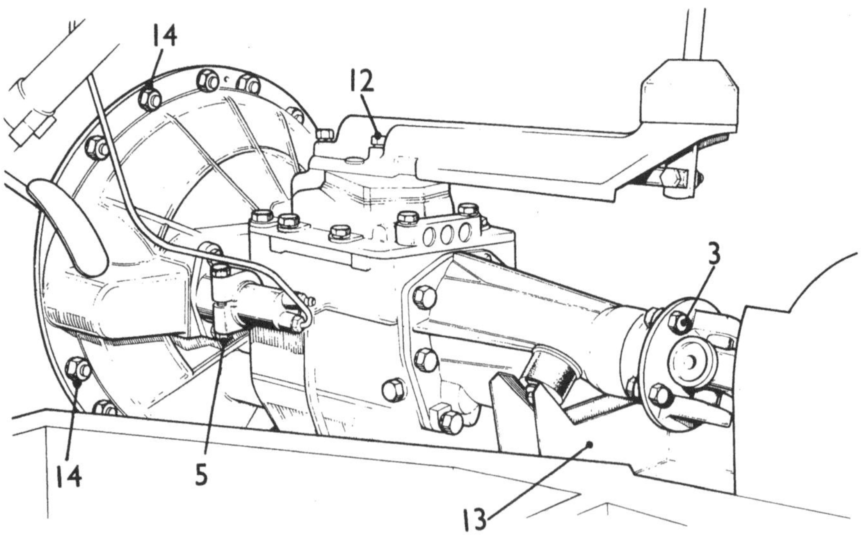

NT2071

GEARBOX ASSEMBLY

– Overhaul **37.20.04**

Service tools: RG 421, S4235A, S4235A-2, S4221A, S4221A-19, S4221A-19/1, S4221A-19/2 S4221A-19/3, S144, A145, S314/1, Needle-roller retaining tube.

1. Remove the gearbox 37.20.01.

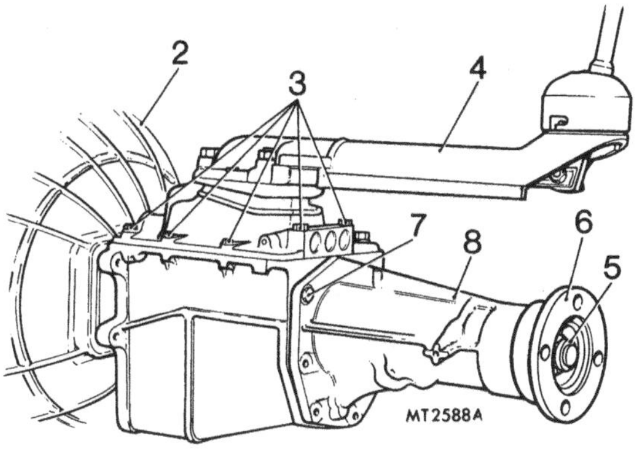

Dismantling

2. Remove the bell housing 37.12.07.
3. Take out the bolts.
4. Lift off the top cover and extension and the joint washer.
5. Using Tool RG 421 unscrew the flange nut and remove the washer.
6. Withdraw the flange.
7. Unscrew the bolts.
8. Withdraw the rear extension and joint washer.
9. Unscrew the retaining bolt.
10. Withdraw the reverse idler spindle and distance tube.
11. Insert the needle-roller retaining tube (corresponding to the dimensions given) and eject the layshaft spindle to the rear. Allow the layshaft cluster to drop to the bottom of the gearbox.
12. Using Tool No. S4235A-2, withdraw the constant pinion assembly.
13. Remove the top gear baulk ring.
14. Remove the circlip retaining the speedometer drive gear.
15. Remove the speedometer drive gear and ball.
16. Remove the snap-ring from the mainshaft ballrace.
17. Remove the circlip from the mainshaft ballrace.
18. Fit the abutment plate, Tool No. S4221A-19.
19. Fit Tool No. S4221A and adaptor S4221A-19/1, to the annular groove in the mainshaft centre ballrace.
20. Withdraw the ballrace.

continued

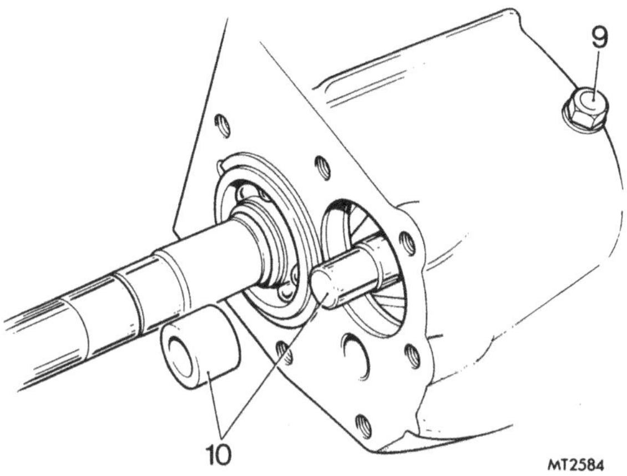

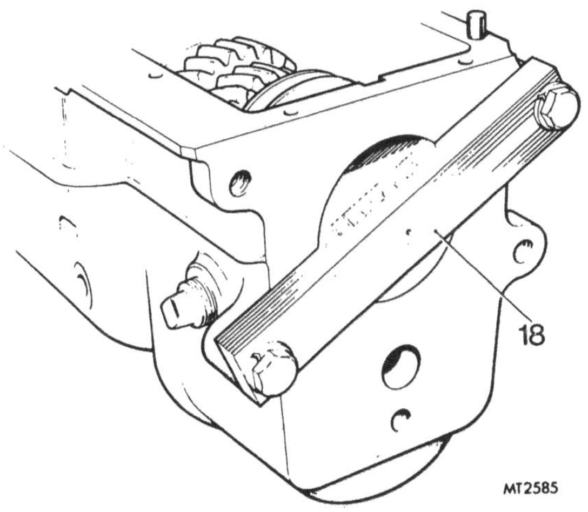

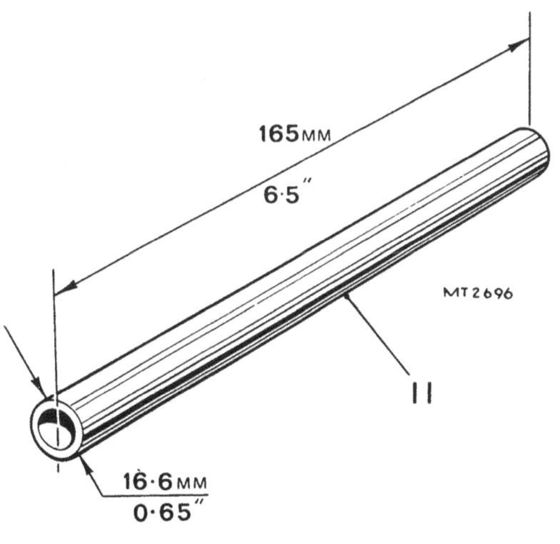

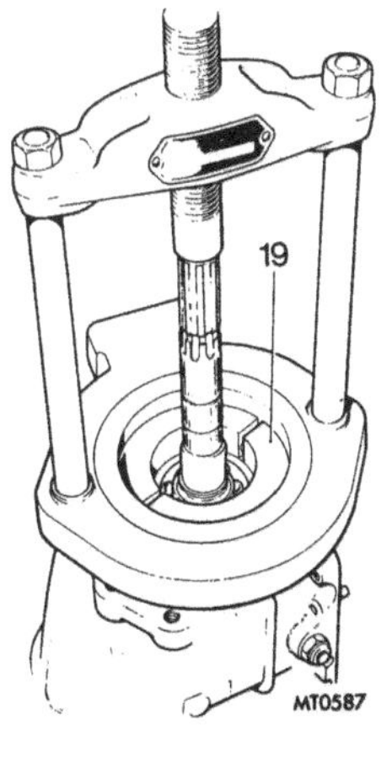

[**NOTE:** In repairing the Mk. III transmission, see the exploded view given following the table of contents for this section (page 218).]

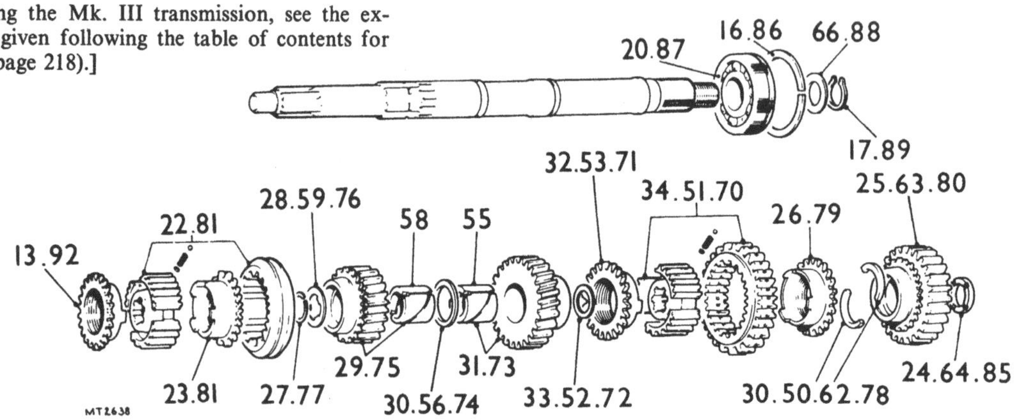

21. Tilt the mainshaft and remove it from the gearbox.
22. Remove the 3rd/top synchro unit.
23. Remove the 3rd gear baulk ring.
24. From the rear of the mainshaft, remove the thrust washer.
25. Remove the 1st speed gear.
26. Remove the 1st gear baulk ring. [except Mk. III].
27. **Remove the circlip. Where the third gear mainshaft thrust washer has three lugs use Tool S.144. Where the washer has six lugs use Tool S.144A.**
28. Withdraw the washer.
29. Remove the 3rd speed gear and bush.
30. Remove the thrust washer.
31. Remove the 2nd speed gear and bush.
32. Remove the 2nd gear baulk ring.
33. Remove the thrust washer.
34. **Remove the [2nd gear synchro unit (Mk. III) or the] 1st/2nd synchro unit.**
35. **Remove the split collars. [except the Mk. III].**
36. Lift out the layshaft cluster.
37. Take out the layshaft thrust washers.
38. Remove the reverse idler gear.
39. Unscrew the nut.
40. Remove the reverse actuator and pivot pin.

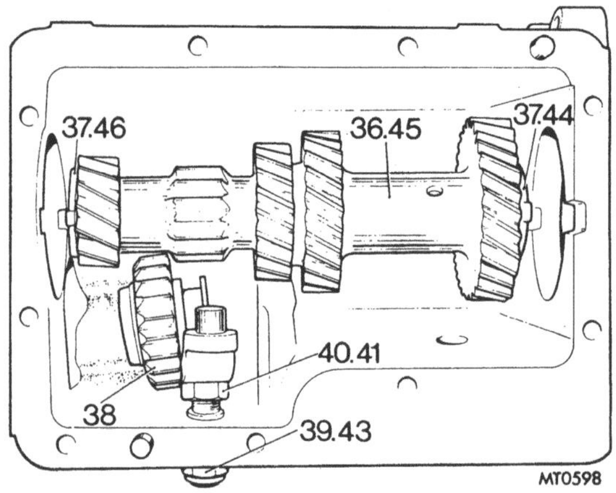

Reassembling

41. Screw the pivot pin into the reverse actuator until one full thread protrudes through the lever boss.
42. Fit the actuator and pivot pin into the gear casing.
43. Fit the plain washer and tighten the nut.
44. Smear the front face of the layshaft front thrust washer with grease and stick it in position in the gear casing. Insert one end of the layshaft spindle through the casing to centralize the washer.
45. Lower the layshaft cluster assembly into the gearbox.
46. Fit the rear thrust washer in position.
47. Insert the layshaft spindle and eject the needle-roller retaining tube.
48. Measure the layshaft cluster end-float. Adjust the layshaft end-float to 0·007 to 0·013 in (0·18 to 0·33 mm), by selective use of thrust washers, repeating 44 to 48 as necessary. Do not remove metal from the bronze face of the thrust washers.
49. Insert the needle-roller retaining tube, eject the layshaft spindle and allow the cluster to drop to the bottom of the gearbox.

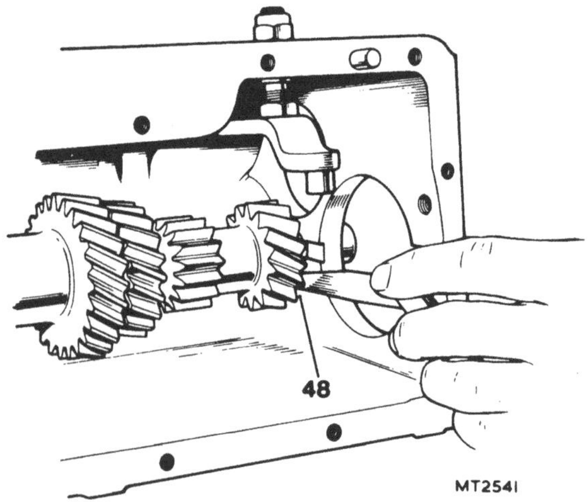

continued

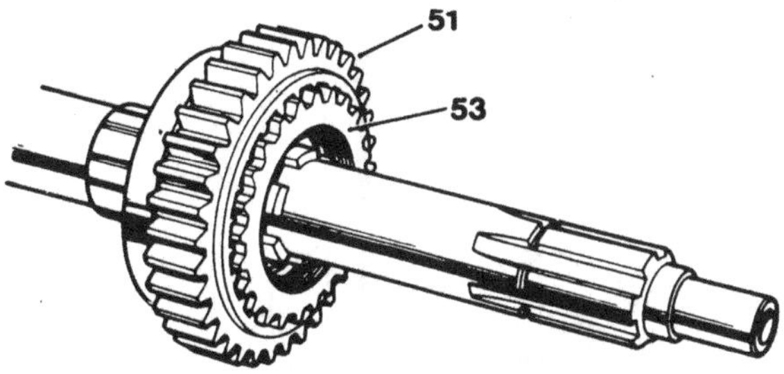

MT2542

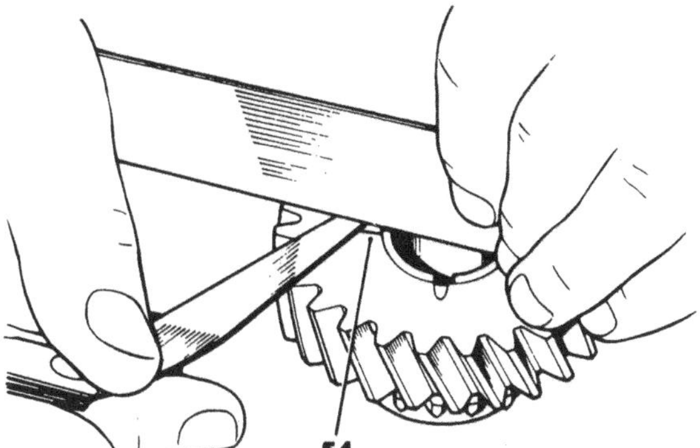

MT2537

50. Fit the split collars to the mainshaft [except Mk. III].
51. **Slide the [2nd gear synchro unit (Mk. III) or the] 1st/2nd synchro unit onto the shaft.**
52. Refit the thrust washer.
53. Fit the 2nd gear baulk ring.
54. Check the end-float of the second gear on the bush. Reduce the length of the bush to reduce the end-float, renew the bush to increase the end-float.
55. Fit the second gear bush to the shaft.
56. Refit the thrust washer.
57. Check the end-float of the 3rd gear on the bush. (Ref. 54).
58. Fit the 3rd gear bush to the shaft.
59. Refit the washer.
60. Secure the assembly using a discarded half circlip.
61. Measure the end-float of the bushes on the mainshaft and adjust by selective use of thrust washers until an end-float of 0·000 to 0·006 in (0·00 to 0·15 mm) is **obtained. [Mk. III end-float should be 0·004 to 0·010 in (0·1016 to 0·256 mm).]**

[NOTE: Instructions 62 through 68 do not apply to Mk. III transmissions.]

62. **[Dismantle the mainshaft.] Fit the split collars to the mainshaft.**
63. Fit the 1st speed gear.
64. Fit the thrust washer.
65. Fit a discarded bearing inner race to distance tube 0·784 to 0·750 in (18·99 to 19·05 mm) long.
66. Measure the thickness of the circlip washer and assemble to the shaft.
67. Fit a discarded half circlip.
68. Measure the 1st speed gear end-float and determine the thickness of the circlip washer required to provide an end-float of 0·000 to 0·002 in (0·00 to 0·05 mm). Dismantle the mainshaft.
69. Assemble components to the mainshaft as follows:
70. **[2nd gear sychro unit (Mk. III) or] first/second synchro unit.**
71. Second gear baulk ring.
72. Thrust washer.
73. Second speed gear and bush.
74. Thrust washer.
75. Third speed gear and bush.

continued

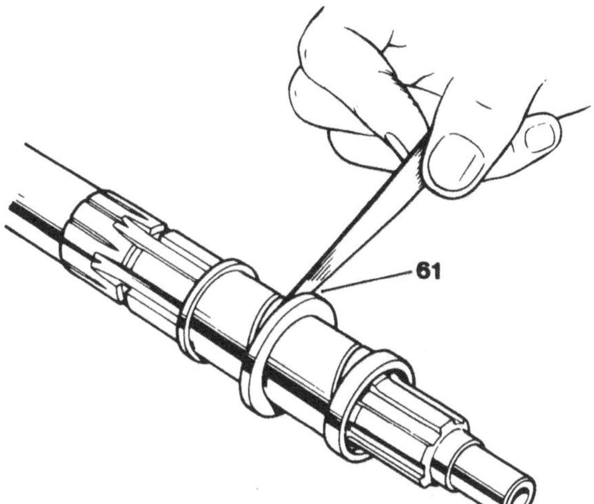

MT2540

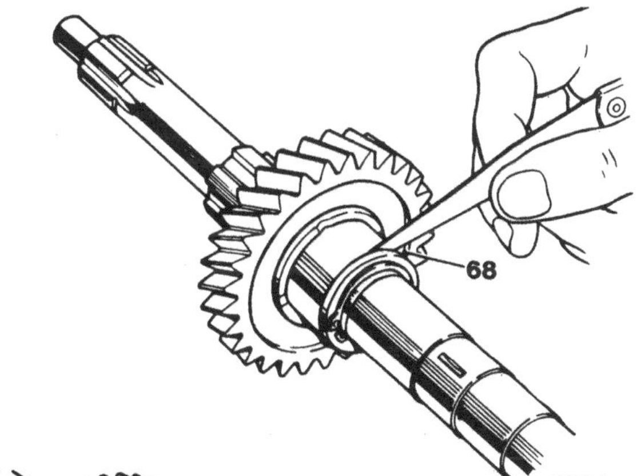

MT2536

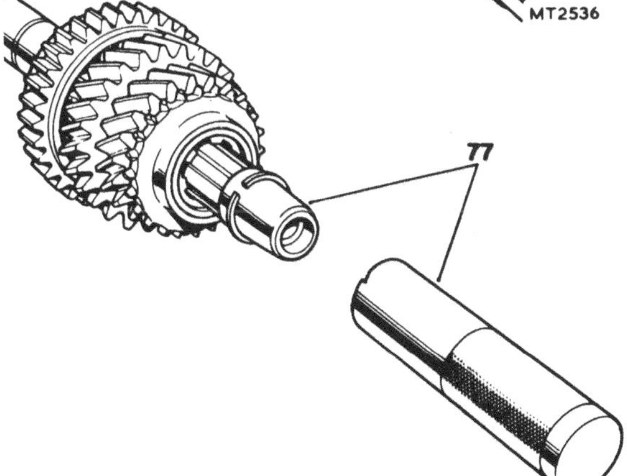

MT2538

76. Washer.
77. Circlip, offset outermost (locate lip of clip in recess of tool No. S145)
78. Split collars **[except Mk. III].**
79. First gear baulk ring **[except Mk. III].**
80. First speed gear.
81. Third/top synchro unit.
82. Position the reverse idler gear in the casing.
83. Place the mainshaft assembly in the gearbox.
84. Fit the abutment plate to the gearbox and mount in a vice.
85. Fit the thrust washer over the mainshaft.
86. Assemble the snap-ring to the ball race.
87. Place the ball race over the mainshaft and drive into position using Tool No. S314/1 and adaptor S4221A-19/3½.
88. Fit the circlip washer selected (Ref. 68).
89. Fit a new circlip.
90. Refit the speedometer drive gear, ball and circlip.
91. Remove the gearbox from the vice and take off the abutment plate.
92. Place the top gear baulk ring in the 3rd/top synchro unit.
93. Refit the constant pinion assembly using Tool No. S314/1 and adaptor S4221A-19/3, ensuring that the top gear baulk ring is correctly located.
94. Invert the gearbox and align the layshaft cluster and thrust washers.
95. Insert the layshaft spindle from the gear and eject the needle-roller retaining tube.
96. Position the reverse idler gear and fit the spindle and distance tube. Fit the locating bolt.
97. Locate the washer on the end of the mainshaft.
98. Refit the rear extension assembly and a new joint washer.
99. Fit the bolts and washers.
100. Replace the drive flange.
101. Fit and tighten the nut and washer and secure with **splitpin [where applicable].**
102. Refit the bell housing assembly and a new joit washer 37.12.07.
103. Fit and tighten the bottom bolt and copper washer.

[NOTE: Before you install the gearbox cover, thoroughly lubricate all the gears and other transmission internal parts with a liberal application of gear oil. Doing so will prevent undue wear when the transmission is first returned to service. Hand-turn the mainshaft so that the lubricant works into the synchronizer assemblies and the bearings.]

104. Select 1st gear on the top cover and the gearbox.
105. Fit the top cover and a new joint washer.
106. Fit and tighten the bolts and washers.

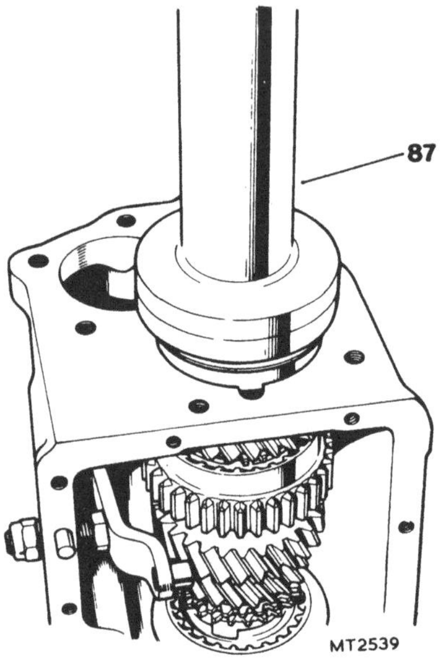

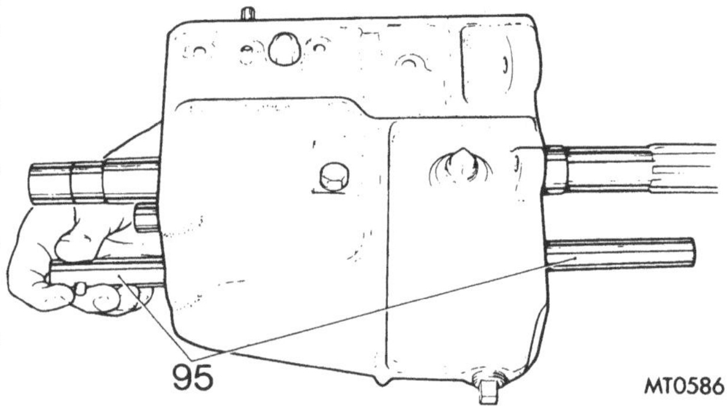

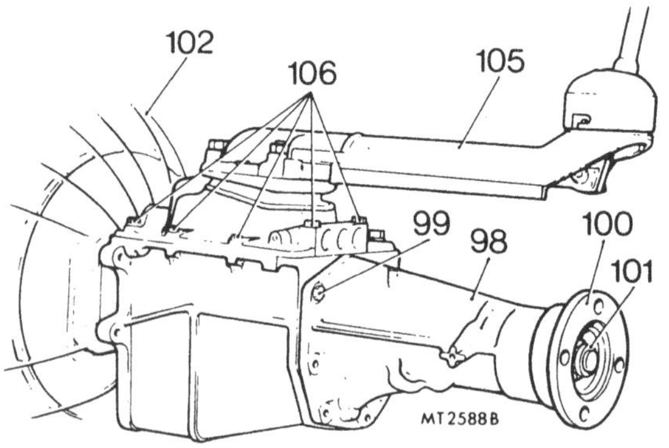

SYNCHRONIZER ASSEMBLIES

– Overhaul **37.20.08**

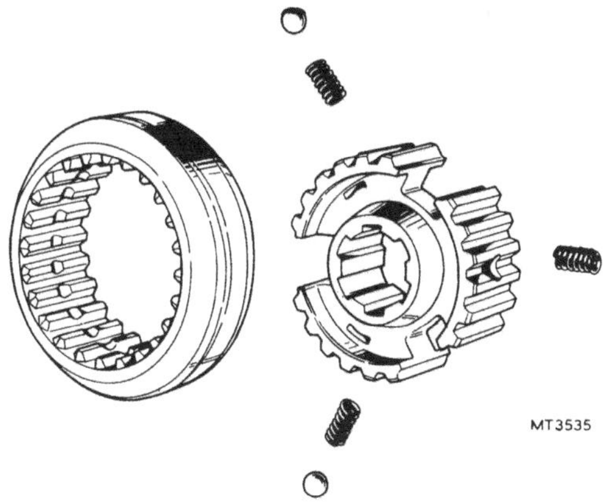

Service tools: spring balance and adaptor.

1. Remove the gearbox assembly 37.20.01.
2. Dismantle the gearbox 37.20.04.

Dismantling

3. Place the assembly in a small box to prevent losing the spring loaded balls, and press the hub through the sleeve.

Reassembling

4. Assemble the synchro springs and balls to the synchro hub and retain them in position with petroleum jelly.
5. Carefully press the sleeve onto the hub.
6. Using the spring balance and adaptor, check the axial release loads, which should be:
   [2nd (Mk. III) or 1st/2nd] 19 to 21 lb (8·6 to 9·5 kg).
   3rd/top . . . . . 19 to 21 lb (8·6 to 9·5 kg).
   If the release loads differ from the specified figures, fit new springs and/or add shims under the springs until the correct loading is achieved.
7. Reassemble the gearbox 37.20.04.
8. Refit the gearbox 37.20.01.

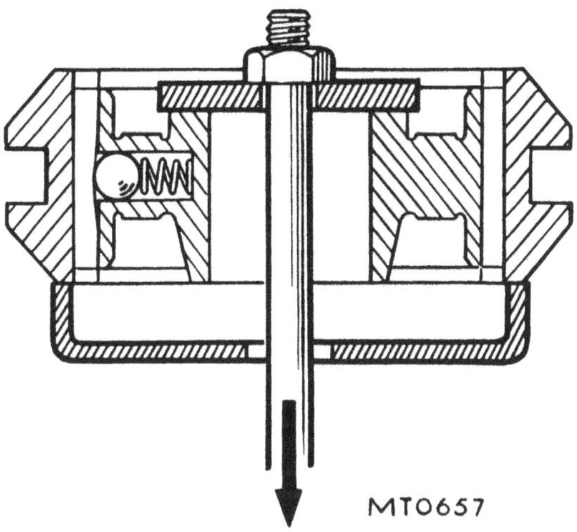

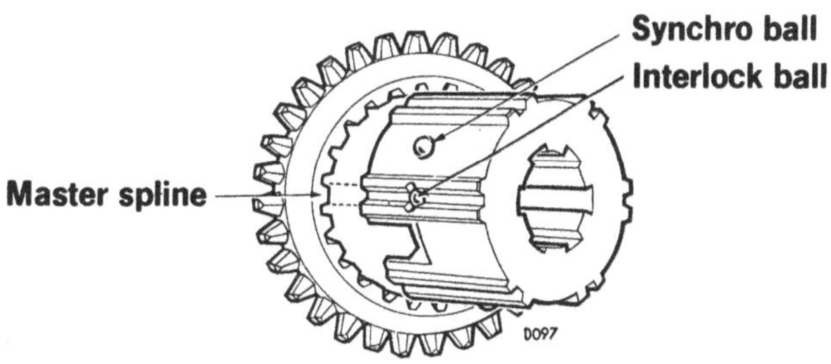

Mk. III 2nd gear sychronizer assembly

LAYSHAFT CLUSTER

– Overhaul 37.20.29

1. Remove the gearbox 37.20.01.
2. Remove the layshaft cluster 37.20.04.

Dismantling

3. Withdraw the needle-roller retaining tube.
4. Shake out the needle-rollers.
5. Prise out the retaining rings.

Reassembling

6. Using a shouldered mandrel, drive the retaining rings into the end bores of the layshaft cluster to the depth shown.
7. Using heavy grease to retain them, refit the needle-rollers.
8. Insert the needle-roller retaining tube.
9. Refit the layshaft cluster 37.20.04.
10. Refit the gearbox 37.20.01.

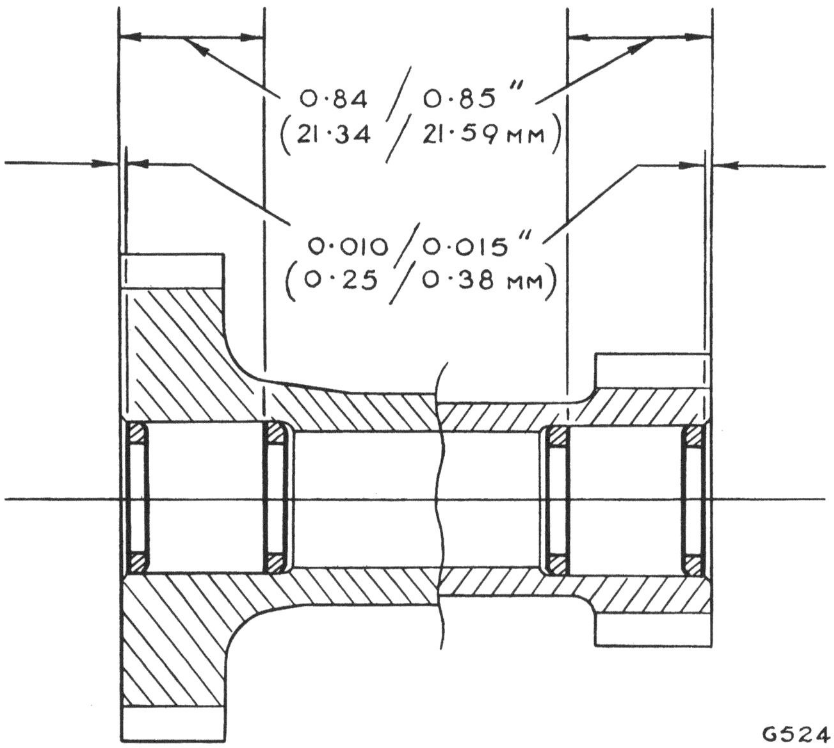

G524

CONSTANT PINION

– Overhaul **37.20.34**

Service tools S4221A, S4221A-19/1

1. Remove the gearbox 37.20.01.
2. Remove the constant pinion 27.20.04 operation 12.

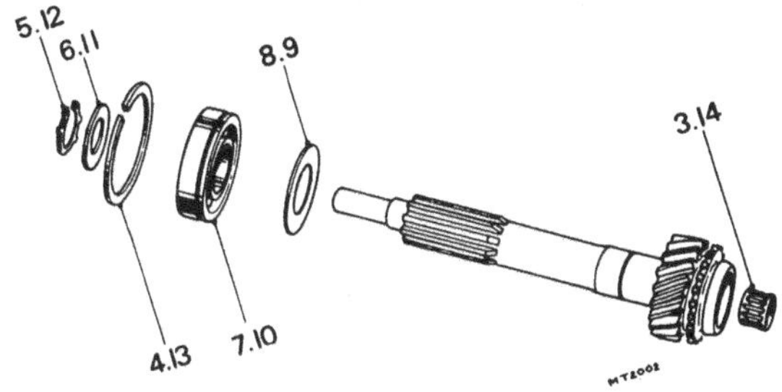

Dismantling

3. Remove the roller bearing.
4. Remove the snap ring.
5. Extract the circlip.
6. Remove the washer.
7. Utilizing Tool No. S4221A and adaptor S4221A-19/1, withdraw the bearing.
8. Remove the oil thrower.

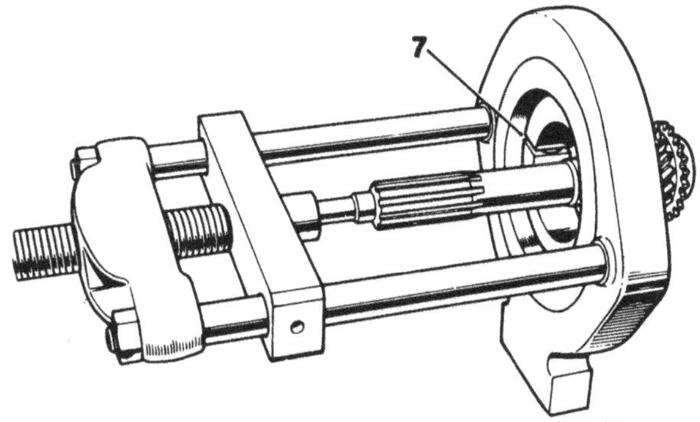

MT3611

Reassembling

9. Refit the oil thrower.
10. Utilizing Tool No. S4221A and adaptor S4221A-19/1, draw the bearing onto the pinion ensuring that the oil thrower is correctly located.
11. Replace the washer.
12. Refit the circlip.
13. Refit the snapring.
14. Replace the roller bearing.
15. Refit the constant pinion 37.20.04 operation 93.
16. Refit the gearbox 37.20.01.

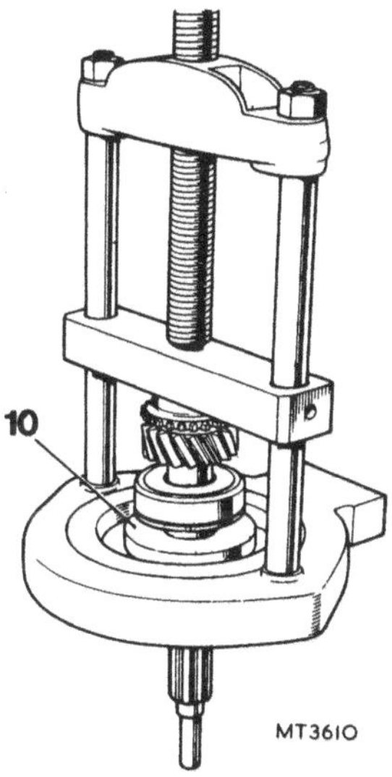

MT3610

REAR OIL SEAL

– Remove and refit **37.23.01**

Removing

1. Remove the propeller shaft assembly.
2. Remove the gearbox drive flange 37.10.01.
3. Prise out the seal.

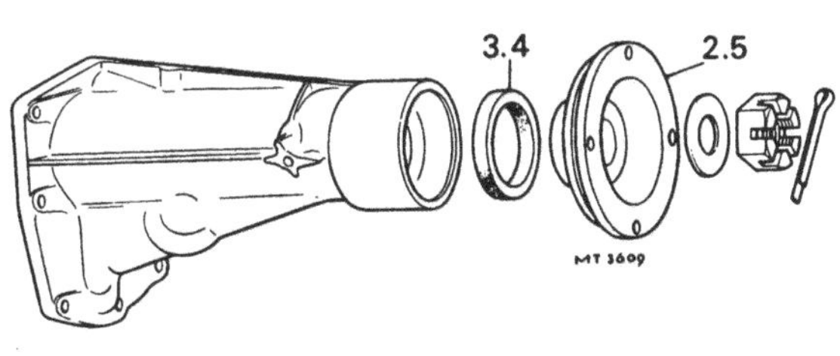

MT 3609

Refitting

4. Drive a new seal into the rear extension.
5. Refit the gearbox drive flange 37.10.01.
6. Refit the propeller shaft.

SPEEDOMETER DRIVE GEAR

– Remove and refit **37.25.01**

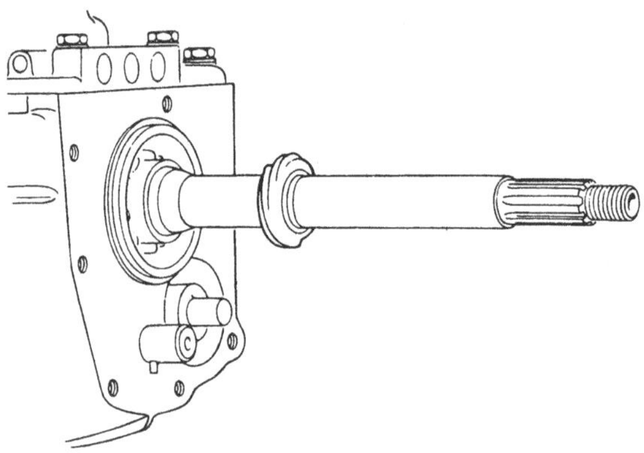

[NOTE: On the Mk. III transmission there is no drive ball and the speedometer drive gear is a press fit on the mainshaft. On such transmissions you must remove the gear using a repair press, or a slide hammer, or a suitable gear puller.]

Removing

1. Remove the rear extension 37.12.01
2. Remove the circlip.
3. Remove the speedometer drive gear and ball.

Refitting

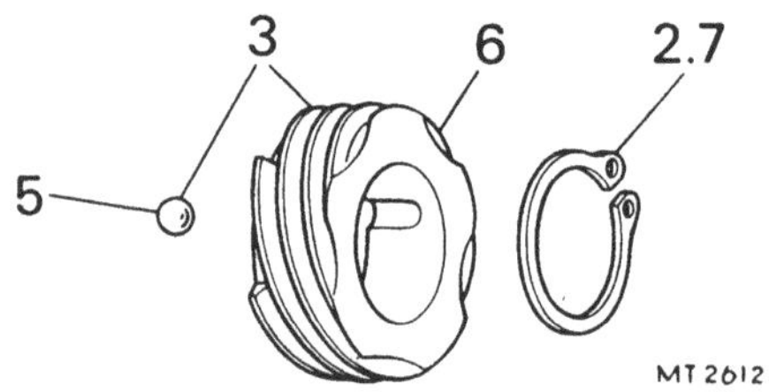

4. Turn the mainshaft until the drive ball detent is uppermost.
5. Locate the drive ball in the detent.
6. Fit the drive gear over the ball.
7. Replace the circlip.
8. Refit the rear extension 37.12.01.

SPEEDOMETER DRIVE GEAR PINION

– Remove and refit **37.25.05**

Removing

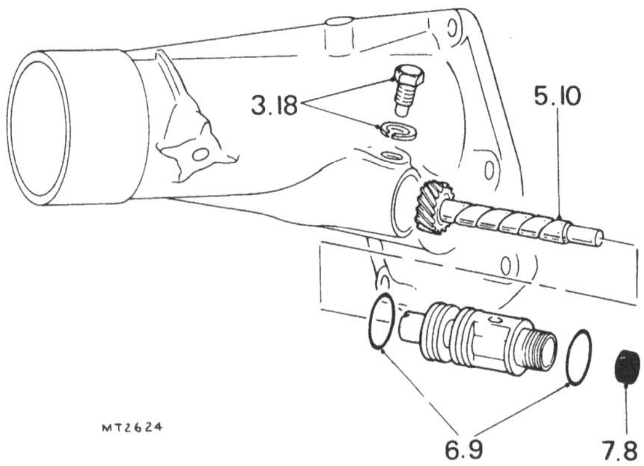

1. Remove gearbox tunnel cover 76.25.07.
2. Disconnect the speedometer cable from the gearbox.
3. Unscrew and remove the peg bolt and washer.
4. Withdraw the speedometer pinion and housing.
5. Withdraw the pinion from the housing.
6. Remove the 'O' ring.
7. Prise out the seal.

Refitting

8. Fit a new seal into the housing.
9. Fit a new 'O' ring.
10. Fit the pinion into the housing.
11. Refit the assembly to the gearbox.
12. Fit and tighten the peg bolt and washer.
13. Reconnect the speedometer cable.
14. Refit gearbox tunnel cover 76.25.07.

OVERDRIVE OPERATIONS

[CAUTION: Please read the instructions thoroughly before you attempt overdrive repairs. If you lack the necessary skills or tools we suggest you leave such work to an Authorized Dealer or other qualified shop. Before taking a removed overdrive in for repair, thoroughly clean its exterior but do not undertake any disassembly.]

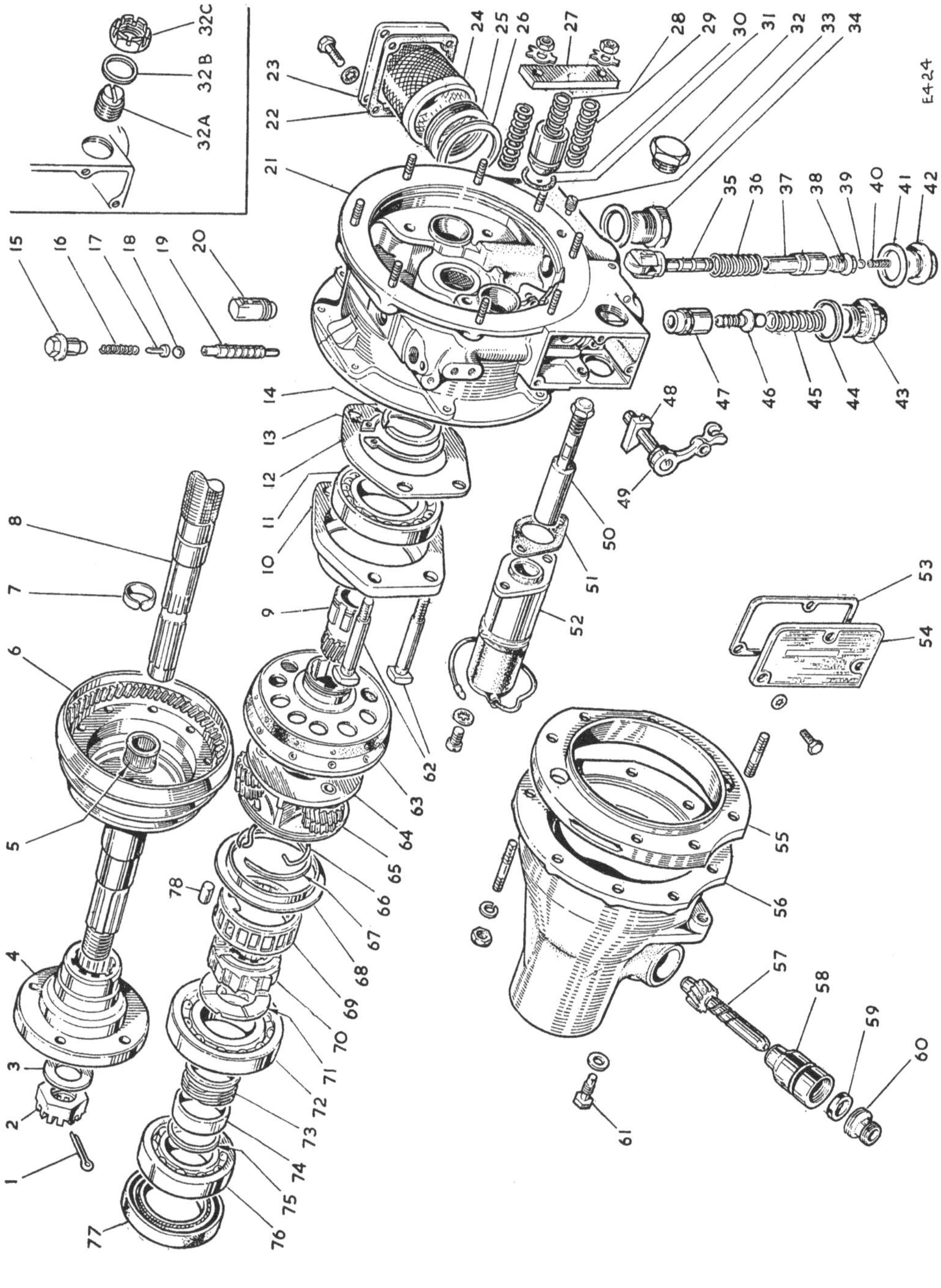

KEY

1	Split pin	22	Gasket	40	Spring	61	Locating screw
2	Nut	23	Cover plate	41	Washer	62	Bolts
3	Washer	24	Filter	42	Plug	63	Cone clutch
4	Coupling flange	25	Magnetic rings	43	Plug	64	Planet carrier assembly
5	Bush	26	Rubber/Steel washer	44	Washer	65	Planet gear
6	Annulus	27	Bridge piece	45	Spring	66	Spring
7	Spring	28	Bias spring	46	Relief valve plunger	67	Circlip
8	Main shaft	29	Clutch return spring	47	Relief valve body	68	Oil thrower
9	Sungear	30	Piston	48	Cam	69	Cage
10	Thrust ring	31	Piston 'O' ring	49	Operating lever	70	Inner member
11	Thrust bearing	32	Plug	50	Solenoid plunger	71	Thrust washer
12	Retaining plate	32A	Adjuster screw } These items replace items 32 on later units	51	Gasket	72	Front bearing
13	Circlip	32B	Locknut }	52	Solenoid	73	Speedometer drive gear
14	Circlip	32C	Fibre washer }	53	Gasket	74	Distance piece
15	Plug	33	Pump locating screw	54	Cover plate	75	Spacer
16	Spring	34	Plug	55	Brake ring	76	Rear bearing
17	Plunger	35	Pump plunger	56	Rear casing	77	Oil seal
18	Ball	36	Return spring	57	Speedometer pinion	78	Roller
19	Operating valve	37	Pump body	58	Speedometer pinion bush	79	Woodruff key
20	Lubrication bush	38	Non-return valve body	59	Seal	80	Ring spring
21	Front casing	39	Ball	60	Screwed end	81	Cam

LAYCOCK DE NORMANVILLE OVERDRIVE

The overdrive is an additional gear unit, mounted on the rear face of the gearbox in place of the normal extension. When in operation, the unit provides a higher overall gear ratio than is available with the standard transmission. Reduced engine speed, resulting from the higher ratio, will reduce fuel consumption, increase engine life, and ensure greater driving comfort, providing the unit is used correctly.

The overdrive is operated by an electrical solenoid, controlled by a switch mounted on the steering column. An inhibitor switch, fitted in the electrical circuit, prevents engagement of overdrive in reverse, first and second gears.

Suggested minimum engagement speeds are:
Top gear . . . . 40m.p.h.
Third gear . . . . 30m.p.h.
Maximum disengagement speeds are:
Top gear . . . . At driver's discretion
Third gear . . . . 70 m.p.h.

Disengagement of the overdrive at a speed higher than stated may cause damage from 'over-revving'.

WORKING PRINCIPLES

Overdrive gears

The epicyclic gear train of the unit consists of a central sungear, meshing with three planet gears which in turn mesh with an internally toothed annulus.

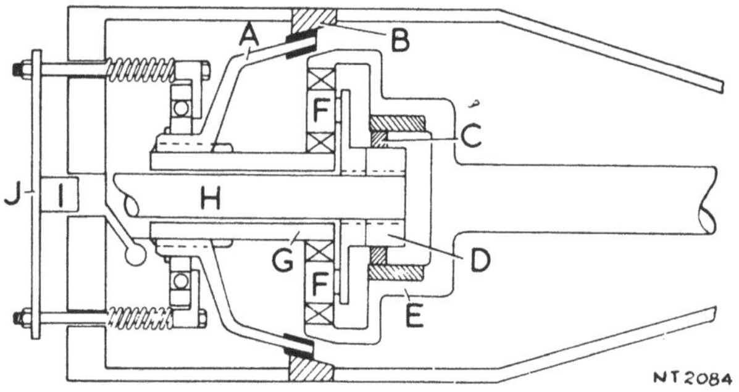

Overdrive disengaged (Fig 1)

A cone clutch (A), mounted on the externally splined extension of the sungear (G) is spring-loaded, by four clutch springs (L), via a thrust ring (K) and bearing (M), against the annulus (E) thus locking the gear train and permitting over-run and reverse torque to be transmitted.

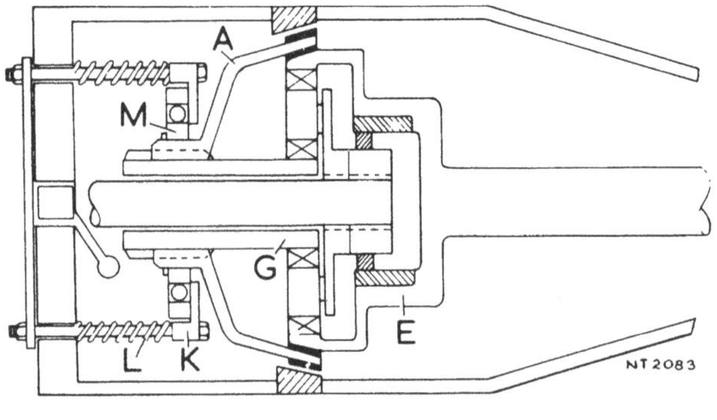

Overdrive engaged (Fig 2)

When overdrive is selected, two hydraulically operated pistons (I) acting against bridge pieces (J), move forward and, overcoming the spring pressure, cause the cone clutch (A) to engage the brake ring (B) with sufficient load to hold the sungear (G) at rest. The planet carrier (D) can now rotate with the input shaft (H) causing the planet gears (F) to rotate about their own axis to drive the annulus at a faster speed than the input shaft, this being allowed by the free-wheeling action of the uni-directional clutch (C).

HYDRAULIC SYSTEM

Hydraulic pressure is developed by a plunger pump, cam operated, from the input shaft. The pump draws oil through a wire mesh filter and delivers it to the operating valve. A relief valve, incorporated in the system, controls the working pressure.

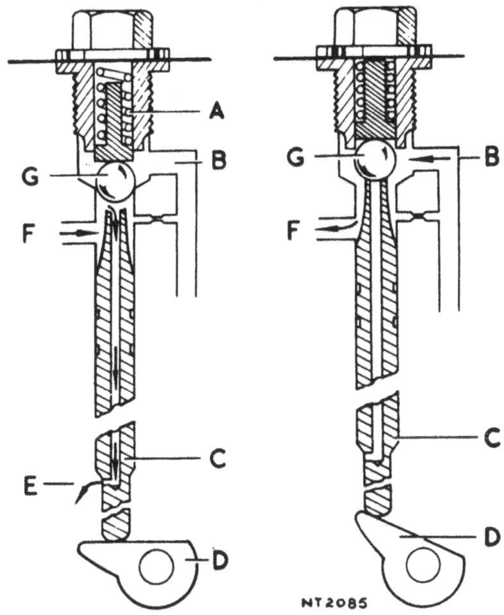

Operating valve (Fig 3)

In direct drive position, the ball valve (G) is seated in the casing thereby isolating the supply (B) from the operating cylinders (F).

When overdrive is selected, a solenoid causes cam (D) to rotate lifting the ball from its seat in the casing, and sealing the top of the valve, thus directing oil under pressure from port (B) to the operating cylinders (F).

When the valve is returned to the direct drive position, oil from the operating cylinders is exhausted down the hollow valve stem through the restrictor (E).

LUBRICATION

Being interconnected, the gearbox and overdrive unit have a common oil level, indicated by a plug on the side of the gearbox. When draining the oil, remove the overdrive unit drain plug and gearbox drain plug. Access to the gauze filter, which must be removed and cleaned prior to refilling with oil, is effected by removing plate (1) (Fig 4) retained by four setscrews.

Spill oil, from the relief valve, is diverted through drilled passages to a bush in the front casing, then into the mainshaft and along a central drilling to the rear bearing in the annulus. From the bearing, oil is passed, due to centrifugal force, through the uni-directional clutch to an oil thrower, from which it is picked up by a catcher on the planet carrier and then to the planet gears via the hollow bearing pins.

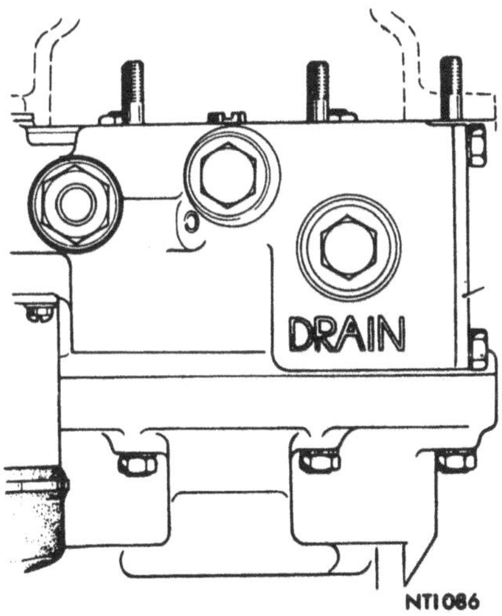

NOTE: All gearbox and overdrive units fitted to new cars are filled with a special oil, formulated to give all necessary protection to new gears. Under normal circumstances, this oil should not be changed, but may be topped up with any of the approved oils. If a new unit is fitted, or parts of an existing unit renewed, the unit should be replenished with new special oil, supplied with a new unit, or ordered separately from the Spares Division.

Should difficulty be experienced in obtaining the special oil, use one of the approved lubricants. ON NO ACCOUNT SHOULD ANTI-FRICTION ADDITIVES BE PUT INTO THE OIL.

After refilling the gearbox and running the car for a short distance, re-check and top up the oil level to replace the oil which has been distributed around the hydraulic system. Always use clean oil and take great care to prevent the entry of foreign matter when any part of the casing is opened.

FAULT FINDING 'D' TYPES	
SYMPTOM	POSSIBLE CAUSE
Overdrive does not engage	1. Insufficient oil in gearbox. 2. Electrical system not working. 3. Solenoid operating lever out of adjustment. 4. Insufficient hydraulic pressure due to pump non-return valve incorrectly seating (probably dirt on seat). 5. Insufficient hydraulic pressure due to sticking or worn relief valves. 6. Pump not working due to choked filter. 7. Pump not working due to damaged pump roller or cam. 8. Leaking operating valve due to dirt on ball seat. 9. Damaged parts within the unit requiring removal and inspection.
Overdrive does not disengage	NOTE: IF OVERDRIVE DOES NOT DISENGAGE DO NOT REVERSE THE CAR OTHERWISE EXTENSIVE DAMAGE MAY RESULT. 1. Fault in electrical control system. 2. Solenoid sticking. 3. Blocked restrictor jet in operating valve. 4. Solenoid operating lever incorrectly adjusted. 5. Sticking clutch. 6. Damaged gears, bearings, or sliding parts within the unit.
Clutch slip in overdrive	1. Insufficient oil in gearbox. 2. Solenoid lever out of adjustment. 3. Insufficient hydraulic pressure due to pump non-return valve incorrect seating. (Probably dirt on seat). 4. Insufficient hydraulic pressure due to sticking or worn relief valve. 5. Operating valve incorrectly seated. 6. Worn or glazed clutch lining.
Clutch slip in reverse or free wheel condition on over-run	1. Solenoid operating lever out of adjustment. 2. Partially blocked restrictor jet in operating valve. 3. Worn or burnt inner clutch lining. NOTE: Before removing any of the valve plugs it is essential to operate the solenoid several times in order to release all hydraulic pressure from the system. To do this, engage top gear, switch on the ignition and operate the overdrive control switch several times.

OIL FILTER
–Remove and refit **40.10.01**

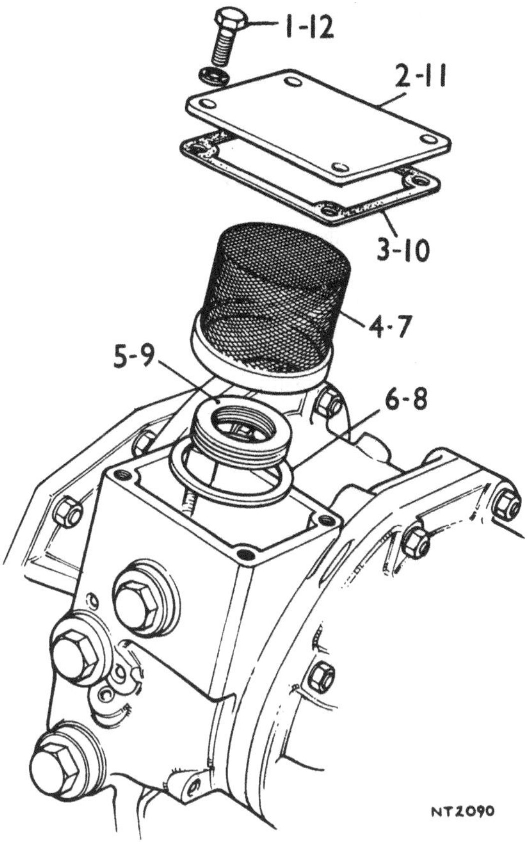

Removing

1. Remove four screws.
2. Remove the cover plate.
3. Carefully remove the cork washer.
4. Lift out copper gauze filter.
5. Remove three small rubber/steel bonded sealing washers.
6. Remove one large rubber/steel bonded sealing washer.

Refitting

7. Clean filter.
8. Place large rubber/steel bonded sealing washer in filter housing.
9. Place small washers in filter and place in housing.
10. Place cork gasket on filter housing.
11. Place plate back on housing.
12. Replace four screws and star washers and tighten.

OIL SEAL

–Remove and refit **40.15.01**

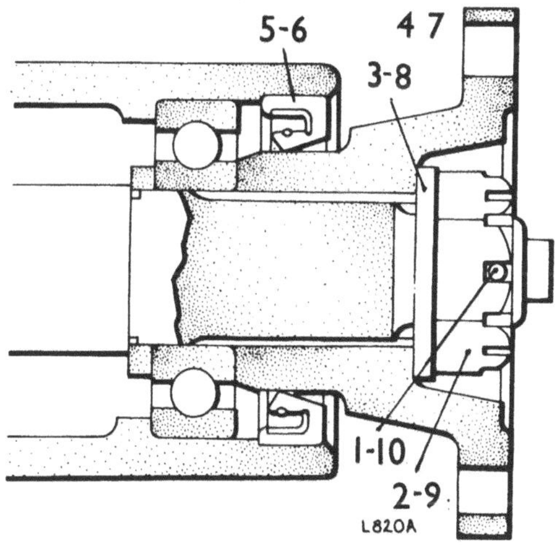

Removing

1. Remove the cotter pin.
2. Remove the nut.
3. Remove the washer.
4. Remove drive flange.
5. Prise out rear oil seal with a driver or similar tool.

Refitting

6. With the use of a large socket tap a new seal into place.
7. Refit the driver flange.
8. Refit the washer.
9. Refit the nut.
10. Replace and fit a new cotter pin.

Triumph Spitfire Mk IV Manual. Part No. 545254. Issue 1.

OPERATING VALVE

–Remove and refit **40.16.01**

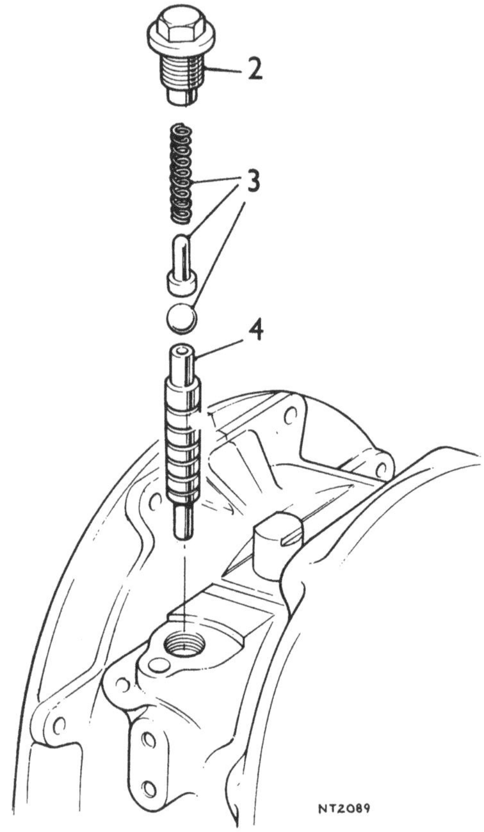

Removing

1. Operate the solenoid several times to release the hydraulic pressure.
2. Unscrew the valve plug.
3. Remove the spring, plunger and ball with a magnet taking care to avoid damage to the valve seat.
4. Remove the operating valve by inserting a length of stiff wire down the centre and drawing it up.
5. Ensure that the small hole at the bottom of the valve is clear, the hole provides a passage for oil exhausted from the operating cylinders when the valve is moved to the direct drive position.

Refitting

6. Reverse operations 1 to 5.
   If the ball does not seat properly:-
7. Place the ball on a block of wood.
8. Position the valve on the ball and give the valve a sharp gentle tap.
9. Clean the valve seat in the casing.
10. Locate the ball on its seat and very gently tap the ball.
    TAPPING THE BALL TOO HARD WILL CLOSE THE MOUTH OF THE VALVE SEAT AND PREVENT VALVE REASSEMBLY.

NON RETURN VALVE

–Remove and refit **40.16.10**

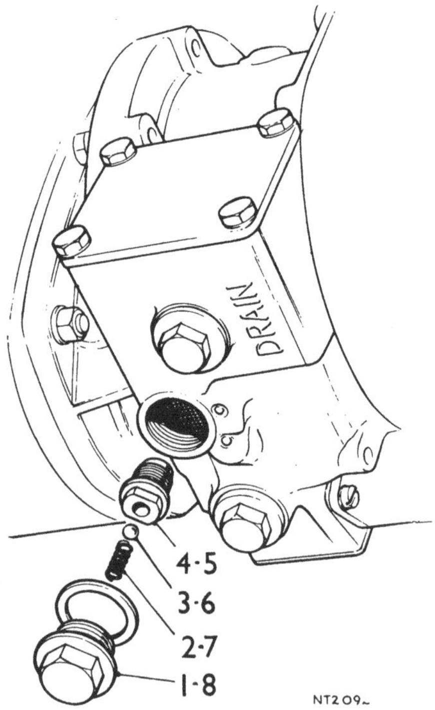

Removing

1. Remove the pump plug.
2. Remove the non return valve spring.
3. Remove the ball.
4. Unscrew the non return valve body using tool L213.

Refitting

5. Refit the non return valve body using tool L213.
6. Re-position the ball.
7. Re-position the non return valve spring.
8. Refit the pump plug ensuring that the spring locates correctly in the plug recess.

OIL PUMP

–Remove and refit **40.18.01**

Removing

1. Remove the non return valve body 40.16.10
2. Remove the grub screw from housing.
3. Using tools L183A and adaptor L183A-2A, remove the pump body using tool L213, by screwing the spindle into the pump body, positioning the adaptor against the casing and screwing the wing nut down which will remove the body.
4. Remove the spring.
5. Remove the pump plunger.

Refitting

6. Refit the pump plunger.
7. Refit the spring.
8. Refit the pump body by locating the assembly within the orifice in the front casing locating the flat of the plunger roller fork against the thrust button situated below the centre bush. Press the pump home using tool L206A until the annular groove in the pump body is in alignment with the locating screw orifice.
9. Insert the dowelled locating screw and tighten, ensuring that the dowel locates in the groove.
10. Refit the non return valve body 40.16.10.

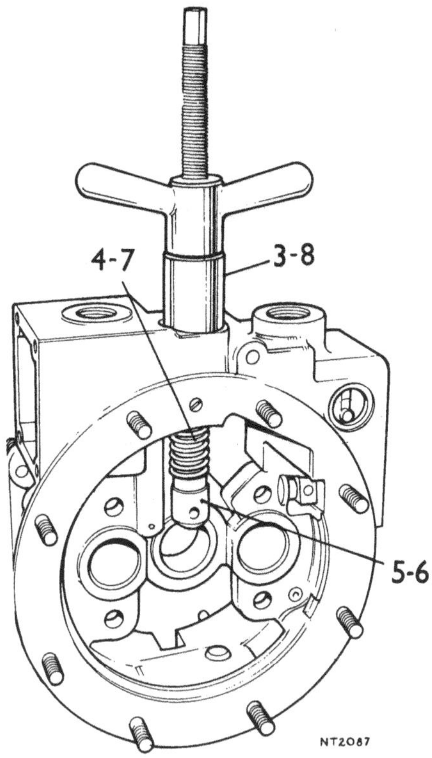

OIL PUMP

–Overhaul **40.18.04**

1. Remove the oil pump 40.18.01
2. Replace the spring which fits over the pump plunger.
3. Replace the spring, ball and copper seal in the non return valve.
4. Refit the oil pump 40.18.01.
5. To check the pump is working, jack up the rear wheels of the car securely and remove the operating valve plug.
6. Start the engine.
7. Engage top gear.
8. Run the engine slowly.
9. Watch for oil being pumped into the valve chamber.
10. If none appears the pump is not functioning.
11. In this case the valve should be cleaned and reseated.
12. To reseat remove the valve body using tool L213.
13. Clean the valve body.
14. Tap the ball sharply onto its seal.
15. A flow of oil does not necessarily indicate that the hydraulic pressure is correct.

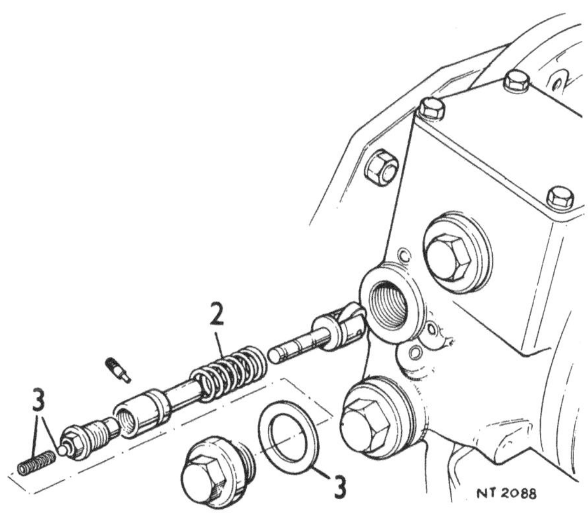

OVERDRIVE ASSEMBLY

Hydraulic pressure test 40.20.01

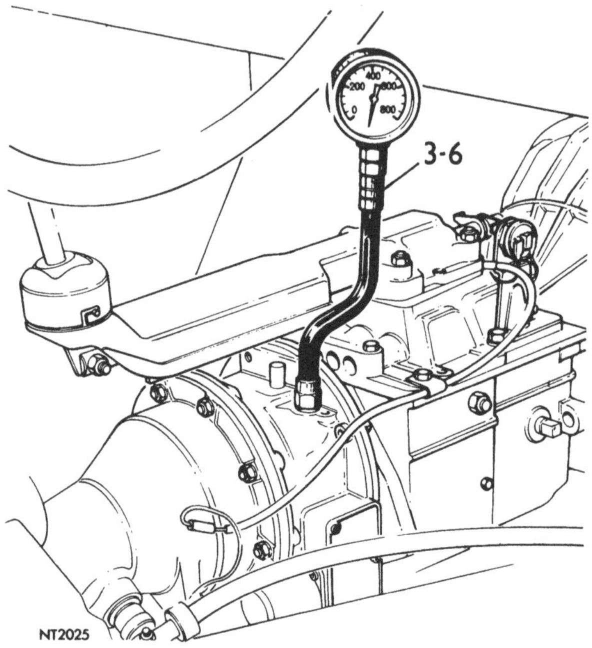

1. Release the hydraulic pressure by switching on the ignition, engaging third or top gear and operating the overdrive switch several times.
2. Remove the operating valve plug.
3. Replace the valve plug with the hydraulic test equipment (churchill tool L188).
4. Jack up the rear wheels of the car securely.
5. Start the engine and run up to at least 20 m.p.h. on the speedometer.
6. Check the hydraulic pressure in the overdrive which should be 540 – 560 lb/sq. in. (37·962 – 39·368 kg/cm).
7. Lack of pressure when overdrive is selected may indicate the pump return valve requires cleaning and reseating and/or the relief valve and filter cleaning.

OVERDRIVE ASSEMBLY

–Remove and refit 40.20.07

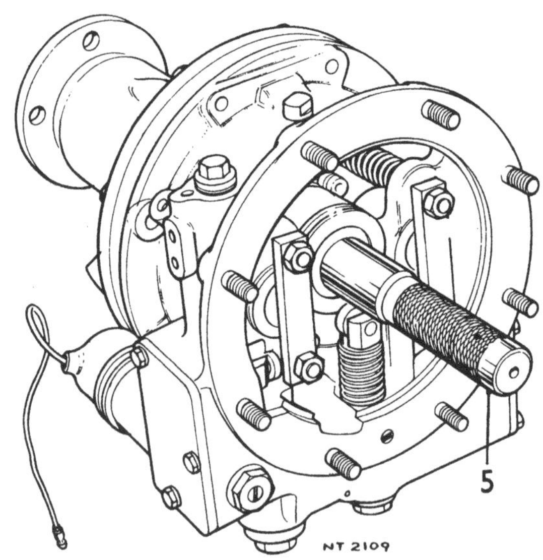

Removing

NOTE: If possible it is advantageous to jack the rear wheels of the car off the ground and run in an overdrive condition before removing the overdrive, this eases removal from gearbox.

1. Disconnect the connectors from the gearbox overdrive switch and solenoid.
2. Remove the gearbox, operation No. 37.20.01.
3. Remove the nuts and spring washers securing the overdrive to the adaptor flange.
4. Carefully remove the overdrive.

Refitting

5. Line up the splines of the planet carrier and the uni-directional clutch by turning the lower set of splines anti-clockwise. Check the final alignment with a dummy drive shaft (churchill tool L201), if available.
6. Turn the shaft to position the cam with its highest point uppermost. The lowest point will then coincide with the overdrive pump roller. Carefully fit overdrive. If the overdrive fails to meet the adaptor plate face by approximately 5/8" it means that the planet carrier and uni-directional clutch splines have become mis-aligned. In this case remove the unit and re-align the splines.

OVERDRIVE ASSEMBLY

–Overhaul **40.20.10.**

1. Remove gearbox from car 30.20.01.
2. Remove overdrive from gearbox 40.20.07.
3. To prevent damage or faulty operation resulting from foreign matter, scrupulous cleanliness must be observed during all service operations.
4. Prepare a clean area in which to lay out the dismantled unit, and clean containers to receive the smaller parts.
5. With the front casing uppermost, secure the unit in suitably protected vice jaws.
6. Release the four bridge retaining nuts and washers.
7. Remove the piston springs.
8. Loosen the two solenoid securing screws to prevent the rubber solenoid cover fouling during front casing removal.
9. Progressively loosen the eight nuts securing the front casing, this ensures the gradual release of the clutch spring loading.
10. If the brake ring remains with the rear casing tap gently to remove.
11. Remove the four clutch return springs.
12. Withdraw the clutch sliding member complete with thrust bearing, thrust ring retaining plate and sungear.
13. Remove the operating valve 40.16.01.
14. Remove the relief valve by removing the plug at the bottom of the front casing adjacent to the solenoid housing cover plate.

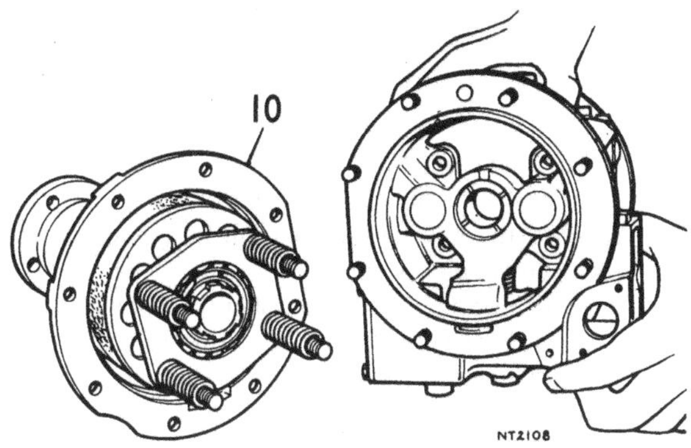

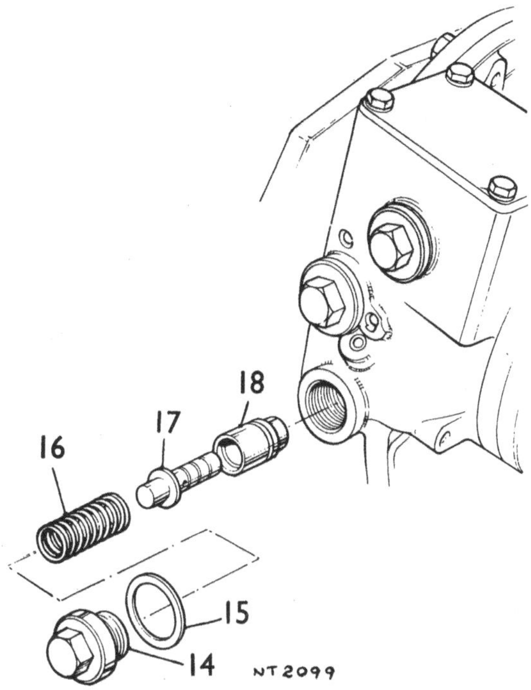

15. Remove the washer.
16. Remove the spring.
17. Remove the releif valve plunger.
18. Remove the relief valve body, this can be withdrawn using a piece of stiff wire shaped into a hook form, into the hole in the side of the body and pulling out.
19. Remove the pump 40.18.01.
20. Remove the filter 40.10.01.
21. Remove the operating pistons from their respective housings.
22. Remove the sungear retaining clip from its groove in the sungear extension.

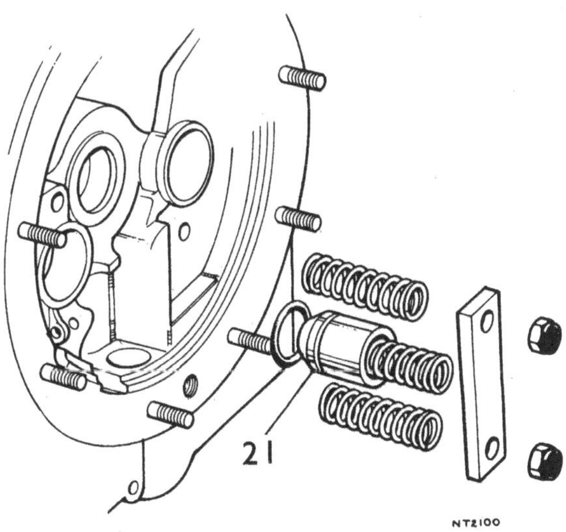

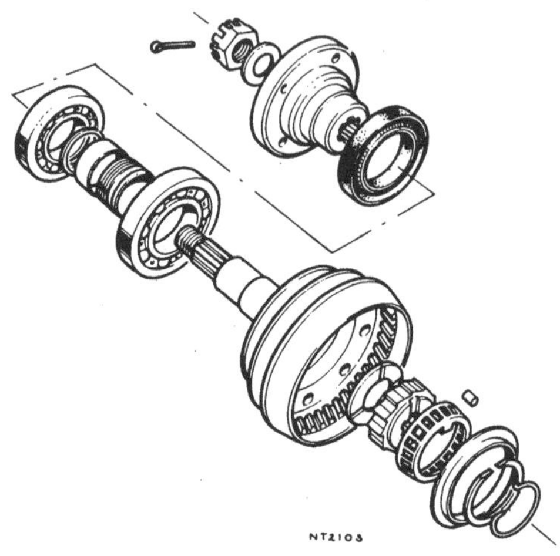

23. Withdraw the sungear.
24. Remove the thrust bearing retaining plate.
25. Remove the bearing circlip from its groove on the cone clutch hub.
26. Press the hub from the bearing and thrust ring.
27. Extract the bearing from the thrust ring, if this is necessary the thrust ring MUST be replaced.
28. Remove the solenoid 40.22.04.
29. Inspect the gear teeth on the planet carrier assembly for damage and wear and check for excessive movement indicating needle bearing or retaining pin wear.

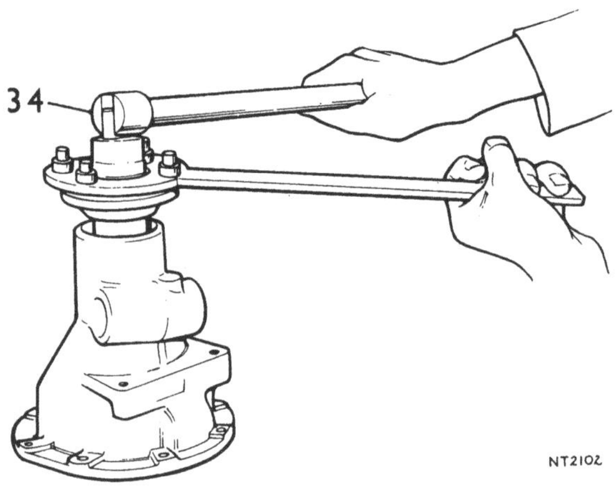

30 If necessary renew the complete carrier assembly.
31. Remove the speedometer bush locating screw.
32. Remove the pinion from the rear casing.
33. Remove the split pin.
34. Remove the nut holding the flange with tool RG421.
35. Remove the washer.
36. Remove the flange.
37. Press the annulus forward out of the rear case.
38. The rear bearing oil seal will remain in situ.
39 Remove the front bearing.
40. Remove the speedo drive gear.
41. Remove the distance piece.
42. Remove the spacer.

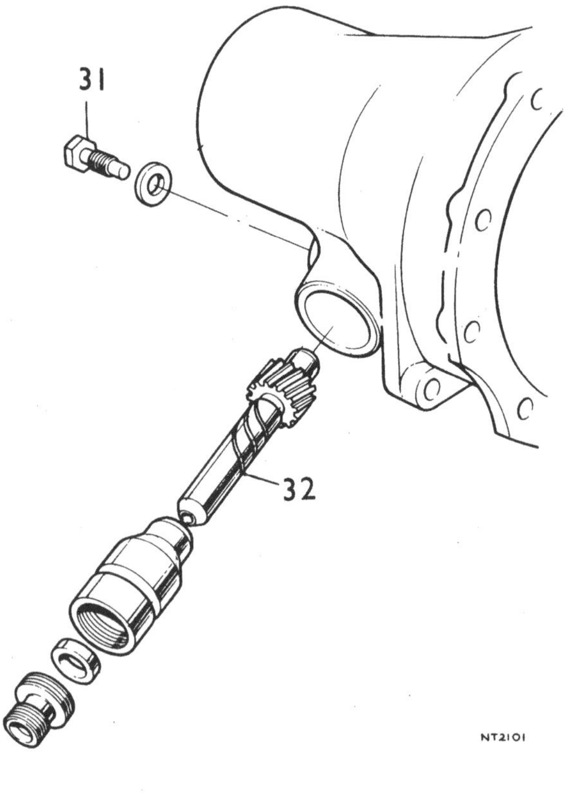

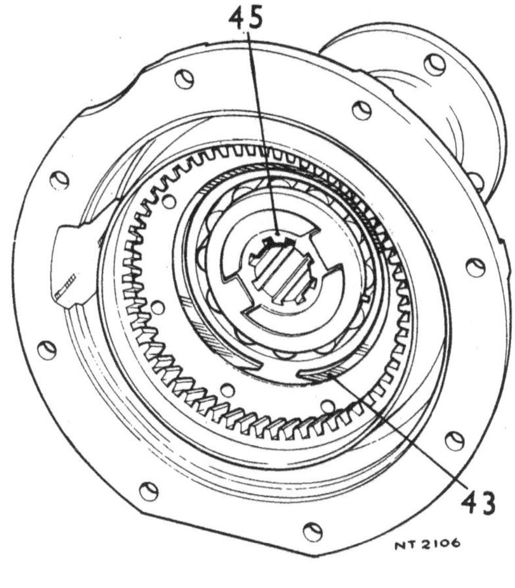

43. Remove the circlip.
44. Remove the brass oil thrower ring.
45. Withdraw the uni-directional clutch from the annulus.
46. Remove the bush from the annulus.
47. Remove oil seal.
48. Remove rear bearing with tool L206A.
49. Clean all components ready for reassembly.

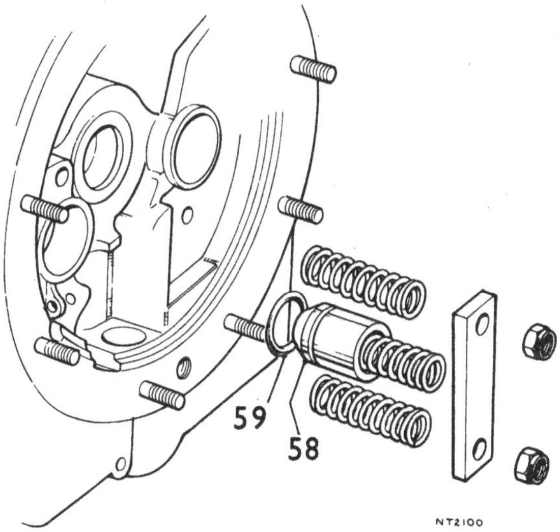

50. Inspect and renew all faulty gaskets, seals and tab washers.
51. Refit operating valve 40.16.01.
52. Insert the relief valve plunger in the relief valve body.
53. Locate the assembly within its orifice at the base of the front casing.
54. Insert the spring locate it on the boss of the plunger.
55. Secure the assembly with the relief valve blanking plug.
56. Reassemble and fit oil pump 40.18.01.
57. Reassemble and fit filter 40.10.01.
58. Replace the operating pistons with the open end of the piston bore facing forward.
59. Carefully ease the sealing rings into the bores during this operation.

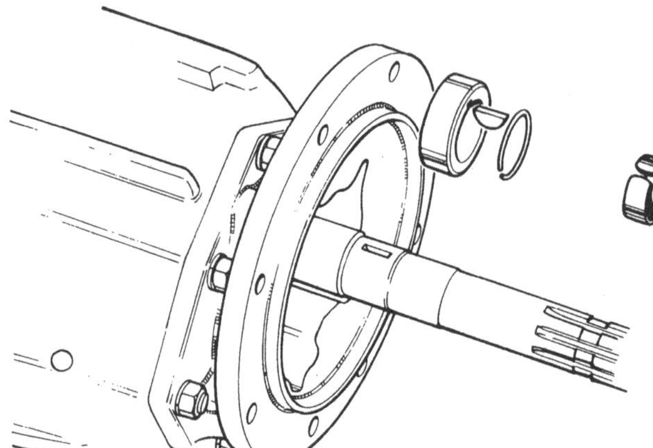
NT2110

60. Locate the front bearing over the annulus tail shaft.
61. Press into position against the locating shoulder at the reat of the annulus.
62. Position the speedometer drive gear.
63. Position the distance piece.
64. Position if fitted the spacing washer on the tail shaft.
65. Fit assembly to rear casing.

NOTE: Where new parts have been fitted, make a dimensional check between the distance piece and the abutment shoulder at the rear bearing.

66. Fit spacing washers as required to give a ·005 in – ·010 in (·1270 – ·254 mm) end float between the rear bearing and the casing.
67. Press the rear bearing onto the rail shaft and into the rear casing simultaneously.
68. Fit the oil seal.
69. Press the rear coupling flange onto the tailshaft.
70. Fit washer.
71. Refit and tighten the castellated nut to (70 – 80 lbs. ft).
72. Line up the castellations with the hole in the tailshaft.
73. Fit split pin.

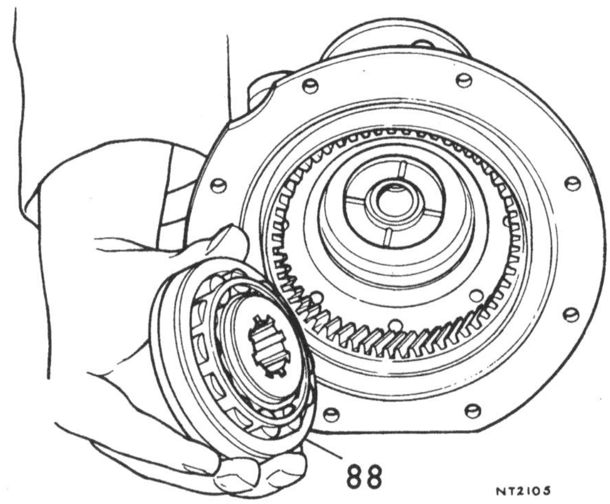

NT2105

74. Insert the speedometer drive pinion.
75. Insert the bush.
76. Rotate the annulus to engage the gear.
77. Align the bush and casing holes.
78. Fit the dowelled locating screw.
79. Insert the bush in the centre of the annulus.
80. Fit the spring in the roller cage of the uni-directional clutch engaging one end of the cage.
81. Insert the inner member.
82. Engage the opposite end of the spring on the hole provided.
83. Ensure that the slots of the inner member engage the tongues of the cage.
84. Place the assembly front face down in the assembly tool L178.
85. Fit the rollers.
86. Check that the spring rotates the cage to drive the rollers up the inclined faces of the inner member.
87. Refit the thrust washer.
88. Refit the uni-directional clutch, transferring the clutch direct from the assembly tool.
89. Fit the brass oil thrower ring.
90. Secure the thrower ring with the circlip.
91. Rotate the gears until the etched lines on the gear and carrier coincide.

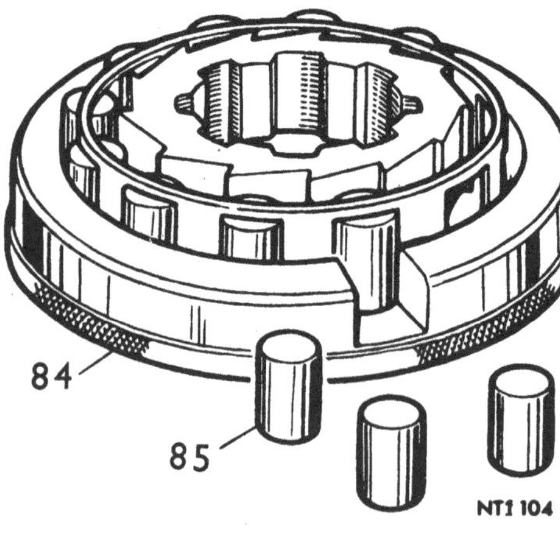

NT1 104

NOTE: On one of the three gears the etched line occurs on the same tooth as the centre pop mark.

Two gears have etched lines but all three have centre pop marks.

Align the etched lines on the two gear teeth with the corresponding lines on the carrier.

Align the centre pop mark on the remaining unetched gear with the corresponding line on the carrier.

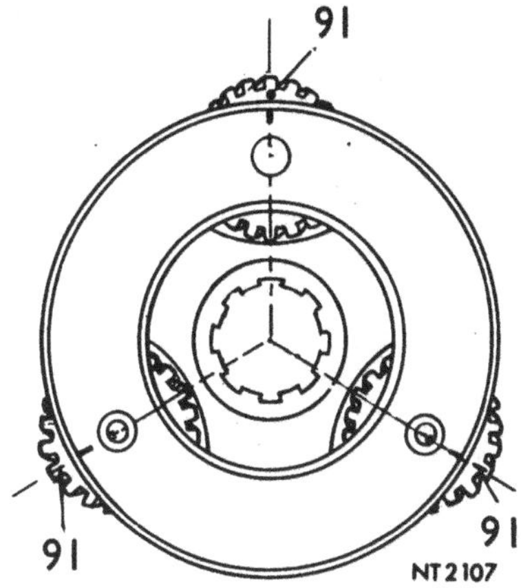

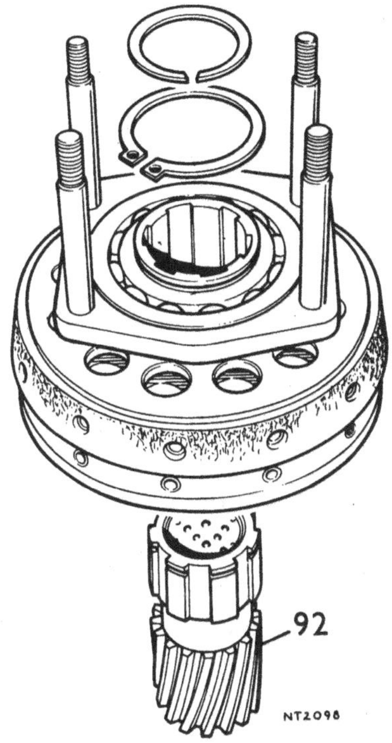

92. Insert the sungear.
93. Re-check the etched lines for alignment.
94. Position the assembly within the annulus.
95. Remove the sungear.
96. Press the thrust bearing into the thrust ring.
97. Fit the four bolts ensuring the heads are correctly positioned.
98. Press the assembly on the cone clutch hub.
99. Secure with a circlip.
100. Fit the retaining plate.
101. Insert the sungear in the splined bore of the cone clutch.
102. Secure with circlip.
103. Locate the assembly within the annulus.
104. Fit the four clutch return springs.
105. Coat both faces of the brake ring with suitable jointing compound.
106. Position the brake ring on the rear face of the front case.
107. Ensure the kidney shaped slot in the brake ring is located at the bottom.
108. Fit the front casing to the rear casing.
109. Clutch spring pressure will now be felt and it will now be necessary to exert a slight pressure to bring the two casings together sufficiently to start the nuts.
110. Ensure that the spring washers are under the nuts before tightening.

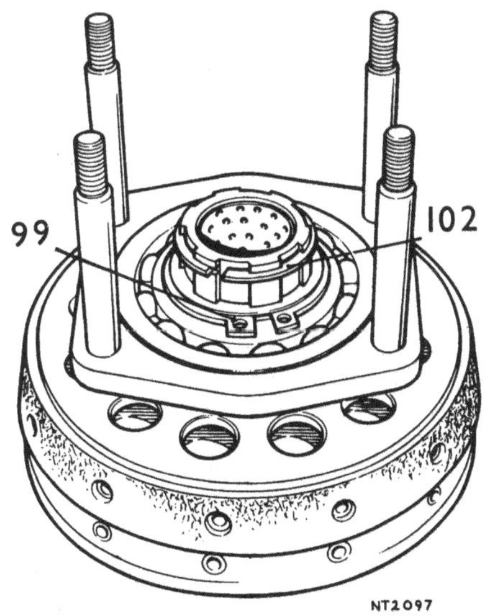

111. Tighten diametrically opposed nuts until the two faces meet.
112. Locate the piston springs with the piston bores.
113. Fit the bridge pieces.
114. Tighten the lock nuts.
115. Position the solenoid plunger in the fork of the
116. operating lever.
Screw on the adjusting nut.
117. Fit solenoid 40.22.04.
118. Adjust and test solenoid 40.22.01.
119. Refit overdrive to gearbox 40.20.07.
120. Refit the gearbox to the car 37.20.01.
121. Connect gearbox overdrive switch.
122. Connect the solenoid.

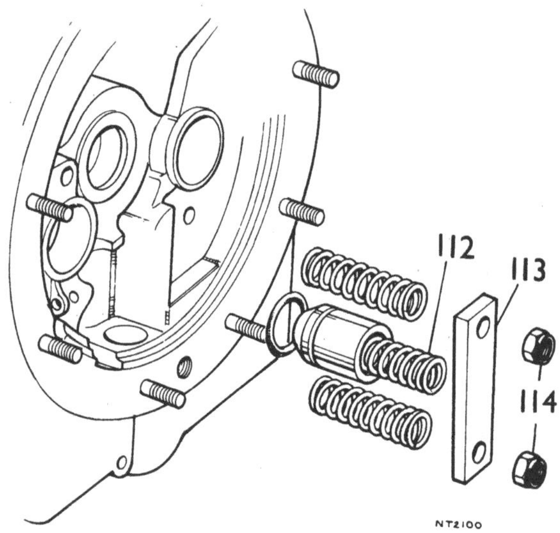

NT2100

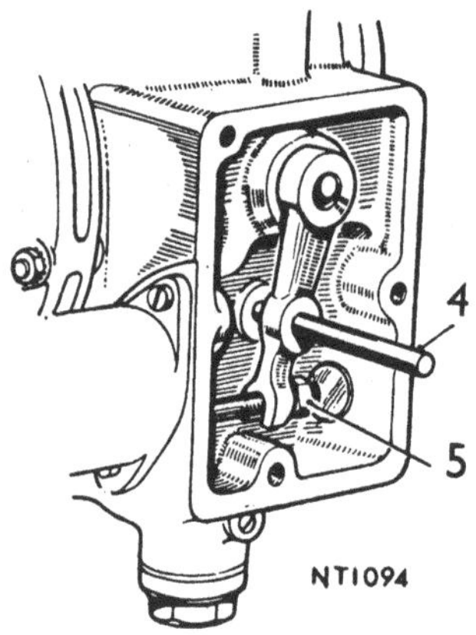

NT1094

OVERDRIVE SOLENOID

–Test and adjust **40.22.01**

1. Remove the three screws and star washers.
2. Remove the cover plate.
3. Remove the gasket.
4. Move the operating lever until a 3/16 in (4·762 mm) setting pin pushed through the hole in the lever aligns with the hole in the casing.
5. With the solenoid energised screw the adjusting nut until it just contacts the operating lever.
6. Remove the setting pin.
7. De-energise the solenoid.
8. Energise the solenoid.
9. Re-check the alignment of the holes.
10. Check the current consumption is approximately 2 amps.
11. A reading of 20 amps indicates the solenoid plunger is not moving far enough to switch from the solenoid operating coil to the holding coil of the solenoid. If this is the case the operating lever must be readjusted (operations 5 to 9).
12. Continuous high current will cause premature solenoid failure.

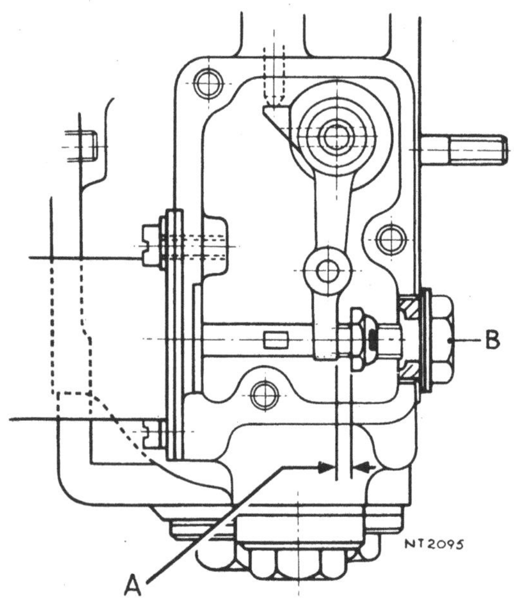

NT2095

NOTE: When the solenoid is not energised ensure a clearance exists at A, this can be adjusted at B.

OVERDRIVE SOLENOID

–Remove and refit **40.22.04**

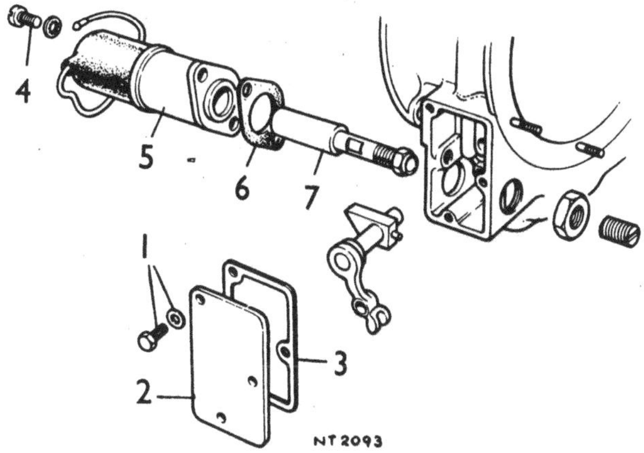

Removing

1. Remove three screws and star washers.
2. Remove cover plate.
3. Remove gasket.
4. Unscrew the two solenoid retaining screws.
5. Remove the solenoid.
6. Remove the joint washer.
7. Remove the plunger.

Refitting

8. Reverse instructions 1 to 7 ensuring that the solenoid is tested and adjusted operation No. 40.22.01.

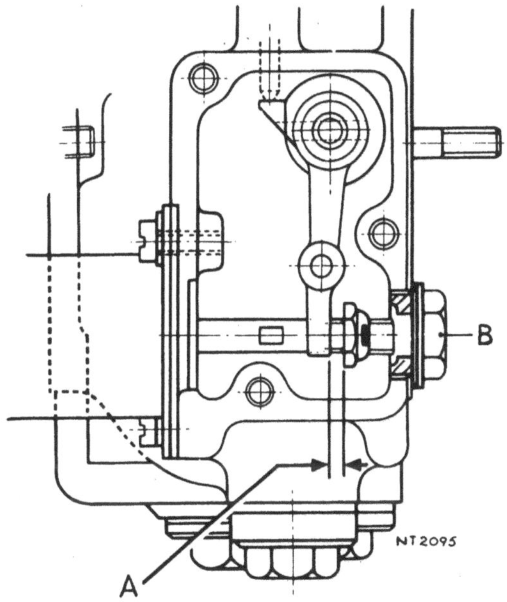

OVERDRIVE SOLENOID 40.22.05

– Remove, refit, test and adjust

1. Remove and refit, operation No. 40.22.04.
2. Test and adjust, operation No. 40.22.01.

PROPELLER AND DRIVE SHAFT OPERATIONS

REAR DRIVE SHAFT

– Remove and refit **47.10.01**

Rear drive shaft and hub bearings – remove and refit **47.10.18**

Special tools S109C, S4221A, S300A.

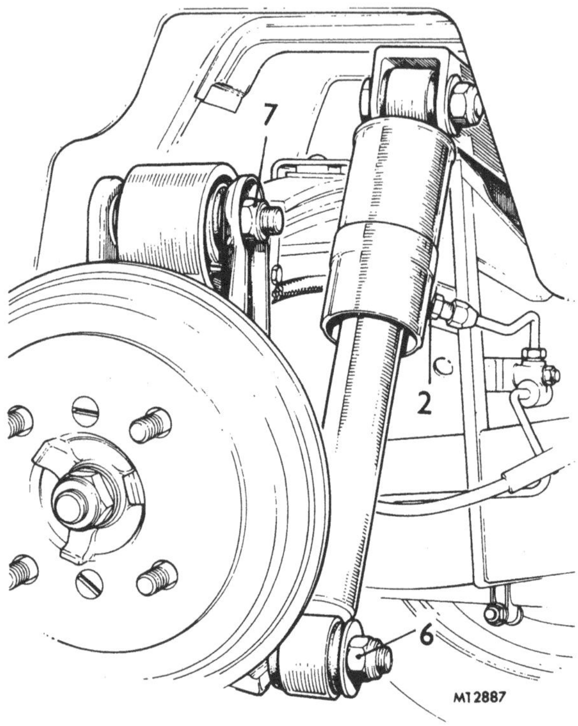

Removing

1. Jack up the rear of the car, support it on chassis stands and remove the road wheel.
2. Disconnect the flexible brake hose from the chassis bracket and pipe union and wheel cylinder.
3. Disconnect the hand brake cable from the brake back plate lever and release the return spring.
4. Using a jack under the vertical link to relieve the damper load, remove the bolt securing the radius arm to the vertical link bracket. 64.35.28.
5. Remove the four bolts and nyloc nuts securing the drive shaft universal joint to the inner drive shaft flange.
6. Remove the nyloc nut and washer securing the damper lower attachment to the vertical link and pull the damper clear of its mounting pin.
7. Remove the jack from beneath the vertical link and whilst supporting the brake drum by hand, remove the nut and bolt securing the road spring to the vertical link.
8. Withdraw the drive shaft from the chassis together with the brake, hub and vertical link assembly.

[NOTE: In order to overhaul the universal joint which is at the inner end of the rear drive shaft, follow the procedure described for propeller shaft universal joints 47.15.18.]

[CAUTION: Please read the instructions thoroughly before you attempt universal joint or wheel bearing repairs. If you lack the necessary skills or tools we suggest you leave such work to an Authorized Dealer or other qualified shop. Before taking a removed drive shaft in for repair, thoroughly clean its exterior but do not undertake any disassembly.]

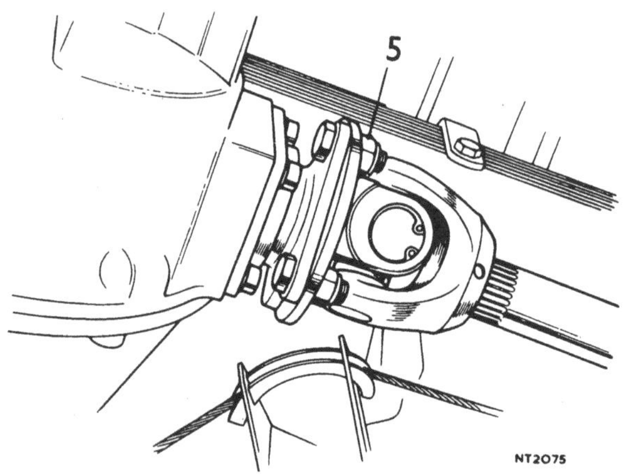

9. Hold the assembly in a vice and remove the brake drum and the hub nut.
10. Using special tool S109C withdraw the hub and key.
11. Remove the nyloc nut and withdraw the bolt securing the trunnion to the vertical link assembly and separate the trunnion from the vertical link.
12. Bend back the lock tabs and remove the four bolts securing the grease trap and brake back plate to the hub bearing housing and remove:
    a. The grease trap.
    b. The back plate.
    c. The seal housing complete with seal.
    d. The joint washer.
13. Using a suitable press or special tool S4221A with suitable packing, press the drive shaft from the hub bearing housing. The flinger may be left in position if it is not damaged and is a satisfactory tight fit on the shaft.
14. From the hub bearing housing press or drift out the inner oil seal and needle roller bearing.

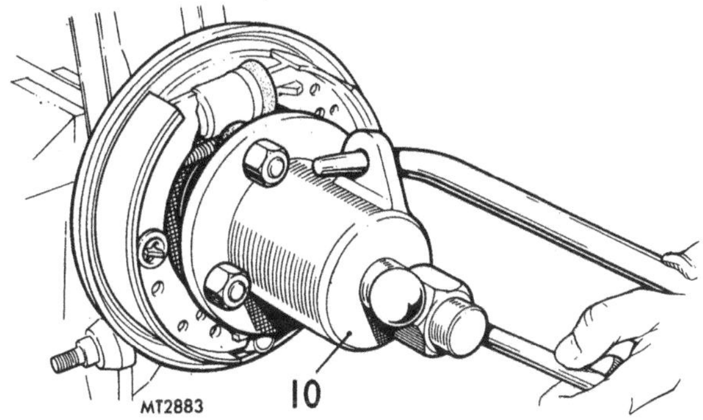

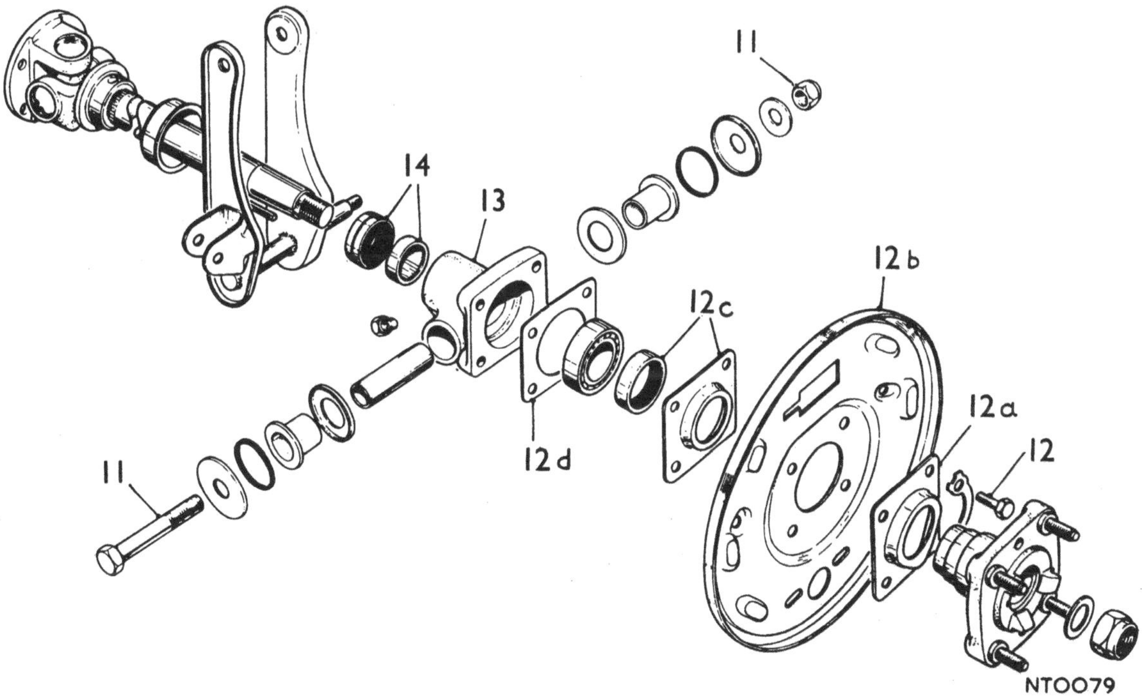

Refitting

15. Press the needle roller bearing (lettered end trailing) into the hub bearing housing to a depth of 0·5 in (12·7 mm) from the housing face using special tool No. S300A.
16. Drift the inner oil seal into the housing with the sealing lips trailing.

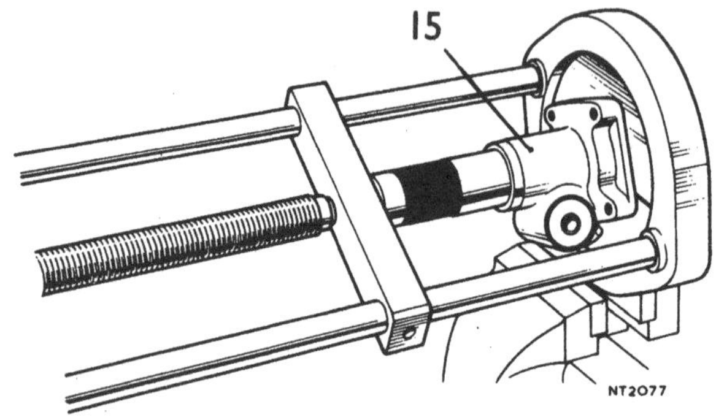

17. If previously removed, drive the flinger on to the axle shaft with a suitable tool to a position 5·740 to 5·760 in (145·8 to 146·3 mm) from the end of the shaft indicated by dimension A.
18. Pack the needle rollers with grease and fit the housing to the axle shaft taking care not to damage the oil seal and needle rollers.

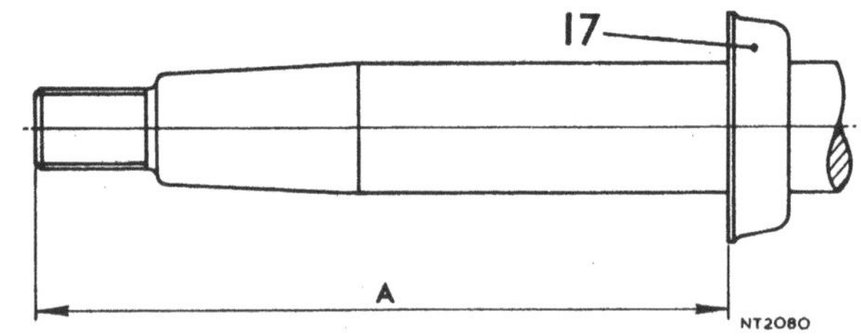

19. Hold the axle shaft in the soft jaws of a vice, pack the bearing housing and ball race with grease and drift the ball race on to the shaft, using special tool No. S304.

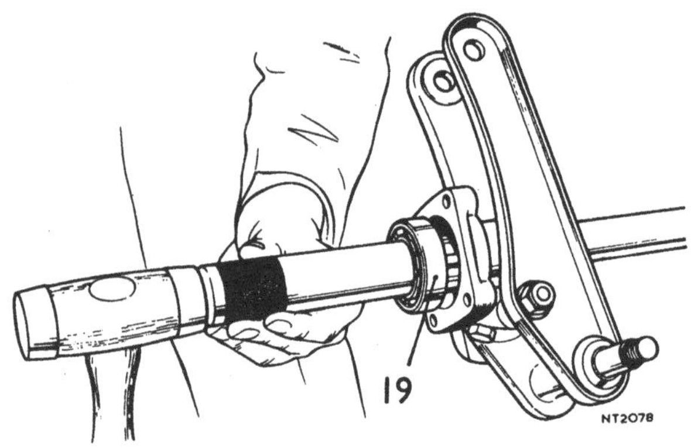

20. Press the outer oil seal into the seal housing with the sealing lip trailing i.e. lip facing towards the bearing.
21. Smear a new paper joint with grease and place it in position on the bearing housing face.
22. Assemble the seal housing together with the brake back plate and grease trap and secure to the bearing housing with new lock plates and the four bolts. Ensure that the following is observed:
    a. The back plate is fitted with the wheel cylinder at the top.
    b. The grease trap is fitted with the duct at the bottom.
23. Insert the key into the key-way and check that the taper on the axle shaft and in the hub is clean and free from burrs. Fit the hub and secure with a plain washer and a new nyloc nut and tighten to 90 to 120 lb ft. (12·4 to 16·6 kgf m).
24. Offer up the vertical link and secure it to the trunnion and drive shaft assembly with the special bolt and a new nyloc nut tightening to 38 to 48 lbf ft. (5·2 to 6·6 kgf m). (Nut to the rear).
25. Fit the vertical link and drive shaft assembly to the road spring eye leaving the securing nyloc nut and bolt slack at this stage (nut to the rear).
26. Jack up the assembly under the vertical link and secure the damper to its lower attachment pin with a new nyloc nut and tighten to 28 to 38 lbf. ft (3·9 to 5·2 kgf. m).
27. Attach the radius arm to the vertical link and tighten to 24 to 32 lbf. ft (3·3 to 4·4 kgf m). 64.35.28.
28. Fit the axle shaft universal coupling to the final drive inner axle shaft flange and secure with bolts and new nyloc nuts. Remove the supporting jack.
29. Fit the handbrake cable to the brake back plate lever and reconnect the return spring.
30. Fit the brake drum and reconnect the flexible hose to the chassis bracket and pipe union and to the wheel cylinder.
31. Bleed the hydraulic brake system 70.25.02.
32. Fit the road wheel and remove the car from the chassis stand.
33. Roll the car to settle the suspension and finally tighten the spring eye bolt to 38 to 48 lbf. ft (5·2 to 6·6 kgf. m).

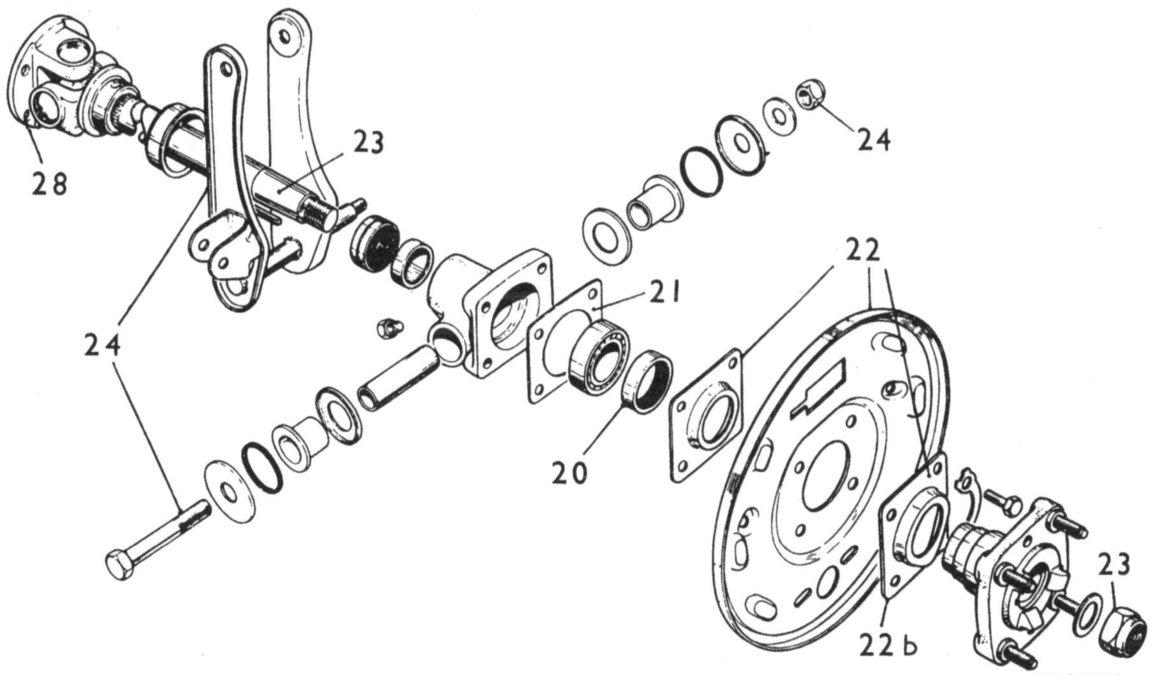

REAR DRIVE SHAFT TRUNNION BUSHES

– Remove and refit **47.10.19**

Removing

1. Jack up the rear of the car and remove the road wheel.
2. Jack up under the vertical link to relieve the damper load and disconnect the damper from its lower attachment.
3. Disconnect the radius rod from the vertical link.
4. Lower the jack under the vertical link.
5. Remove the special trunnion nut and bolt.
6. Push the vertical link assembly towards the final drive unit to expose the trunnion and remove the following:
   - *a.* two outer shims
   - *b.* two rubber seals
   - *c.* one distance tube
   - *d.* two nylon bushes
   - *e.* two inner shims

Refitting

7. Assemble the components to the trunnion in the following order:
   - *a.* Fit the two inner shims (smaller diameter)
   - *b.* Press in the two nylon bushes
   - *c.* Fill the space between the two nylon bushes with grease
   - *d.* Fit the rubber seals
   - *e.* Fit the distance tube
   - *f.* Fit the outer shims (larger diameter)
8. Pull the vertical link assembly back into its correct position and secure the trunnion assembly to the vertical link with the special bolt and nyloc nut.
9. Jack up under the vertical link and reconnect the damper to its lower attachment pin on the vertical link.
10. Reconnect the radius rod to the vertical link and lower the jack under the vertical link.
11. Refit the road wheel and lower the jack.
12. Remove the lubrication plug and grease the hub – see maintenance section.

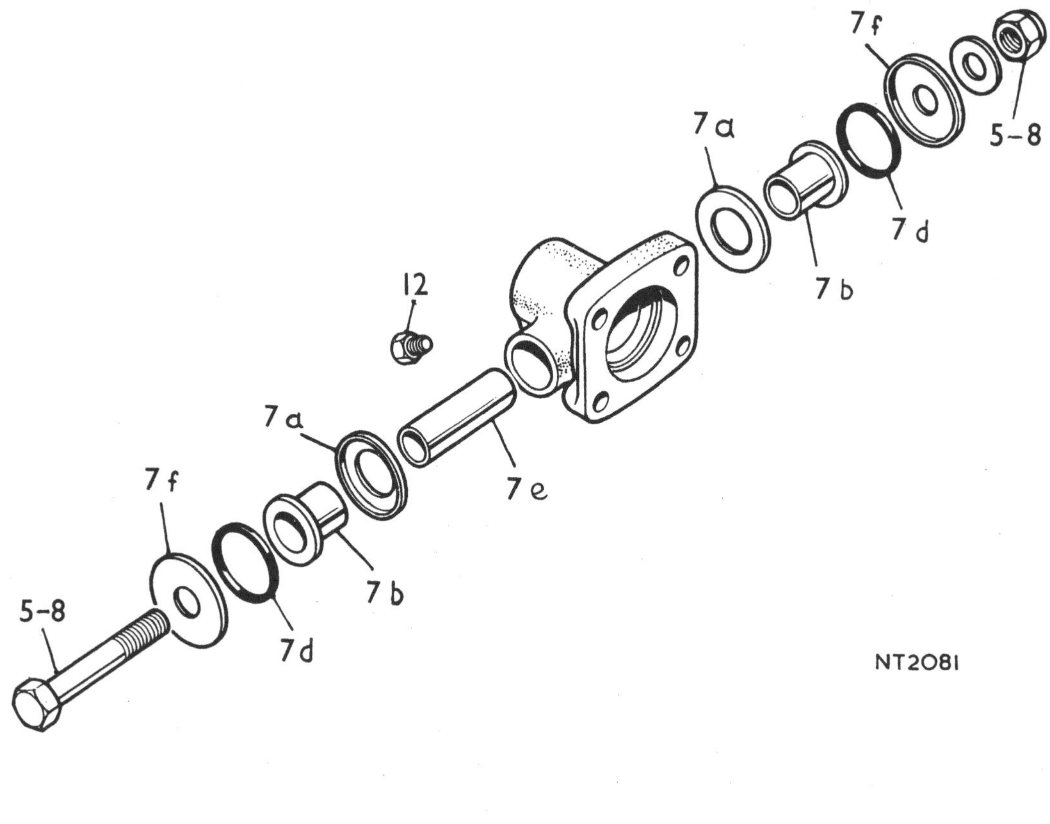

PROPELLER SHAFT

– Remove and refit **47.15.01**

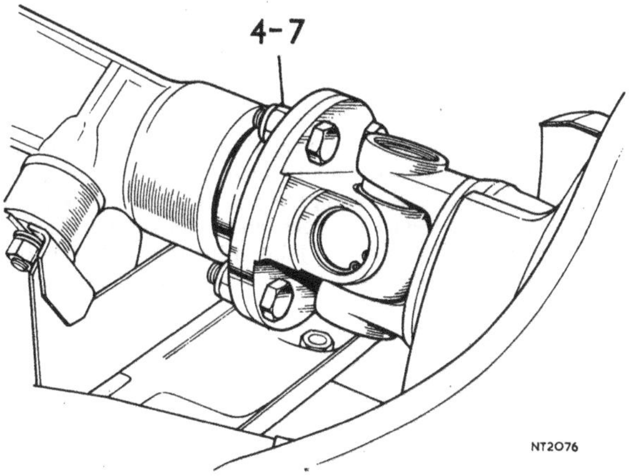

NT2076

Removing

1. Place the car on a ramp or over a pit.
2. Remove the fascia support bracket 76.46.09.
3. Remove the gearbox cover 76.25.07.
4. Mark for reassembly the relationship of the gearbox driving flange to the universal joint flange and remove the four securing nuts and bolts.
5. Working beneath the car, mark for reassembly as in instruction 4 and remove the four nuts and bolts securing the rear end of the propeller shaft to the final drive flange and remove the propeller shaft from the car.

NOTE: Strap or frictionless type propeller shafts may be fitted as standard alternatives.

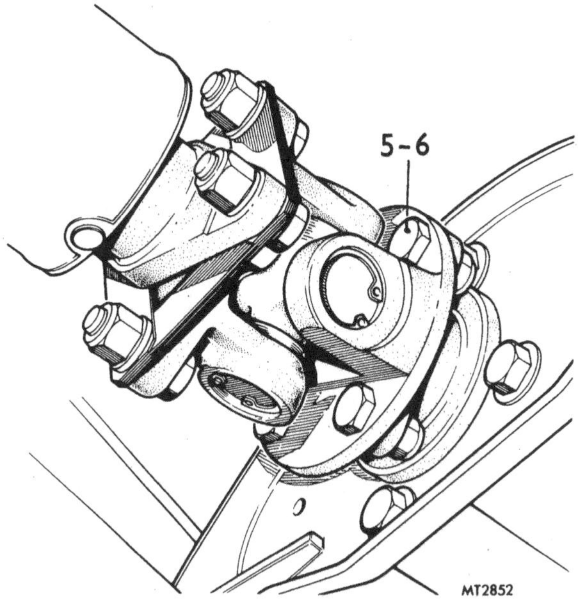

MT2852

Refitting

6. Offer up the propeller shaft to the final drive flange so that the identification marks line up and secure in position using new nyloc nuts. Tighten the nuts and bolts to 26 to 34 lbf. ft (3·6 to 4·7 kgf. m).
7. Working inside the car offer up the propeller shaft to the gearbox flange so that the bolt holes line up and the identification marks coincide. Fit the bolts with new nyloc nuts and tighten as in instruction 6.
8. Reverse instructions 1 to 3.

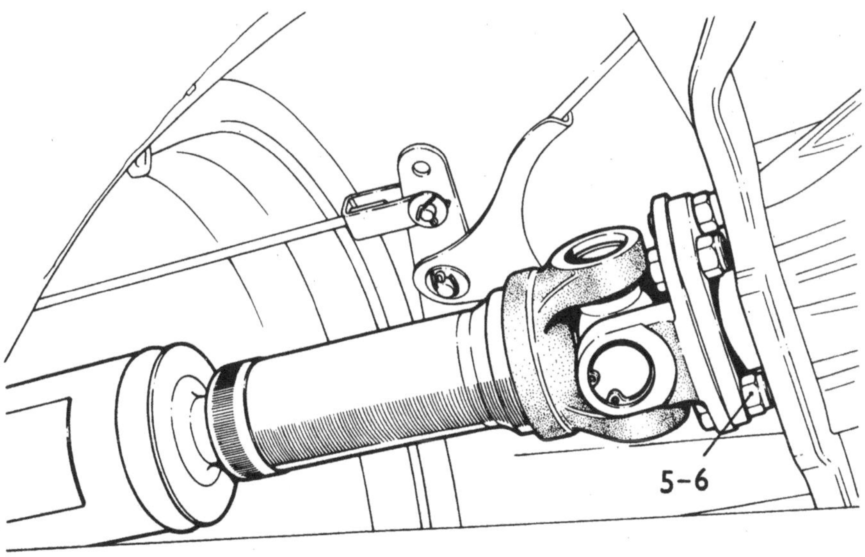

MT3853

UNIVERSAL JOINT

– Overhaul 47.15.18

[CAUTION: Please read the instructions thoroughly before you attempt universal joint repairs. If you lack the necessary skills or tools we suggest you leave such work to an Authorized Dealer or other qualified shop. Before taking a removed shaft and joint in for repair, thoroughly clean its exterior but do not undertake any disassembly.]

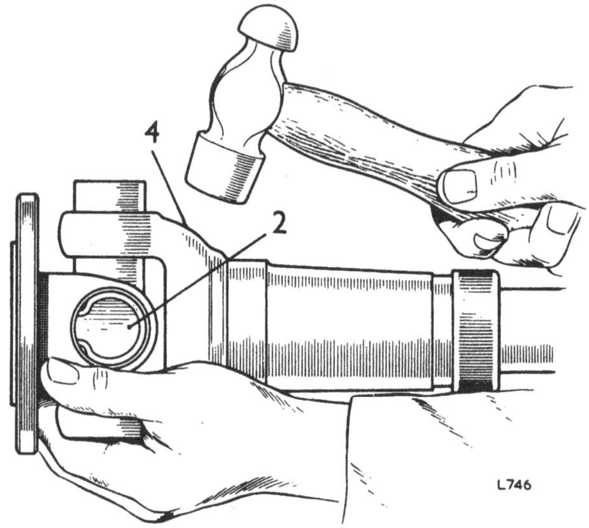

L746

Dismantling

1. Remove the propeller shaft 47.15.01.
2. Remove the paint, rust etc. from the vicinity of the bearing cups and circlips.
3. Remove the circlips.
4. Tap the yokes to eject the bearing cups.
5. Withdraw the bearing cups and spider.

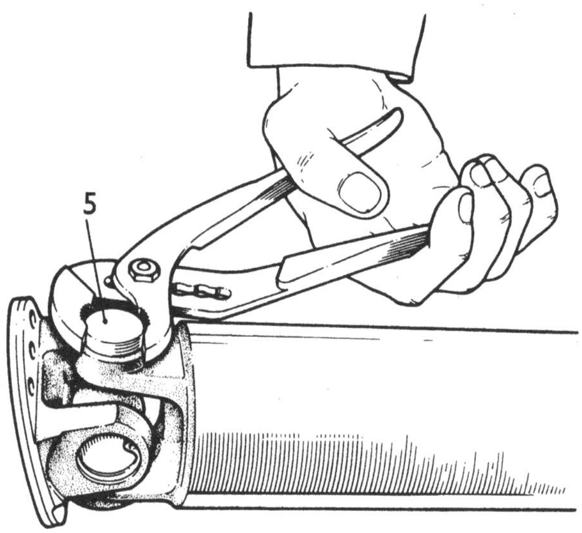

MT2850

Reassembling

6. Remove the bearing cups from the new spider.
7. Ensure that the cups contain approved lubricant (one third full) and that the needle bearings are complete and in position.
8. Fit the spider to the propeller shaft yoke.
9. Engage the spider trunnion in the bearing cup and insert the cup into the yoke.
10. Fit the opposite bearing cup to the yoke and carefully press both cups into position, ensuring that the spider trunnion engages the cups and that the needle bearings are not displaced.
11. Using two flat faced adaptors of slightly smaller diameter than the bearing cups press the cups into the yokes until they reach the lower land of the circlip grooves. Do not press the bearing cups below this point or damage may be caused to the cups and seals.
12. Fit the circlips.
13. Refit the propeller shaft 47.15.01.

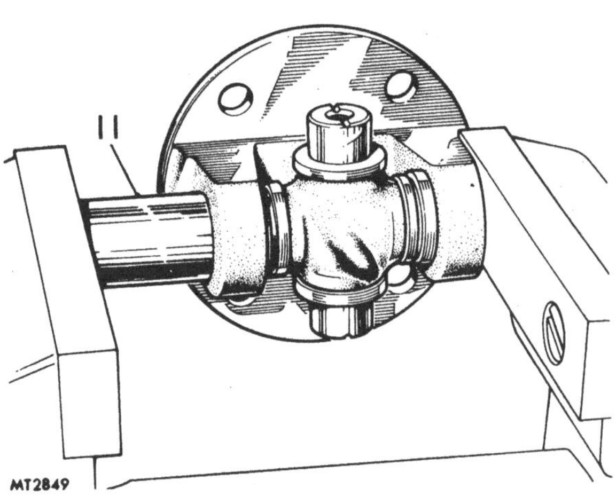

MT2849

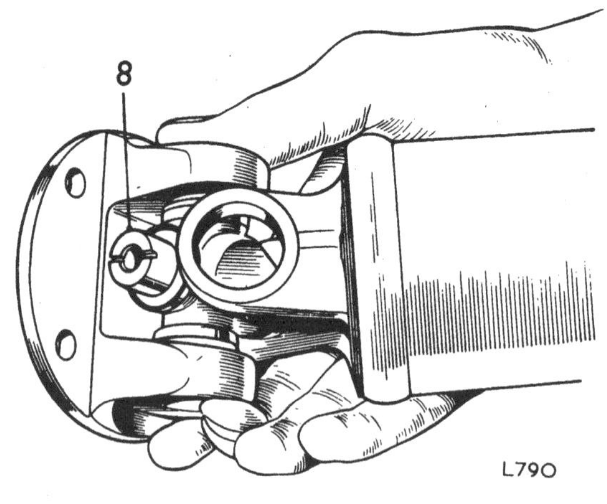

L790

FINAL DRIVE OPERATIONS

INNER DRIVE SHAFT BEARING

– Remove and refit **51.10.04**

Inner drive shaft oil seal – remove and refit **51.10.05**

Special tools S4221A and adaptor set S4221A-7B.

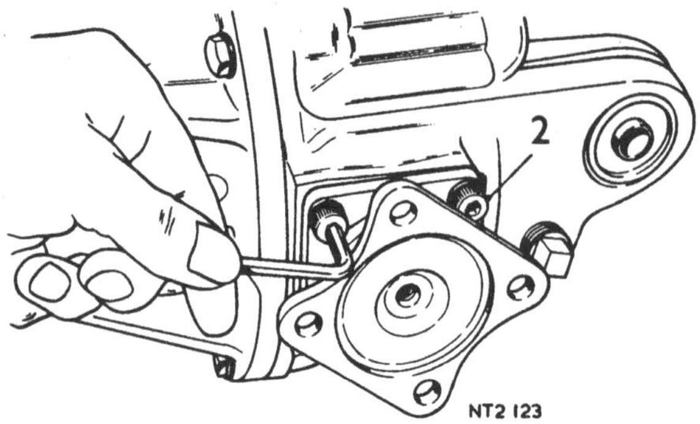

Removing

1. Remove the rear drive shaft assembly – instructions 1 to 8, 47.10.01.
2. Using a hexigon socket key remove the four socket headed screws securing the seal housing plate to the final drive casing. **NOTE**: if both shafts are being removed identify for reassembly.

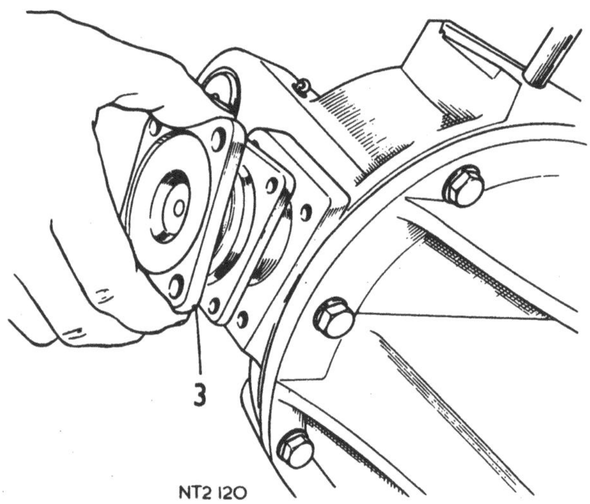

3. Place a suitable receptacle under the drive shaft to catch the oil and withdraw the inner drive shaft complete with bearing and oil seal housing plate.

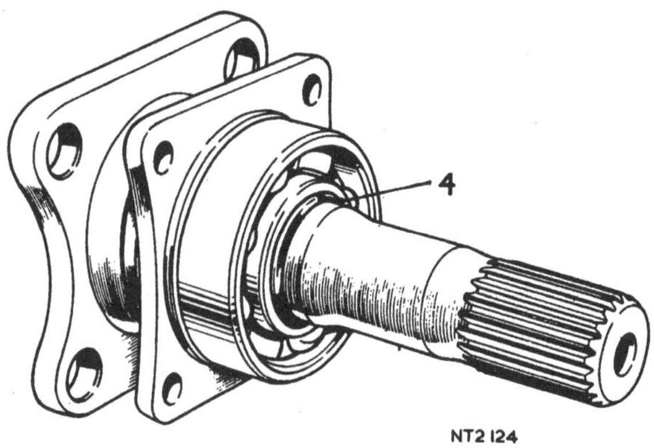

4. Remove the circlip locating the bearing.

5. Using special tool S4221A and S4221A-7B remover, press the inner drive shaft from the bearing.
6. Remove the oil seal housing plate and drift out the oil seal.

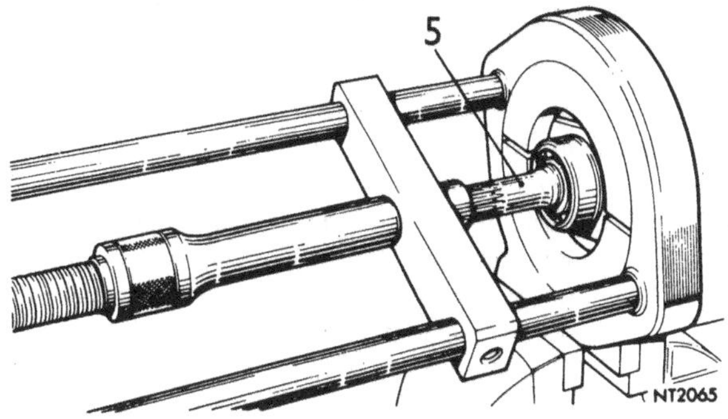

Refitting

7. With the lip of the oil seal leading press or drift a seal into the housing plate.
8. With the sealing lip trailing assemble the housing plate to the inner drive shaft taking care not to damage the seal when passing it over the splines.

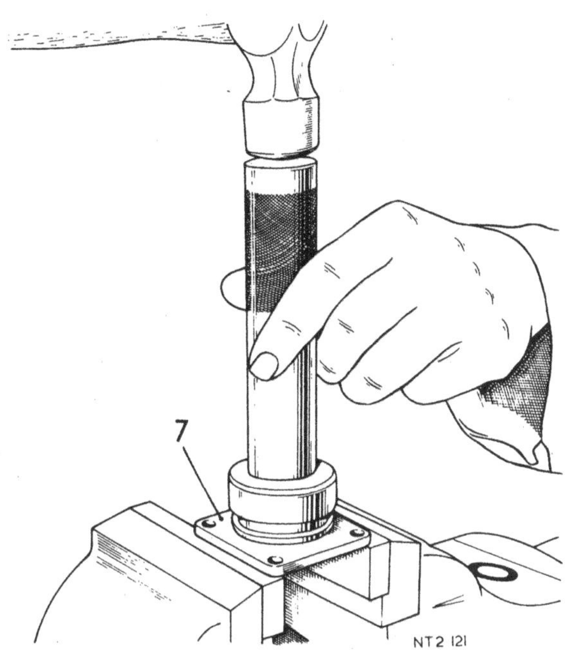

9. Using special tool S4221A and bearing replacer adaptor press the inner drive shaft through the bearing, until the bearing contacts the locating shoulder on the shaft, and the circlip groove is fully exposed.
10. Fit the circlip ensuring that it is properly located in the groove around its entire circumference.
11. Clean the housing plate mating faces and apply a sealing compound, clean and lubricate the shaft splines and insert the assembly into the final drive casing. Oscillate the drive shaft to enable the splines to locate in the internal splines of the sun gear before finally pushing the assembly home.
12. Secure the assembly to the final drive casing with four socket headed screws and washers.
13. Refit the rear drive shaft assembly 47.10.01.
14. Refil the final drive unit with a recommended E.P. oil – see maintenance.

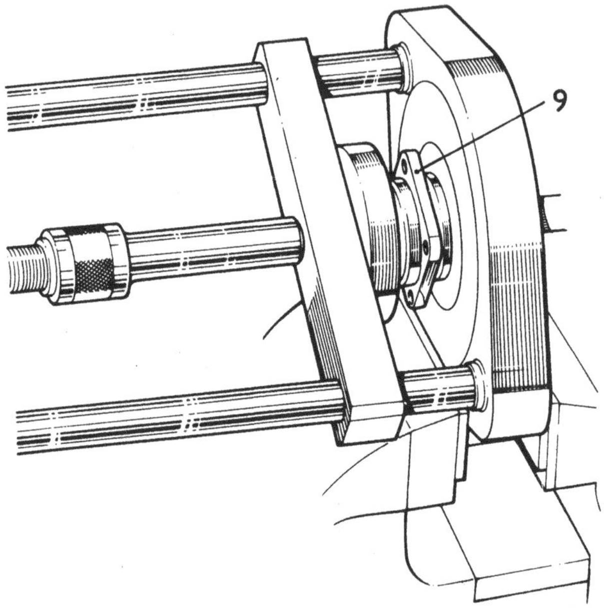

NT2066

DIFFERENTIAL ASSEMBLY

– Remove and refit 51.15.01

[CAUTION: Please read the instructions thoroughly before you attempt final drive repairs. If you lack the necessary skills or tools we suggest you leave such work to an Authorized Dealer or other qualified shop. Before taking a removed final drive unit in for repair, thoroughclean its exterior but do not undertake any disassembly.]

Removing

1. Remove the final drive unit from the car 51.25.13.
2. Clean and degrease the unit.
3. Identify for reassembly and remove the inner drive shafts – instructions 2 and 3, 51.10.04.
4. Remove the pinion nut and driving flange – instructions 4 and 5, 51.20.01.
5. Remove the four bolts and shakeproof washers securing the final drive mounting plate to the differential housing and remove the plate. 51.25.33
6. Remove the eight bolts and spring washers securing the differential housing to the final drive casing and withdraw the differential assembly.

Refitting

7. Clean the differential and final drive casing mating faces, apply sealing compound and fit a new paper gasket.
8. Offer up the differential assembly to the final drive casing and secure with the eight bolts and spring washers and tighten evenly to 15 to 20 lbf. ft (2·1 to 2·8 kgf. m).
9. Refit the inner drive shafts – instructions 12 and 13, 51.10.04.
10. Refit the final drive mounting plate to the assembly.
11. Refit the pinion driving flange and nut – instructions 8 and 9, 51.20.01.
12. Fit the final drive unit to the car – 51.15.01 and refil the unit with a recommended grade of E.P. oil.

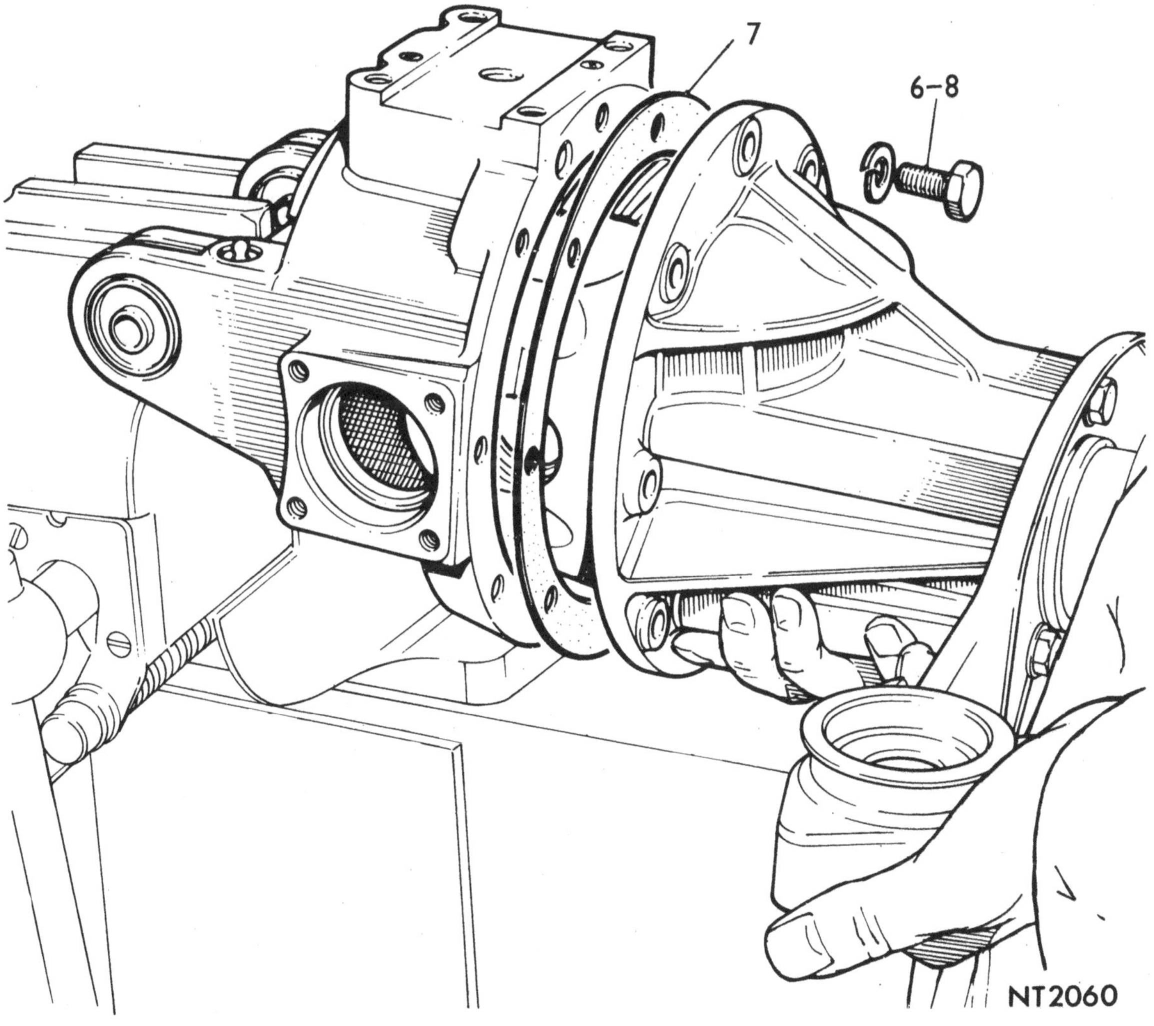

DIFFERENTIAL ASSEMBLY

– Overhaul 51.15.07

Service tools: S101, S101-1, S4221A, S4221A-8C, S4221A-17, S337 (or RG 421), 18G 134, 18G 134 DH, 18G 191, 18G191 M.

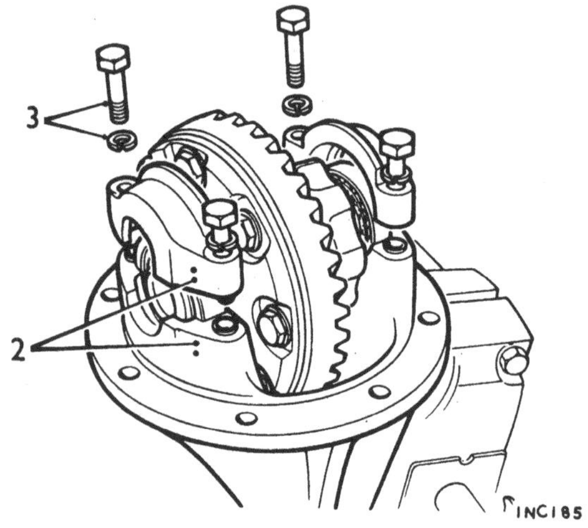

Dismantling

1. Remove the differential assembly. 51.15.01.
2. Place the unit upright in a vice and mark one bearing cap and adjacent side of the differential carrier to ensure the bearing caps are refitted in their original positions.
3. Remove the bearing cap retaining bolts and spring washers and remove the bearing caps.

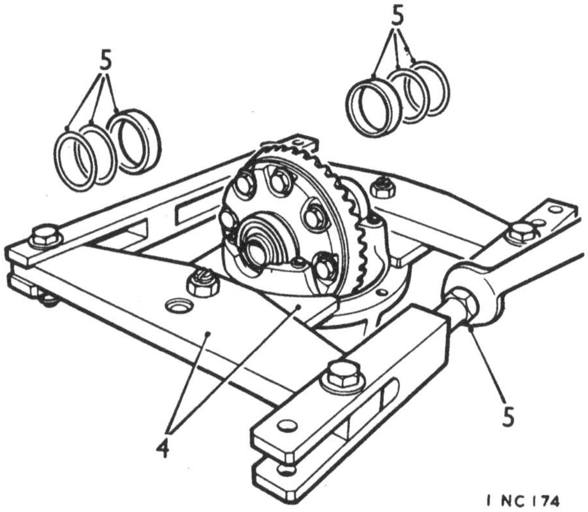

4. Assemble the spreader S101 and adaptor S101-1 to the casing.
5. Stretch the unit case by tightening the turnbuckle three to four flats until the differential carrier can be levered out and the bearing shims and caps removed.

IMPORTANT: To avoid damaging the case, do not spread any more than is necessary. Each flat on the turnbuckle is numbered to provide a check on the amount turned. The maximum stretch is 0·008 in (0·20 mm). Do not lever against the stretcher.

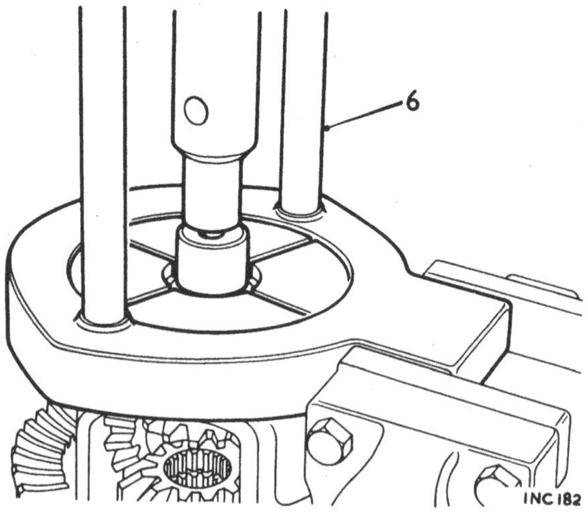

6. Remove the gear carrier bearing cups, using 18G 47 C and 18G 47 BD or S4221A and S4221A-8C.

continued

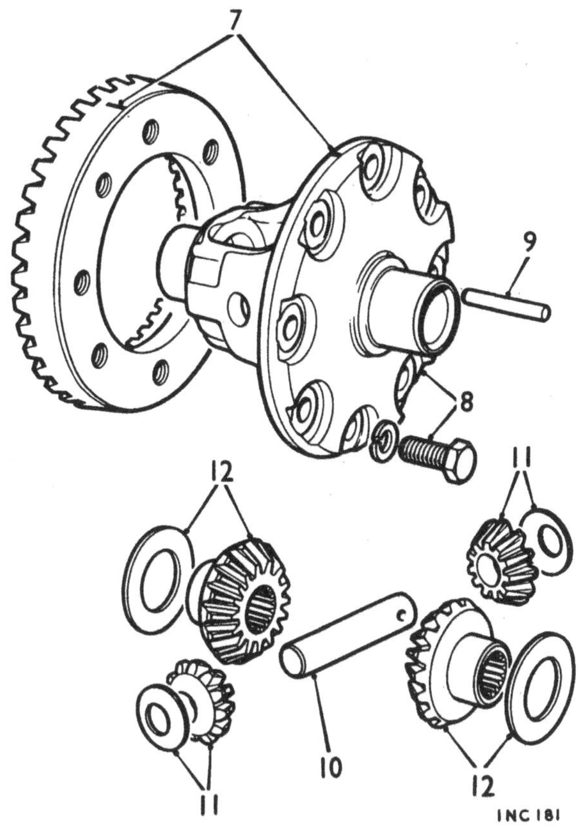

7. Mark the crown wheel and differential carrier to ensure correct replacement.
8. Remove the crown wheel retaining bolts and spring washers and remove the crown wheel.
9. Drive out the differential pinion pin locking peg.
10. Drive out the differential pinion pin.
11. Turn the differential gear wheels by hand until the differential pinions are opposite the openings in the differential gear case: remove the differential pinions and their selective thrust washers.
12. Remove the differential gear wheels and their thrust washers.

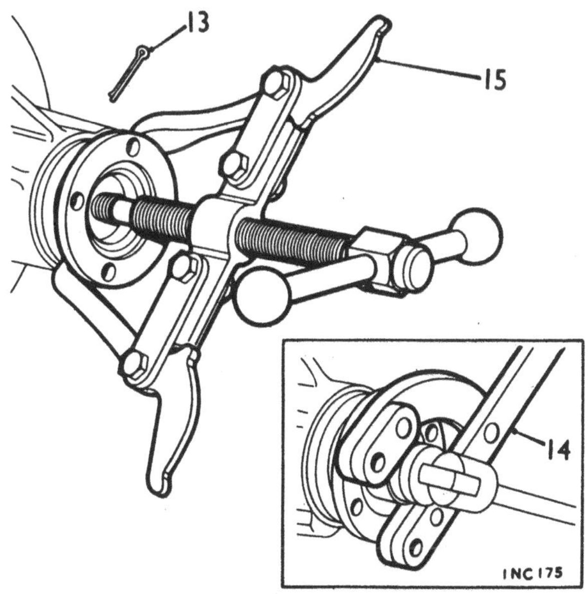

13. Remove the drive flange nut split pin.
14. Restrain the drive flange, using RG421 or S337, and remove the drive flange nut.
15. Remove the drive flange.

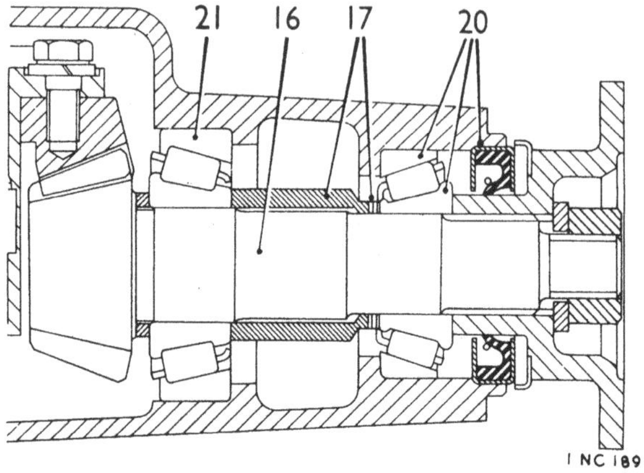

16. Press out the pinion.
17. Remove the pinion bearing shims and pinion bearing spacer.

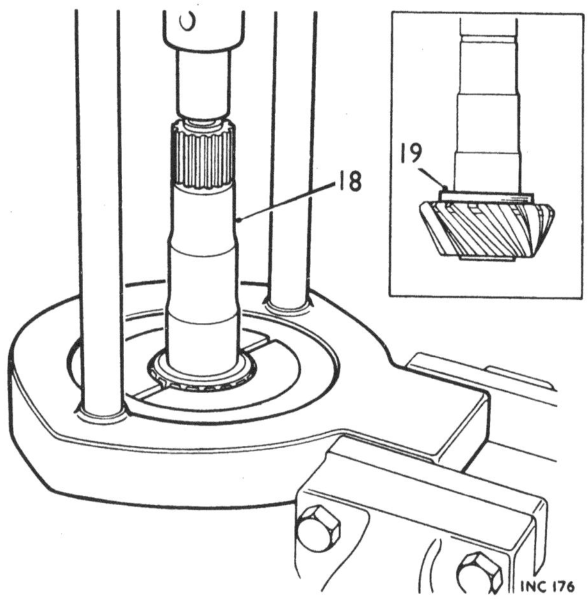

18. Using S4221A and adaptor S4221A-17, remove the inner bearing from the pinion.
19. Remove the pinion head washer.
20. Drift out the pinion outer bearing cup, bearing and oil seal.
21. Drift out the pinion inner bearing cup.

Inspection

22. Clean all components.
23. Renew all worn or damaged parts.
24. The crown wheel and pinion must only be replaced as a matched pair. The pair number is etched on the outer face of the crown wheel and the forward face of the pinion.
25. If one differential bearing is defective replace both differential bearings. If one pinion bearing is defective replace both pinion bearings.

continued

Reassembling

Setting the crown wheel position

26. Fit the differential bearing cones to the gear carrier, using 18G 134 and adaptor 18G 134 DH.

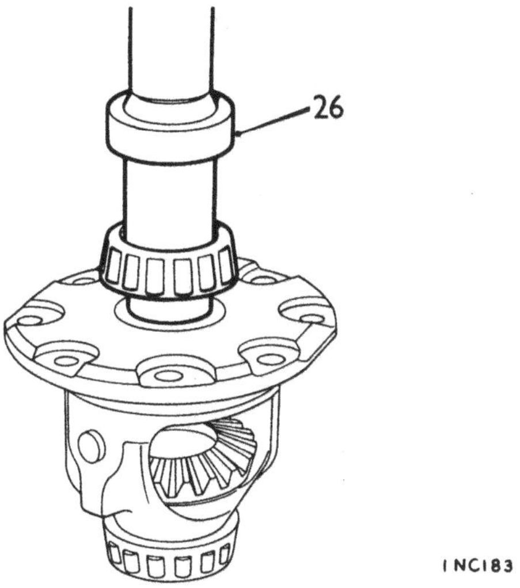

27. Fit the two differential gears to the carrier, ensuring each thrust washer is correctly positioned.

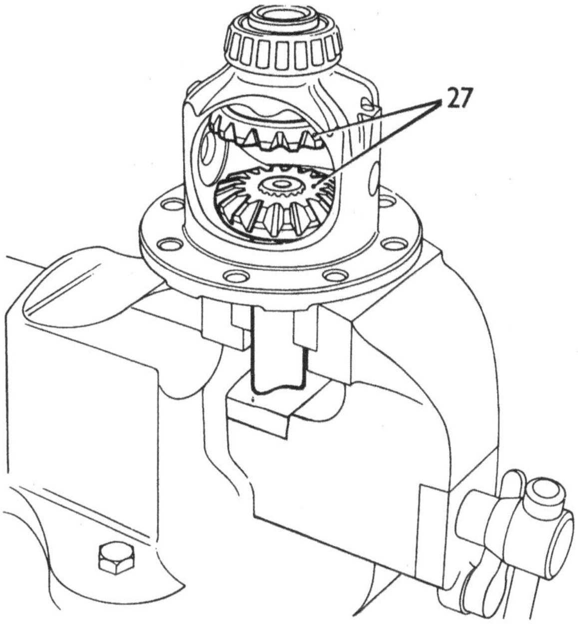

28. Position the two pinion gears, one each side of the carrier, and mesh them with the differential gears.
29. Using an axle shaft inserted in a differential gear, turn the gears and ensure the two pinions rotate into mesh and align awith the pinion pin hole in the carrier.
30. Fit the pinion pin.
31. Press each pinion in turn firmly into mesh with the differential gears and assess the required pinion thrust washer thickness.
32. Remove the pinion pin and the two pinions.
33. Select a thrust washer of the thickness required for each pinion. Eight thrust washers are available in 0·002 in (0·05 mm) steps from 0·027 in to 0·041 in (0·685 mm to 1·03 mm).
34. By selection from the above range of pinion shims, reduce the end-float to give ZERO backlash. Note that at zero backlash the assembly will be tight and difficult to rotate. Lubricate prior to final assembly.
35. Fit the pinion pin locking peg and secure by peening the metal of the differential carrier.

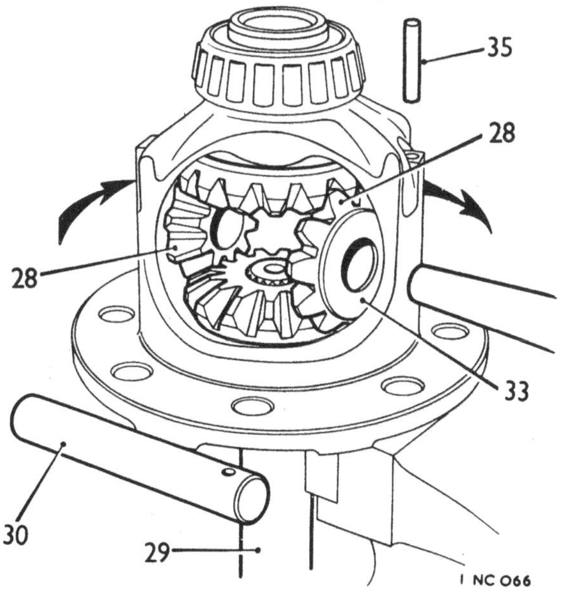

continued

36. Clean the crown wheel mounting and gear carrier face and fit the crown wheel and secure with the bolts and spring washers.
37. Tighten the bolts.
38. Assemble the carrier bearing cups to the bearings and position the assembly in the case; do not fit the bearing shims.
39. Fit and secure the bearing caps as originally marked.
40. Using 18G 191 mounted on the axle spreader adaptor plate S101-1 with the plunger operating squarely on the rear of the crown wheel. Rotate the carrier and check the 'run-out'. Maximum 'run-out' must not exceed 0·003 in (0·076 mm).
41. Remove the bearing caps.
42. Press the differential bearings cups onto the bearings and move the carrier assembly to one side of the case. Zero the dial gauge and move the carrier assembly fully in the opposite direction. The indicated movement, which should be noted, is the **Total side-float** and is referred to as **Dimension 'A'**.
43. Remove the differential assembly from the case.

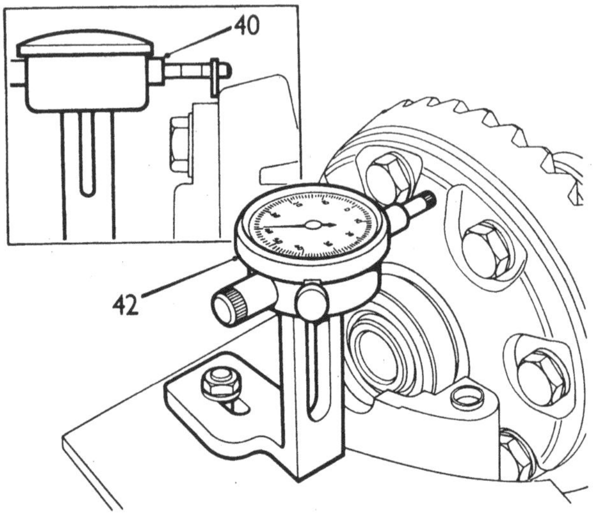

INC 177

Setting the pinion position

44. Drift the pinion inner and outer bearing cups into the casing.

45. Fit the pinion inner bearing to the dummy pinion 18G 191 M.
The standard pinion head spacer 0·077 in (1·95 mm) is incorporated in the dummy pinion.
46. Oil the bearings and fit the dummy pinion, outer bearing, tool spacer, washer and nut.

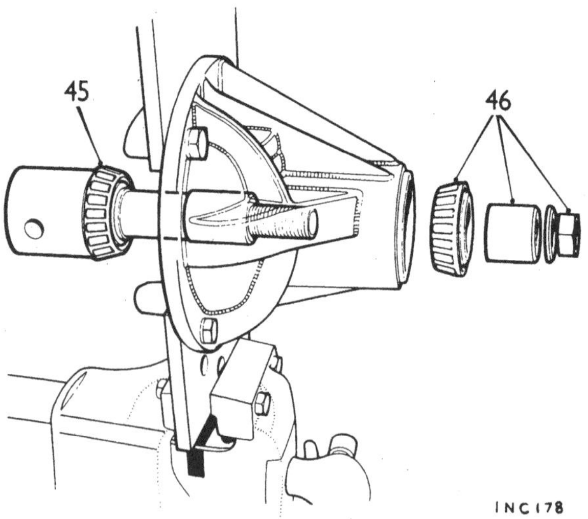

INC 178

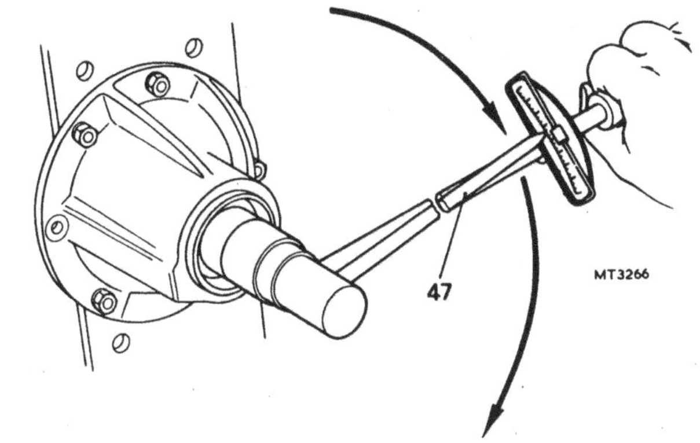

47. Tighten the nut gradually until a bearing pre-load of 15 to 18 lbf. in (0·17 to 0·21 kgf. m) is obtained. This can be measured using a lbf. in (kgf. m) scale torque wrench and suitable size socket.

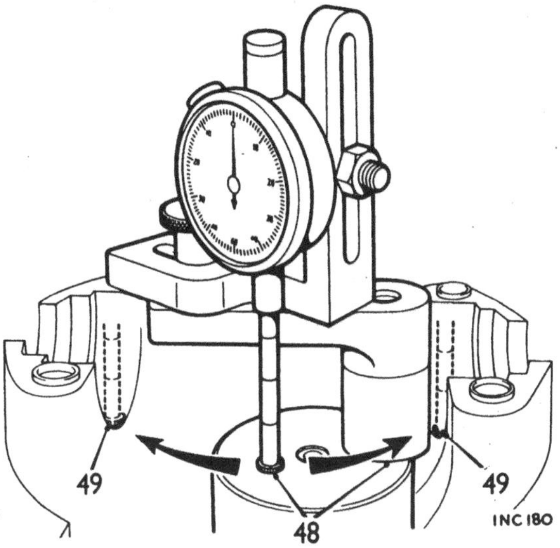

48. Clean the dummy pinion head. Position the dial gauge foot of 18G 191 on the dummy head and zero the gauge onto the head.
49. Move the gauge foot over the centre of one differential bearing bore. Note the indicated measurement. Repeat for the opposite bearing bore.
50. **Add the two measurements figures and divide by two.**
51. Twenty-two **pinion head washers** are available ranging in size from 0·075 to 0·096 in (1·91 to 2·44 mm).
52. Remove the dummy pinion 18G 191 M
53. Remove the inner bearing from the dummy pinion.

Calculating pinion head washer size

Example

Sum of each bore measurement divided by two	0·002 in	(0·05 mm)
Plus dummy pinion washer allowance from 18G 191 M	0·077 in	(1·95 mm)
Required size of pinion head washer	**0.79 in**	**(2·00 mm)**

NOTE: Whilst etched +, –, or 'N' markings will be found on the pinion face these should be ignored since they are taken into consideration in the design and method of using the dummy pinion.

Setting the pinion bearing pre-load

54. Fit the correct **pinion head washer** to the pinion.
55. Fit the inner bearing, using S4221A and adaptor S4221A-17.
56. Oil the bearing and fit the pinion to the casing.
57. Fit the bearing spacer, chamfered end towards the drive flange.
58. Fit the shims: oil and fit the outer bearing.
59. Fit the drive flange washer and nut.
60. Gradually tighten the nut, **do not exceed** 90 lbf. ft (12·4 kgf. m), checking the bearing pre-load during tightening operations.
61. Rotate the flange to settle the bearings, and check the pre-load using S98A. Pre-load should be set at 15 to 18 lbf. in (0·17 to 0·21 kgf. m) at 90 lbf. ft (12·4 kgf. m) torque on the flange nut.
    If the pre-load is high, increase shim thickness.
    If the pre-load is low, decrease shim thickness.
62. Four shims are available ranging in size from 0·003 to 0·030 in (0·076 to 0·762 mm). Note: 0·001 shim thickness equals approximately 4 lbf. in (0·046 kgf. m) pre-load.
63. Remove the drive flange nut washer and flange nut.
64. Soak the new oil seal in oil for **one hour** and fit the seal.
65. Fit the drive flange, washer and flange nut.
66. Tighten the drive flange nut to 90 lbf. ft (12·4 kgf. m) using RG421 or S337 to retain the flange.
67. Lock the nut, using a new split pin.

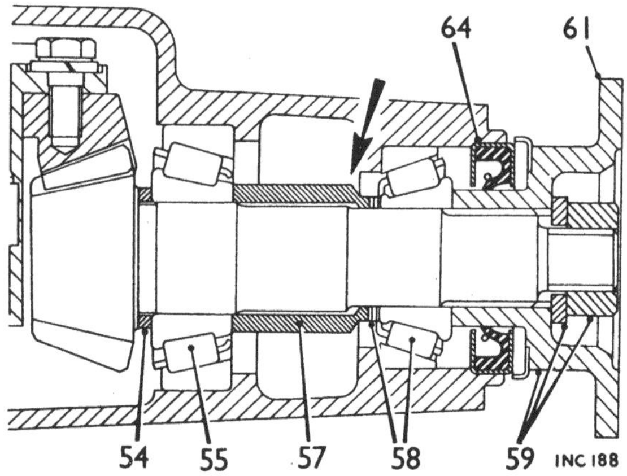

INC 188

Setting the backlash

68. Place the bearing cups on the differential bearings and fit the differential carrier in the case.
69. Position dial gauge 18G 191 on the axle spreader adaptor plate S101-1, crown wheel side.
70. Move the crown wheel fully into mesh with the pinion and zero the gauge on the rear of the crown wheel.
71. Move the crown wheel and carrier in the opposite direction until the bearing crown-wheel-side is butted in its housing. The indicated measurement, which should be noted, is the 'IN-OUT' of mesh clearance.

continued

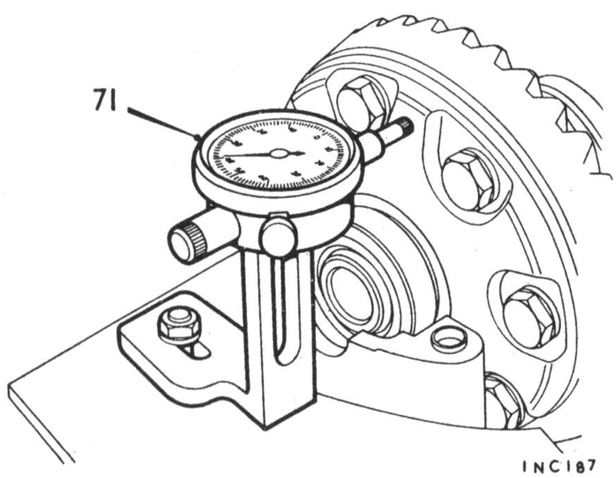

INC187

Setting the crown wheel backlash

In-Out of mesh clearance **(from operation 71)**	0·025 in	(0·63 mm)
Minus backlash (see DATA)	0·005 in	(0·13 mm)
Required crown wheel side shim pack	**0·020 in**	**(0·50 mm)**
Total side-float **(dimension 'A', operation 42)**	0·060 in	(1·52 mm)
Minus crown wheel side shim pack	0·020 in	(0·50 mm)
Required shim pack opposite crown wheel	**0·400 in**	**(1·02 mm)**

Pre-load: add **0·002 in (0·051 mm)** to each shim pack calculated above.

72. Fit S101 and S101-1, and stretch the unit case by tightening the turnbuckle three or four flats.
73. From the calculations made in 76 and 77, select the required shim pack from those available· 0·003 in (0·0762 mm), 0·005 in (0·127 mm), 0·010 in (0·25 mm), 0·20 in (0·50 mm).
74 Fit the differential assembly to the case and fit the shim packs. Slacken S101 and remove.
75. Refit the bearing caps as originally marked, and secure using the bolts and spring washers.
76. Rotate the pinion to settle the differential bearings. Position dial gauge service tool 18G 191 on the adaptor plate and position the foot on the crown wheel gear. Measure the total backlash at several positions, which must be 0·004 to 0·006 in (0·10 to 0·15 mm).

NOTE: A movement of 0·002 in (0·05 mm) shim thickness from one differential bearing to the other will vary the backlash by approximately 0·002 in (0·05 mm).

77. Tighten the bearing cap bolts.
78. Remove S101-1.
79 Refit the differential assembly. 51.15.01.

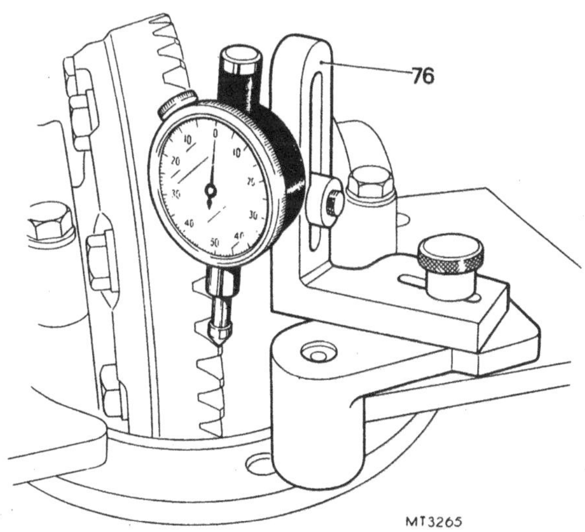

continued

DATA

Differential bearing shims	0·003 in (0·076 mm)	
	0·005 in (0·127 mm)	
	0·010 in (0·254 mm)	
	0·020 in (0·508 mm)	
Differential case, maximum stretch	0·008 in (0·20 mm)	
Differential pinion gears thrust washer	8, in 0·002 in (0·05 mm) steps	
Sizes	0·027 in (0·685 mm)	0·035 in (0·889 mm)
	0·029 in (0·737 mm)	0·037 in (0·940 mm)
	0·031 om (0·787 mm)	0·039 in (0·991 mm)
	0·033 in (0·838 mm)	0·043 in (1·092 mm)
Crown wheel run-out	Max. 0·003 in (0·076 mm) 0·004 to 0·006 in (0·102 to 0·152 mm)	
Optimum setting	0·005 in (0·127 mm)	
Pinion bearing pre-load	15 to 18 lbf. in (0·17 to 0·21 kgf. m)	
Pinion head washer sizes:		
STD:	0·077 in (1·956 mm)	
Alternatives:		

0·075 in (1·905 mm)	0·082 in (2·083 mm)	0·087 in (2·210 mm)
0·0765 in (1·930 mm)	0·0825 in (2·095 mm)	0·0885 in (2·247 ,,)
0·078 in (1·981 mm)	0·083 in (2·108 mm)	0·090 in (2·286 mm)
0·079 in (2·007 mm)	0·084 in (2·134 mm)	0·0915 in (2·323 mm)
0·0795 in (2·019 mm)	0·085 in (1·259 mm)	0·0935 in (2·337 mm)
0·080 in (2·032 mm)	0·0855 in (2·171 mm)	0·0945 in (2·400 mm)
0·081 in (2·057 mm)	0·086 in (2·184 mm)	0·096 in (2·438 mm)

Pinion bearing shims sizes	0·003 in (0·076 mm)
	0·005 in (0·127 mm)
	0·010 in (0·254 mm)
	0·030 in (0·762 mm)

PINION OIL SEAL

– Remove and refit **51.20.01**

Special tool S337 or RG421

Removing

1. Place the car on a ramp or over a pit.
2. Disconnect the exhaust pipe at its rear mounting and the silencer at the two attachment points.
3. Disconnect the rear end of the propeller shaft from the pinion driving flange – instruction 5, 47.15.01.
4. Withdraw the split pin from the pinion securing nut and remove the nut, using special flange holder tool S337 or RG421.
5. Place a receptacle under the drive flange to collect oil and remove the flange – instruction 15, 51.15.07.
6. Lever out the old oil seal.

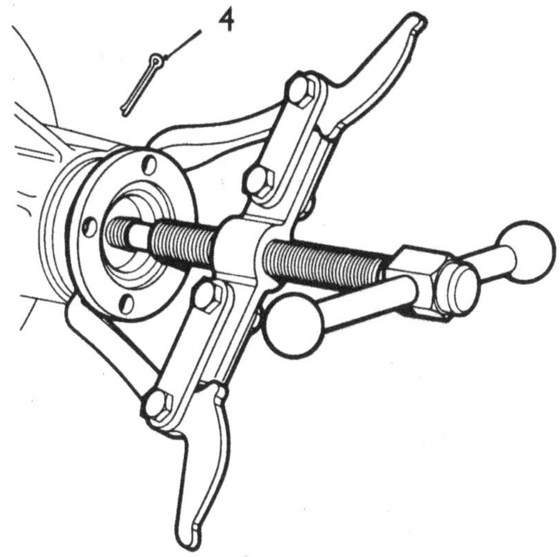

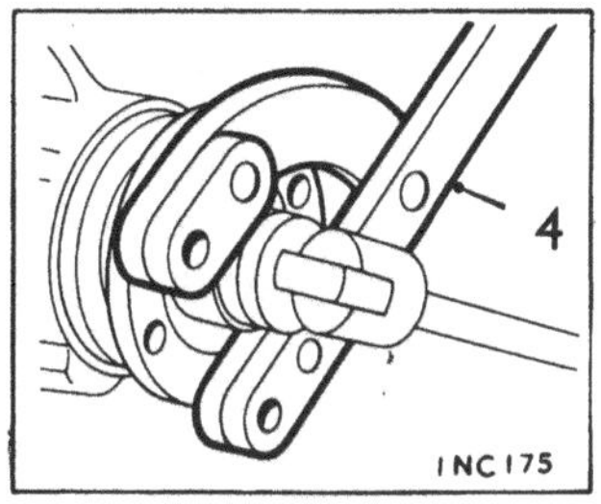

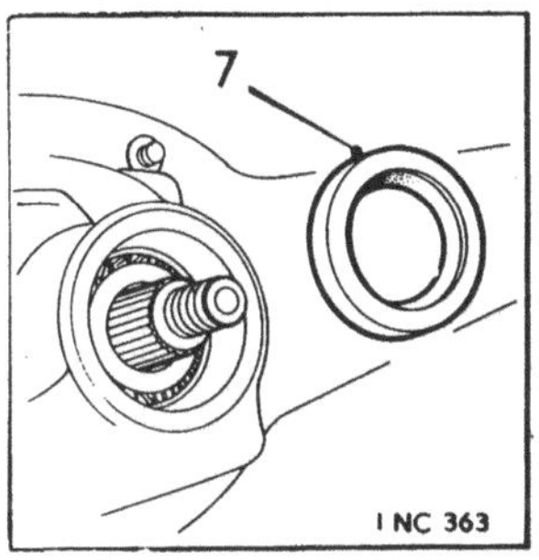

Refitting

7. Immerse a new seal in light oil for one hour before fitting.
8. With the sealing lip leading, drive in the seal.
9. Refit the driving flange and secure with the pinion nut and tighten to 90 to 120 lbf. ft (12·4 to 16·6 kgf. m).
10. Fit a new split pin.
11. Reconnect the rear end of the propeller shaft to the driving flange – instruction 6, 47.15.01.
12. Reverse instructions 1 and 2.
13. Recharge the final drive unit with a recommended E.P. oil.

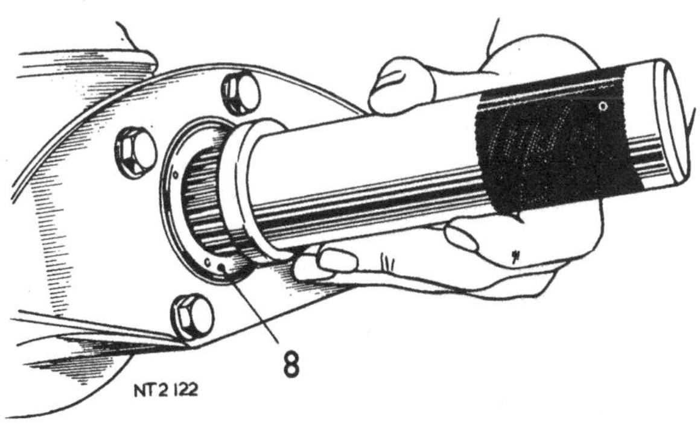

FINAL DRIVE UNIT

– Remove and refit 51.25.13

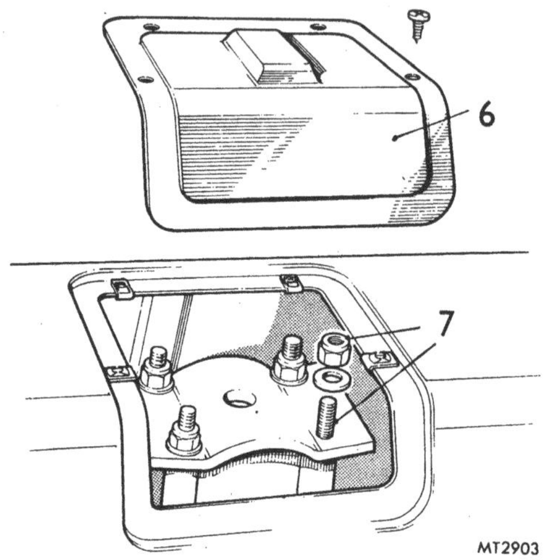

1. Place the car on a ramp or over a pit.
2. Jack up the car on both sides under the rear body jacking points.
3. Disconnect the exhaust pipe at its rear mounting and remove the silencer.
4. Disconnect the propeller shaft from the pinion driving flange. 47.15.01.
5. Disconnect the two rear drive shaft universal joint couplings from the inner drive shafts 47.10.01.
6. Lower the ramp and remove the rear trim panel and rear spring access cover.
7. Remove the four nuts and studs securing the rear spring to the final drive assembly.

[NOTE: There are six nuts and a flat retaining plate on Mk. III models, all spring leaves being rigidly attached to the final drive units.]

8. Raise the ramp and remove the final drive rear mounting nut and special bolt.
9. Whilst supporting the final drive unit remove the two nyloc nuts securing the final drive front mounting plate to the chassis. 51.25.34.
10. Manoeuvre the final drive unit clear of the chassis taking care to keep the unit upright to avoid oil spillage from the stud holes.

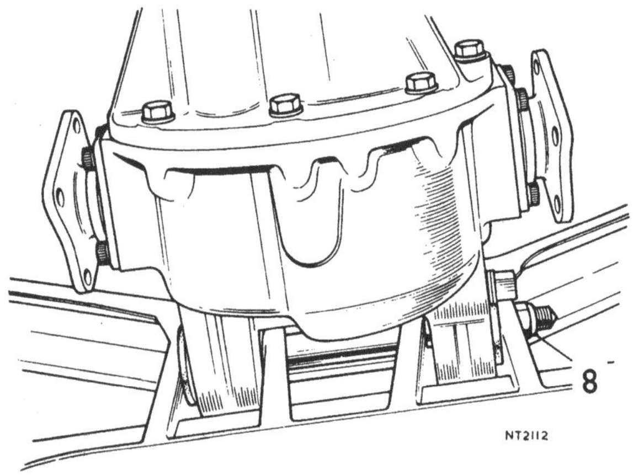

Refitting

11. Offer up the final drive unit to its rear mounting location ensuring that the two rubber washers are positioned correctly on the outside of the mounting lugs. Secure the rear mounting with the special bolt and nyloc nut.
12. Connect the final drive front mountings, ensuring the rubbers are correctly positioned i.e. the upper rubber with the stepped end facing downwards. Fit the lower rubber and special washer and secure with the nyloc nuts. Tighten to 26 to 34 lbf. ft (3.6 to 4.7 kgf. m). It should be possible to revolve the lower rubbers with the fingers when the nuts are tightened to this torque figure. 51.25.34.
13. Reconnect the rear drive shaft couplings to the inner drive shafts.
14. Reconnect the propeller shaft to the pinion flange.
15. Reconnect the exhaust pipe to its rear mounting and fit the silencer.
16. Lower the ramp and position the spring so that the dowel on the bottom leaf locates in the hole in the final drive spring platform. Refit the studs and secure with the nyloc nuts and plain washers and tighten to 26 to 34 lbf. ft (3·6 to 4·7 kgf. m).
17. Lower the body jacks fit the rear spring access cover plate and rear trim panel.
18. Check the final drive oil level – see maintenance – and if necessary top up with a recommended E.P. oil.
19. Remove the car from the ramp.

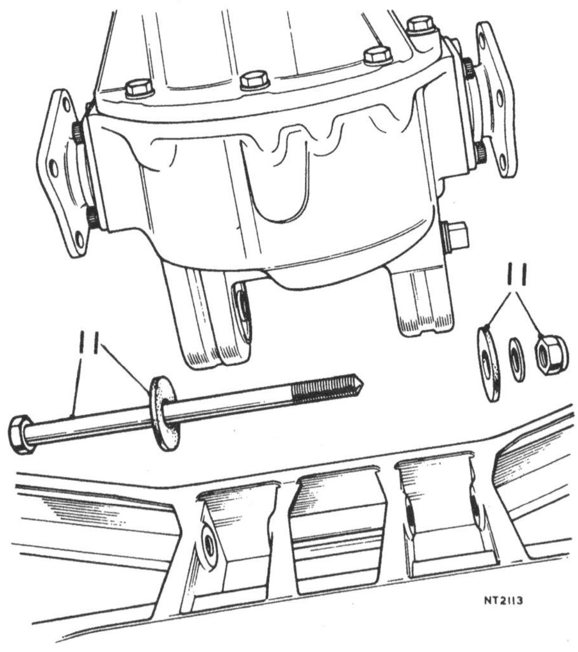

FINAL DRIVE UNIT MOUNTING PLATE

– Remove and refit 51.25.33

Removing

1. Place the car on a ramp or over a pit.
2. Jack up the rear of the car at the two rear jacking points until the wheels are clear of the ramp or ground.
3. Disconnect the exhaust pipe at its rear mounting and the silencer at its two attachment points.
4. Mark for reassembly and disconnect the rear end of the propeller shaft from the pinion flange, 47.15.01.
5. Remove the pinion nut and flange – instructions 4 and 5, 51.20.01.
6. Remove the four bolts and shakeproof washers securing the mounting plate to the differential housing.
7. Remove the nyloc nut, special washer and lower rubber from both sides of the mounting plate and manoeuvre the plate off the captive studs and remove from the car. 51.25.34.

Refitting

8. Place the mounting plate in position and refit the pinion driving flange and nut – instructions 8 and 9 51.20.01.
9. Refit the mounting plate and rubbers to the chassis – instructions 8 to 12, 51.25.34.
10. Reconnect the exhaust pipe and silencer.
11. Remove the car from the ramp.

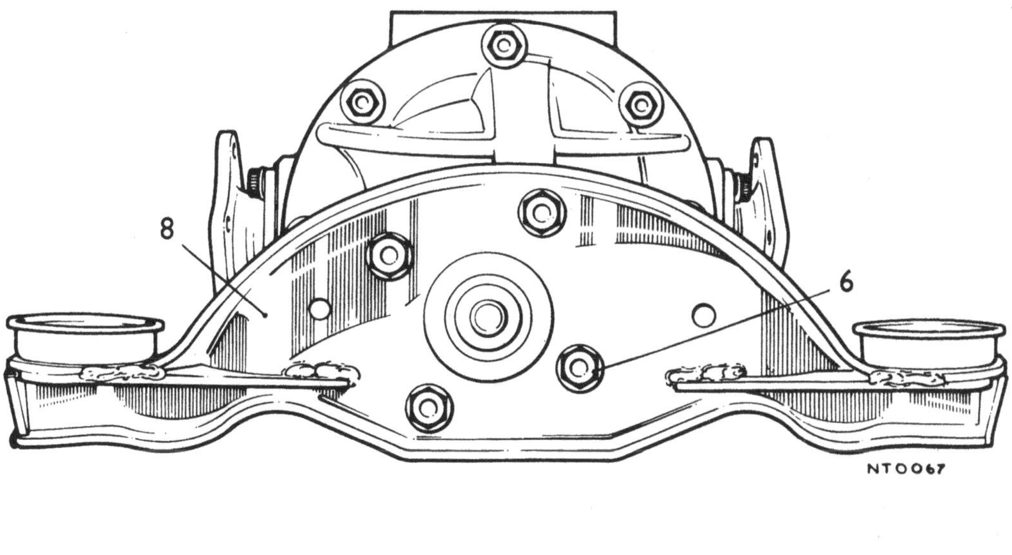

FINAL DRIVE UNIT MOUNTING PLATE RUBBERS

– Remove and refit 51.25.34

Removing

1. Place the car on a ramp or over a pit.
2. Disconnect the exhaust pipe at its rear mounting and the silencer at its two attachment points.
3. Jack up the rear of the car under both sides of the chassis until the rear wheels are clear of the ramp or ground.
4. Mark for reassembly and disconnect the rear end of the propeller shaft from the pinion flange. 47.15.01.
5. Remove the four bolts and shakeproof washers securing the final drive mounting plate to the differential casing, 51.25.33.
6. Remove the two nyloc nuts (one each side) securing the mounting to the chassis, together with the special washer and lower rubber.
7. Using a suitable lever prise the mounting plate downwards until the upper mounting rubber can be removed from the captive stud. Repeat instructions on the other side.

Refitting

8. Prise the mounting plate downwards as in instruction 7 sufficiently to enable the replacement upper rubber to be fitted over the captive stud. Ensure that the stepped end of the rubber is facing downwards and locates squarely in the mounting plate cup. Repeat this instruction on the opposite side.
9. Fit the lower rubber and special washer and loosely secure – at this stage – with the nyloc nut. Repeat this instruction on the other side.
10. Secure the mounting plate to the differential casing with the four bolts and shakeproof washers.
11. Reconnect the rear end of the propeller shaft to the pinion flange ensuring that the assembly marks line up.
12. Lower the jack and finally tighten the two nyloc mounting nuts to 26 to 34 lbf. ft (3·6 to 4·7 kgf. m). It should be possible to revolve the lower mounting rubbers with the fingers when the nuts are tightened to this torque figure.
13. Reverse instructions 1 and 2.

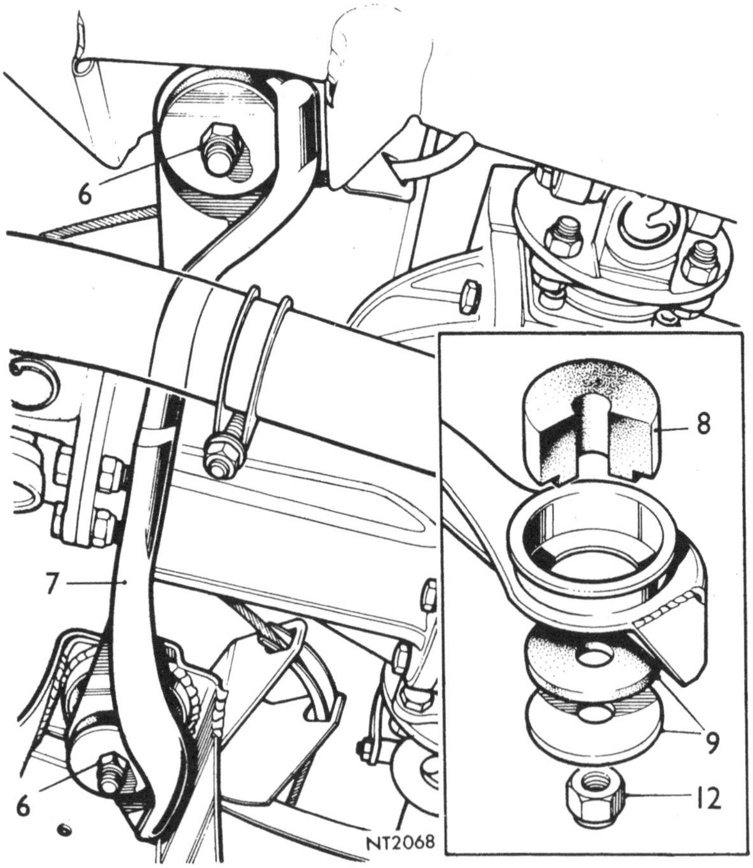

STEERING OPERATIONS

STEERING RACK AND PINION

– Remove and refit 57.25.01

Removing

1. Remove the nyloc nut and plain washer securing the outer tie rod to left and right hand steering arms.
2. Release both outer tie rod ends from the steering arms.
3. Remove the pinch bolt clamping the steering rack pinion to the lower end of the flexible coupling.
4. Remove the four nyloc nuts and plain washers securing the steering rack 'U' bolts to the chassis. Withdraw the two reinforcement plates.
5. Remove the 'U' bolts complete with the rack clamp brackets.
6. Slide rack forward to disengage the rack pinion splines from the flexible coupling.
7. Withdraw the steering rack.

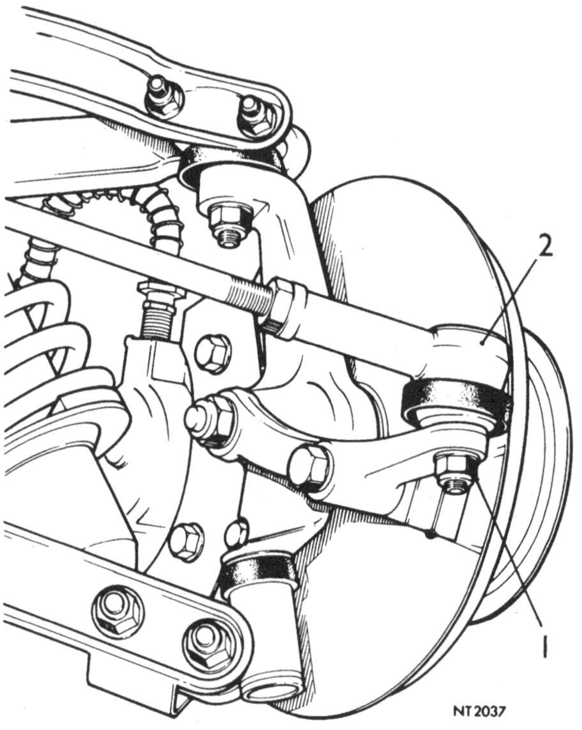

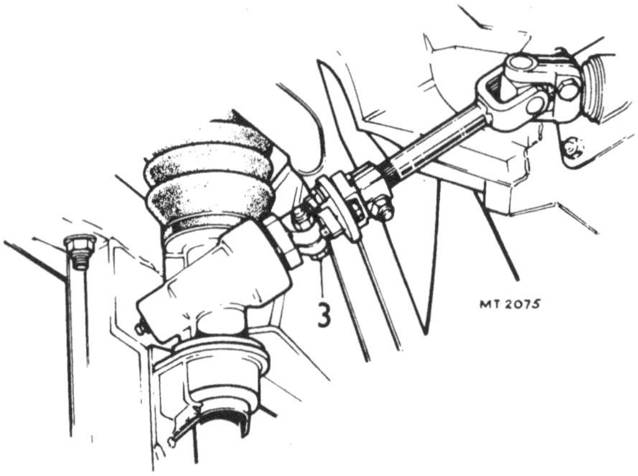

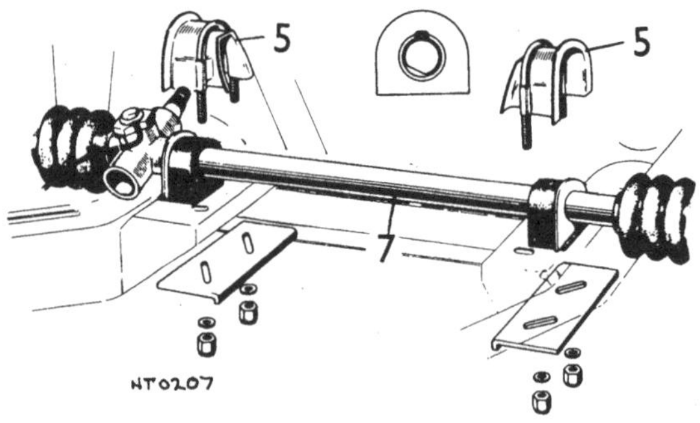

Refitting

8. Remove the centre plug from rack plunger and align dimple in the rack shaft with plug aperture. In this position the rack shaft is centralised.
9. Fit plug to plunger ensuring bonding strap is not omitted.
10. Carefully slide rack into position in car taking care not to disturb position of rack shaft.
11. With steering wheel held in straight ahead position engage rack pinion shaft splines in the flexible coupling and fit and tighten the pinch bolt in the flexible coupling.
12. Ensure the rack mounting rubbers are correctly positioned on inboard side of rack flanges with lip under the straight face on the flange.
13. Fit 'U' bolts and clamp brackets.
14. Fit the reinforcement plates to the 'U' bolts ensuring angled lips are inboard with lip projecting downward.
15. Fit the plain washers and nyloc nuts to 'U' bolts taking care not to omit the bonding strap at the pinion end of rack.
16. With angled lips of the reinforcement plates in contact with the inboard edges of the chassis brackets and the mounting rubbers compressed tighten the 'U' bolt securing nuts.
17. Fit tie rod outer ends to steering arms.

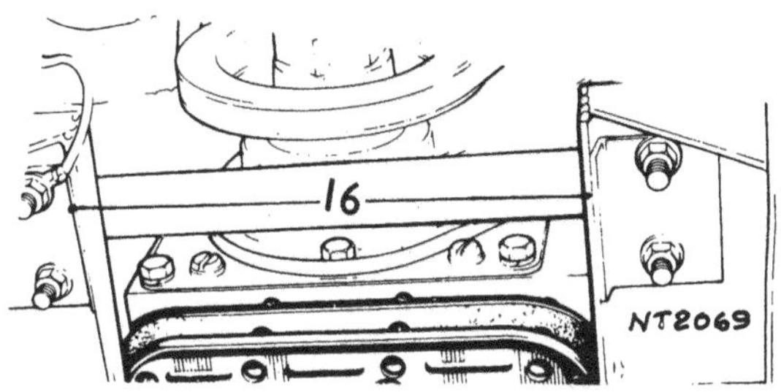

STEERING RACK GAITERS

– Remove and refit 57.25.02

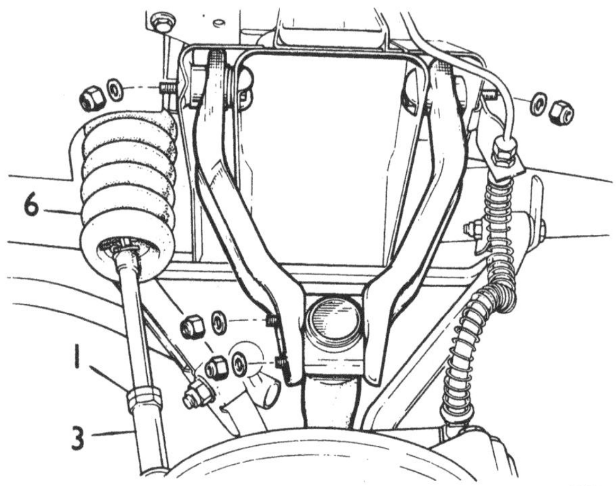

Removing

1. Slacken the locknut securing tie rod outer ball joint.
2. Remove the nut and washer securing the tie rod outer ball joint to the steering arm.
3. Release the outer ball joint from the steering arm.
4. Unscrew the outer ball joint and locknut from the tie rod.
5. Remove the inner and outer clips retaining the gaiter to the rack and tie rod respectively.
6. Withdraw the gaiters.
7. Repeat above instructions on opposite tie rod.

Refitting

8. Lubricate the tie rod inner ball joints with fresh grease.
9. Slide the new gaiters along the tie rods into position on the rack and secure with clip.
10. Screw locknut and tie rod outer ball joint assembly on tie rod and locate them as near as possible to original locations.
11. Connect outer ball joint to steering arm and fit plain washer and nut.
12. Repeat instructions 8 to 11 on opposite tie rod.
13. Check front wheel toe-in and adjust as necessary, ensuring tie rods are equally adjusted.
14. Secure the outer ball joint assemblies with their respective locknuts.
15. Adjust the outer end of both gaiters so that they are capable of accommodating movement of the tie rods when moving from lock to lock.

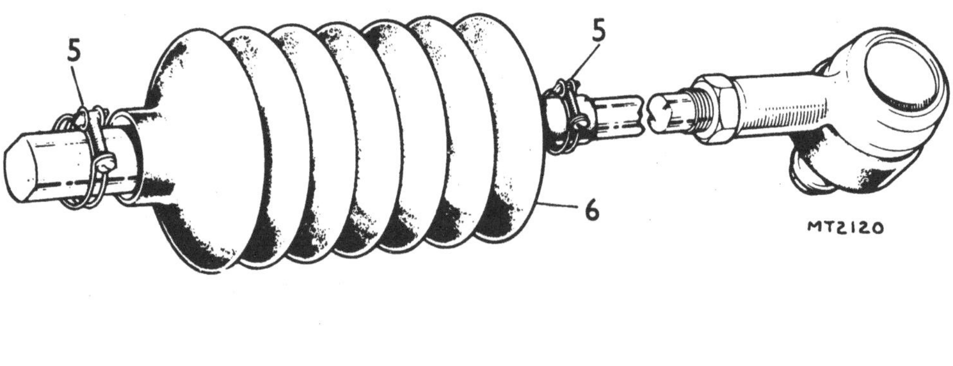

STEERING RACK AND PINION

– Overhaul **57.25.07**

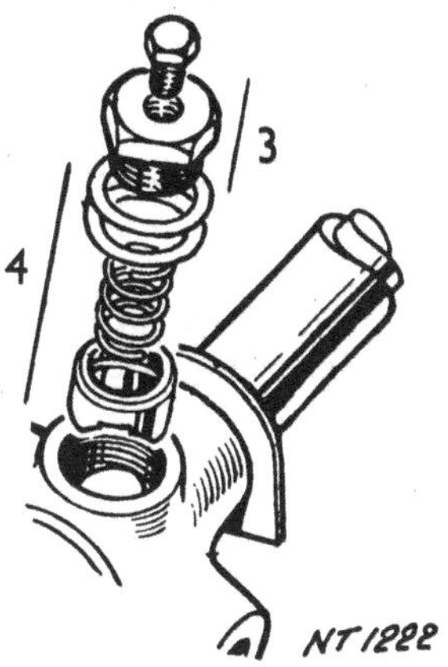

1. Remove the rack from the car 57.25.01.

Dismantling

Rack plunger

2. Slacken the plug securing the rack plunger to the rack housing.
3. Remove the plug and shim pack.
4. Withdraw the spring, washer and plunger.

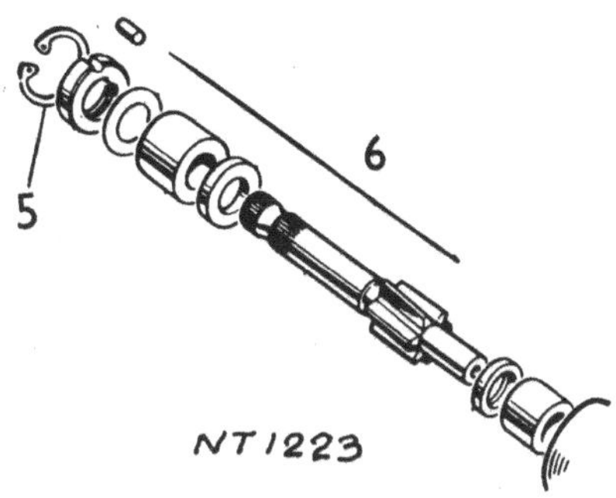

Pinion

5. Remove the circlip retaining the pinion to the rack housing.
6. Withdraw pinion complete with plug-end, locating pin, shims, bush and upper thrust washer.

Tie rods and rack shaft

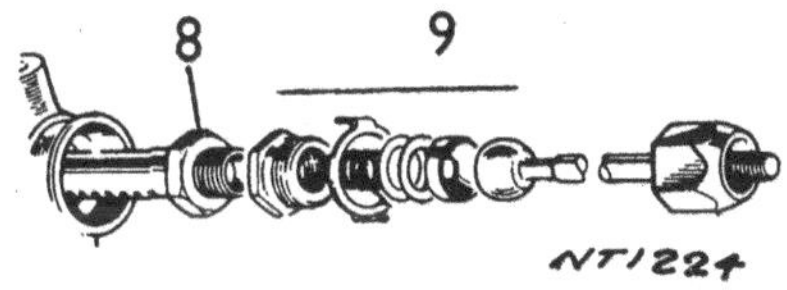

7. Release the clips securing the gaiters to the rack housing and tie rods and slide the gaiters clear of the rack.
8. Slacken the locknuts securing inner ball joint adaptors to the rack shaft.
9. Unscrew and remove the inner ball joint adaptors and tie rods.
10. Remove the locknuts and withdraw the rack shaft.

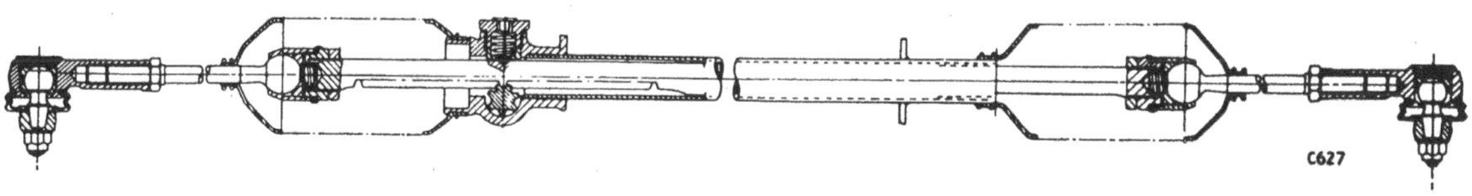

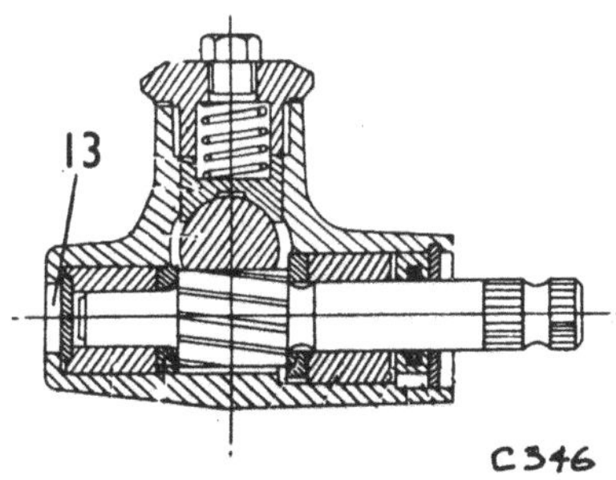

Rack housing bush

11. Remove the bush from the rack housing.

Pinion housing

12. Extract the pinion lower thrust washer.
13. Drive or press out the pinion end cover and the lower bush.

Assembling

Rack housing bush

14. Fit a new bush to the rack housing.

Pinion housing

15. Fit the end cover and lower bush to the pinion housing ensuring that the recessed end of the bush is adjacent to the end cover.

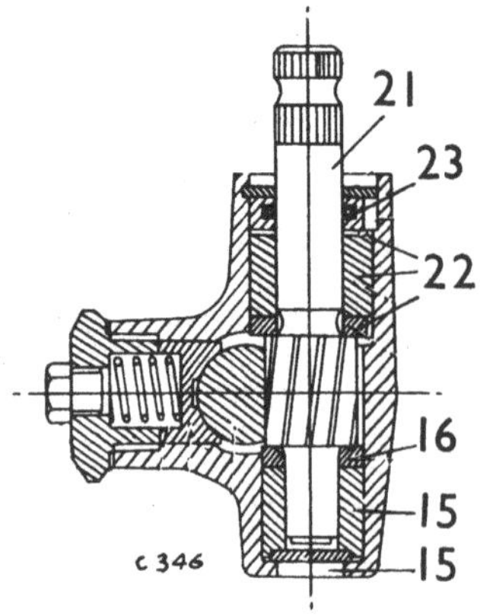

Rack shaft and tie rods

16. Insert pinion lower thrust washer in position above pinion lower bush ensuring that the internal fillet faces away from the bush.
17. Insert rack shaft in housing and fit locknuts.
18. Fit both tie rods and adaptors and secure with locknuts. (torque 80 lbf/ft. 11·06 kgf/m).
19. Pack tie rod inner ball joints and gaiters with approximately 2 oz (56 gms.) of grease each side and secure gaiters to rack.

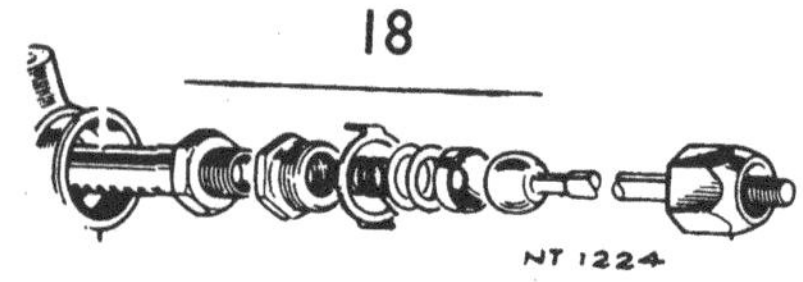

Pinion

20. Rotate rack shaft until teeth will permit pinion engagement.
21. Fit pinion to pinion housing.
22. Fit upper thrust washer, bush and shims.
23. Fit a new internal 'O' ring to the plug-end.
24. Fit circlip and check pinion end movement. End movement must not exceed 0·010 in (0·254 mm). Adjust shim pack as necessary.

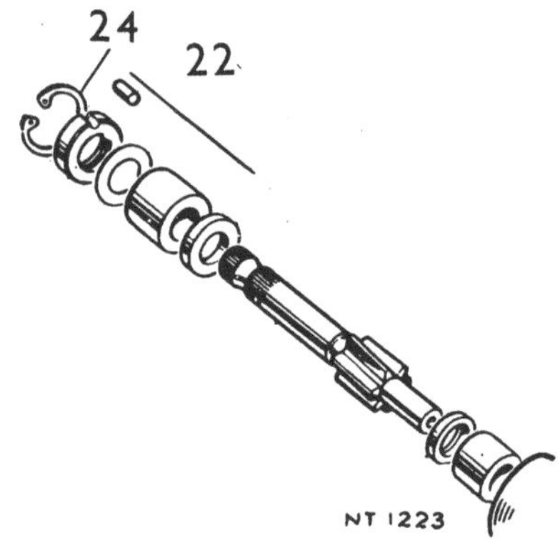

Rack plunger

25. Fit plunger.
26. Fit the damper plug, shims and spring.
27. Check the rack shaft for side movement (90° to shaft axis). Side movement should be within 0·004 to 0·008 in (0·1016 to 0·2032 mm). Adjust as required by adding or removing shims as necessary.

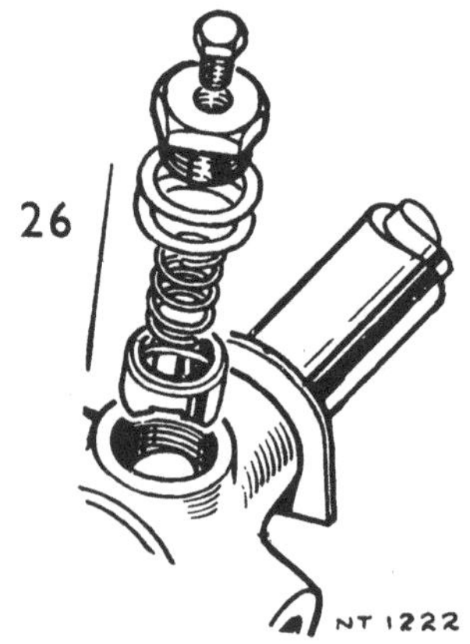

STEERING RACK DAMPER

– Remove and refit 57.35.10

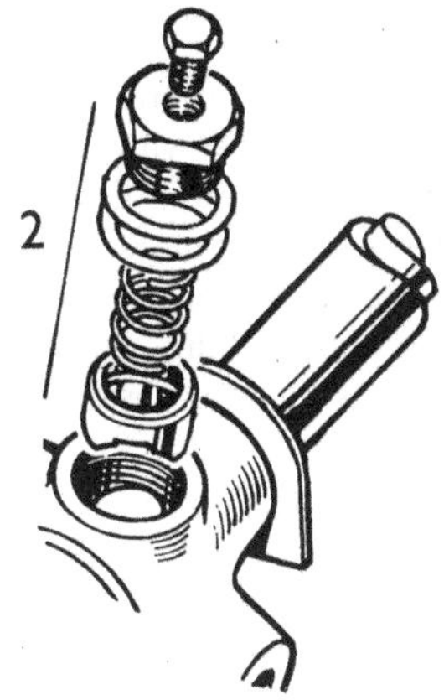

Removing

1. Disconnect the bonding strap at the damper plug.
2. Unscrew the damper plug and remove plug, shims, spring washer and plunger.

Refitting

3. Fit plunger and washer.
4. Fit the damper plug, shims and spring.
5. Check the rack shaft for side movement (90° to shaft axis) side movement should be within 0·004 to 0·008 in (0·1016 to 0·2032 mm). Adjust as required by adding or removing shims as necessary. Ensure bonding strap is reconnected.

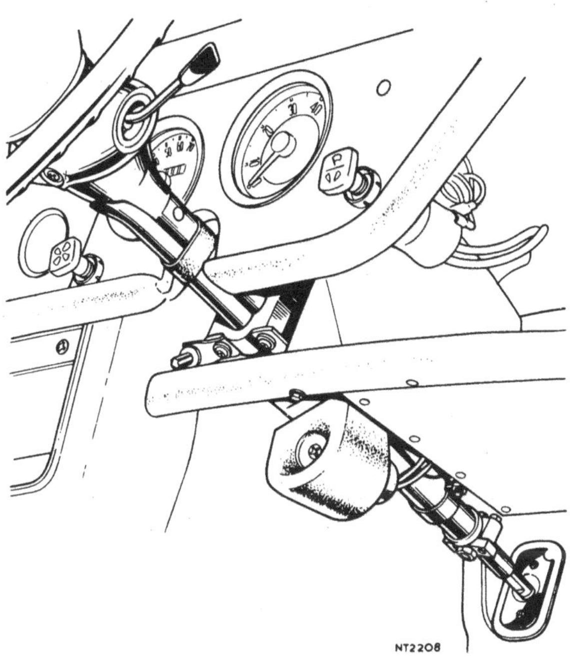

STEERING COLUMN ASSEMBLY

– Remove and refit 57.40.01

Removing

1. Disconnect the battery.
2. Remove the driver's parcel shelf.
3. Remove the pinch bolt securing the steering mast to the flexible coupling.
4. Remove the nuts, spring and plain washers from the two bolts securing the steering column forward bracket.
5. Withdraw forward support housing and felt liner.
6. Disconnect connections for horn, trafficators and lights.
7. Disconnect electrical plug from steering column lock.
8. Remove the two cap screws securing the steering column rear bracket and withdraw both clamp halves and upper screwed plate.
9. Withdraw steering column complete with tie-bar.

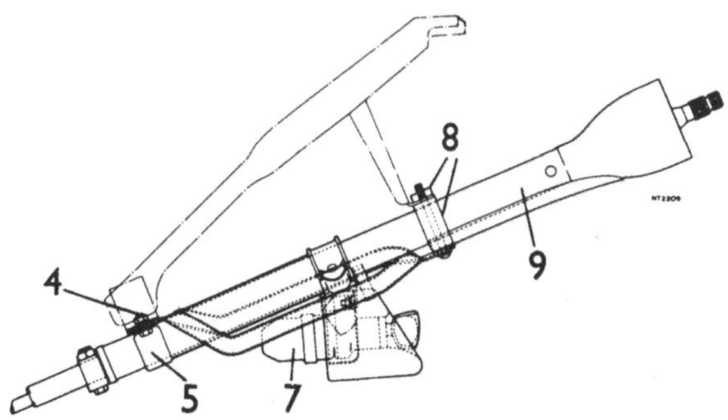

Refitting

10. Locate road wheels in straight ahead position.
11. Hold steering wheel with two spokes horizontal and the third spoke towards drivers seat and engage lower end of steering mast in splines of flexible coupling.
12. Fit and tighten pinchbolt in flexible coupling.
13. Slide cardboard tube rearwards and fit felt over steering mast housing ensuring ends of felt are below mast housing.
14. Engage forward ends of tie-bar in forward mounting bolts.
15. Fit felt housing and engage forward mounting bolts. Fit plain and spring washers and nuts to mounting bolts.
16. Ensure spring clip is fitted on mast housing to align with rear clamp bracket.
17. Fit upper and lower clamp halves to mast housing.
18. Place screwed plate in bracket above clamp halves.
19. Enter cap bolts through tie bar, upper and lower clamp halves, and mounting bracket and engage screwed plate. Tighten cap bolts.
20. Tighten bolts and nuts securing forward bracket.
21. Make good connections to horn, lights, trafficators and steering lock.
22. Fit drivers parcel tray.
23. Connect battery.

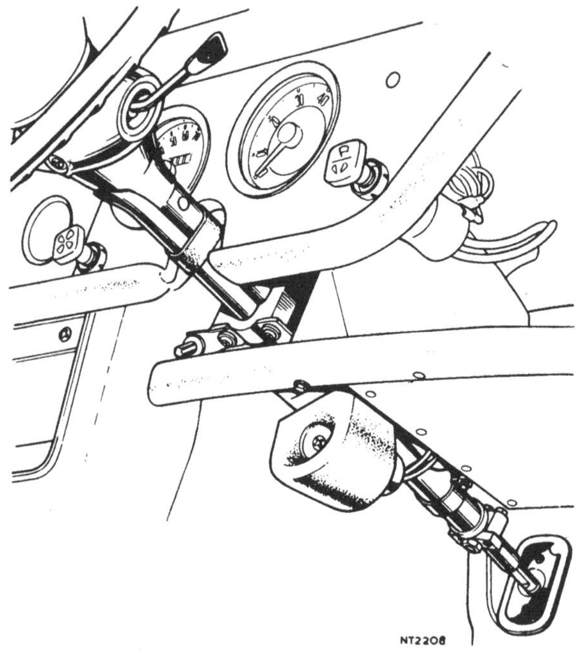

STEERING COLUMN SAFETY CLAMP

– Remove and refit 57.40.07

Removing

1. Slacken and remove the two bolts and spring washers securing clamp halves to steering mast.
2. Withdraw clamp.
3. Slacken locknut and pinch screw.

Refitting

4. Offer up clamp halves to steering mast ensuring the straight member sits across the flat on the steering mast.
5. Engage bolts complete with spring washers.
6. Slide clamp rearwards until it is against the steering mast housing. Ensure plastic thrust washer is located between clamp bracket and mast housing.
7. Tighten clamp bolts to 6 to 9 lbf/ft (0·8 to 1·2 kgf/m).
8. Tighten grubscrew to 18 to 20 lbf/ft (2·5 to 2·8 kgf/m).
9. Tighten locknut.

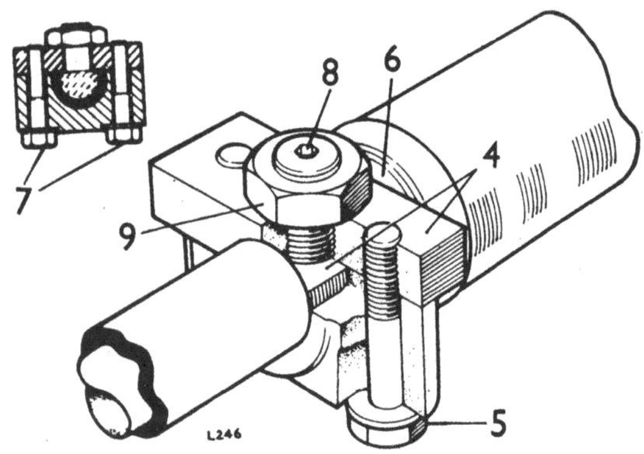

Triumph Spitfire Mk IV Manual. Part No. 545254. Issue 1.

STEERING COLUMN UNIT

– Overhaul **57.40.10**

1. Remove the steering column from the car 57.40.01

Dismantling

2. Remove the steering wheel 57.60.01
3. Remove the lights and trafficator, switches. 86.55.54, 86.65.17.
4. Unlock the steering column lock.
5. Remove the two set bolts and spring washers securing the safety clamp and temove the safety clamp and thrust washer
6. Withdraw the lower steering mast.
7. Withdraw the cap from the lower end of the steering column tube and remove the tubular cardboard distance piece.
8. Move the upper steering mast axially towards the upper end of the column tube, and remove the trafficator cam.
9. Move the steering mast axially towards the lower end of the column tube until the column is felt to butt against the column tube lower bush.
10. Depress the rubber locating buttons on the column lower bush and withdraw the steering mast and the lower bush.
11. Remove the lower bush from the mast.
12. Remove the upper bush from the column tube.

Assembling

13. Fit a new upper bush to the column tube ensuring that it is entered with the rubber dots on the end face of the bush leading. That is with the rubber dots towards the bottom of the column tube.
14. Enter the steering mast from the bottom of the column tube ensuring that the upper bush is not displaced and that the rubber locating dowels remain aligned with the holes in the column tube.
15. Enter the lower bush in the column tube ensuring that the rubber dots on the end face of the bush are trailing (dots to bottom of tube) and that the rubber locating dowels engage the holes in the column tube.
16. Fit the trafficator cam.
17. Fit the end cap to the column tube.
18. Fit the lower steering mast.
19. Align the flat on the lower mast with the slot on the upper mast and fit the thrust washer and safety clamp. Do not tighten the safety clamp at this stage. The safety clamp must be tightened after the column is installed in the car and the position of the lower column determined. (Instructions 6 and 7. 57.40.25).
20. Fit the lights and trafficator switches.
21. Fit the steering wheel.

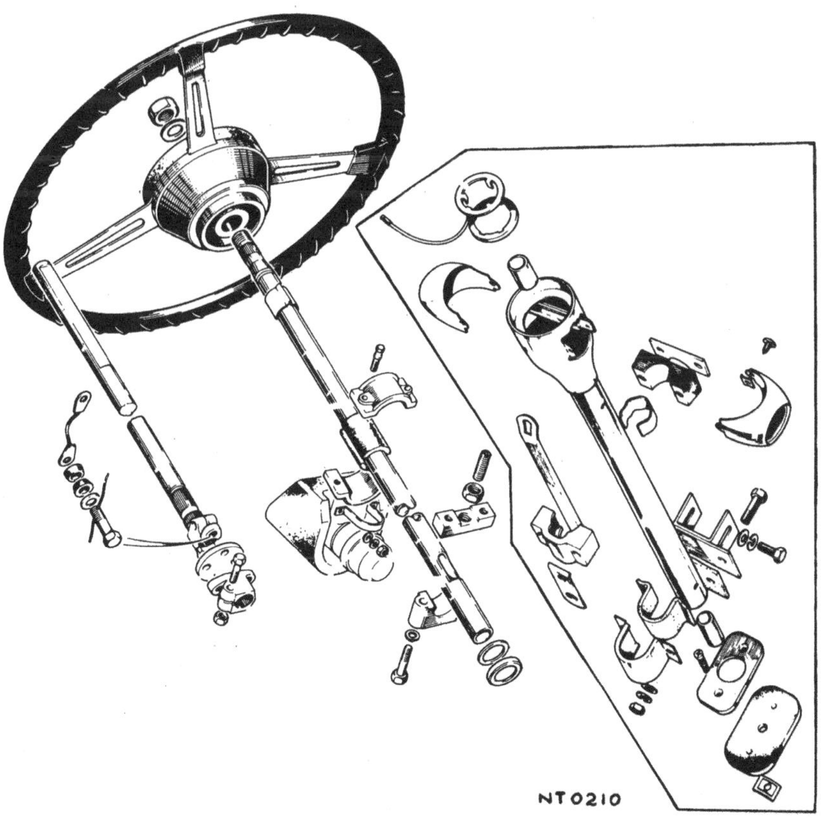

STEERING COLUMN FLEXIBLE COUPLING

– Remove and refit **57.40.25**

Removing

1. Slacken the two bolts on the steering column safety clamp.
2. Remove the pinchbolts securing the flexible coupling to the steering column lower shaft and the rack pinion shaft.
3. Slide lower steering column shaft clear of flexible coupling.
4. Withdraw flexible coupling from pinion shaft.

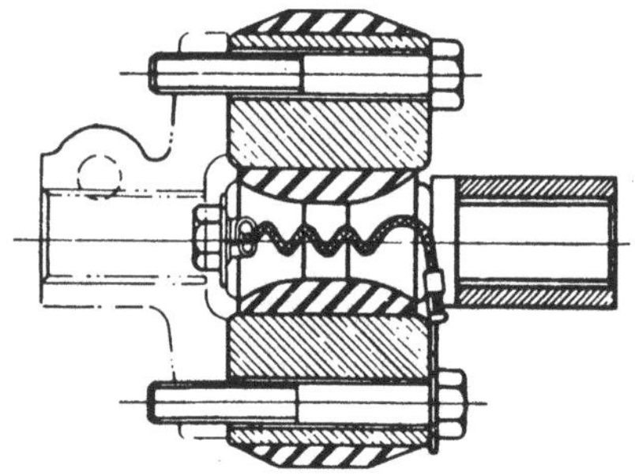

NT 1225

Refittig

5. Engage flexible coupling on pinion shaft and fit and tighten the pinchbolt.
6. With front wheels in straight ahead position and steering wheel in central position engage the lower steering column shaft in flexible coupling. Fit and tighten pinchbolt.
7. Slide safety clamp upwards until it abuts against the steering column housing. Tighten the safety clamp bolts, Torque 6 to 9 lbf/ft (0·8 to 1·2 kgf/m).

STEERING COLUMN LOCK/IGNITION SWITCH

– Remove and refit **57.40.31**

Removing

1. Remove steering column assembly 57.40.01.
2. Remove the two nuts and washers securing steering lock shroud and withdraw shroud and steering column tie-bar.
3. With a centre punch mark the centre of the two shear bolt heads securing the steering lock clamp bracket.
4. Using a small chisel unscrew the shear head bolts.
   or
4a. If instruction 4 proves unsuccessful drill into the shear bolt heads where previously marked by centre punch and using an Easiout extractor unscrew the shear-heads bolts.
5. Remove the steering lock.

Refitting

6. Offer up steering column lock to steering column and locate the dowel in the column drilling.
7. Fit steering lock clamp and secure with two new shear bolts.
8. Evenly tighten shear bolts until heads shear.
9. Install steering column assembly in car 57.40.01.

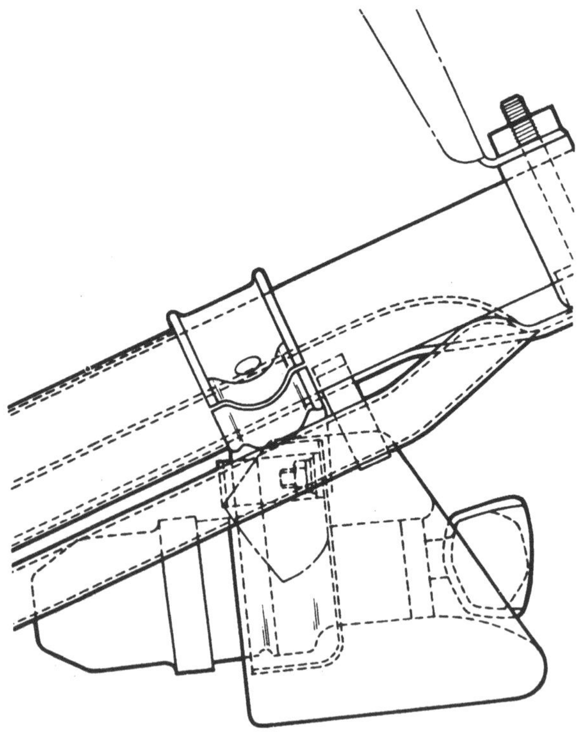

NT 2209 A

TIE ROD BALL JOINT – OUTER

– Remove and refit 57.55.02

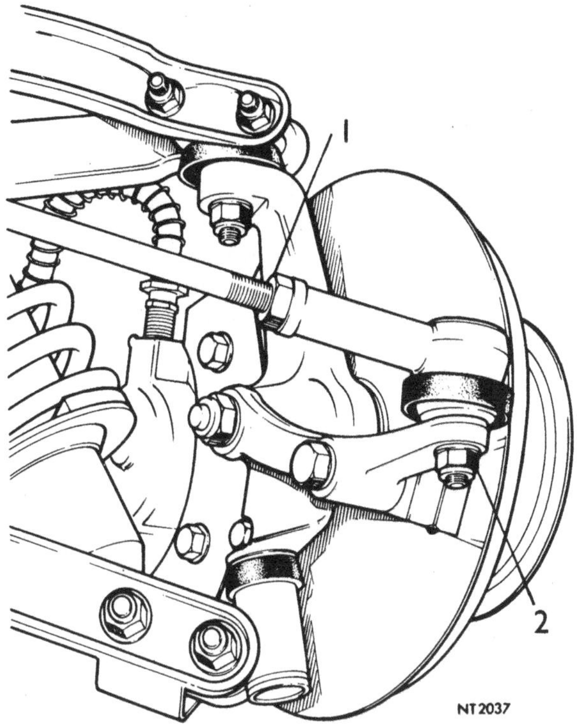

Removing

1. Slacken the locknut securing tie rod outer ball joint to tie rod.
2. Remove the nyloc nut securing ball joint to steering arm. Detach ball joint from steering arm.
3. Unscrew the ball joint housing from tie rod.

Refitting

4. Screw the ball joint housing into position on the tie rod. The distance between tie rod ball joint centres (inner to outer) is 8·72 in (221·5 mm).
5. Connect the ball joint to the steering arm and secure with the nyloc nut.
6. Check and adjust front wheel track as necessary.
7. Tighten the tie rod locknut.

TIE ROD BALL JOINT – INNER

– Remove and refit 57.55.03

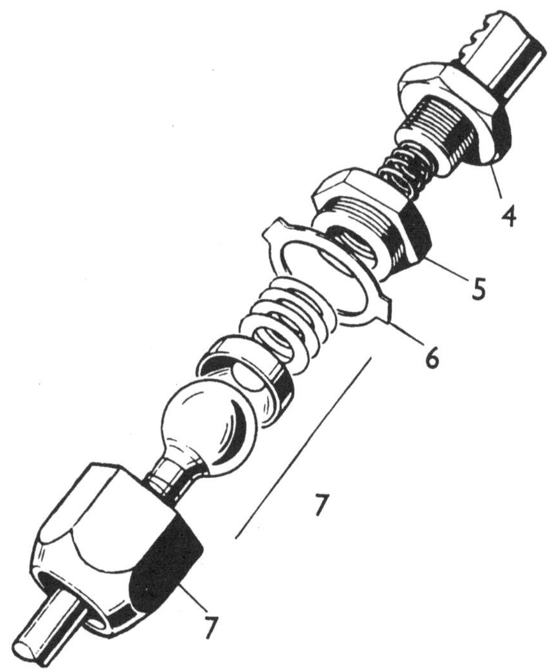

Removing

1. Remove the tie rod outer ball joint and housing.
2. Release the gaiter clips and remove the gaiter.
3. Wipe the grease from the inner ball joint.
4. Slacken the locknut securing inner ball joint adaptor to the steering rack. To prevent stress being transmitted to the pinion the opposite adaptor should be held with a spanner.
5. Unscrew the inner ball joint adaptor from the rack shaft observing the precaution mentioned in 4.
6. Straighten the lock tabs on the adaptor/ball housing washer.
7. Unscrew the adaptor from the ball housing and withdraw spring, shims, ball seat and tie rod from housing.

Refitting

8. Lubricate the ball housing and insert the tie rod.
9. Fit the ball seat, shim(s), new tab washer and adaptor.
10. Tighten the adaptor, torque 80 lbf/ft. (11.06 kgf/m), and check the tie rod for end-float and articulation. End-float should be within 0·0005 to 0·003 in (0·0127 to 0·0762 mm), there must be no tight spots in articulation. Adjust by adding or removing shims as necessary. Shims available are 0·002,0·004 and 0·010 in (0·0508, 0·1016 and 0·254 mm).
11. Bend over lock washer tabs.
12. Slide the spring into the adaptor.
13. Fit the locknut to the rack shaft.
14. Fit and tighten the adaptor to the rack shaft and secure with the locknut. When tightening the adaptor and locknut employ a spanner at the opposite adaptor to prevent stress to the pinion. Locknut torque should be 80 lbf/ft (11.06 kgf/m).
15. Cover the adaptor with fresh grease.
16. Fit the gaiter and clips.
17. Fit the tie rod outer end.
18. Check and adjust front wheel toe-in as necessary.

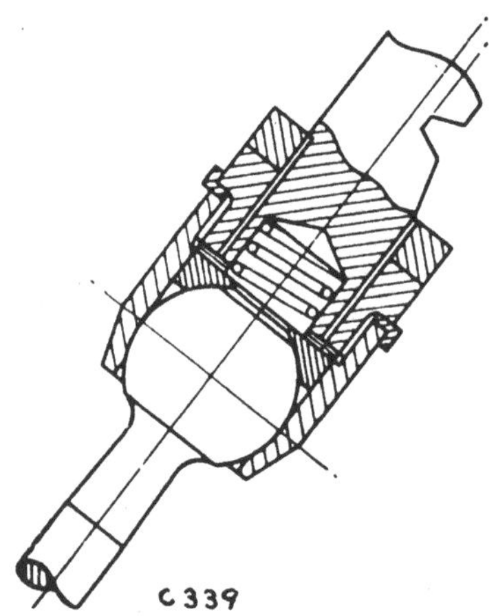

Triumph Spitfire Mk IV Manual. Part No. 545254. Issue 1.

STEERING WHEEL

– Remove and refit **57.60.01**

Removing

1. Prise off steering wheel pad.
2. Prise out horn push.
3. Withdraw the horn brush connection.
4. Remove the nut and washer securing steering wheel hub to steering mast.
5. To ensure that the steering wheel hub will be replaced in its original spline location scribe both hub centre and top of steering mast.
6. Using a suitable extractor withdraw the steering wheel.

Refitting

7. Fit steering wheel and hub to steering mast. If the steering wheel was withdrawn without spline location being marked, set the front wheels to the straight ahead position and centralise the steering wheel.
8. Reverse instructions 1 to 4.

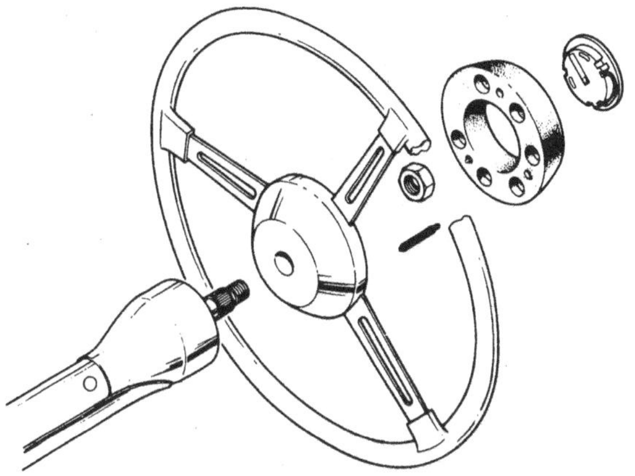
NT2226.

STEERING WHEEL HUB

– Remove and refit **57.60.02**

Removing

1. Remove steering wheel and hub from car 57.60.01.
2. Remove the six bolts and washers securing the steering wheel spokes to the hub.

Refitting

3. Reverse instructions 1 and 2.

STEERING GEOMETRY

57.65.00

Only two adjustments are possible: Front wheel alignment and Camber angle.

DATA	Kerb condition	Laden condition (2 up)
Camber	3° Positive ±1°	2° Positive ±1°
Castor	4°±1 °	4½° ±½°
King pin inclination	5¾° ±1°	6¾° ±¾°
Wheel alignment	1/16 - 1/8 in Toe-in (1·59 - 3·18 mm)	0 - 1/16 in Toe-in (0 - 1·59 mm).

FRONT WHEEL ALIGNMENT

– Check and adjust 57.65.01

Checking

1. Locate the car on level ground and position the front wheels in straight ahead position.
2. Using wheel alignment equipment check the front wheels for toe-in. Front wheel toe-in should be within 1/16 to 1/8 in (1.59 to 3.18 mm).

Adjusting

3. Slacken the outer clips on the rack gaiter.
4. Slacken the locknuts at the tie rod outer ends.
5. Shorten or extend both tie rods an equal amount to obtain the required toe-in of 1/16 to 1/8 in (1·59 to 3·18 mm).
6. Tighten the locknuts at the tie rod outer ends.
7. Tighten the gaiter clips.

NOTE: that both tie rods should be adjusted equally. Differences in tie rod lengths will result in incorrect wheel angles on turns.

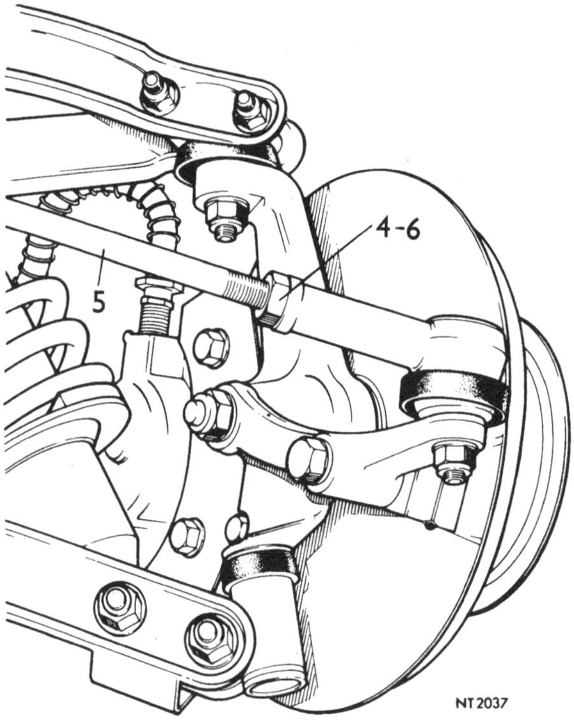

CAMBER ANGLE

– Check and adjust 57.65.05

Front wheel camber should be within the following limits:

Kerb condition	3° Positive ±1°
Laden condition (2 up)	2° Positive ±½°

Adjusting

1. Jack up the car and support chassis on stands.
2. Slacken the two nuts securing the lower wishbone bracket studs to the inner side of the chassis.
3. Remove or add shims equally to both lower wishbone brackets as necessary. Add shim(s) to go negative: remove shim(s) to go positive. One shim equals approximately 1°. Repeat as necessary on opposite wishbone.
4. Tighten the nuts securing the lower wishbone mounting bracket studs.
5. Remove the chassis stands.

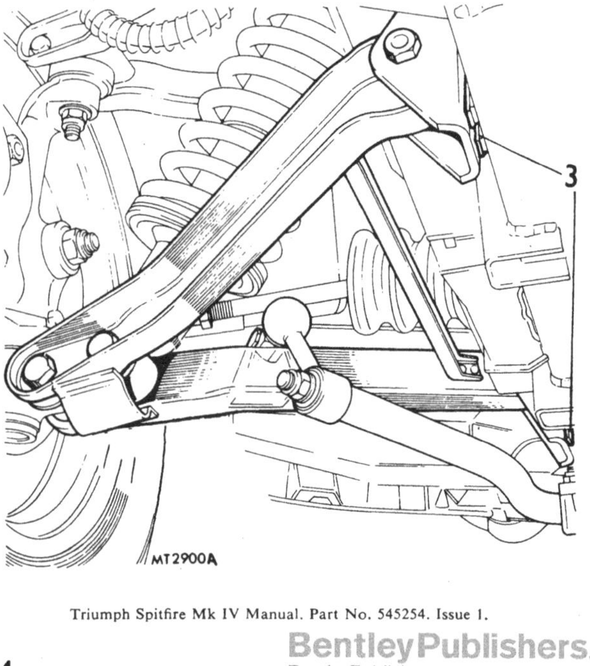

57.65.01
57.65.05

FRONT SUSPENSION OPERATIONS

ANTI-ROLL BAR

– Remove and refit **60.10.01**

Removing

1. Remove nyloc nut and plain washer securing anti-roll bar link to left and right hand lower wishbones.
2. Release anti-roll bar links from wishbones.
3. Remove the nyloc nuts and plain washers securing anti-roll bar 'U' bolts to left and right hand chassis brackets.
4. Withdraw 'U' bolts and bush housing bottom halves.
5. Withdraw anti-roll bar.

Refitting

6. Reverse instuctions 1 to 5.

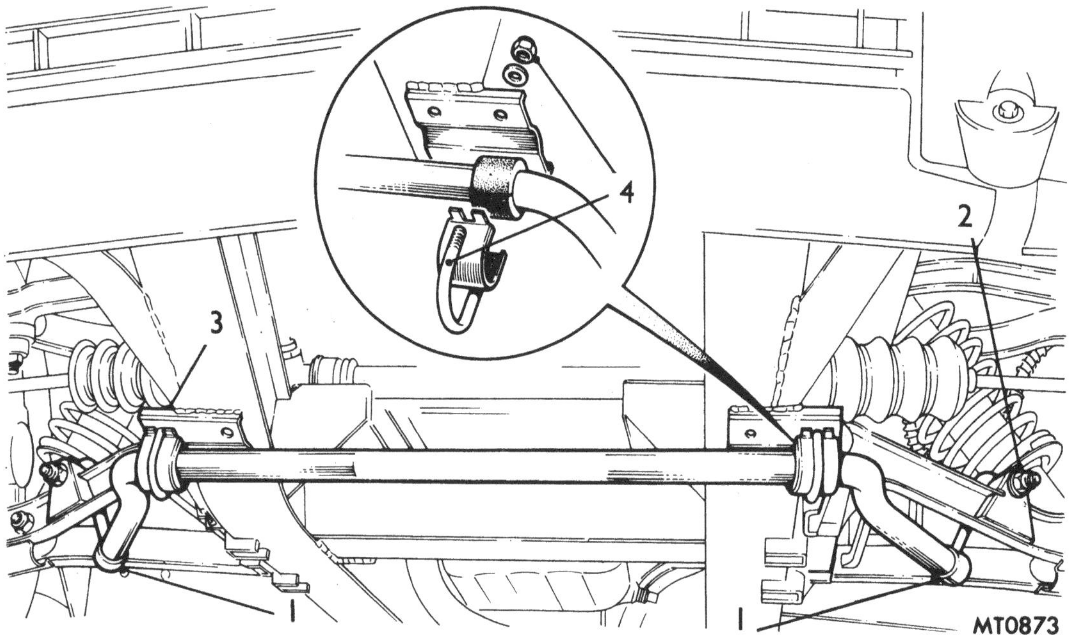

ANTI-ROLL BAR LINK

– Remove and refit **60.10.02**

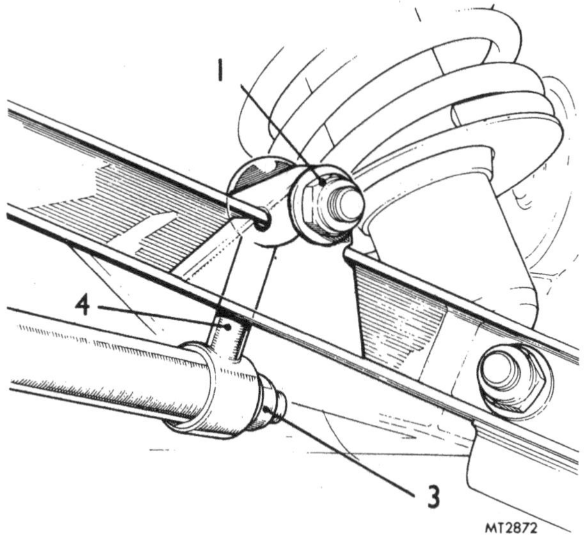

Removing

1. Remove the nyloc nut and plain washer securing the anti-roll bar link to the lower wishbone.
2. Withdraw the anti-roll bar link from the lower wishbone.
3. Remove the nyloc nut and plain washer securing the lower end of link to the anti-roll bar.
4. Withdraw the anti-roll bar link.

Refitting

5. Reverse instructions 1 to 4.

ANTI-ROLL BAR RUBBERS

– Remove and refit **60.10.04**

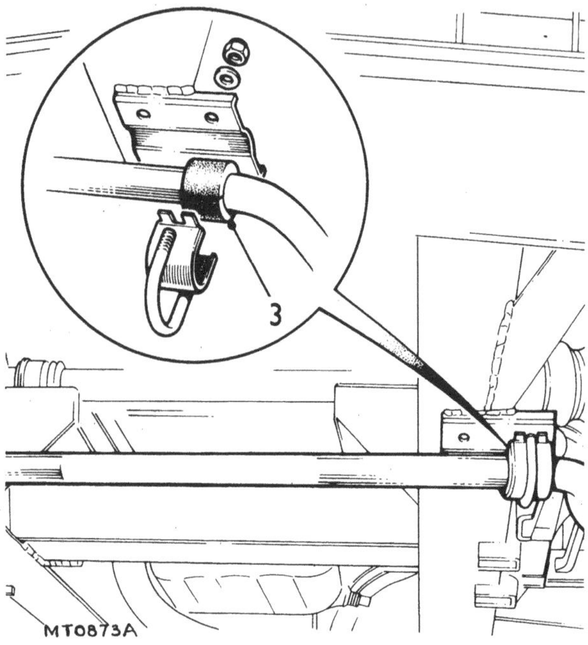

Removing

1. Remove anti-roll bar 60.10.01.
2. Remove the anti-roll bar links from anti-roll bar.
3. Slide off anti-roll bar bushes.

Refitting

4. Slide new bushes into position on anti-roll bar.
5. Fit the anti-roll links to the anti-roll bar.
6. Connect the anti-roll bar link to the lower wishbone.
7. Offer up the anti-roll bar to its mounting brackets and align the rubber bushes.
8. Fit the bush housings and 'U' bolts.
9. Secure the 'U' bolts with nyloc nuts and plain washers.

BALL JOINT

– Remove and refit **60.15.02**

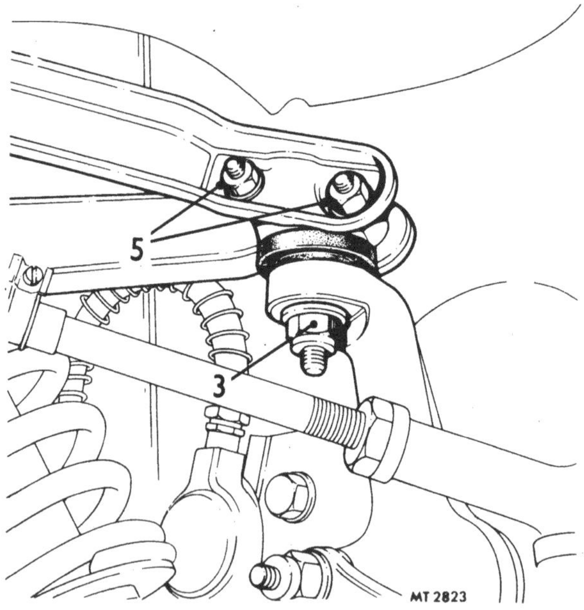

Removing

1. Jack up car and support chassis on stand(s).
2. Remove road wheel.
3. Remove nyloc nut and plain washer securing ball joint shank to vertical links.
4. Release ball joint shank from vertical link.
5. Remove the two bolts and nuts securing ball joint to upper wishbone.
6. Withdraw ball joint.

Refitting

7. Reverse instuctions 1 to 6.

TRUNNION

– Remove and refit **60.15.03**

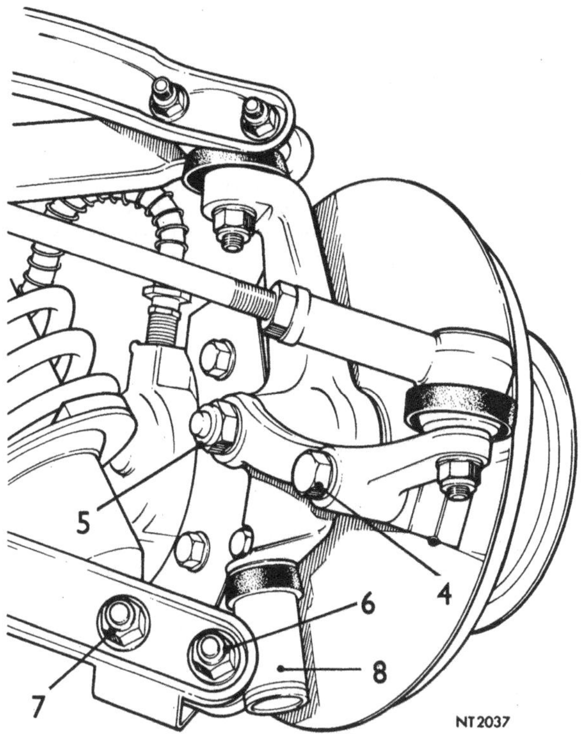

Removing

1. Jack up the car and support the chassis on stand(s).
2. Remove the road wheel.
3. Remove the hub 60.25.01/02.
4. Remove the bolt securing the steering arm and disc shield to the vertical link.
5. Slacken the nut and lift off the disc shield.
6. Remove the nut and bolt securing the trunnion to the lower wishbone.
7. Slacken the nut and bolt securing the shock absorber to the lower wishbone.
8. Detach the trunnion from the lower wishbone and unscrew the trunnion.

Refitting

9. Reverse instructions 1 to 8, ensure that the trunnion outer washers and dust seals are correctly positioned.
10. Lubricate the trunnion, page 09-1.

TRUNNION

– Overhaul **60.15.13**

Dismantling

1. Remove the trunnion 60.15.03.
2. Remove the two washers.
3. Extract the two dust seals.
4. Press out the distance piece.
5. Extract the two nylon bearings and remove the two washers.

Reassembling

6. Lubricate the nylon bearings with a zinc base grease.
7. Reverse instructions 1 to 5.

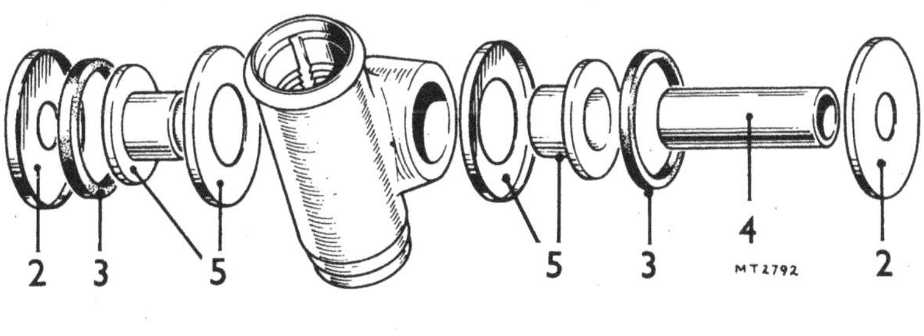

FRONT ROAD SPRING

– Remove and refit **60.20.01**

Service Tools: S4221A, S4221A-5, S4221A-18.

[WARNING: Please read the instructions thoroughly before you attempt spring repairs. If you lack the necessary skills or tools we suggest you leave such work to an Authorized Dealer or other qualified shop. Severe personal injury may result if you undertake disassembly without a proper spring compressing tool.]

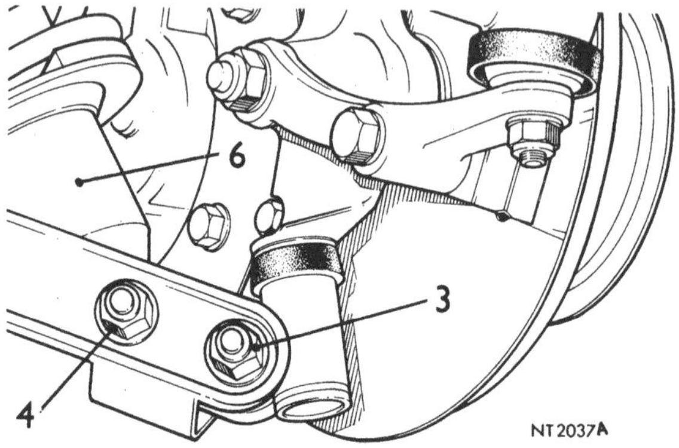

Removing

1. Jack up the car and support chassis on stand(s).
2. Remove front wheel.
3. Slacken the bolt and nut securing the steering trunnion to lower wishbone.
4. Remove the bolt, nut and washer securing bottom of damper to lower wishbone.
5. Remove three nyloc nuts and washers securing road spring pad to bracket.
6. Release damper from lower wishbone and withdraw damper and spring assembly complete.
7. Using Tool S4221A and adaptor S4221A-5 and S4221A-18 compress the road spring until the upper mounting flange is released of spring loading.
8. Remove locknut and nut securing damper rod to mounting flange.
9. Withdraw mounting rubbers, mounting rubber seats and mounting flange from damper rod.
10. Release spring tension and remove tool.
11. Remove spring from damper.

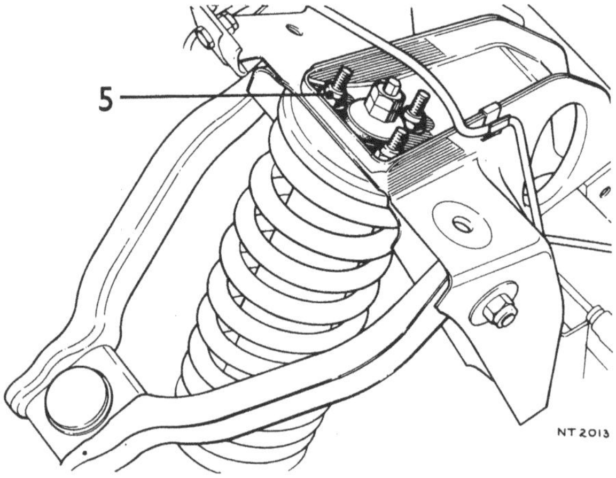

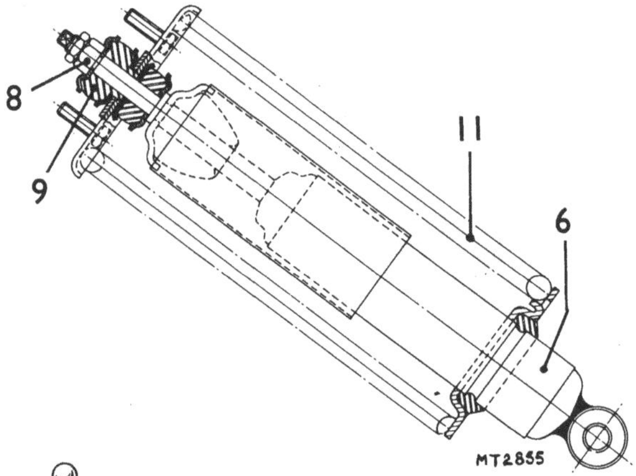

Refitting

12. Fully extend damper rod and locate road spring on damper flange.
13. Using Tool S4221A and adaptors 5 to 18, compress road spring until the free end of the spring is below the damper rod end.
14. Fit the mounting rubbers, mounting rubber seats and mounting flange.
15. Fit and tighten nut and locknut.
16. Remove spring and damper from the tool.
17. Offer up spring and damper assembly to vehicle and engage mounting flange studs in upper bracket.
18. Fit plain washers and nyloc nuts to mounting flange studs. Tighten nuts.
19. Engage bottom end of damper in lower wishbone and secure with bolt, washer and nut.
20. Tighten bolt and nut securing lower swivel to wishbone.
21. Fit road wheel and remove jack.

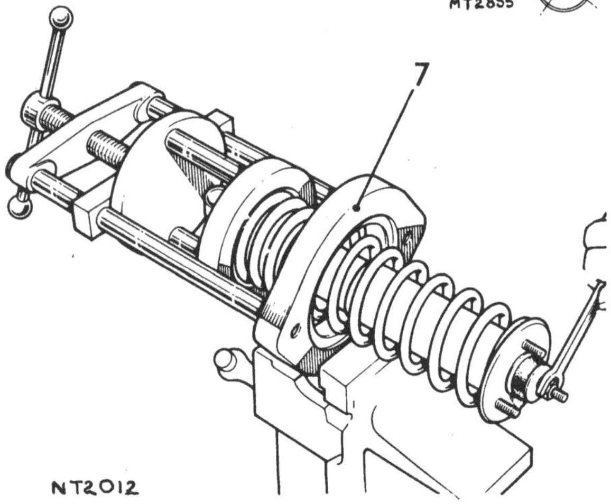

FRONT HUB

– Remove and refit **60.25.01/02**

Removing

1. Jack up car and support chassis on stand(s).
2. Remove front wheel 74.20.01.
3. Remove the two bolts and spring washers securing brake caliper to vertical link.
4. Remove caliper shield and withdraw caliper from vertical link and disc. Ensure that weight of caliper is not supported by brake hose.
5. Remove hub cap.
6. Remove split pin, slotted nut and washer from stub axle.
7. Withdraw front hub complete with bearings and oil seal.

Refitting

8. Partially pack hub with fresh grease.
9. Locate the oil seal in the hub and enter the hub and bearings on the stub axle.
10. Fit the washer and slotted nut to the stub axle.
11. Tighten the slotted nut to obtain hub end-float of 0·002 to 0·008 in (0·2032 to 0·508 mm).
12. Secure the slotted nut with a new split pin.
13. Partially fill the hub cap with fresh grease and fit the cap to the hub.
14. Fit the caliper and shield to the vertical link and secure with two bolts and spring washers.
15. Fit the road wheel and remove the stand(s).

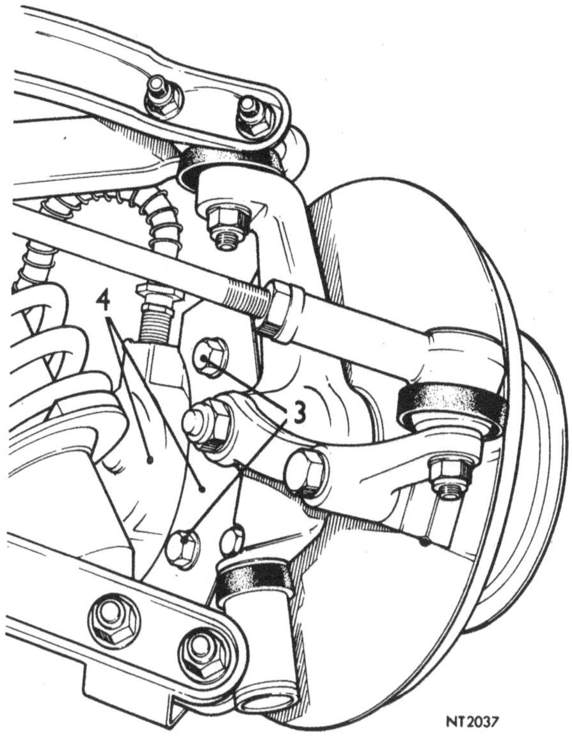

NT2037

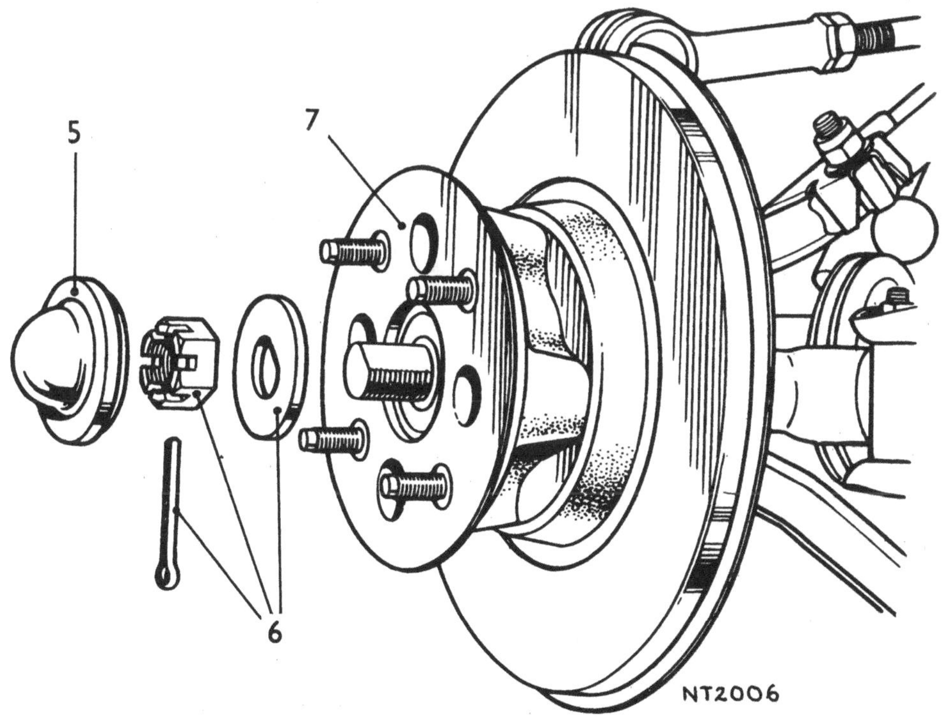

NT2006

FRONT HUB

– Overhaul **60.25.07/08**

Dismantling

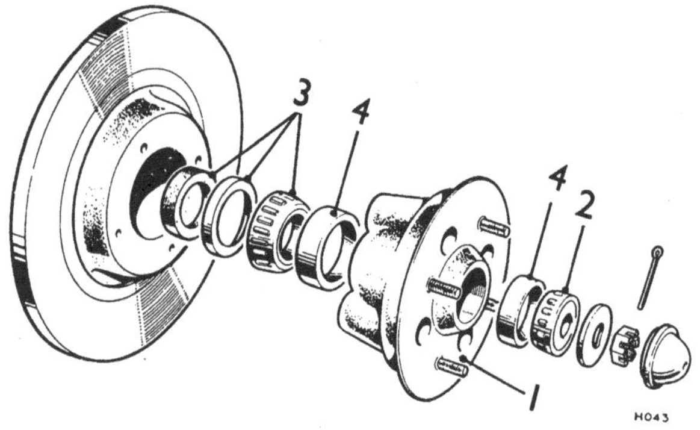

1. Remove the front hub 60.25.01/02.
2. Withdraw the outer bearing.
3. Withdraw the oil seal, inner bearing shield and inner bearing.
4. Carefully and evenly extract the inner and outer bearing tracks. The removal of the bearing tracks is not advised unless renewal is intended.
5. Thoroughly clean all components.

Reassembling

6. Examine all components and renew as necessary.
7. Fit the inner and outer bearing tracks to the hub.
8. Partially fill the hub with fresh grease and fit the inner bearing.
9. Fit the inner bearing shield.
10. Lubricate the new felt seal and enter the seal in the hub.
11. Fit the outer bearing.
12. Install the hub on the stub axle 60.25.01/02.

FRONT HUB BEARING END-FLOAT

– Check and adjust **60.25.13.**

Checking

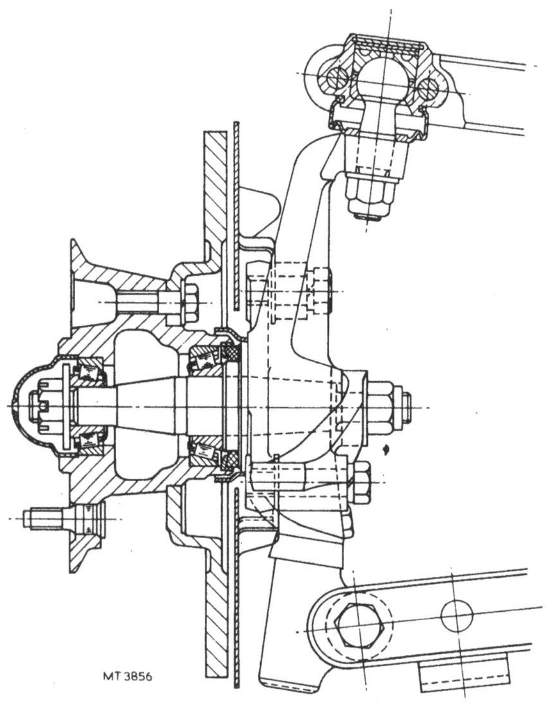

1. Jack up the car and remove the front wheel.
2. Remove the front brake pads.
3. Check hub bearing end-float. A correctly adjusted hub will have end-float within 0·002 to 0·008 in (0·2032 to 0·508 mm). A dial gauge can be used for checking purposes.

Adjusting

4. Remove the hub cap.
5. Remove the split pin from the slotted nut at the end of the stub axle.
6. Tighten or slacken the slotted nut as necessary to obtain the required end-float (item 3 above.)
7. Fit a new split pin.
8. Clean the hub cap and partially fill with fresh grease.
9. Fit the hub cap.
10. Fit the front brake pads.
11. Fit the front wheel and remove the jack.

FRONT HUB BEARINGS

– Remove and refit **60.25.14**

As operation 60.25.07.

FRONT HUB OIL SEAL

– Remove and refit **60.25.15**

Removing

1. Remove the front hub 60.25.01/02.
2. Withdraw the oil seal.

Refitting

3. Partially pack the hub with fresh grease.
4. Lubricate the new hub seal and enter the seal in the hub.
5. Fit the hub to the stub axle 60.25.01/02.

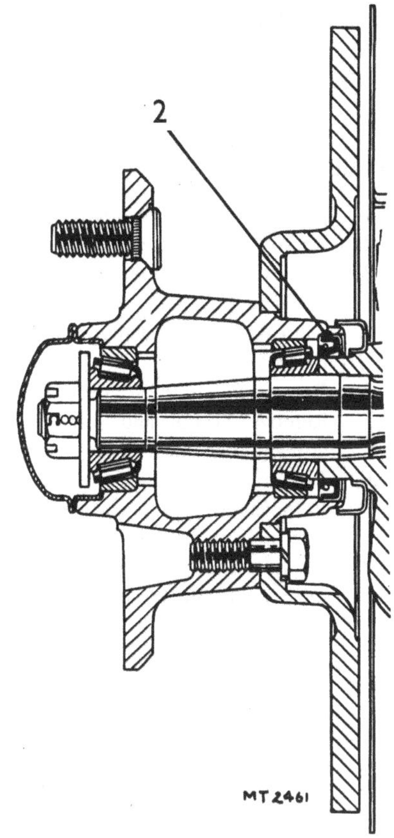

FRONT HUB STUB AXLE

– Remove and refit **60.25.22**

Removing

1. Remove the vertical link 60.25.23.
2. Press out the stub axle from the vertical link.

Refitting

3. Reverse instructions 1 and 2.
4. Fit the vertical link and hub assembly 60.25.23.

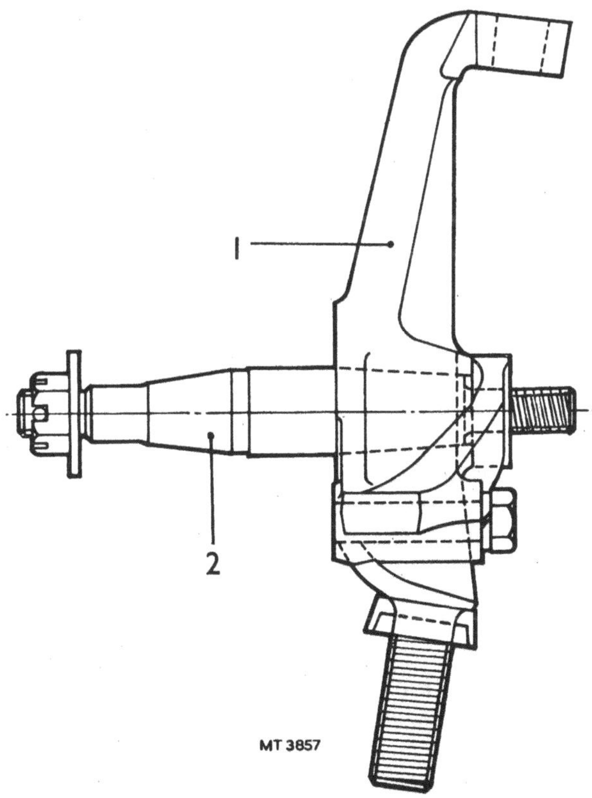

VERTICAL LINK

– Remove and refit 60.25.23

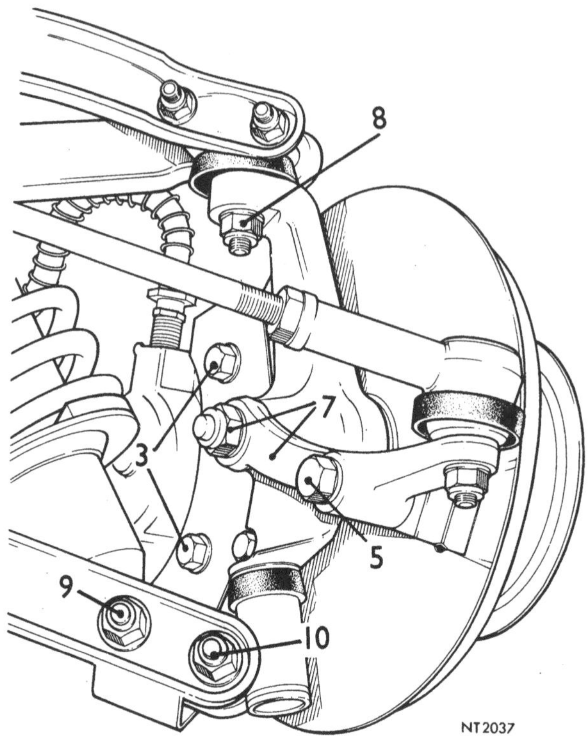

NT2037

Removing

1. Jack up the car and support chassis on stand(s).
2. Remove front wheel.
3. Remove two bolts and spring washers securing caliper shield and caliper to vertical link and withdraw shield and caliper. Do not permit caliper to hang suspended by brake hose.
4. Remove hub and disc (items 5-7, 60.25.01/02).
5. Remove the bolt and spring washer securing steering arm and disc shield to vertical link.
6. Remove disc shield.
7. Remove the nyloc nut and plain washer securing steering arm to stub axle and withdraw steering arm.
8. Remove the nyloc nut and plain washer securing upper ball joint to vertical link and release ball shank from vertical link.
9. Slacken the nut and bolt securing bottom end of damper to lower wishbone.
10. Remove the nut, washer and bolt securing lower swivel to lower wishbone.
11. Withdraw vertical link.

Refitting

12. Reverse instructions 1 to 11.

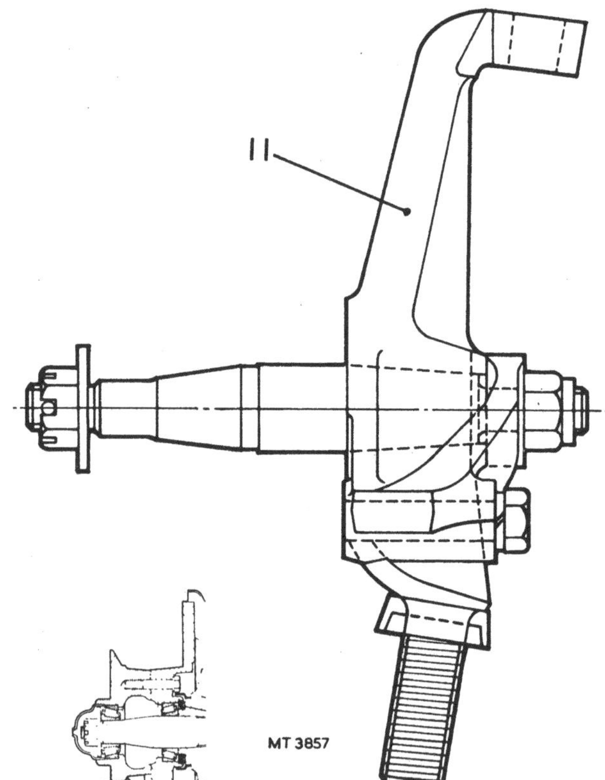

FRONT HUB WHEEL STUDS

– Remove and refit 60.25.29

Removing

1. Jack up the car and remove the front wheel.
2. Tap the wheel stud towards the disc.
3. Withdraw the stud. Wheel studs should not be removed except for purposes of renewal.

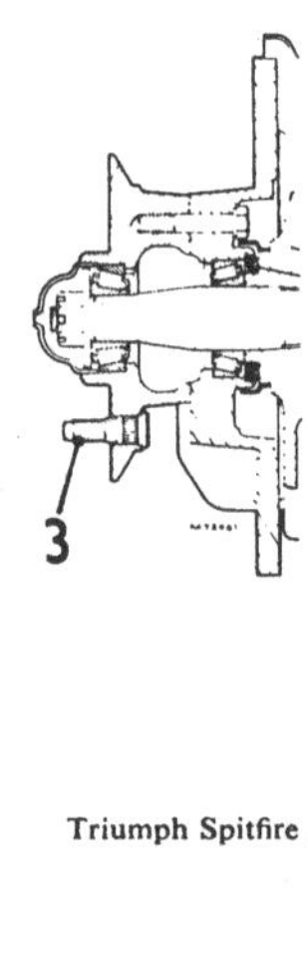

MT 3857

Refitting

4. Ensure that the mating countersunk faces of both stud and hub are clean.
5. Enter the stud from the rear of the hub flange.
6. Using suitable packing (e.g. a short length of steel tubing) draw the stud into position.
7. Remove the nut and packing.
8. Fit the road wheel and remove the jack.

FRONT DAMPER

– Remove and refit **60.30.02**

As operation 60.20.01.

FRONT DAMPER BUSHES

– Remove and refit **60.30.07**

1. Remove damper and road spring assembly from the car (items 1 to 6 operation 60.20.01).

Upper mountings – remove and refit.

Service Tools: S4221A, S4221A-5, S4221A-18

2. Using above tools, release spring load from damper upper mountings.
3. Remove locknut and nut securing damper rod to mounting flange.
4. Withdraw mounting rubbers, mounting rubber seats and mounting flange from damper rod.
5. Fit new mounting rubbers, mounting rubber seats, and mounting flange in order illustrated in operation 60.20.01.
6. Fit and tighten damper rod nut and locknut.
7. Release spring and damper assembly from Service Tool.

Lower bush – remove and refit

8. Press out bush and sleeve from damper eye.
9. Press in new bush and sleeve ensuring both are centralised.
10. Install damper and road spring assembly on car (60.20.01).

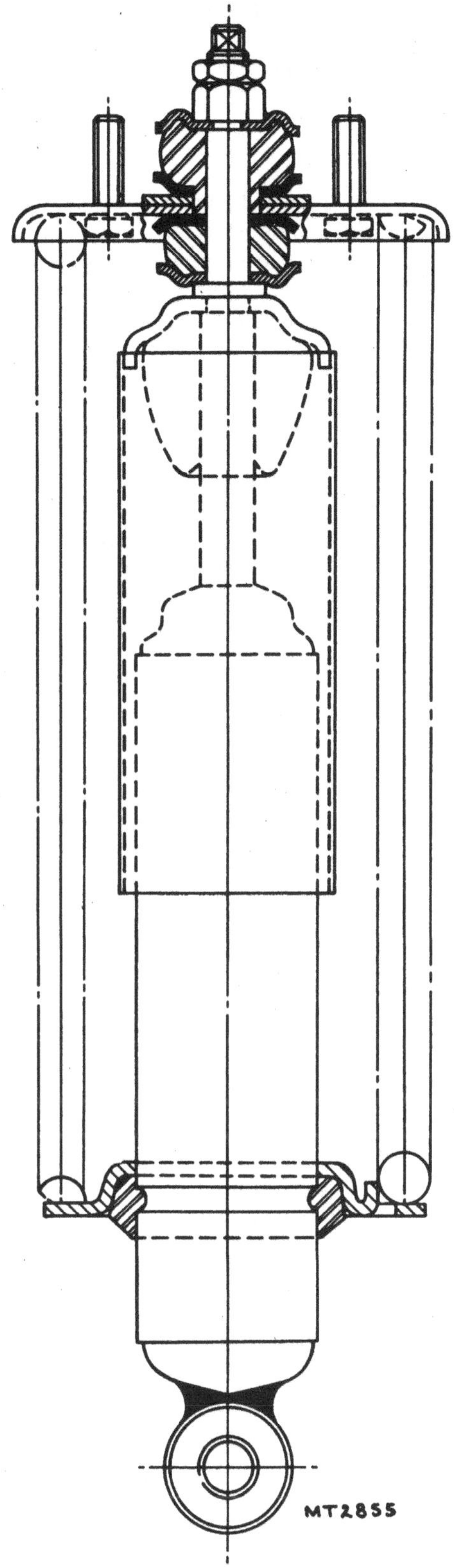

WISHBONE – UPPER

– Remove and refit **60.35.01**

Removing

1. Jack up car and support chassis on stand(s).
2. Remove front wheel.
3. Remove front spring and damper assembly 60.20.01.
4. Remove two nyloc nuts, plain washers and bolts securing upper ball joint to outer end of wishbone.
5. Remove two nyloc nuts, plain washers and bolts securing inner ends of wishbone to bracket.
6. Withdraw upper wishbone.

Refitting

7. Reverse instructions 1 to 6. Do not tighten wishbone inner fulcrum bolts until car is static laden.
8. Fit front wheel and remove stand(s).
9. Tighten wishbone inner fulcrum bolts.

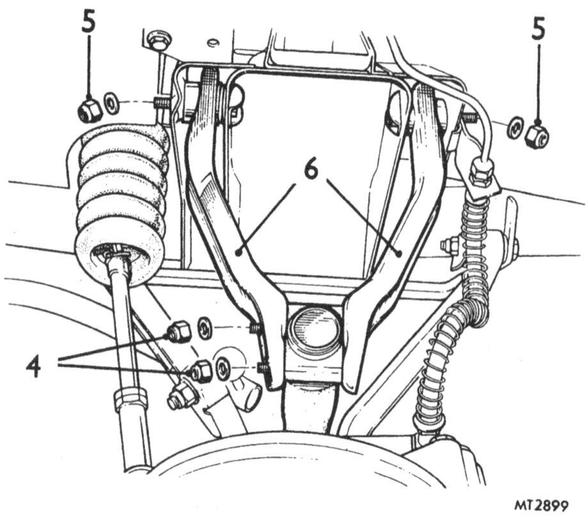

WISHBONE – LOWER

– Remove and refit **60.35.03**

Removing

1. Jack up car and support chassis on stand(s).
2. Disconnect anti-roll bar link from lower wishbone.
3. Remove two bolts and nyloc nuts securing bottom end of damper and lower swivel bearing to outer end of lower wishbone.
4. Remove two nyloc nuts and bolts securing inner ends of wishbone to brackets.
5. Withdraw lower wishbone.

Refitting

6. Reverse instructions 1 to 5. Do not tighten wishbone inner fulcrum bolts until car is static laden.
7. Fit front wheel and remove stand(s).
8. Tighten wishbone inner fulcrum bolts.

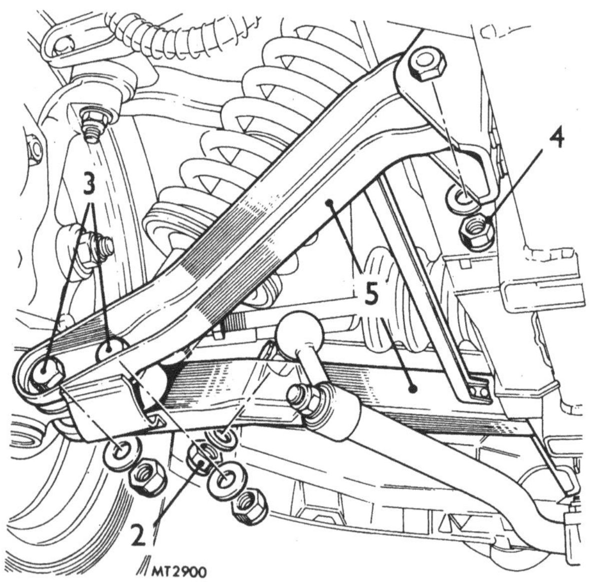

REAR SUSPENSION OPERATIONS

REAR HUB

– Remove and refit **64.15.01**

Service Tool: S109C

Removing

1. Jack up the car and remove the road wheel.
2. Release the handbrake and remove the brake drum.
3. Remove the nyloc nut and plain washer securing the rear hub to the drive shaft.
4. Using Tool S109C, withdraw the rear hub.

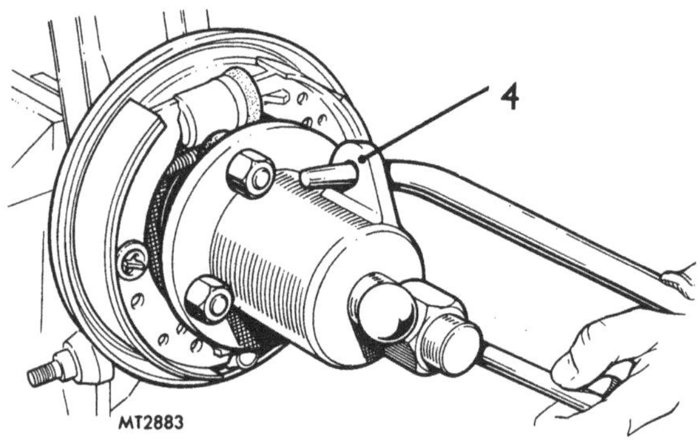

Refitting

5. Ensure that the mating tapered faces, of the drive shaft and hub are clean.
6. Enter hub on drive shaft, engaging the key in the keyway.
7. Fit the plain washer and the nyloc nut. Tighten the nut. Use a tyre lever or similar across the hub studs to apply reverse torque.
8. Fit the brake drum.
9. Fit the road wheel and remove the jack.

REAR HUB WHEEL STUDS

– Remove and refit **64.15.26**

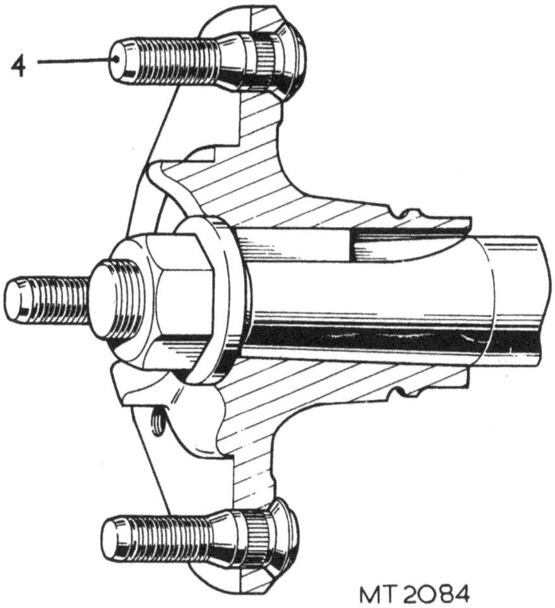

Removing

1. Jack up the car and remove the road wheel.
2. Remove the brake drum.
3. Rotate hub until rear face of wheel stud is clear of brake shoe return springs.
4. Tap out wheel stud towards the brake backplate. Wheel studs should not be removed unless renewal is intended.

Refitting

5. Enter the stud squarely from the rear of the hub flange ensuring that the tapered faces on both hub and studs are clean.
6. Using suitable packing (e.g. a short length of tubing and washers) inserted over the wheel stud draw the stud into position. If new studs are a loose fit in the hub flange, a new hub should be fitted.
7. Fit the brake drum and road wheel and remove the jack.

REAR ROAD SPRING

– Remove and refit 64.20.01

Removing

1. Jack up the car and support the chassis on stands.
2. Remove the road wheels.
3. Remove the two nyloc nuts, plain washers and bolts securing the road spring to the vertical links.
4. Remove the two nyloc nuts and plain washers. Disconnect the lower end of the dampers from the vertical links.
5. Remove the rear squab trim pad – six screws and cup washers.
6. Remove the panel – two bolts.
7. Remove the four nyloc nuts securing the spring bracket to the differential casing.

[NOTE: There are six nuts and a flat retaining plate on Mk. III models, all spring leaves being rigidly attached to the final drive unit.]

8. Using two nuts, remove the four studs from the differential casing.
9. Ease the spring ends clear of the vertical links and withdraw the spring from the car.

Refitting

10. Slide the spring into position with the spring clips stamped 'FRONT' facing the front of the car and the ground edge of the main leaf facing rearwards.
11. Locate the spring centre bolt in the recess in the differential casing.
12. Reverse 1 to 8. Do not fully tighten bolts (3) until the car is lowered to the ground.

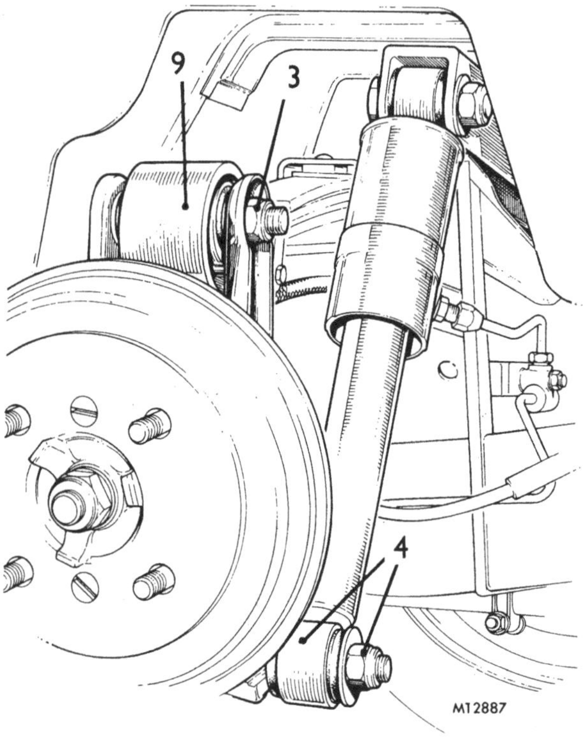

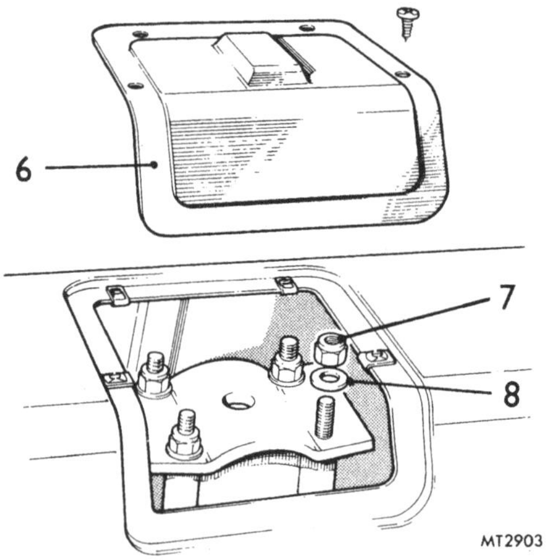

REAR ROAD SPRING

– Overhaul **64.20.04**

Dismantling

1. Remove the spring 64.20.01.
2. Withdraw the spring plate.
3. Compress the spring in a vice and remove the spring bracket and rubber pad, one nut, bolt, spring washer and spacer. Release the spring from the vice.

[NOTE: The rear road spring for the Mk. III Spitfire is assembled into a conventional rigid spring unit as shown in the illustration. There is no bracket or rubber pad, the spring being mounted directly onto the final drive unit by a flat steel retaining plate and six studs.]

4. Remove the four nuts, bolts and spacers from the spring clips.
5. Lift off the top three leaves.
6. Slightly bend the leg of each spring clip on the safety leaf and withdraw the main leaf.

Reassembling

7. Clean the spring leaves and lubricate with graphite grease.
8. Insert the main leaf into the safety leaf with the ground edge facing the plain, unstamped legs of the spring clips.
9. Reverse 1 to 5, ensuring that all eight rubber buttons are correctly located in the recessed tips of the spring leaves.

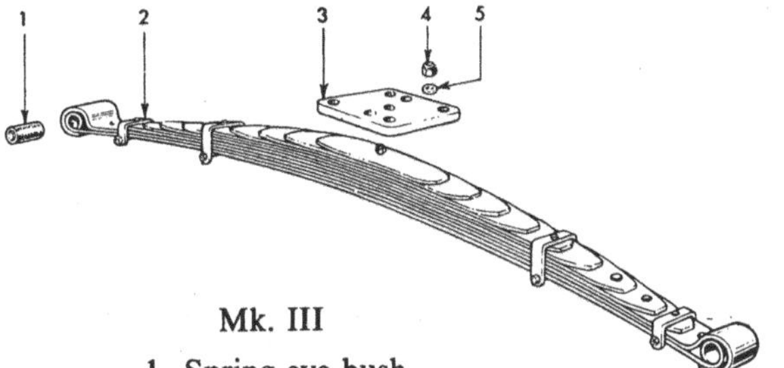

Mk. III

1. Spring eye bush
2. Road spring
3. Retaining plate
4. Nut
5. Washer

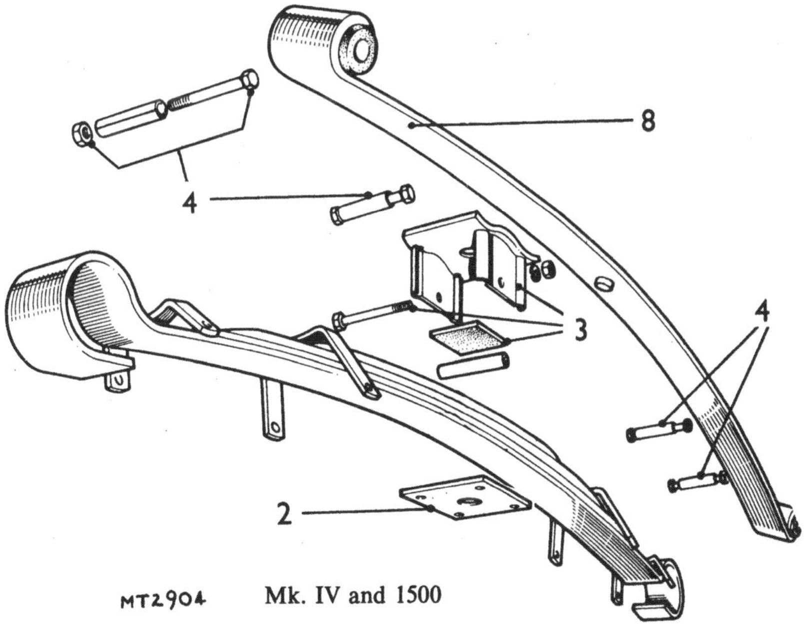

MT2904 Mk. IV and 1500

REAR WHEEL ALIGNMENT

– Check and adjust 64.25.17

Checking

Rear Suspension Data

Kerb condition:
- Camber . . . . 1° negative ±1°
- Toe-out . 1/32 in – 3/32 in (0·79 – 2·38 mm)

Laden condition (2 up):
- Camber . . . . . . 3·3/4° negative ±1°
- Toe-in . . . . 0 in – 1/16 in (0 to 1·59 mm)

Camber. No provision is made for adjustment to camber angle. Camber angles found to be outside the above limits are generally indicative of a weak or incorretly set spring, or wear in the vertical link bushes.

Toe-in/Toe-out. Rear wheel toe-in/toe-out dimensions vary as given according to the load condition of the vehicle. In addition rear wheel alignment is also influenced by the condition of the rear spring and wear factors in the vertical link linkage. It is important therefore before altering rear wheel alignment, that the causes necessitating adjustment are understood.

Adjusting

1. Locate car on level ground.
2. Using suitable equipment, check for alignment.
3. Adjust as necessary by adding to or removing from the shims fitted between the radius rod front brackets and the body. Add shims to increase toe-out remove shims to reduce toe-out.

In contrast to front wheel track adjustment where alteration is made equally to both tie rods to maintain balanced steering lock angles, adjustment to rear wheel track is individual to either wheel and is determined with alignment with its respective front wheel to which it has a tolerance of 0 to 1/32 in (0 to 0·79 mm) toe-in – kerb condition or 1/64 in – 3/64 in (0·39 to 1·19 mm) toe-out – laden condition (2 up).

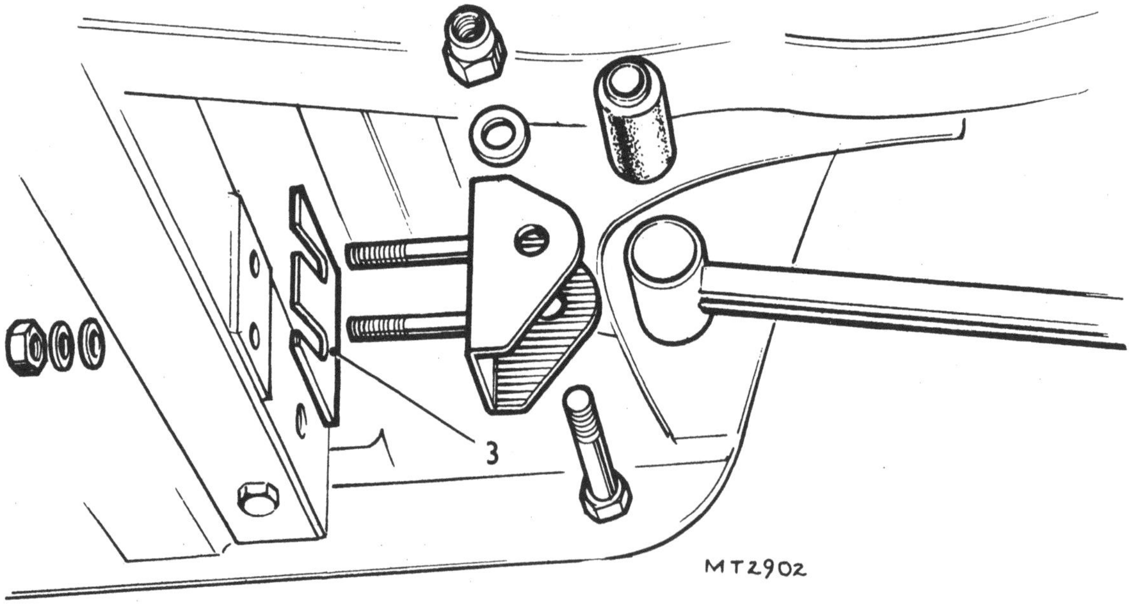

REAR DAMPER

– Remove and refit 64.30.01

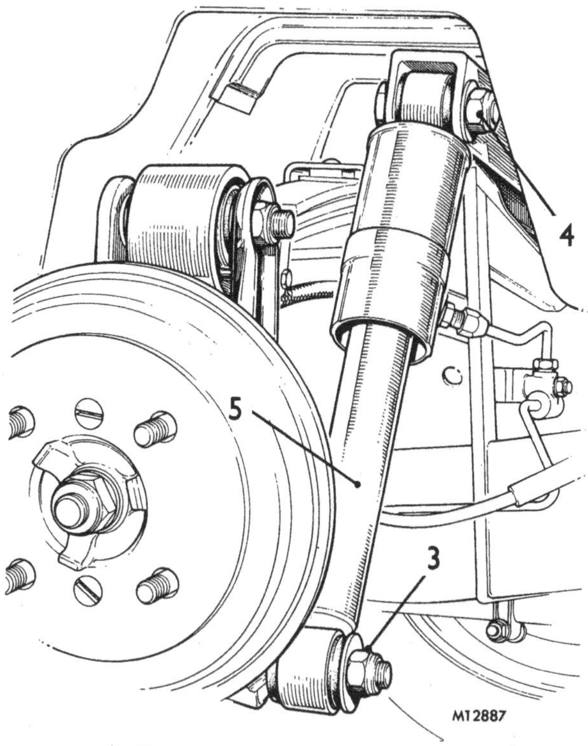

Removing

1. Jack up the car and support the chassis on stands.
2. Remove the rear wheel.
3. Remove the nyloc nut and plain washer securing the lower end of damper to the vertical link.
4. Remove the nyloc nut, plain washer and bolt securing the upper end of damper to the chassis.
5. Withdraw the damper.

Refitting

6. Reverse instructions 1 to 5.

VERTICAL LINK

– Remove and refit 64.35.03

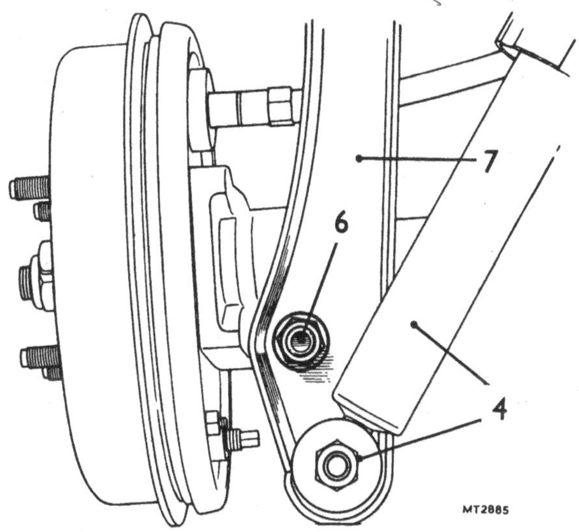

Removing

1. Jack up the car and support the chassis on stands.
2. Remove the rear wheel.
3. Remove the nyloc nut, plain washer and bolt securing the radius rod to the bracket on the vertical link. Detach the radius rod from the vertical link.
4. Remove the nyloc nut and plain washer securing lower end of damper to vertical link. Detach damper.
5. Remove the nyloc nut, plain washer and bolt securing the upper end of the vertical link to the road spring.
6. Remove the nyloc nut, plain washer and bolt securing the vertical link to the inner hub.
7. Remove the vertical link.

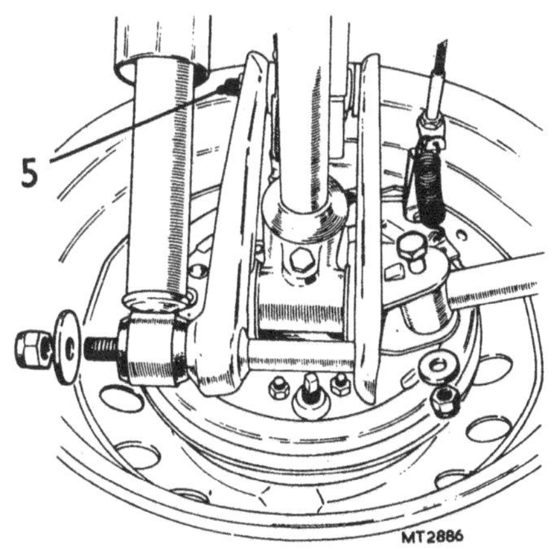

Refitting

8. Offer up the vertical link to the car and engage the lower end of the link in the damper bush.
9. Ensure inner hub pivot bush seals and covers are intact and in position and fit the vertical link to the inner hub. Secure in position with bolt, plain washer and nyloc nut.
10. Fit bolt, plain washer and nyloc nut to upper end of vertical link and road spring.
11. Fit plain washer and nyloc nut to secure the damper to the vertical link.
12. Connect the radius rod to the vertical link bracket and secure with bolt, plain washer and nyloc nut.
13. Fit the road wheel.
14. Remove the chassis stands.

RADIUS ROD

– Remove and refit 64.35.28

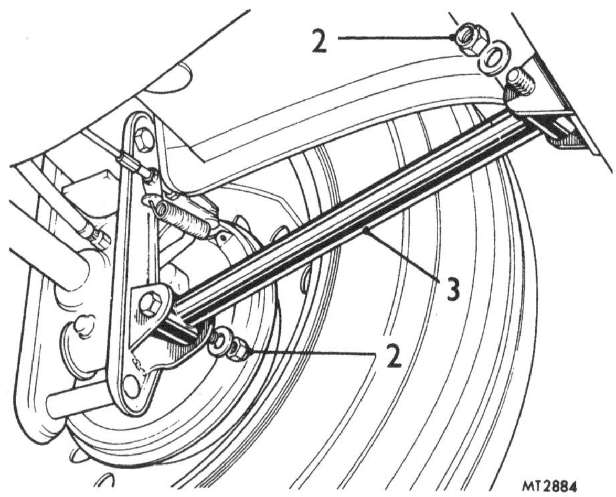

Removing

1. Jack up the car, support the chassis on stands and remove the rear wheel.
2. Remove the nyloc nuts, plain washers and bolts securing the radius rod ends to the body and the vertical link.
3. Withdraw the radius rod.

Refitting

4. Reverse instructions 1 to 3.

[NOTE: After you have replaced either a radius rod or a radius rod bushing, you should always check the rear wheel alignment 64.25.17 and, if necessary, adjust the toe.]

RADIUS ROD BUSHES

– Remove and refit 64.35.29

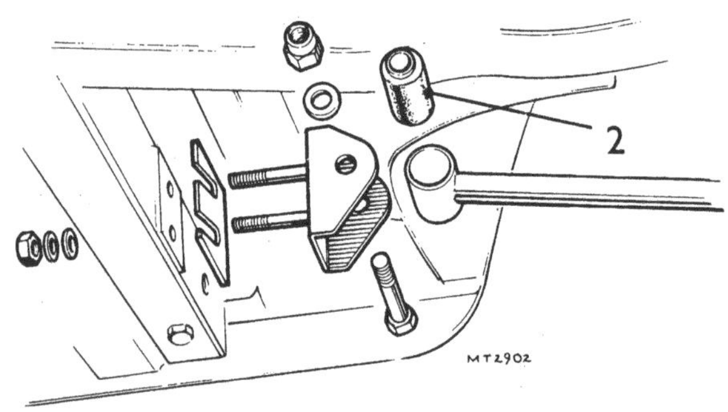

Removing

1. Remove the radius rod 64.35.28.
2. Press out bushes.

Refitting

3. Press in new bushes.
4. Fit the radius rod to the car.

BRAKE OPERATIONS

***Pressure Differential Warning Actuator**

REAR BRAKE DRUM

– Remove and refit **70.10.03**

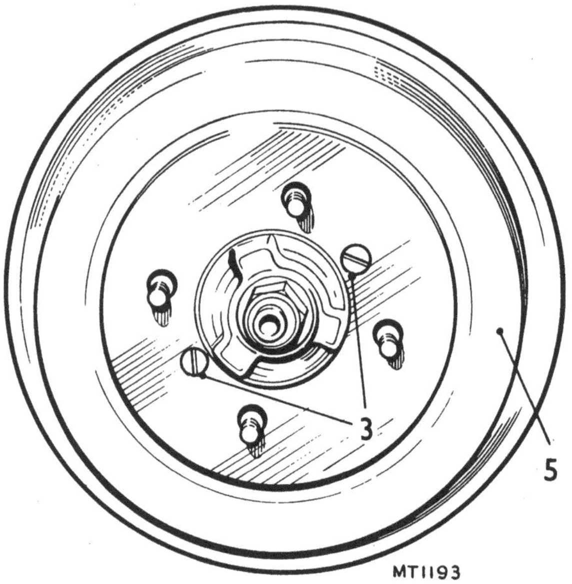

Removing

1. Jack up the car and support the chassis on stands.
2. Remove the rear wheel.
3. Remove two countersunk screws securing the rear brake drum to the hub.
4. Release the handbrake.
5. Withdraw the brake drum.

Refitting

5. Reverse instructions 1 to 5.

FRONT DISC

– Remove and refit **70.10.10**

Removing

1. Jack up the car and support the chassis on stands.
2. Remove the front wheel.
3. Remove two bolts securing the caliper shield and caliper to the vertical link.
4. Withdraw the caliper ensuring that strain is not imposed on the brake hose.
5. Remove the front hub 60.25.01.
6. Remove four bolts and spring washers securing the disc to the front hub.
7. Withdraw the disc.

Refitting

8. Offer up the disc to the front hub ensuring that the mating faces are clean.
9. Fit and tighten the four spring washers and bolts.
10. Locate the felt oil seal in the hub.
11. Fit the hub and disc to the stub axle.
12. Fit the caliper and shield to the vertical link.
13. Fit the road wheel and remove the chassis stands.

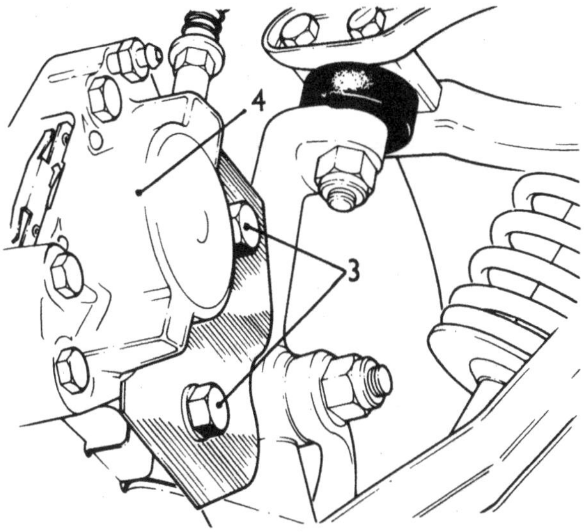

NT2007

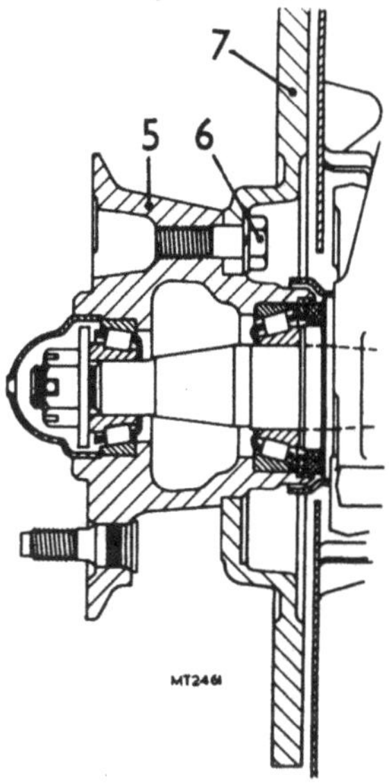

FRONT DISC SHIELD

– Remove and refit **70.10.18**

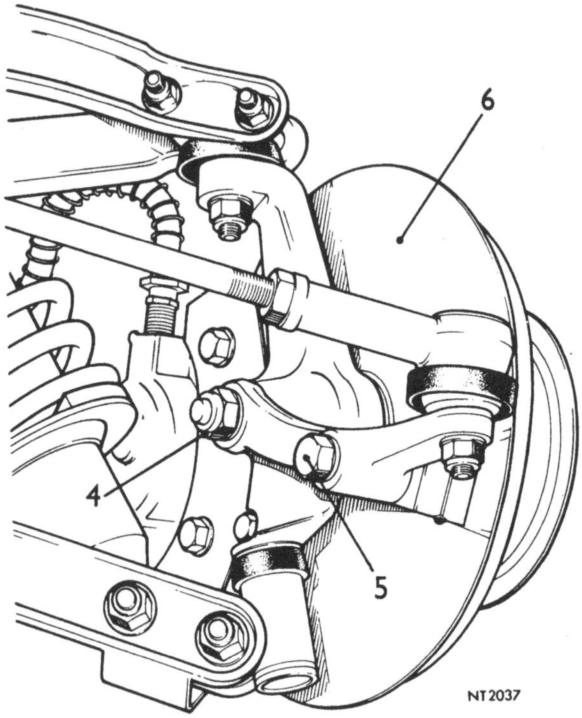

Removing

1. Jack up the car and support the chassis on stands.
2. Remove the front wheel.
3. Remove the brake caliper and the front hub 60.25.01.
4. Slacken the nyloc nut securing the steering arm to the stub axle.
5. Remove the bolt and spring washer securing the steering arm to the vertical link.
6. Withdraw the disc shield.

Refitting

7. Reverse instructions 1 to 6.

REAR BACKPLATE

– Remove and refit **70.10.26**

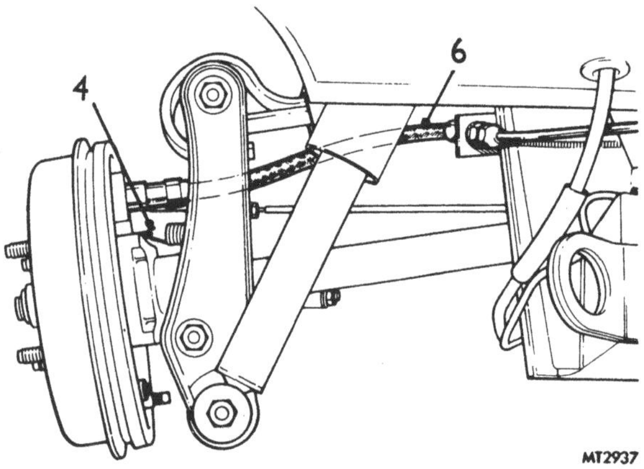

Removing

1. Jack up the car and support the chassis on stands.
2. Remove the rear wheel.
3. Remove the rear brake drum and hub 64.15.01
4. Detach the handbrake cable return spring from the bracket on the backplate.
5. Remove the clevis pin securing the handbrake cable to the brake shoe operating lever.
6. Remove the rear flexible brake hose.
7. Straighten tabs on locking plates securing the backplate bolts.
8. Remove the backplate securing bolts.
9. Withdraw the backplate and bearing shield.

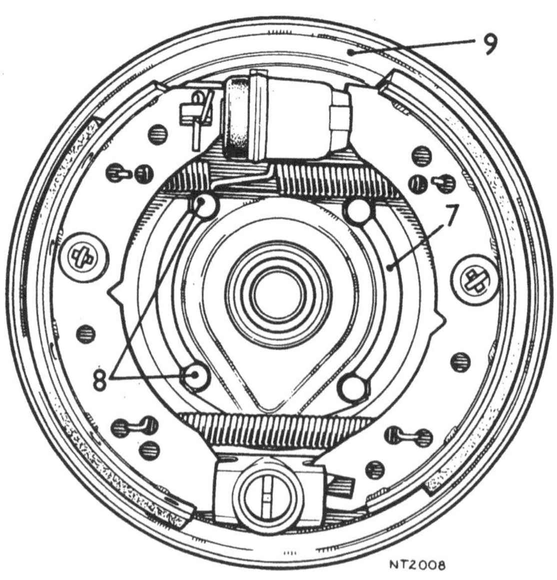

Refitting

10. Reverse instructions 1 to 9.
11. Bleed the brakes.

BRAKE HOSE – FRONT

– Remove and refit

Left hand	**70.15.02**
Right hand	**70.15.03**

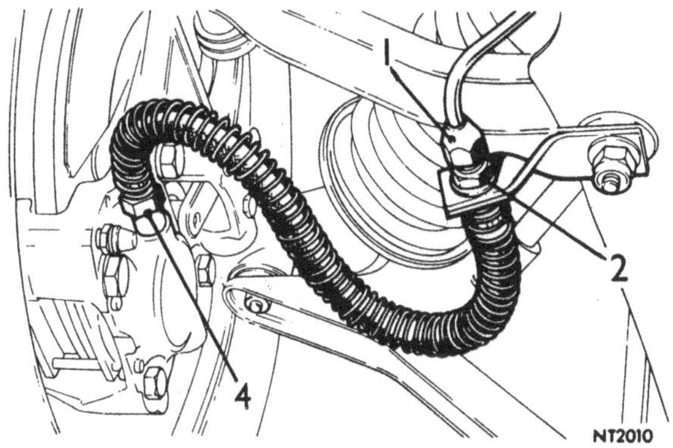

Removing

1. Disconnect the brake pipe from the inboard end of the flexible hose.
2. Using two spanners remove the nut and washer securing the inboard end of the flexible hose to the front suspension bracket.
3. Withdraw the inboard end of the flexible hose from the suspension bracket.
4. Unscrew the outboard end of the flexible hose from the brake caliper.

Refitting

5. Reverse instructions 1 to 4, ensure that the hose is neither kinked nor twisted when installed.
6. Bleed the brakes.

BRAKE HOSE – REAR

– Remove and refit

Left hand	**70.15.17**
Right hand	**70.15.18**

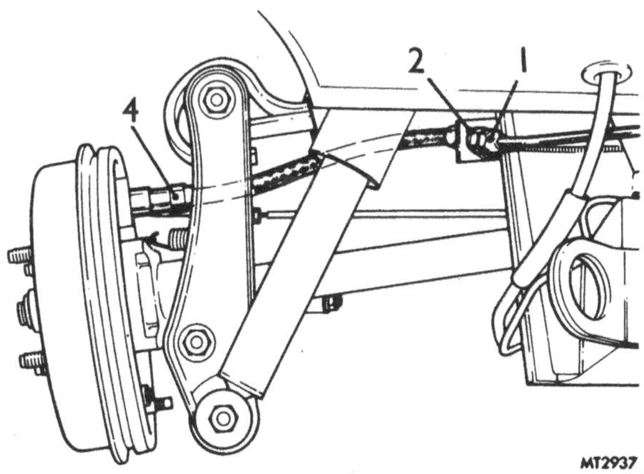

Removing

1. Disconnect the union and brake pipe from the inboard end of the flexible hose.
2. Using two spanners remove the nut and washer securing the flexible hose to the chassis bracket.
3. Withdraw inboard end of hose from the chassis bracket.
4. Unscrew the outboard end of the hose from the brake backplate.

Refitting

5. Reverse instructions 1 to 4. Ensure that the hose is neither kinked nor twisted when installed.
6. Bleed the brakes.

CONNECTOR – 2-WAY

– Remove and refit **70.15.32**

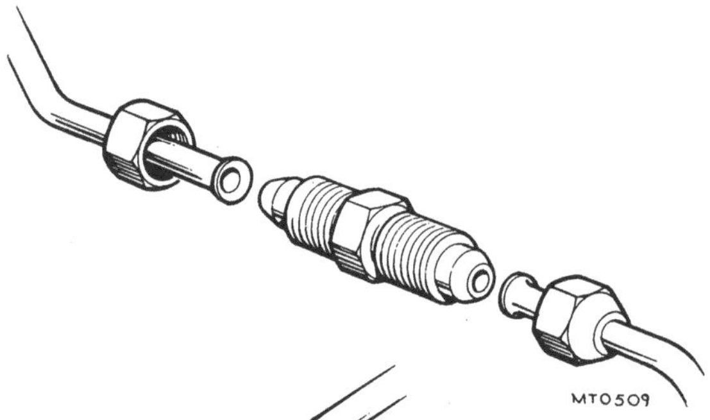

Removing

1. Disconnect the unions and brake pipes from he connector.
2. Withdraw the connector.

Refitting

3. Reverse instructions 1 and 2.
4. Bleed the brakes.

CONNECTOR

– Remove and refit

3-way	**70.15.34**
4-way	**70.15.35**

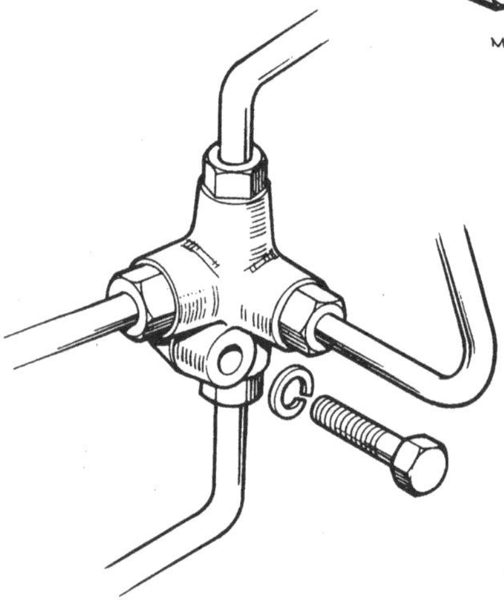

MTO 509

Removing

1. Disconnect unions and brake pipes from the connector.
2. Remove the nut, spring washer and bolt securing the connector to the chassis.
3. Remove the connector.

Refitting

4. Reverse instuctions 1 to 3.
5. Bleed the brakes.

P.D.W.A. UNIT

– Remove and refit **70.15.36**

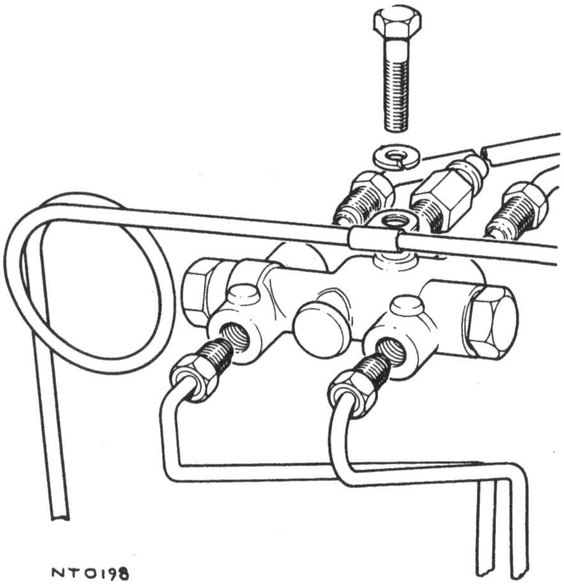

NT0198

Removing

1. Release the snap connector from P.D.W.A. unit.
2. Disconnect the four brake pipe unions at P.D.W.A. unit.
3. Remove the bolt securing P.D.W.A. unit to car.
4. Withdraw the P.D.W.A. unit.

Refitting

5. Reverse instructions 1 to 4.
6. Bleed the brakes.
7. Check operation of brake circuit warning light and ensure that the P.D.W.A. shuttle is centralised.

70.15.32 70.15.34/35
70.15.36

BRAKE PIPES

– Remove and refit

Pipe – Master cylinder to 4-way connector	70.20.01
Pipe – 4-way connector to L.H. front hose	70.20.02
Pipe – 4-way connector to R.H. front hose	70.20.03
Pipe – 2-way connector to 3-way connector	70.20.14
Pipe – 3-way connector to L.H. rear hose	70.20.15
Pipe – 3-way connector to R.H. rear hose	70.20.16
Pipe – 4-way connector to 2-way connector	70.20.27

[CAUTION: Never attempt to straighten a brake pipe which has been bent or dented. Doing so may weaken the pipe structurally. Damaged brake pipes should always be replaced.]

Removing

1. Disconnect unions at both ends of pipe.
2. Release pipe from securing clips (where fitted).

Refitting

3. Reverse instructions 1 and 2.
4. Bleed the brakes.

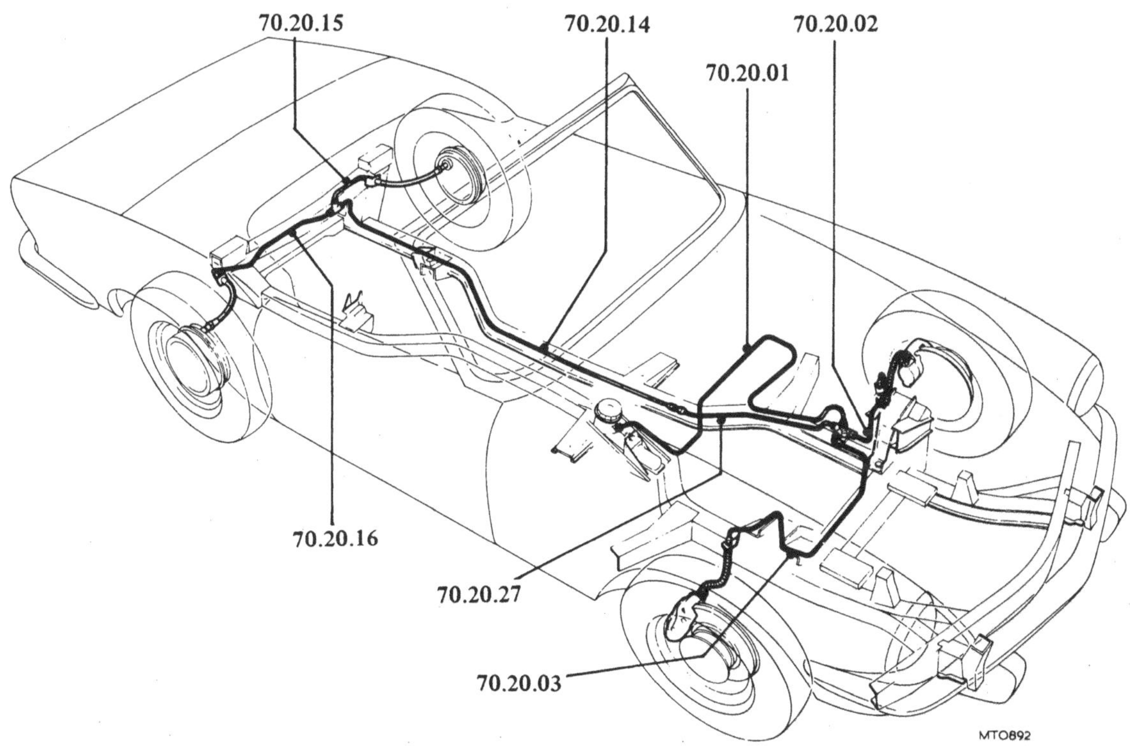

BRAKES [except U.S.A.]

– Bleed **70.25.02**

Do not allow the fluid in the reservoir to fall below half capacity. When topping up the reservoir do not use aerated fluid discharged from the system.

1. Release the handbrake.
2. Attach a rubber or plastic tube of suitable bore to the bleed nipple of the front caliper farthest from the master cylinder. Allow the free end of the tube to hang submerged in brake fluid in a transparent container.
3. Slacken the bleed nipple.
4. Depress the brake pedal fully and follow with three rapid strokes. Allow the pedal to return. Repeat until fluid, free from air bubbles issues from the bleed tube.
5. Depress the brake pedal and tighten the bleed nipple.
6. Release the brake pedal and remove the tube and container.
7. Repeat instructions 2 to 6 on opposite front caliper.
8. Bleed both rear wheel cylinders in similar manner commencing with the longest pipe run (i.e. commence with left hand cylinder on right hand steer models and right hand cylinder on left hand steer models).
9. Remove the bleed tube and container.

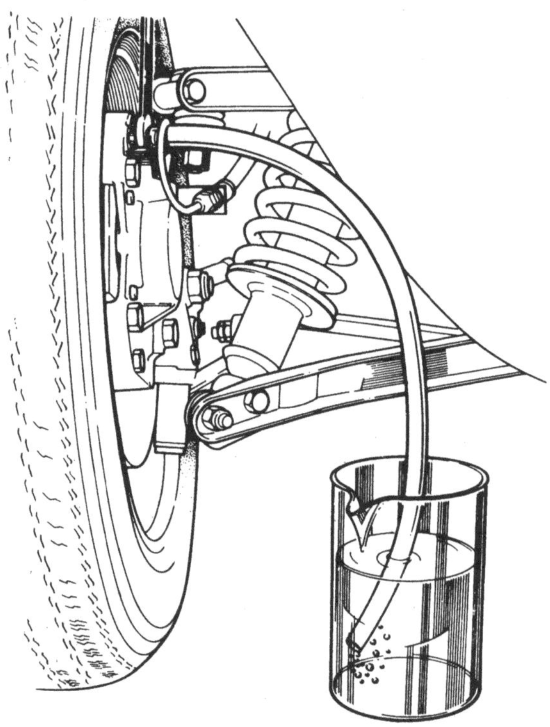

NT2011

BRAKES [U.S.A. only]

– Bleed (Tandem type master cylinder) **70.25.02**

Do not allow the fluid in the reservoir to fall below half capacity. When topping up the reservoir do not use aerated fluid discharged from the system.

1. Release the handbrake.
2. Attach a bleed tube to the nipple of the rear wheel cylinder farthest from the master cylinder. Allow the free end of the tube to hang submerged in brake fluid in a transparent container.
3. Slacken the bleed nipple and depress the brake pedal using light pedal pressure. Do not use the full travel of the brake pedal as this may cause the shuttle of the P.D.W.A. switch to be moved out of centre.
   Allow the brake pedal to return to its idle position and again depress lightly. Continue until discharge from the bleed tube is free of air.
4. Hold the pedal depressed, tighten the nipple and remove the bleed tube.
5. Repeat instructions 2 and 3 on the opposite rear wheel cylinder.
6. Attach the bleed tube to the front caliper farthest from the master cylinder.
7. Depress and release the brake pedal until air bubbles cease to issue from the bleed tube.
8. Hold the pedal depressed, tighten the nipple and release the brake pedal.
9. Attach the bleed nipple to the opposite caliper and repeat instructions 7 and 8.
10. Check that the P.D.W.A. shuttle is centred by switching on ignition and observing the warning light. Re-centre the shuttle as necessary.

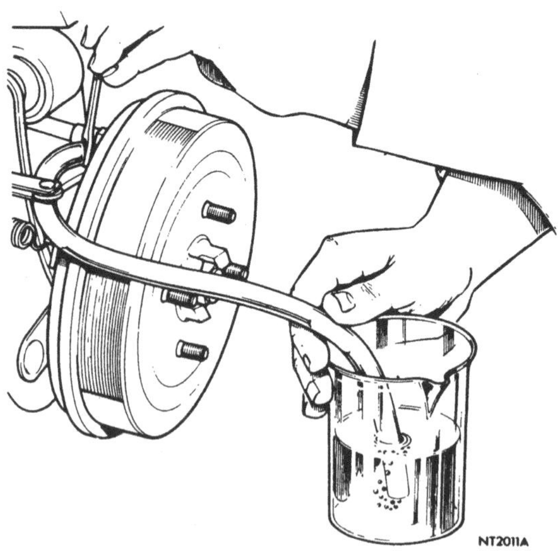

NT2011A

BRAKES

– Adjust **70.25.03**

Front

The front disc brakes are self-adjusting.

Rear

1. Release the handbrake and raise the rear wheels clear of ground contact.
2. Rotate the square-ended adjusters on rear of backplate in clockwise direction until the wheel is locked.
3. Rotate the adjuster anti-cloclwise until the wheel spins freely.
4. Repeat instructions 3 and 4 on opposite rear wheel.
5. Remove the jack.

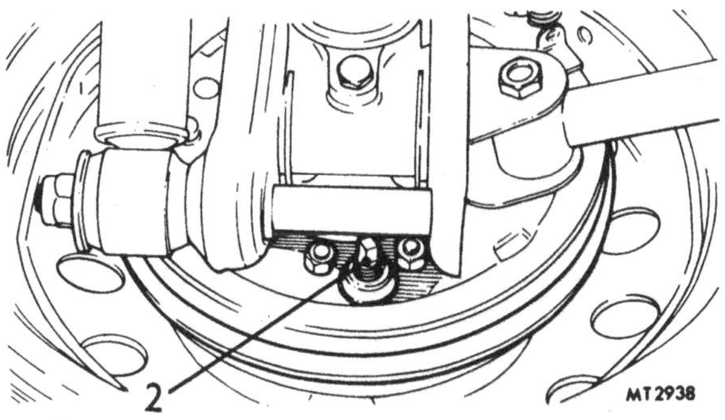

MASTER CYLINDER[except U.S.A.]

– Remove and refit **70.30.01**

Removing

1. Remove the clevis pin securing the master cylinder push rod fork to the brake pedal.
2. Disconnect the brake pipe at the master cylinder. Seal the master cylinder fluid outlet and the brake pipe to prevent ingress of foreign matter.
3. Remove the two bolts and spring washers securing the master cylinder to the mounting bracket.
4. Withdraw the master cylinder.

Refitting

5. Reverse instructions 1 to 4.
6. Bleed the brakes.

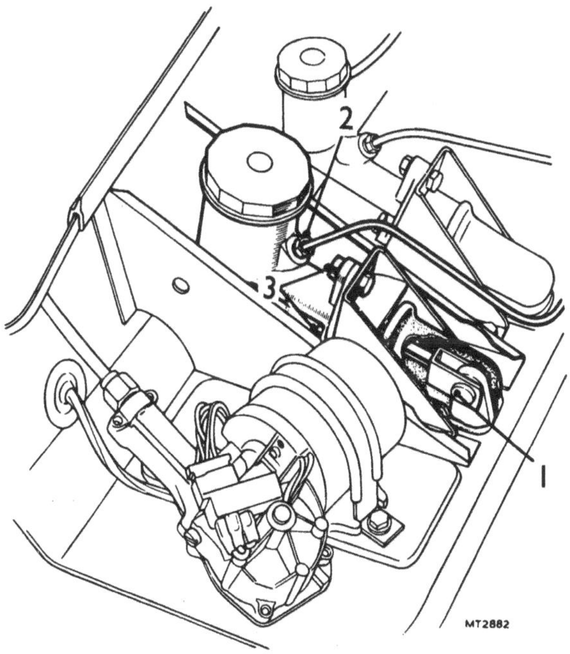

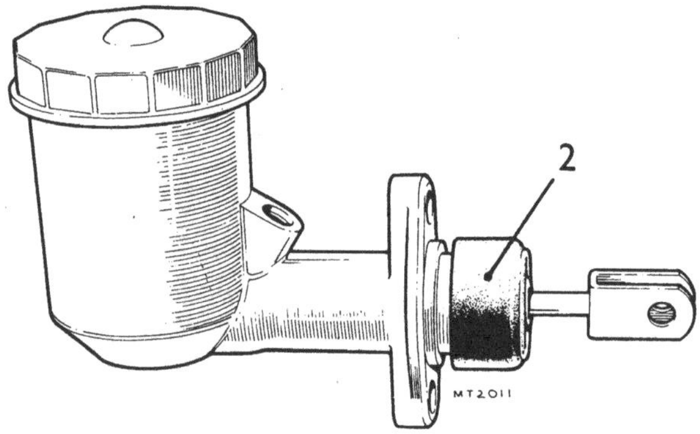

MASTER CYLINDER [except U.S.A.]

[CAUTION: Wash hydraulic cylinder components in brake fluid only. Never use gasoline or other petroleum-based solvents since they are damaging to rubber parts in the system.]

– Overhaul **70.30.02**

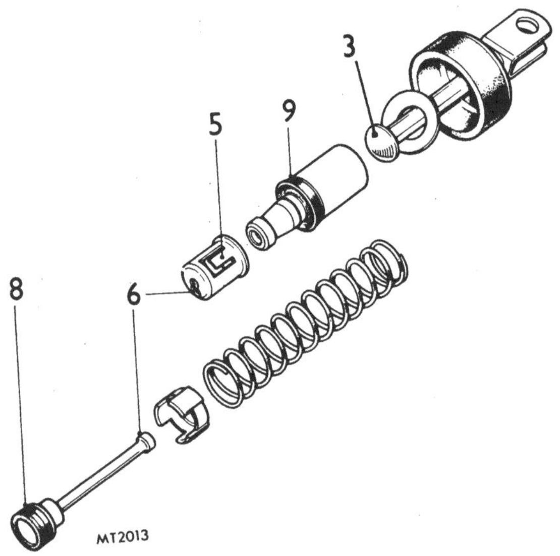

Dismantling

1. Drain the fluid reservoir.
2. Slide the rubber dust cover along the push rod to expose end of master cylinder.
3. Remove the circlip and withdraw the push rod and dished washer.
4. Withdraw the piston and seal assembly.
5. Straighten the prong of the spring thimble and remove the thimble from the piston.
6. Release the valve stem from the keyhole slot in the thimble.
7. Remove the spring and slide the valve seal spacer along the valve stem.
8. Remove the valve seal from the valve stem.
9. Remove the seal from the piston.
10. Clean and examine all components.

Assembling

11. Fit a new seal to the piston.
12. Fit a new seal to the valve stem.
13. Assemble the spacer, spring and thimble to the valve stem.
14. Fit the thimble to the piston and carefully depress the thimble prong.
15. Lubricate the master cylinder bore with clean brake fluid and insert the piston and seal assembly.
16. Fit the push rod and dished washer and secure with the circlip.
17. Fit a new rubber dust cover.

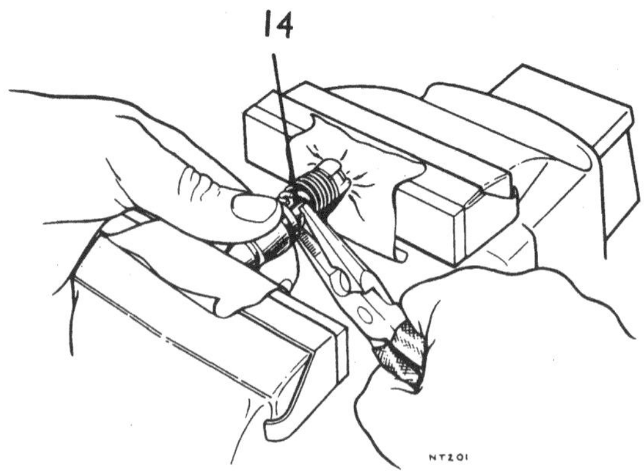

MASTER CYLINDER (TANDEM TYPE) [U.S.A. only]

– Remove and refit 70.30.08

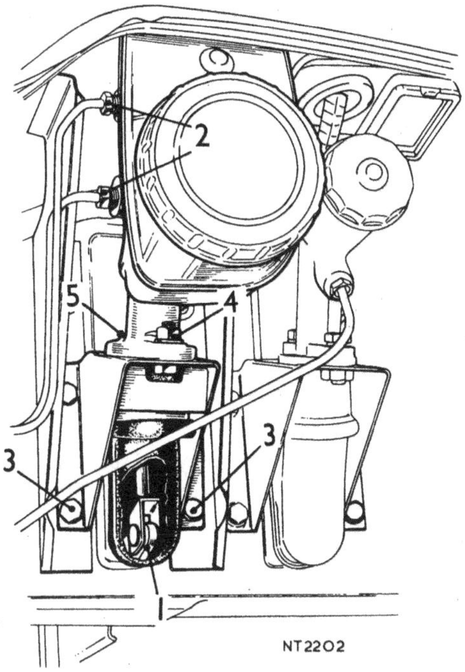

NT2202

Removing

1. Raise the rubber boot to expose the master cylinder/brake pedal linkage and remove the split pin, plain washer, clevis pin and the rubber boot.
2. Disconnect the brake pipes at the master cylinder. Seal the master cylinder fluid outlets and the brake pipes to prevent ingress of foreign matter.
3. Remove the four bolts and spring washers securing the master cylinder mounting bracket to the scuttle and withdraw the mounting bracket and the master cylidner.
4. Remove the nut, spring washer and the bolt securing the master cylinder flange to the mounting bracket.
5. Remove the bolt and spring washer securing the master cylinder flange to the mounting bracket.

Refitting

6. Reverse instructions 1 to 5.
7. Bleed the brakes.

MASTER CYLINDER (TANDEM TYPE) [U.S.A. only]

– Overhaul 70.30.09

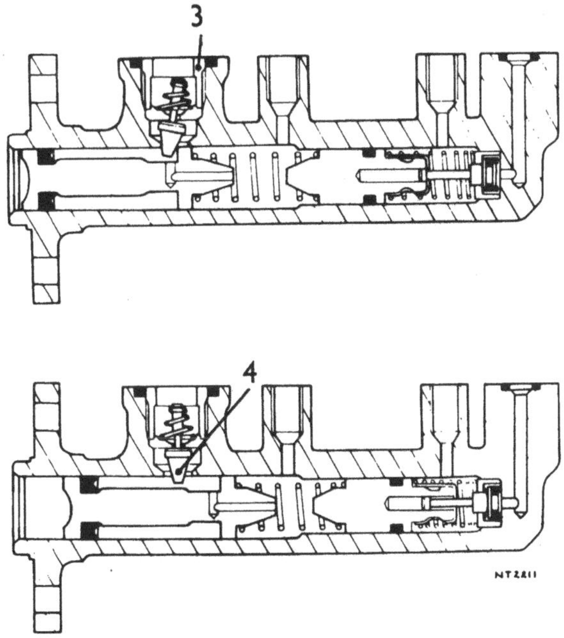

NT2411

1. Remove the master cylinder, detach mounting bracket and drain the brake fluid. 70.30.08.

Dismantling

2. Remove the reservoir (four screws).
3. Remove the internal nut securing the tipping valve to the master cylinder body.
4. Apply slight pressure on the push rod to allow the tipping valve to seat and withdraw the tipping valve.
5. Slide the dust cover clear of the cylinder to expose the circlip retaining the push rod and remove the circlip.

6. Withdraw the push rod.
7. Withdraw the primary and secondary plungers complete with seals and springs.
8. Separate the plungers and the intermediate spring.
9. Prise up the leaf of the spring retainer and remove the spring and centre valve sub-assembly from the secondary plunger.
10. Withdraw the valve spacer, spring washer and valve seal from the valve head.
11. Remove the seals from the primary and secondary plungers.
12. Thoroughly clean all components.
13. Examine the cylinder bore for wear, scoring and corrosion. If damage or wear is evident a new tandem cylinder must be obtained.

Assembling

14. Fit new seals to the primary and secondary plungers.
15. Fit a new valve seal, smaller diameter leading, to the valve head.
16. Place the spring washer on the valve stem ensuring that the convex side of the washer is adjacent to the valve.
17. Fit valve spacer, legs leading.
18. Fit the spring retainer to the valve stem, keyhole leading.
19. Slide the secondary spring over the spring retainer and offer up the secondary plunger.
20. Place the secondary plunger and vale assembly between the protected jaws of a vice. To ensure cleanliness, clean paper should be placed on the vice jaws. Compress the spring and using a small screwdriver, press the spring retainer hard against the secondary plunger. Holding the retainer in this position depress the leaf of the spring retainer against the plunger. Remove the plunger and valve assembly from the vice and check that the retainer is firmly located on the plunger.
21. Fit the intermediate spring between the primary and secondary plungers.
22. Generously lubricate the cylinder bore and plungers with clean brake fluid and insert the plungers in the cylinder.
23. Fit the push rod and the retaining circlip and rubber boot.
24. Depress the push rod and insert the tipping valve. Fit seal and securing nut and tighten nut to a torque of 35 to 40 lbf. ft (4·8 to 5·5 kgf. m).
25. Fit the reservoir.

[CAUTION: Wash hydraulic cylinder components in brake fluid only. Never use gasoline or other petroleum-based solvents since they are damaging to rubber parts in the system.]

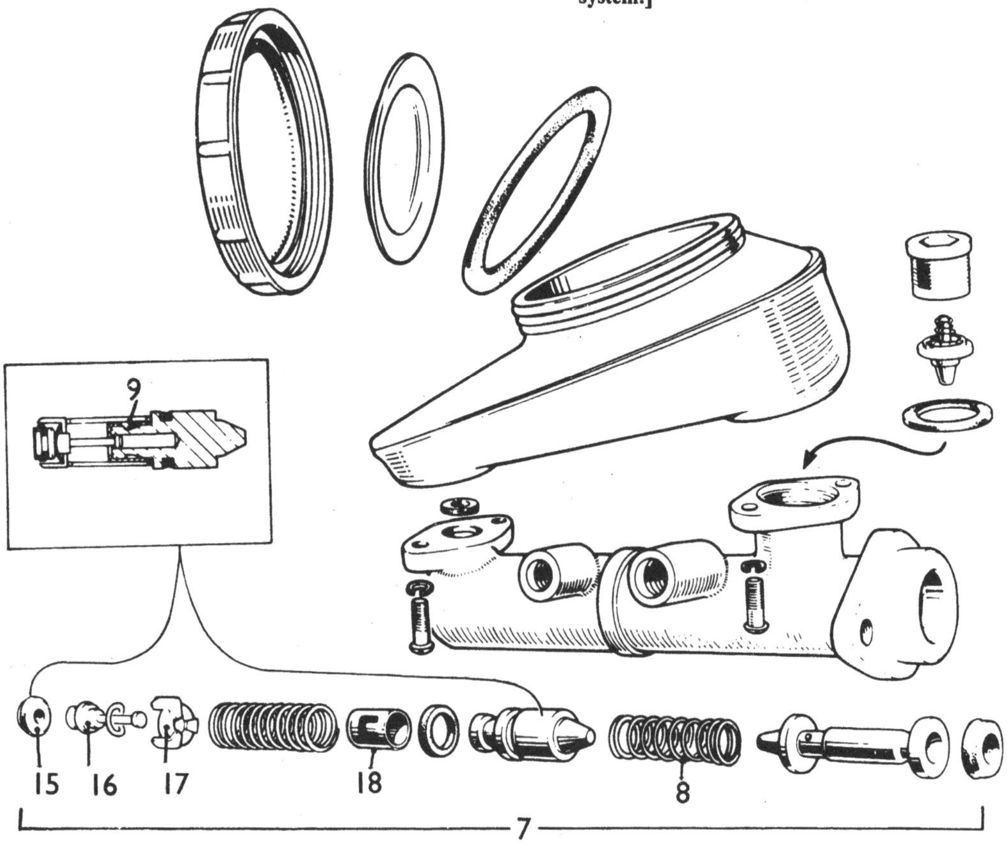

BRAKE PEDAL ASSEMBLY

– Remove and refit **70.35.01**

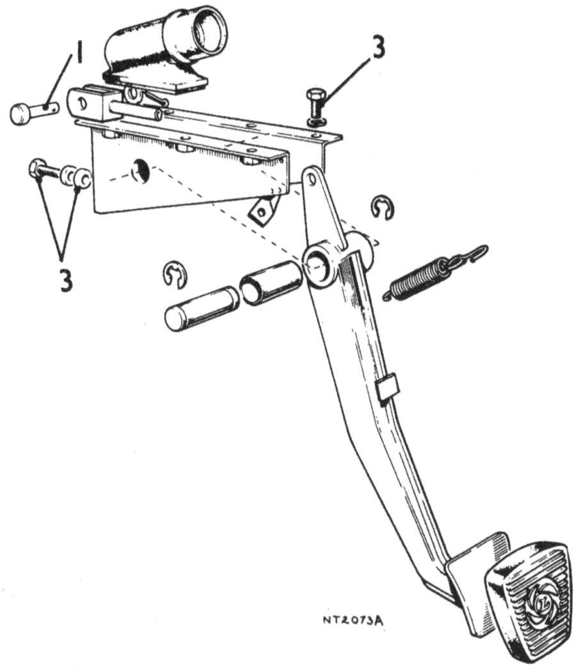

Removing

1. Remove the clevis pin securing master cylinder push rod fork to brake pedal.
2. Remove the terminal connectors from the brake light switch.
3. Remove the bolts, nuts and spring washers seeuring the brake pedal bracket to the scuttle.
4. Withdraw the brake pedal and bracket assembly.

Refitting

5. Reverse instructions 1 to 4.

HANDBRAKE LEVER

– Remove and refit **70.35.08**

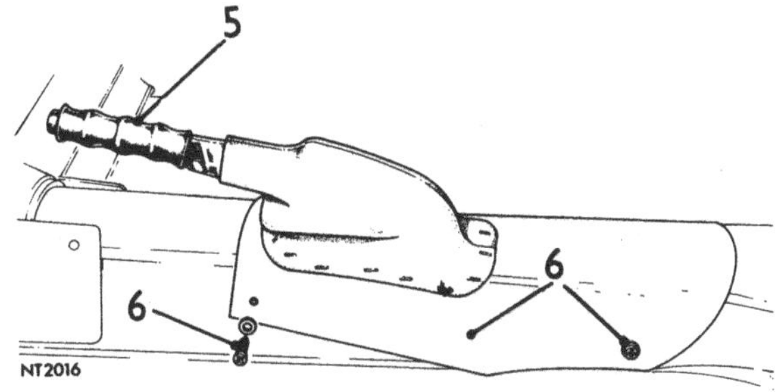

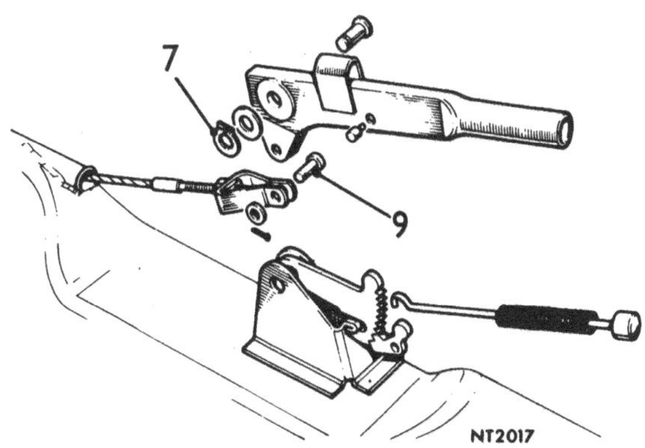

Removing

1. Remove the driver's and passenger's seats complete with runners. 76.70.04/05.
2. Remove the safety belt buckle brackets from tunnel.
3. Remove the safety belt buckles from the floor of car.
4. Remove the carpet.
5. Remove the hand grip from the handbrake lever.
6. Remove the four screws securing the handbrake cover and slide the cover clear of the handbrake.
7. Remove the circlip securing handbrake fulcrum pin to bracket.
8. Remove the fulcrum pin and lift the handbrake clear of the bracket.
9. Remove the split pin and clevis pin securing the handbrake lever to the cable fork and withdraw the handbrake lever.

Refitting

10. Reverse instuctions 1 to 9.

HANDBRAKE PAWL AND RATCHET

– Remove and refit 70.35.09

Removing

1. Remove the handbrake lever. 70.35.08.
2. File the riveted end of the pawl fulcrum pin until the end of the pin is flush with the handbrake lever.
3. Depress the handbrake button and retain in this position with tape.
4. Remove the pawl fulcrum pin and withdraw the pawl and ratchet.

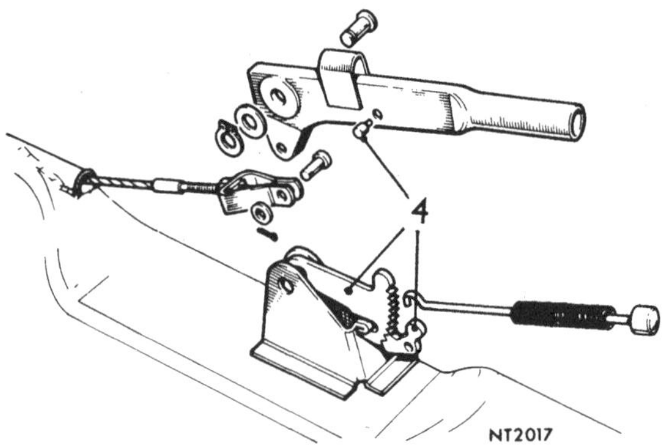

Refitting

5. Insert a new ratchet (hooked end downwards) into the handbrake lever.
6. Insert new pawl (teeth downward, facing teeth on ratchet) into handbrake lever.
7. Fit a new fulcrum pin to pawl and rivet the end of pin ensuring that the pawl pivots freely.
8. Engage hooked end of release lever in the pawl and remove the tape from the release button.
9. Secure the handbrake lever to the handbrake cable fork and fit a new split pin to the clevis pin.
10. Align the ratched hole with the hole in the handbrake lever and fit the lever to the car mounting bracket.
11. Fit the fulcrum pin and secure with the circlip.
12. Refit the handbrake cover, handgrip, carpet and seats. 70.35.08.

HANDBRAKE CABLES

Adjust 70.35.10

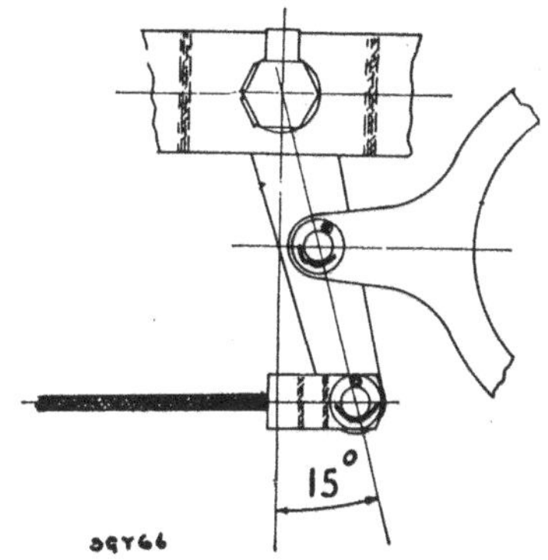

1. Jack up the rear wheels and support the chassis on stands.
2. Release the handbrake.
3. Rotate the rear brake adjusters and lock both rear wheels.
4. Check that the relay lever is positioned as shown. Adjust front and rear cables as necessary to obtain this condition.

Adjusting the front cable

5. Remove the carpet and handbrake lever cover to expose the handbrake lever 70.35.08.
6. Slacken the cable locknut at the rear of the cable fork and rotate the cable as necessary, clockwise to tighten the cable, anti-clockwise to slacken the cable. Tighten the locknut. Note that it may be necessary to first release the rear cable from the brake backplate.
7. Refit the handbrake lever cover, carpets and seats. 70.35.08.

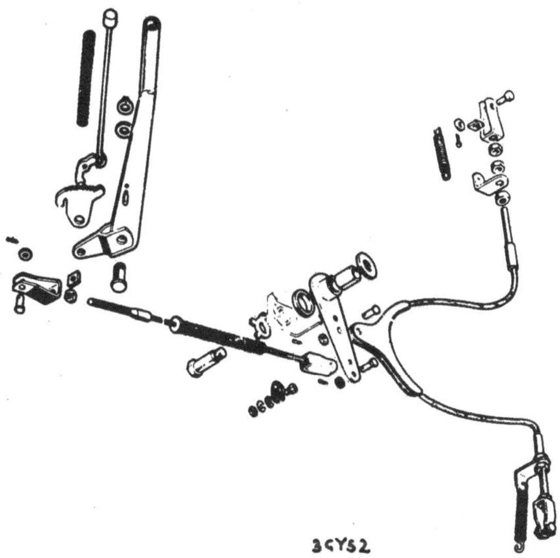

Adjusting the rear cable

8. Disconnect the cable fork ends from the brake backplate lever.
9. Adjust the cable forks equally at both ends of the cable until with slight strain applied to the cable, the clevis pins can be inserted to secure the cable forks to the backplate levers. Do not force the cables taut to insert the clevis pins.
10. Fit the washers and new split pins to the clevis pins.
11. Tighten the cable fork lock nuts.
12. Slacken the brake adjusters until the wheels spin freely.
13. Remove the chassis stands.

HANDBRAKE COMPENSATOR

Remove and Refit 70.35.11

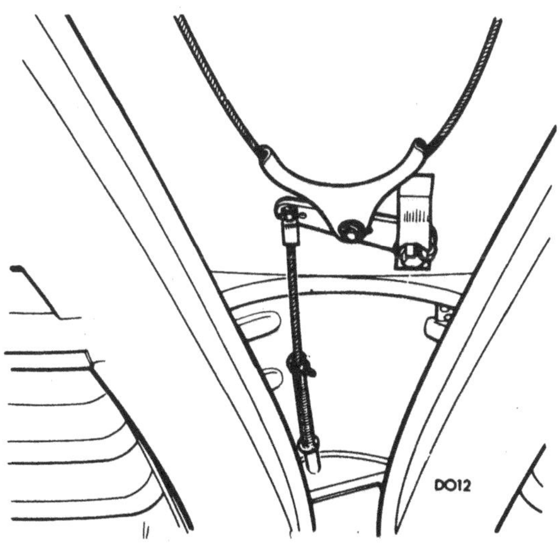

Removing

1. Release the handbrake.
2. Remove the split pin and plain washer securing the compensator to the relay lever and withdraw the clevis pin.
3. Remove the compensator from the rear cable.

Refitting

4. Reverse instructions 1 to 3.

HANDBRAKE CABLES

– Remove and refit **70.35.16**

Removing

Front

1. Remove the handbrake lever 70.35.08.
2. Disconnect the cable fork from the handbrake lever.
3. Remove the front cable fork from the cable.
4. Disconnect the cable from the relay lever.
5. Withdraw the cable from the car.

Rear

6. Disconnect the compensator from the relay lever.
7. Unhook the cable return springs.
8. Disconnect the cable forks from the backplate levers.
9. Remove the cable forks and spring plates from the cable.
10. Withdraw the cable from the car.

Refitting

Front

11. Fit the cable in position on car.
12. Fit the fork and nuts to the front end of the cables.
13. Connect the front end of the cable to the handbrake lever.
14. Fit the handbrake lever to the car.
15. Set the relay lever as shown and connect the cable rear fork to the left hand inboard hole.
16. With the handbrake lever fully 'OFF' adjust the cable length. Do not alter the position of the relay lever.
17. Check that the cable spring is properly positioned and adjust the spring tension as necessary, by means of the rear pinch clamp.

Rear

18. Thread the cable through the cable guides.
19. Fit the nuts, spring plates and forks to both ends of the cable.
20. With the handbrake lever in the 'OFF' position, attach the cable and compensator to the right hand inboard hole of the relay lever.
21. Jack up the rear wheels and lock the rear brakes by means of the adjusters.
22. With the cable pulled by hand, adjust the fork ends so that the clevis pins will slip into place.
23. Secure fork lock nuts and fit split pins to clevis pins.
24. Attach return springs to the spring plates and the backplates.
25. Adjust the position of the spring plates to tension the springs.
26. Slacken the brake adjusters until wheels spin freely.
27. Remove the jack.

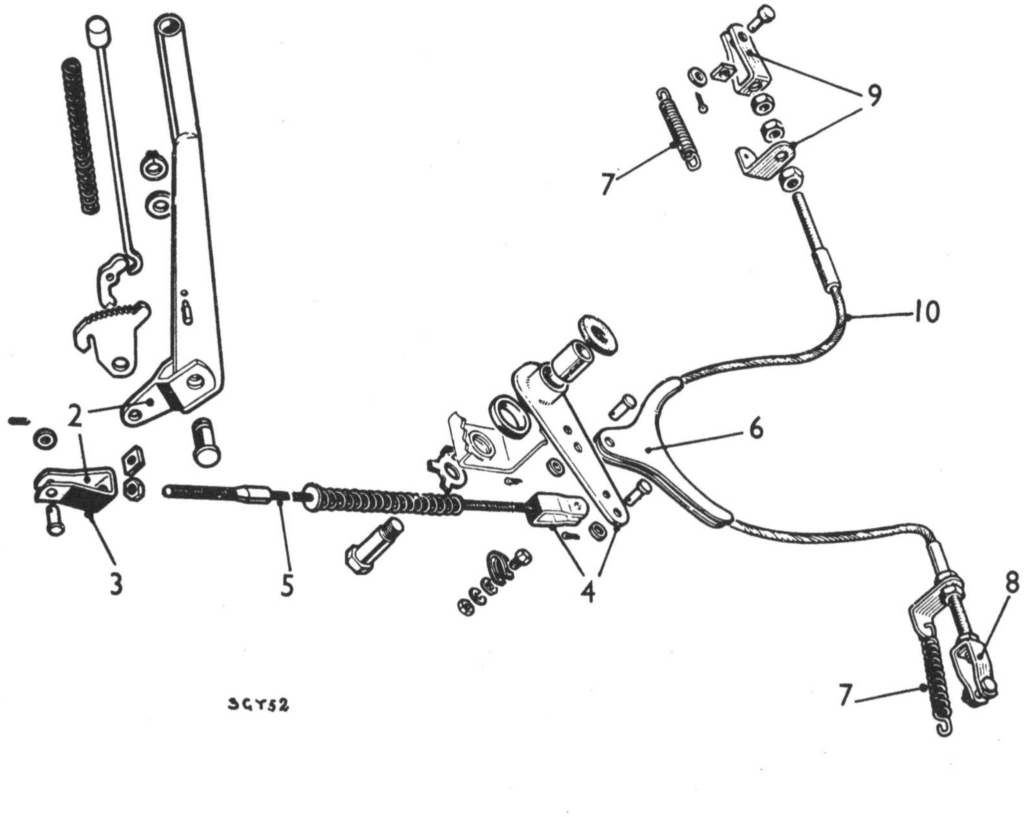

RELAY LEVER

– Remove and refit **70.35.22**

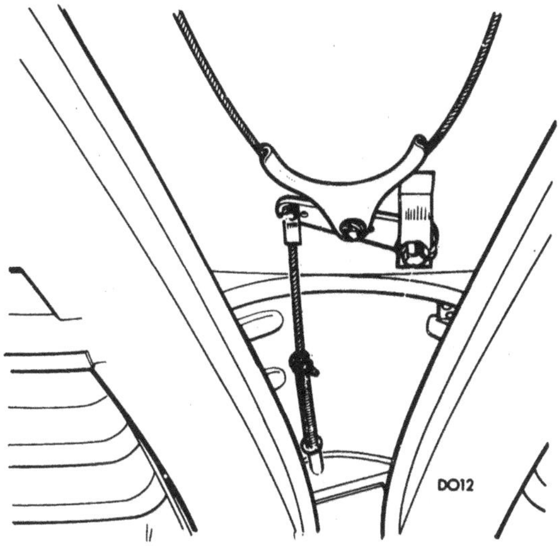

Removing

1. Release the handbrake.
2. Disconnect the compensator from the relay lever.
3. Disconnect the front handbrake cable from the relay lever.
4. Straighten the tab washers securing the relay lever fulcrum bolt.
5. Remove the fulcrum bolt.
6. Withdraw the relay lever complete with rubber and felt seals.

Refitting

7. Lubricate the relay lever bush.
8. With the larger boss of the relay lever downwards position the relay lever and seals in the bracket (felt seal top, rubber seal bottom).
9. Fit the fulcrum bolt and tab washer to the relay lever bracket. Bend the tab washer to retain the bolt.
10. Connect the compensator to the right hand inboard hole in the relay lever.
11. Connect the front brake cable to the left hand inboard hole in the relay lever.

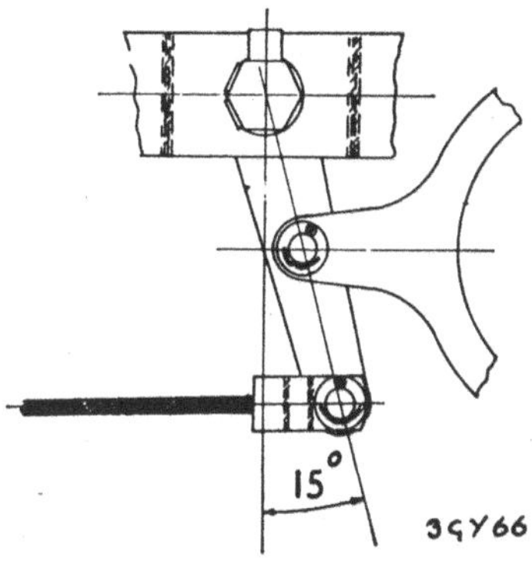

FRONT BRAKE PADS

– Remove and refit **70.40.02**

Removing

1. Remove the front wheel.
2. Withdraw the two spring pins from the brake pad retaining pins.
3. Withdraw the brake pad retaining pins.
4. Lift out the brake pads complete with damping shims.

Refitting

5. If new brake pads are being fitted press the caliper pistons into their respective bores.

[NOTE: Pressing in the pistons will force fluid back to the master cylinder reservoir. Therefore, remove some fluid beforehand to prevent an overflow.]

6. Clean the brake pad locations in the caliper.
7. Fit the brake pads complete with damping shims.
8. Fit the brake pad retaining pins to the caliper and secure with spring pins.
9. Fit the front wheel.

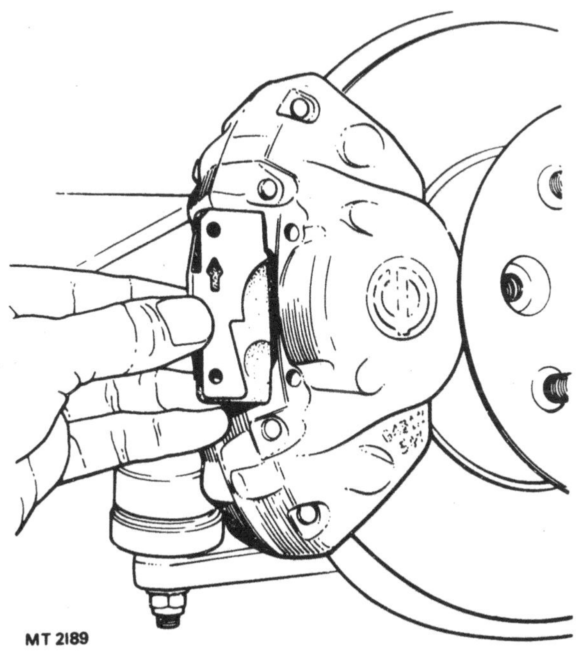

REAR BRAKE SHOES

– Remove and refit **70.40.03**

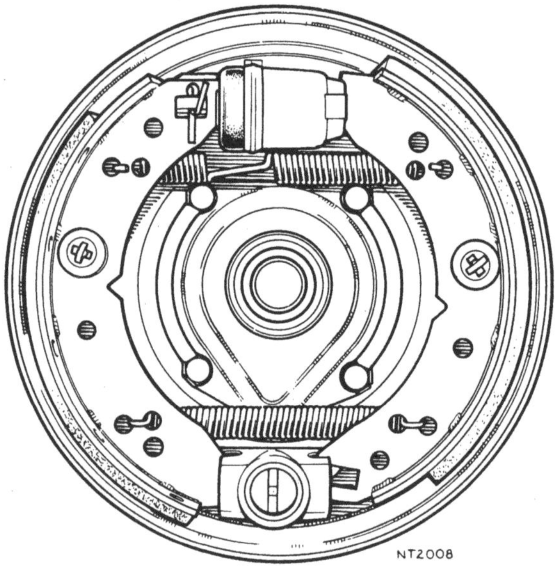

Removing

1. Remove the rear wheel.
2. Remove the brake drum.
3. Remove the split pin from the handbrake lever.
4. Remove the shoe-steady pins, caps and springs.
5. Release the lower end of one shoe from the brake adjuster.
6. Release the upper end of the same shoe from the wheel cylinder.
7. Unhook the brake shoe return springs.
8. Remove the brake shoes.

Refitting

9. Reverse instructions 1 to 8. Note that both brake shoe return springs are fitted inboard.

REAR BRAKE ADJUSTER

– Remove and refit **70.40.17**

1. Remove the rear wheel.
2. Remove the brake drum.
3. Remove the brake shoes.

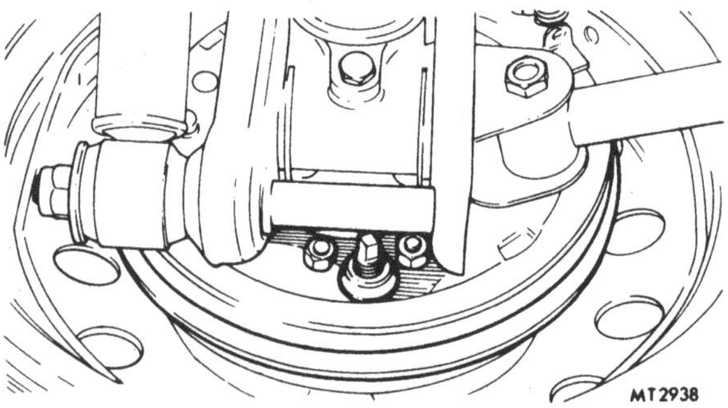

1. Remove the rear wheel.
2. Remove the brake drum.
3. Remove the brake shoes.
4. Remove two bolts and spring washers securing the brake adjuster to the backplate.
5. Withdraw the adjuster.

Refitting

6. Reverse instructions 1 to 5.
7. Adjust the brake.

Triumph Spitfire Mk IV Manual. Part No. 545254. Issue 1.

FRONT CALIPER

– Remove and refit **70.55.02**

Removing

1. Remove the front wheel.
2. Disconnect the fluid supply pipe to the caliper flexible hose.
3. Remove the nut securing the flexible hose to the front suspension bracket and withdraw the hose from the bracket.
4. Unscrew the flexible hose from the caliper.
5. Remove the two bolts and spring washers securing the caliper shield to the vertical link.
6. Withdraw the caliper shield and caliper.

Refitting

7. Reverse instructions 1 to 6.
8. Bleed the brakes.

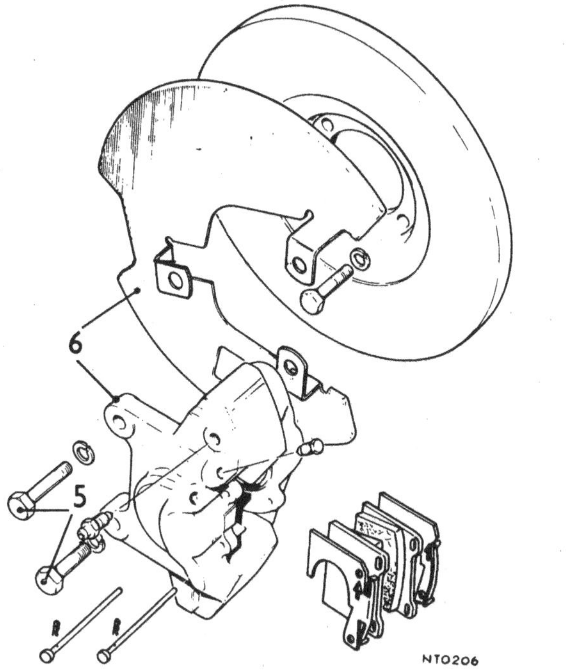

NT0206

FRONT CALIPER

– Renew seals **70.55.13**

[CAUTION: Wash hydraulic cylinder components in brake fluid only. Never use gasoline or other petroleum-based solvents since they are damaging to rubber parts in the system.]

Dismantling

1. Remove the caliper 70.55.02.
2. Remove the spring pins from the brake pad retaining pins.
3. Withdraw the pad retaining pins.
4. Withdraw the brake pads and shims.
5. Remove the cirlip retaining the piston dust covers and withdraw the dust covers.
6. Extract the caliper piston. Piston removal may be facilitated using a low pressure air line. Do not interchange the pistons.
7. Prise out the cylinder seals taking care not to damage the cylinder bore.
8. Thoroughly clean the caliper bores and pistons using brake fluid or methylated spirits. If the pistons or the bores are scored or corroded, a new caliper must be obtained.

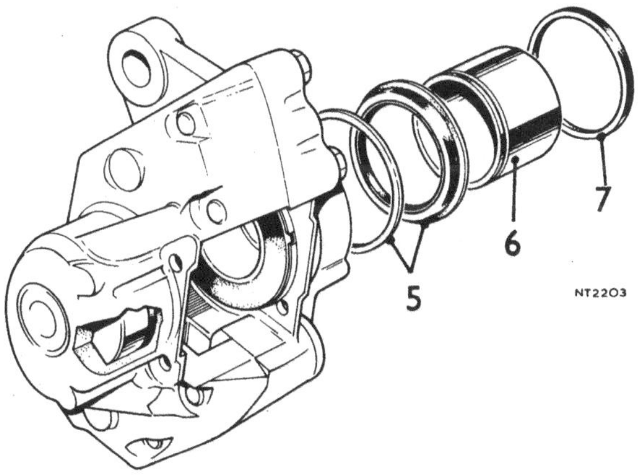

NT2203

Refitting

9. Carefully install new seals in the cylinder bores.
10. Lubricate the bores with clean brake fluid.
11. Evenly enter the pistons into their original locations in the caliper.
12. Fit new dust covers and circlips.
13. Fit the caliper and caliper shield to the vertical link.
14. Fit the flexible hose.
15. Fit the brake pads and shims.
16. Fit the front wheel.
17. Bleed the brakes.

REAR BRAKE CYLINDER

– Remove and refit 70.60.18

Removing

1. Jack up the car and support the chassis on stands.
2. Remove the rear wheel.
3. Release the handbrake and remove the brake drum.
4. Remove the brake shoes.
5. Disconnect the handbrake cable at the brake backplate lever.
6. Remove the flexible brake hose.
7. Remove the wheel cylinder rubber shield at the rear of the backplate.
8. Remove the horse-shoe clip and spring plate securing the wheel cylinder to the backplate.
9. Withdraw the wheel cylinder complete with the handbrake operating lever.

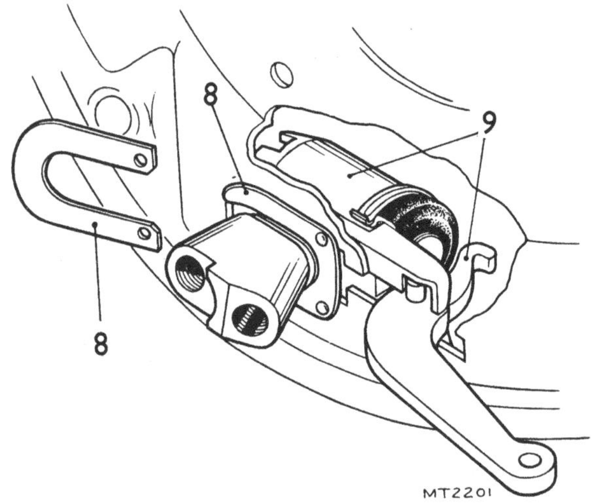

Refitting

10. Ensure that the wheel cylinder mounting face and the backplate are clean.
11. Lightly smear the wheel cylinder mounting face and both sides of the backplate faces with a zinc base grease.
12. Install the handbrake lever in the wheel cylinder.
13. Enter the wheel cylinder and the handbrake lever in the backplate.
14. Install the spring plate at the rear of the wheel cylinder (open end of plate towards rear of car, dimples on plate towards differential).
15. Install the horse-shoe clip (open end of clip towards the front of car). Ensure that the holes in the clip engage the dimples on the spring plate. Check that the wheel cylinder is free to slide on the backplate.
16. Fit the rubber shield.
17. Reverse instructions 1 to 6.
18. Bleed the brakes.

REAR BRAKE CYLINDER

– Overhaul 70.60.26

[CAUTION: Wash hydraulic cylinder components in brake fluid only. Never use gasoline or other petroleum-based solvents since they are damaging to rubber parts in the system.]

1. Remove the clip retaining the rubber boot to the cylinder body.
2. Withdraw the piston complete with rubber boot and seal from the cylinder.
3. Remove the boot and seal from the piston.
4. Examine the piston and cylinder bore for signs of corrosion or scoring. If either are damaged a new brake cylinder must be obtained.
5. Fit a new seal to the piston.
6. Fit a new rubber boot to the piston.
7. Lubricate the cylinder bore, seal and piston with clean brake fluid and insert the piston in the cylinder bore.
8. Fit the boot to the cylinder and secure with the clip.

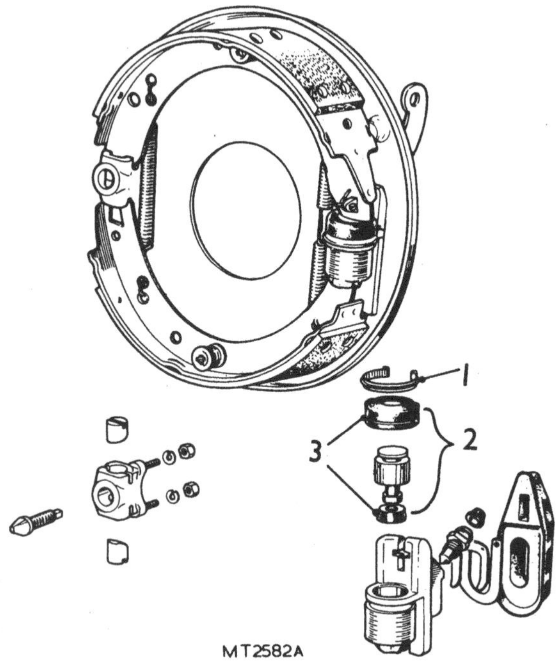

Triumph Spitfire Mk IV Manual. Part No. 545254. Issue 1.

WHEEL AND TYRE OPERATIONS

TYRES – GENERAL

Type	Size	Load Conditions	Pressures			
			Front		Rear	
			lb/in²	kg/cm²	lb/in²	kg/cm²
Cross-ply Tubeless	5.20S – 13	All	21	1·476	26	1·828
Radial ply Tubeless	145SR – 13		21	1·476	26	1·828
Radial ply Tubed*	145SR – 13		21	1·476	26	1·828
Radial ply Tube/Tubeless	**155SR – 13**		21	1·476	26	1·828

***[Tubes must be used whenever radial tires are installed on wire wheels.]**

Tyres of the correct type and dimensions, at the correct cold inflation pressures, are an integral part of the vehicle's design and regular maintenance of tyres contributes not only to safety but to the designed functioning of the vehicle, Road-holding, steering and braking are especially vulnerable to incorrectly pressurized, badly fitted or worn tyres.

Tyres of the same size and type but of different make may have widely varying characteristics. It is therefore recommended that tyres of the same make are fitted to all wheels.

Radial and cross-ply tyres

It is both dangerous and, in the U.K. illegal, to use on the public roads a vehicle fitted with unsuitable combinations of tyres. The following recommendations should therefore be observed:

1. Do not mix radial-ply and cross-ply tyres on the same axle.
2. Do not fit radial-ply tyres to the front wheels and cross-ply tyres to the rear wheels.
3. With suitable tyre pressure adjustments it may be possible to obtain acceptable handling with cross-ply tyres on the front wheels and radial-ply tyres on the rear wheels, but this combination is not recommended.

Size, type, pressures

The pressures recommended (see chart) provide optimum ride and handling characteristics for all normal operating conditions. The pressures should be checked, and adjusted if necessary, once per week. This should be done with the tyres cold. Tyre temperatures and pressures increase when running. Bleeding a warm tyre to the recommended presssures will result in under-inflation which may be dangerous. A slight natural pressure loss occurs with time. If this exceeds 2 lb/in² (0·14 kg/cm²) per week, the cause should be investigated and rectified. It should be noted that it is an offence in the U.K. to use a motor vehicle if a tyre is not so inflated as to make it fit for the use to which the vehicle is being put.

The spare wheel tyre should be maintained at the highest pressure quoted in the chart, and adjusted to the correct pressure for its position when fitted for use.

Wear

All tyres fitted as original equipment include wear indicators in their tread pattern. When the tread has worn to a remaining depth of 0·06 in (1·5 mm) the indicators appear at the surface as bars which connect the tread pattern across the full width of the tyre. It is illegal in the U.K. and certain other countries to continue to use tyres on which the tread is worn to less than 1mm. It should be noted that the properties of many tyres alter progressively with wear. In particular the 'wet grip' and aquaplaning resistance are gradually but substantially reduced. Extra care and speed restriction should therefore be exercised on wet roads as the effective tread depth diminishes. Incorrect wheel alignment will accelerate tyre wear. Fins on the inside or outside edges of the tread pattern are caused by excessive toe-in or toe-out respectively. As fins may also be caused by high cornering speeds or road camber it is better to have the cause ascertained by having the wheel alignment checked (see 'General Specification' for data'.)

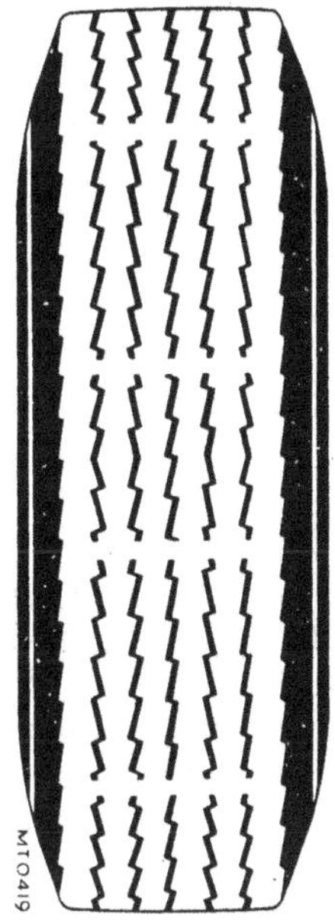

Damage

Excessive local distortion can cause the casing of the tyre to fracture and may lead to premature tyre failure. Tyres should be examined especially for cracked walls, exposed cords, etc. Flints and other sharp objects should be removed from the tyre tread, if neglected they may work through the cover. Any oil or grease which may get onto the tyres should be cleaned off by using fuel sparingly. Do not use paraffin (kerosene) which has a detrimental effect on rubber.

Repairs

Tubeless tyres

A temporary repair can be made to tubeless tyres, using a special kit, provided the puncturing hole is small and confined to the central tread area. The following precautions must, however, be observed.

1. Do not use more than one plug in each hole.
2. Do not use the tyre for high speeds.
3. Ensure that a permanent 'cold patch' or vulcanized repair is made at the earliest opportunity.

Tubes

When repairing tubes, have punctures or injuries vulcanized. Ordinary patches should only be used for emergiences. Vulcanizing is absolutely essential for tubes manufactured from synthetic rubber.

Winter tyres

Winter tyres are designed to give improved traction and braking in mud and snow. Their performance on hard surfaces may, however, be inferior to normal road tyres and extra care is required when using them under normal conditions.

Racing and competition tyres

Should the vehicle be tuned to increase its maximum speed, or be used for racing or competition, consult the respective tyre company regarding the need for tyres of special or racing construction.

Valves

Whenever a new tubeless tyre is fitted, the Schrader snap-in type valve must also be renewed. To facilitate fitting, lubricate the valve with soap solution before using a special tool to snap the valve squarely into an airtight position in the rim hole.

WHEEL AND TYRE BALANCE 74.15.00

Tyre and wheel assemblies should be statically balanced to within 5 ozf. ins. Balance weights are available in ½oz. increments from ½ to 3 oz.

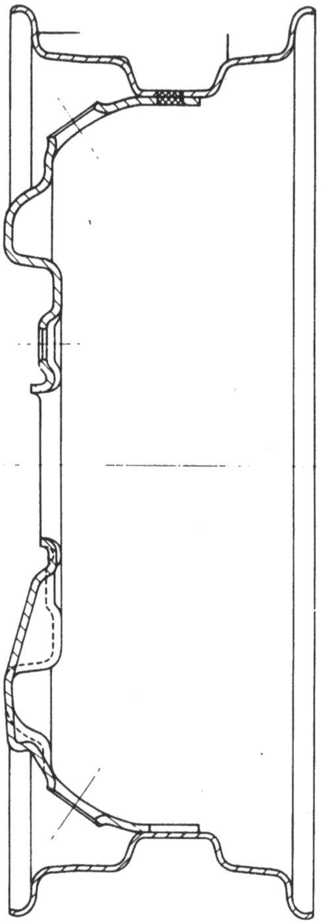

WHEELS

– General 74.20.00

Pressed Steel Wheels

4½J rims. Wheels are located and retained on hubs by four 3/8 ins U.N.F. studs and chrome-plated dome nuts tightened to 38 to 48 lbf. ft (5·2 to 6·6 kgf. m). Embellishment is provided by spring-loaded plastic hub covers fitted under the dome nuts.

Wire Wheels

4½J rims with 16 outer spokes and 32 inner spokes. Wheels are located and retained by four 3/8 in U.N.F. studs and chrome plated dome nuts. tightened to 38 to 48 lbf. ft. (5·2 to 6·6 kgf. m).

It is recommended that the servicing and reconditioning of wire wheels is entrusted to those who are equipped to fulfill this specialist function. It is pointed out that the renewal of a single broken spoke may necessitate extensive re-adjustment to spoke tension throughout the wheel.

The average torque of outer spokes should not be less than 75 lbf. ins (0·86 kgf. m) Average torque of inner spokes should not be less than 60 lbf. ins (0·7 kgf, m). With spokes fully tightened at least two full threads should be visible below the nipple.

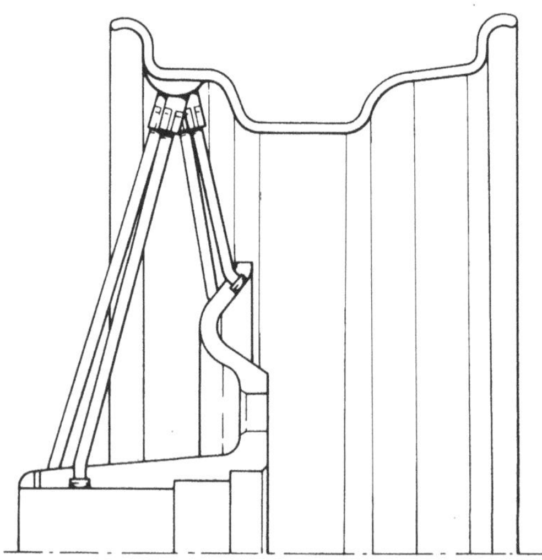

Wheel Tolerances

Wobble

1. Lateral variation measured on the vertical inside face should not exceed 0·050 ins (1·270 mm).

Lift

2. On a truly mounted and revolving wheel the difference between the high and low points measured on either rim ledge should not exceed 0·050 ins (1·270 mm).

BODY OPERATIONS

BODY OPERATIONS — *continued*

Door trim pad – remove and refit	76.34.01
Fascia – centre – remove and refit	76.46.02
– drivers side – remove and refit	76.46.01
– passengers side – remove and refit	76.46.03
Fascia crash pad	76.46.04
Fascia support bracket	76.46.09
Gearbox tunnel cover – remove and refit	76.25.07
Glass – windscreen – remove and refit	76.81.01
Grille – front – remove and refit	76.55.03
Hardtop – remove and refit	76.61.01
Hood assembly – remove and refit	76.61.08
Luggage compartment lid – remove and refit	76.19.01
Luggage compartment lid hinges – remove and refit	76.19.07
Luggage compartment lid lock – remove and refit	76.19.11
Luggage compartment lid lock striker – remove and refit	76.19.12
Luggage compartment lid seal – remove and refit	76.19.06
Parcel shelf – drivers side – remove and refit	76.67.04
Parcel shelf – passengers side – remove and refit	76.67.05
Private lock – remove and refit	76.37.39
Rear compartment trim pad – remove and refit	76.13.20
Seat belt – static – remove and refit	76.73.02
Seat cushion cover – remove and refit	76.70.02
Seat – driver – remove and refit	76.70.04
Seat – passenger – remove and refit	76.70.05
Seat runners – remove and refit	76.70.21

BODY OPERATIONS — *continued*

CHASSIS FRAME

– Alignment check 76.10.02

Whilst severe damage to the chassis is readily detected, less serious damage may cause distortion, that is not visually apparent. If steering and suspension checks indicate a fault which cannot be attributed to anything other than chassis distortion, the chassis frame should be checked for twist and squareness.

Checking for Twist

1. Position the vehicle on a clean and level floor.
2. Remove the underriders 76.22.01
3. Place a jack under each jacking point and remove the road wheels 74.20.01.
4. Adjust the jacks until the following conditions are achieved:
   Points 'A' are 25·29 in (64·23 cm) above the floor.
   Points 'E' are 24·94 in (63·35 cm) above the floor.
   This condition sets the datum 20 in (50·8 cm) above the floor.
   If the height of points 'A' cannot be equalised, the difference in height of points 'A' indicates the amount by which the chassis is twisted.

Checking for squareness

5. Transfer all the lettered points shown to the floor, using a plumb-bob and fine cord.
6. Letter the points on the floor and connect each pair by drawing a line between them.
7. Mark the central point of each line and place a straight edge along these mid-points.
8. Check for squareness.
9. Using a straight edge mark the diagonals as shown.
10. Check for squareness. If the chassis is square then each pair of opposite diagonals must be equal in length and the points of intersection must lie on the same straight line.
11. The extent of lateral chassis distortion is assessed by the amount and direction by which any central point on the transverse line and/or the point of intersection of any pair of diagonals deviates from the centre line.

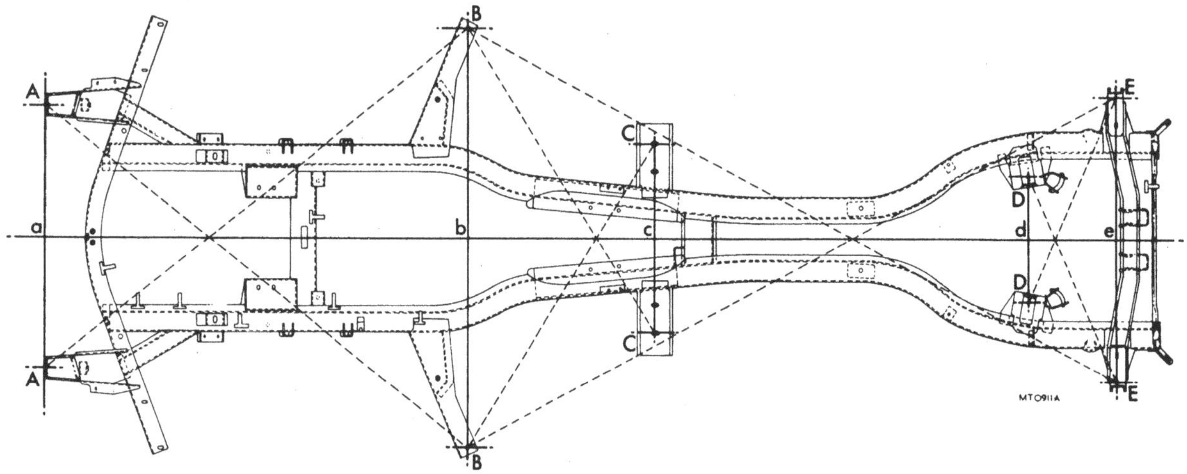

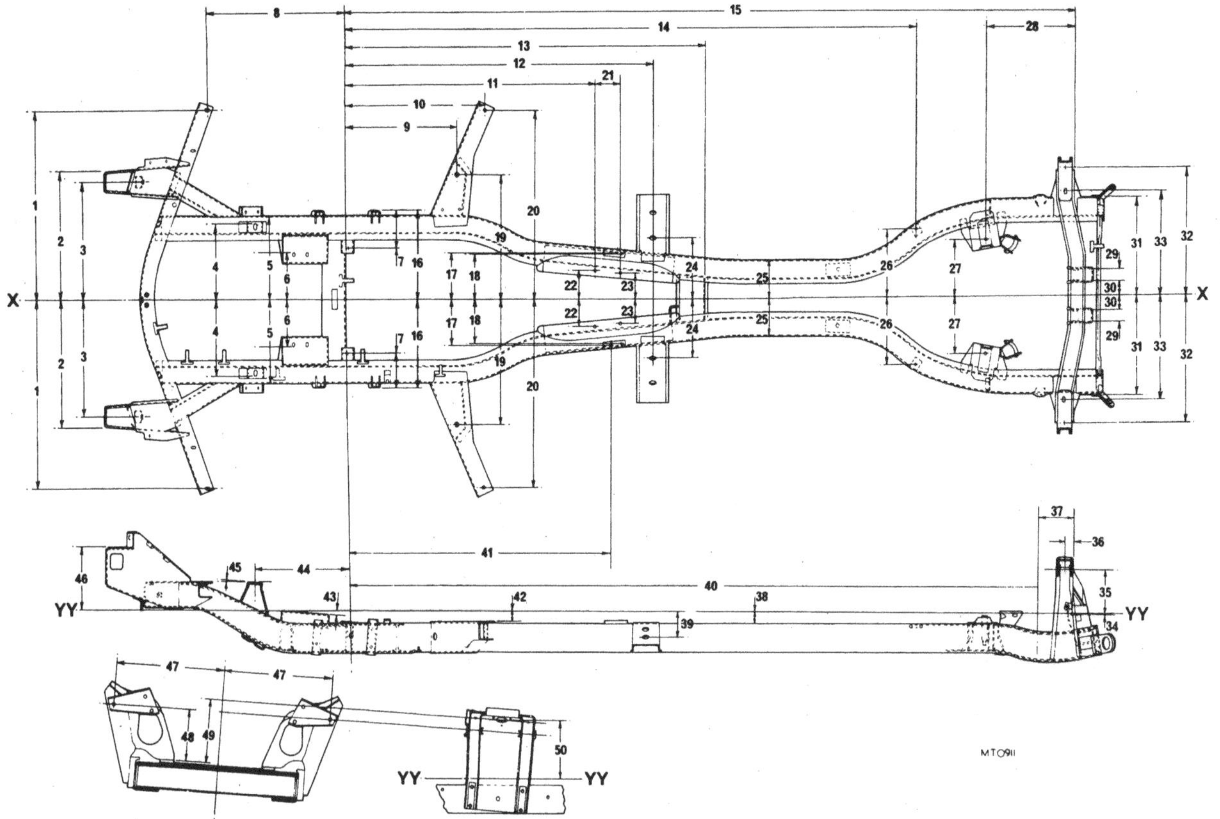

	Inches	Centimetres		Inches	Centimetres		Inches	Centimetres
1	22·15	59·69	19	14·51	36·83	38	1·13	2·87
	22·10	58·42		14·48	36·78	39	2·80	7·11
2	14·56	36·98	20	21·03	53·39	40	83·26	211·48
3	13·78	35·00		20·97	53·27	41	31·45	79·88
	13·72	34·85	21	3·02	7·67		31·39	77·73
4	8·64	21·89		2·98	7·56	42	1·06	2·69
5	9·78	24·84	22	3·23	8·20	43	4½°	4½°
	9·72	24·69		3·12	8·15	44	11·69	29·69
6	5·55	14·05	23	2·91	7·37		11·56	29·36
	5·50	13·97		2·89	7·24	45	4½°	4½°
7	4·32	10·97	24	7·19	18·26	46	5·32	12·76
	4·30	10·92		7·13	18·11		5·26	12·77
8	17·37	44·12	25	4·25	10·80	47	13·06	33·17
	17·25	43·91	26	7·83	19·86		12·94	32·87
9	12·78	32·36		7·80	19·81	48	5·91	15·01
	12·72	32·31	27	6·64	16·87	49	7·31	18·50
10	16·09	40·87		6·61	16·79	50	6·29	15·97
	16·03	40·72	28	10·70	27·18		6·17	15·67
11	29·62	75·13		10·65	27·05			
	29·50	74·77	29	1·51	3·83			
12	36·56	92·86	30	1·63	4·14			
	36·44	92·56		1·61	4·09			
13	42·19	107·16	31*	11·28	28·65			
	41·94	104·83		11·22	28·50			
14	68·22	173·28	32	14·75	37·46			
	68·10	172·97	33	12·00	30·48			
15	87·19	221·46	34	0·15	0·38			
	86·93	220·80	35	5·00	12·70			
16	10·32	26·21		4·88	12·40			
17	5·39	13·69	36	1·12	2·87			
	5·36	13·61		1·00	2·54	*MK. III		
18	5·23	13·28	37	4·06	10·31	31	11·78	29·92
	5·20	13·20					11·72	29·77

REAR COMPARTMENT TRIM PAD

– Remove and refit **76.13.20**

The trim pad is secured to the rear compartment by six screws and cup washers.

BONNET

– Remove and refit **76.16.01**

Removing

1. Disconnect the battery.
2. Disconnect the six headlamp leads.
3. Remove the nut, bolt, spring and plain washers securing the bonnet stay.
4. Lower the bonnet and remove the underriders 76.22.01.
5. Remove the two bolts, spring and plain washers.
6. Lift the bonnet away.

Refitting

7. Reverse instructions 1 to 6, ensuring that the distance pieces and thrust washers are correctly located when refitting bolts 5.
8. Adjust the bonnet if necessary 76.16.02

BONNET

– Adjust **76.16.02**

Adjust the bonnet mountings as follows to achieve a parallel gap of 3/16 in (5 mm) between the bonnet, scuttle and doors:

1. Remove the underriders 76.22.01
2. Slacken the two bolts and move the bonnet forwards or rearwards as required.
3. Retighten the two bolts.
4. Slacken the four screws and move the stop plates forward or rearwards to ensure correct location of the buffers.
5. Retighten the four screws.
6. Slacken the four bolts and raise or lower the bonnet as required.
7. Retighten the four bolts.
8. Slacken the two locknuts and screw the buffer in or out as required.
9. Retighten the two locknuts.
10. Slacken the four bolts and re-position the catch plates to ensure correct engagement of the bonnet catches.
11. Retighten the four bolts.
12. Refit the underriders 76.22.01.

[NOTE: On Mk. III models the bonnet hinges project ahead of the frame. To remove the bonnet, or to adjust the bonnet front to rear, either remove or loosen the pivot bolts. Front height adjustments are made by loosening the two bolts that hold each hinge assembly on the frame. In addition, front to rear adjustments made at the rear of the bonnet are accomplished by moving the buffer inside the bonnet rather than by moving the stop plates on the body — as shown for Mk. IV and 1500 models.]**

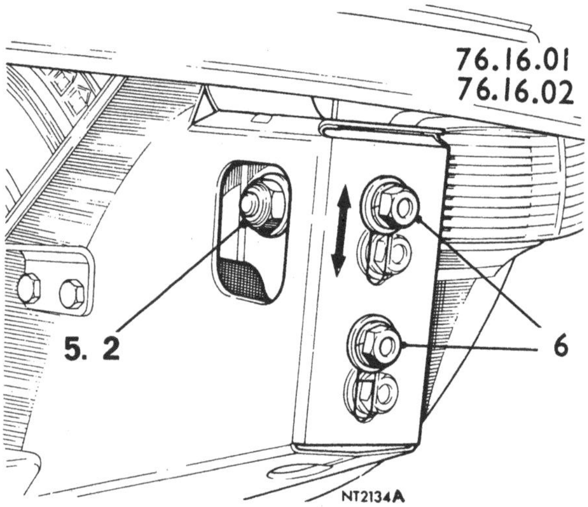

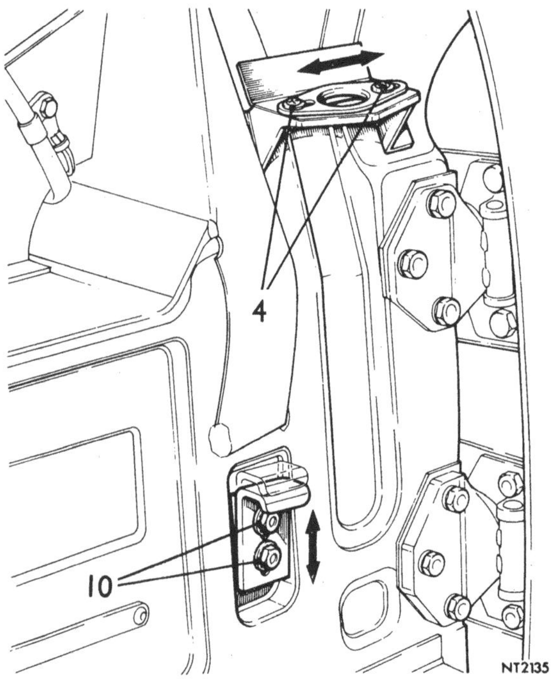

Triumph Spitfire Mk IV Manual. Part No. 545254. Issue 1.

BONNET STAY

– Remove and refit **76.16.14**

Removing

1. Raise and support the bonnet.
2. Remove the nut, bolt, spring washer and plain washers securing the stay to the wheel arch.
3. Remove the nut, bolt and plain washer securing the stay to the chassis. Remove the stay.

Refitting

4. Reverse instructions 1 to 3.

BONNET CATCH

– Remove and refit **76.16.34**

Removing

1. Remove the three nuts, spring washers, plain washers and screws.
2. Withdraw the bonnet catch.

Refitting

3. Reverse instructions 1 and 2.

LUGGAGE COMPARTMENT LID

– Remove and refit **76.19.01**

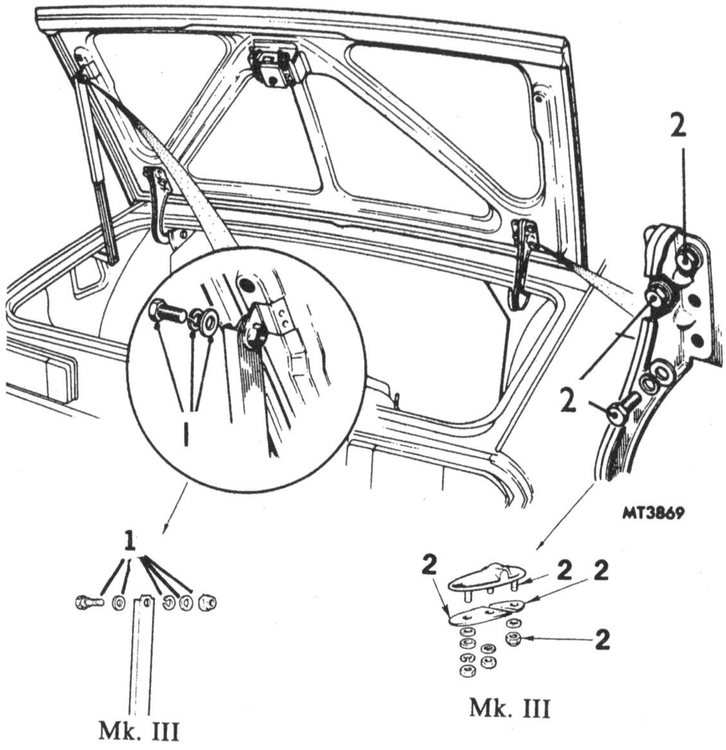

Removing

1. Remove the bolt, spring washer and plain washer securing the support stay to the lid.
2. Remove six bolts, spring washers and plain washers securing the hinges to the lid.

Refitting

3. Reverse instructions 1 and 2, ensuring correct alignment.

LUGGAGE COMPARTMENT LID SEAL

– Remove and refit **76.19.06**

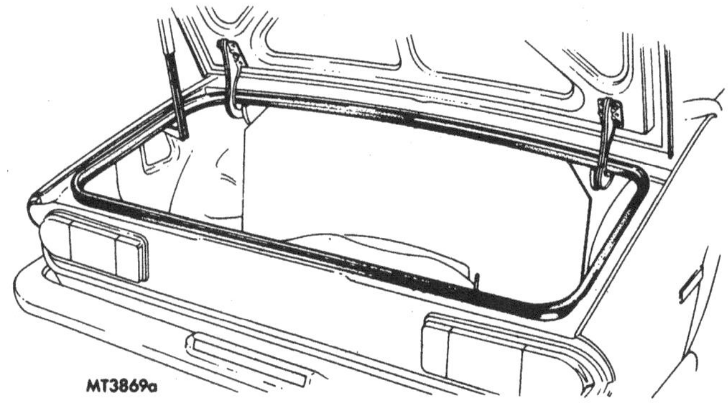

Removing

1. Free the seal from the body, using a suitable blunt tool if necessary.

Refitting

2. Fit the seal, using Seelastik SR51.

LUGGAGE COMPARTMENT LID HINGES

– Remove and refit **76.19.07**

Removing

1. Remove the lid 76.19.01.
2. Remove the four nuts, spring washers and plain washers and lift off the hinges.

Refitting

3 Reverse instructions 1 and 2, ensuring correct alignment of the lid.

LUGGAGE COMPARTMENT LID LOCK

– Remove and refit **76.19.11**

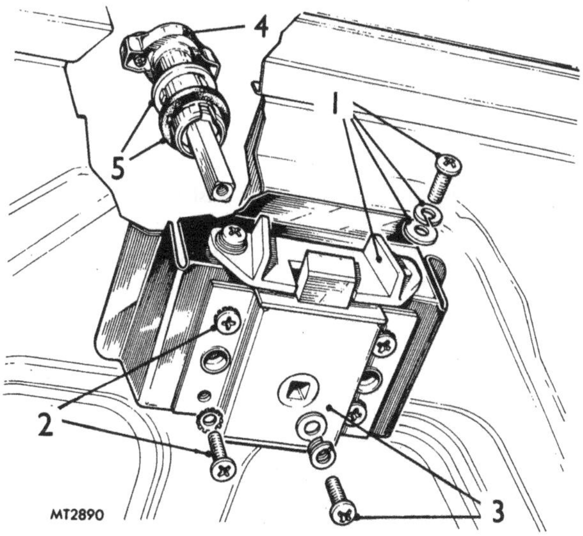

Removing

1. Remove the striker sleeve – two screws, spring washers and plain washers.
2. Remove the four screws and shakeproof washers.
3. Remove the screw, spring washer and plain washer and lift off the latch.
4. Withdraw the lock barrel and shaft.
5. Press out the escutcheon and seating washer.

Refitting

6. Reverse instructions 1 to 5, Check lid alignment and adjust striker sleeve if necessary.

LUGGAGE COMPARTMENT LID LOCK STRIKER

– Remove and refit **76.19.12**

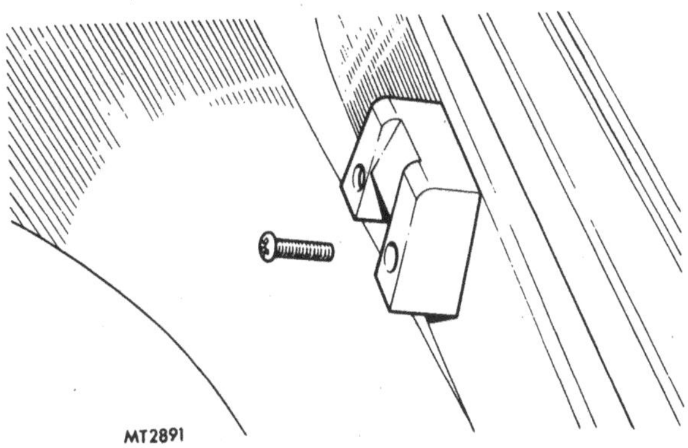

Removing

1. Remove the two screws and lift off the striker.

Refitting

2. Reverse instruction 1, ensuring correct lid closing action and alignment.

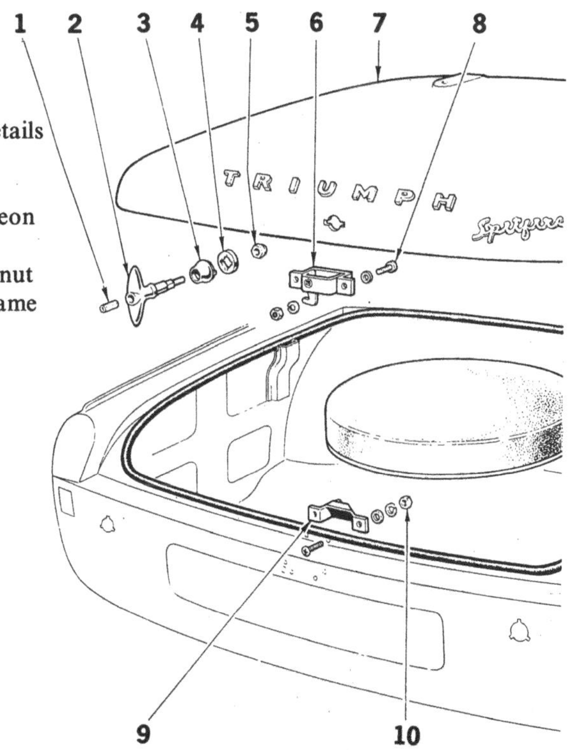

Mk. III lock details
1. Lock
2. Handle
3. Escutcheon
4. Seal
5. Special nut
6. Lock frame
7. Lid
8. Screw
9. Catch
10. Nut

UNDERRIDERS [Mk. IV and 1500]

– Remove and refit 76.22.01

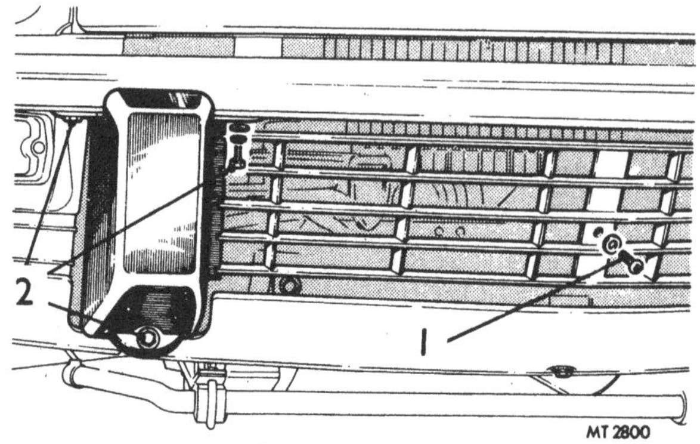

Removing

1. Remove one screw and plain washer.
2. Remove three bolts, spring washers and plain washers.
3. Lift off underrider together with front grille.
4. Repeat operation 2 to remove the second underrider.

Refitting

5. Reverse instructions 1 to 4.

BUMPER – FRONT [Mk. IV and 1500]

– Remove and refit 76.22.08

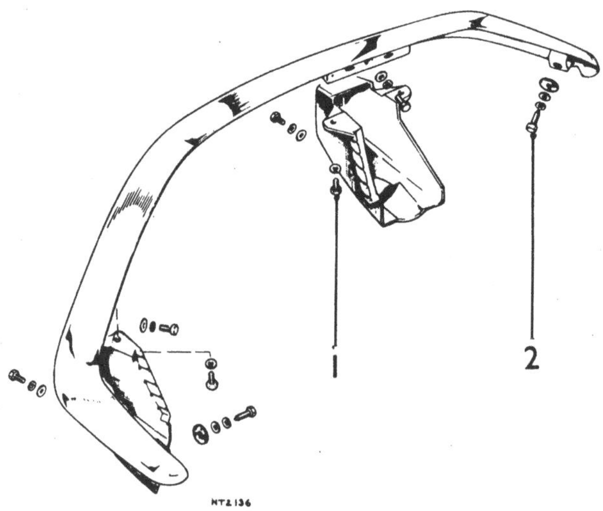

Removing

1. Remove the four bolts, spring washers and plain washers securing underriders to bumper.
2. Raise the bonnet and remove the four bolts, spring washers and plain washers.
3. Lower the bonnet and lift off the bumper.

Refitting

4. Reverse instructions 1 to 3.

BUMPER – REAR [Mk. IV and 1500]

– Remove and refit 76.22.15

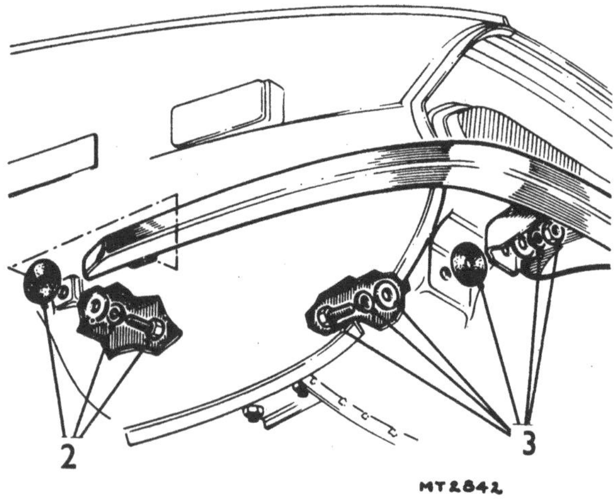

Removing

1. Disconnect the two number plate lamp leads. 86.40.86. Operation 1.
2. Remove the two bolts, spring washers and plain washers.
3. Remove the two nuts, bolts, spring washers and plain washers and lift off the bumper.

Refitting

4. Reverse instructions 1 to 3.

UNDERRIDERS (Mk. III)

—Remove and refit **[76.22.01**

Removing

1. Remove the front bumper. 76.22.08
2. Remove the bolts that hold the underriders to the bumper.
3. Remove the underriders and the P.V.C. mouldings.
4. If necessary, remove the rubber buffers from the underriders.

Refitting

5. Reverse instructions 1 to 4.

BUMPER — FRONT

—Remove and refit **[76.22.08]**

Removing

1. Remove the two bolts which hold the ends of the bumper to the body.
2. Remove the six bolts, 23 and 24 in the illustration given below, that hold the bumper on the brackets.
3. Lift the bumper clear, complete with underriders.

Refitting

4. Reverse instructions 1 to 3.

BUMPER — REAR

—Remove and refit **[77.22.15]**

Removing

1. Working inside the luggage compartment, remove the two bolts, 9 and 14 in the illustration given below, that hold each bumper on the body.
2. Remove the bumpers.
3. If necessary, remove the bolts that hold the spring bars and then remove the spring bars from the luggage compartment.

Refitting

4. Reverse instructions 1 to 3.

Bumper arrangement – Mk. III

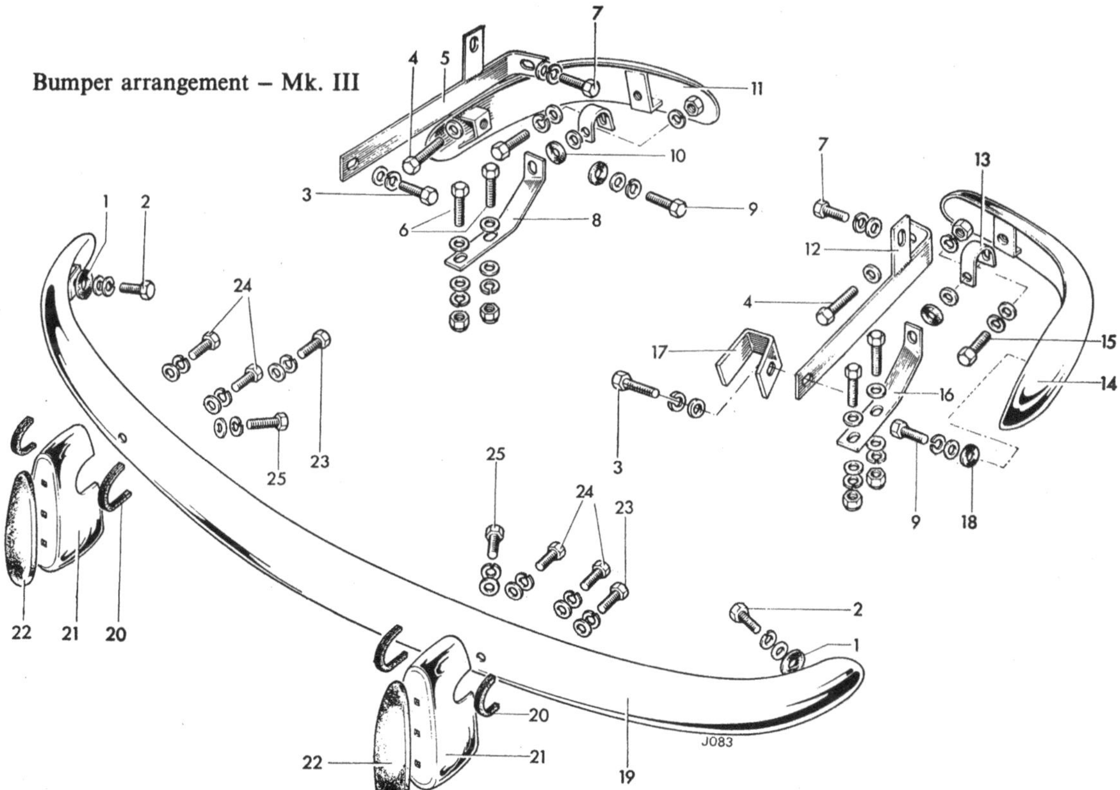

1. Rubber washer
2. Bolt — front bumpers to body
3. Bolt — support bracket
4. Bolt — spring bar
5. Spring bar — upper
6. Bolt — support bar
7. Bolt — spring bar
8. Rear bumper support
9. Bolt — body side
10. Rubber washers
11. Rear bumper — R.H.
12. Spring bar upper
13. Support extension — rear bumper
14. Rear bumper — L.H.
15. Bolt — support extension
16. Rear bumber support
17. Jack stowage bracket
18. Rubber washer
19. Front bumper
20. Moulding — P.V.C.
21. Overrider assembly
22. Rubber buffer — overrider
23. Bolt — lower overrider attachments
24. Bolt — front bumper to body
25. Bolt — overrider to front bumper

GEARBOX TUNNEL COVER

– Remove and refit 76.25.07

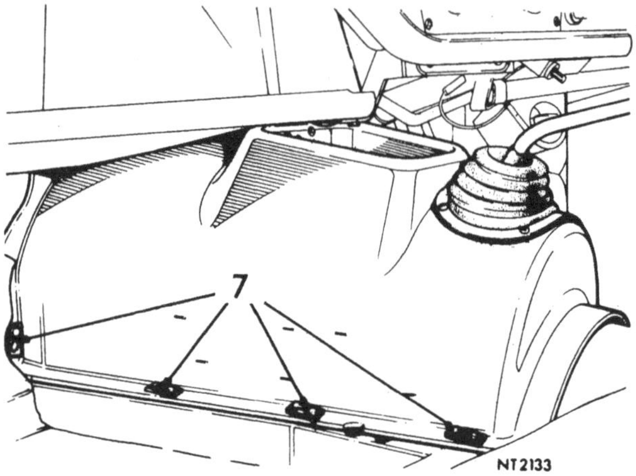

Removing

1. Remove gear lever knob.
2. Remove two screws holding trim pads to gearbox.
3. Remove four bolts holding fascia support bracket to floor.
4. Remove two screws holding fascia support bracket to fascia rail.
5. Remove fascia support bracket.
6. Remove carpet.
7. Remove eleven self tapping screws holding gearbox cover to floor.
8. Remove the gearbox cover.

Refitting

9. Reposition gearbox cover.
10. Refit eleven screws holding gearbox cover to floor.
11. Refit carpet.
12. Reposition fascia support bracket.
13. Refit two screws holding fascia support bracket to fascia rail.
14. Refit four bolts holding fascia support bracket to floor.
15. Refit two trim pads and tighten two self tapping screws.
16. Refit gear lever knob.

DOOR

– Remove and refit **76.28.01**

Removing

1. Isolate the battery.
2. Drill out the rivet securing the check strap.
3. Support the door and remove six bolts, spring washers and plain washers.

Refitting

4. Reverse instructions 1 to 3. Check door closing action and adjust if necessary.

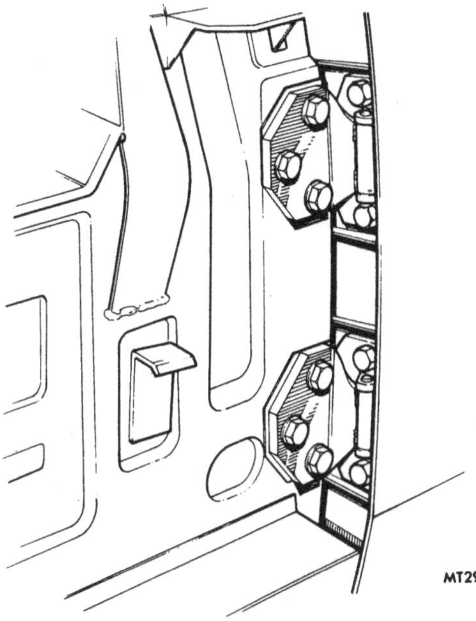

DOOR HINGES

– Remove and refit **76.28.42**

Removing

1. Remove the door 76.28.01.
2. Remove hinges – three bolts, spring washers and plain washers.

Refitting

3. Reverse instructions 1 and 2. Check the door closing action and adjust if necessary.

DOOR GLASS

– Remove and refit **76.31.01**

Removing

1. Remove the trim pad 76.34.01.
2. Remove the spring.
3. Remove the guide packing pieces – two bolts and spring washers.
4. Prise off the two stud retainers and washers securing the regulator arms to the glass frame.
5. Support the glass and detach the regulator arms from the frame.
6. Lower the glass.
7. Detach the inner and outer door waist seals from the clips.
8. Lift out the glass, taking care to avoid scratching it on the seal clips.

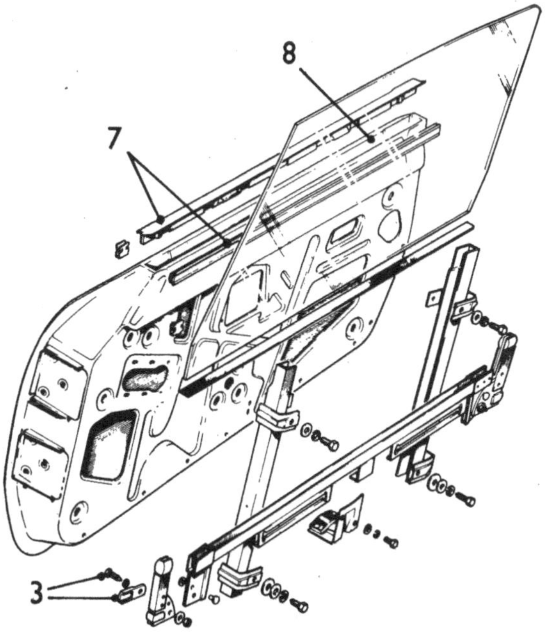

Refitting

9. Reverse instructions 1 to 8.

DOOR GLASS REGULATOR

– Remove and refit 76.31.45

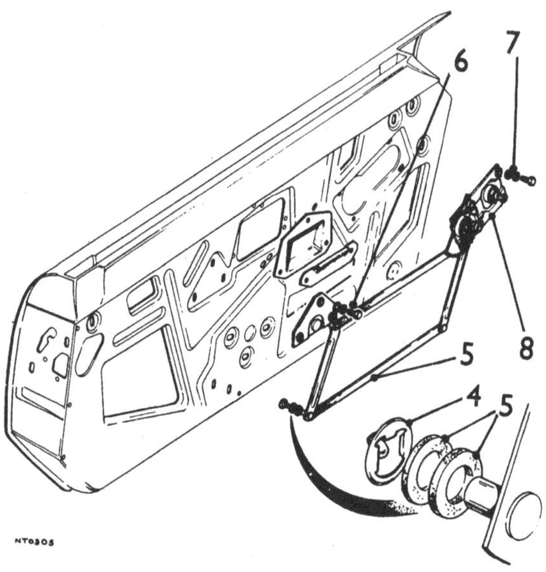

Removing

1. Remove the trim pad 76.34.01.
2. Remove the spring and loosely refit the regulator handle.
3. Raise the door glass and retain it with a small rubber wedge.
4. Detach the regulating arms by prising off the retainer studs.
5. Remove the inter-connecting link and four leather washers.
6. Remove the three bolts, spring washers and plain washers.
7. Remove the four bolts, spring washers and plain washers.
8. Carefully withdraw the regulator assembly.

Refitting

9. Reverse instructions 1 to 8.

DOOR TRIM PAD

– Remove and refit 76.34.01

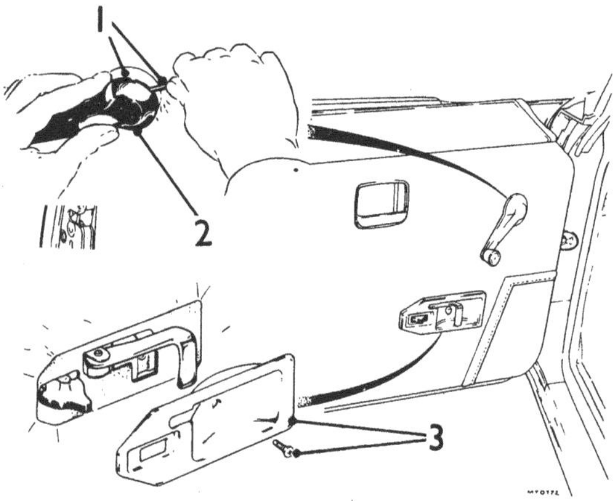

Removing

1. Depress the bezel and press out the pin.
2. Remove the handle and bezel.

[NOTE: The door handle on Mk. III cars is removed in the same manner as the window regulator handle.]

3. Remove the handle plate – 1 screw.
4. Prise off the trim pad – 21 clips.

Refitting

5. Reverse instructions 1 to 4.

DOOR LOCK

– Remove and refit 76.37.12

Removing

1. Remove the trim pad 76.34.01.
2. Release the two linkages.
3. Remove the three screws.
4. Release the linkage.
5. Remove the door lock.

Refitting

6. Reverse instructions 1 to 5.

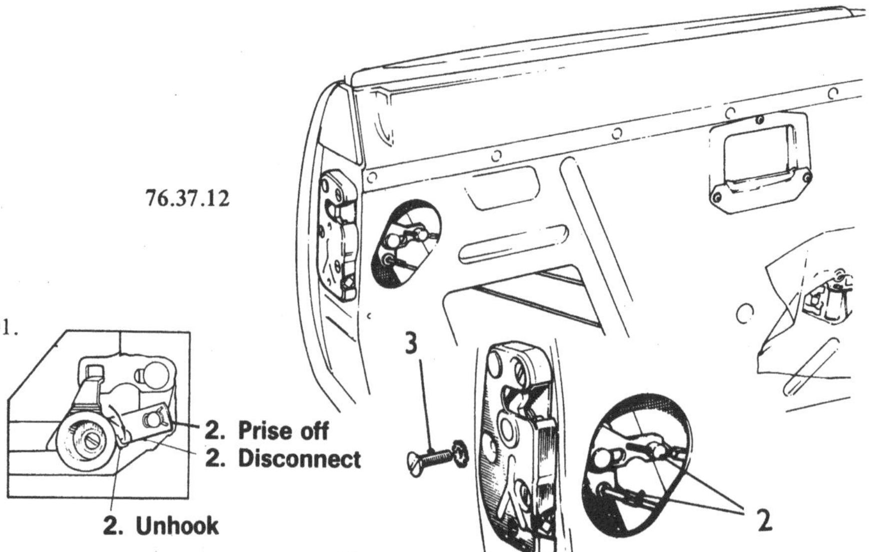

Mk. III

Mk. IV and 1500

Triumph Spitfire Mk IV Manual. Part No. 545254. Issue 1.
Part No. 511243. page 5.517. Issue 1.

DOOR LOCK STRIKER

– Remove and refit 76.37.23

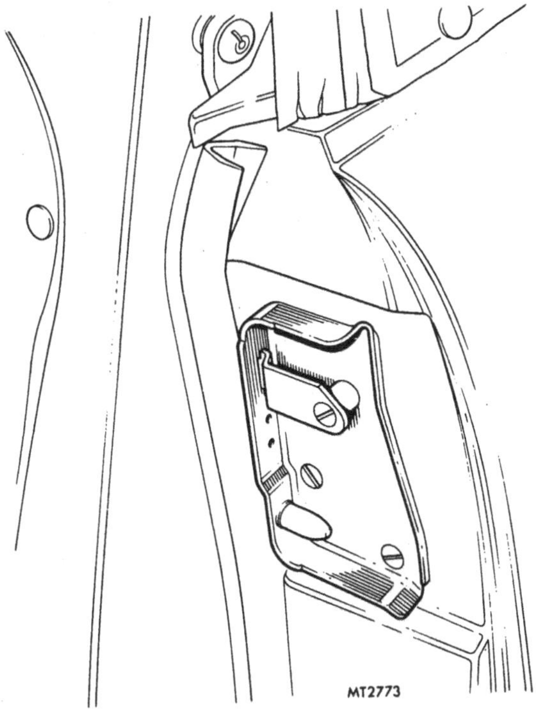

Removing

1. Remove the three screws and lift off the striker.

Refitting

2. Reverse instruction 1, adjusting if necessary to ensure correct door locking action.

DOOR LOCK REMOTE CONTROL

– Remove and refit 76.37.31

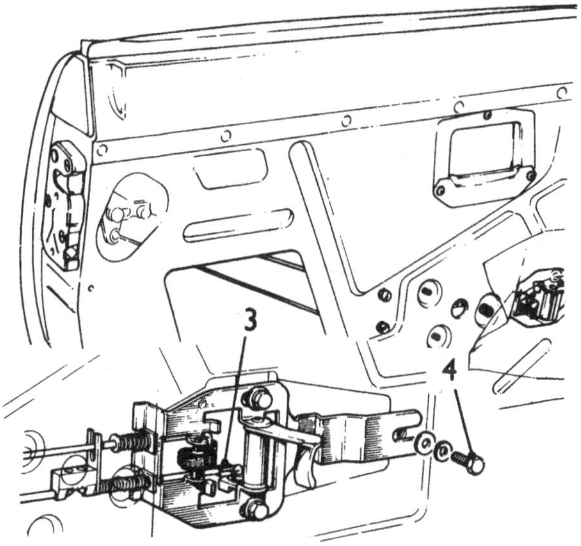

Removing

1. Remove the trim pad 76.34.01.
2. Remove the foam rubber pad.
3. Detach the clip from the control rod.
4. Remove the three bolts. spring washers and plain washers.
5. Disconnect the control rod from the locking lever [except Mk. III].
6. Remove the remote control.

Refitting

7. Reverse instructions 1 to 6, adjusting if necessary to ensure correct operation.

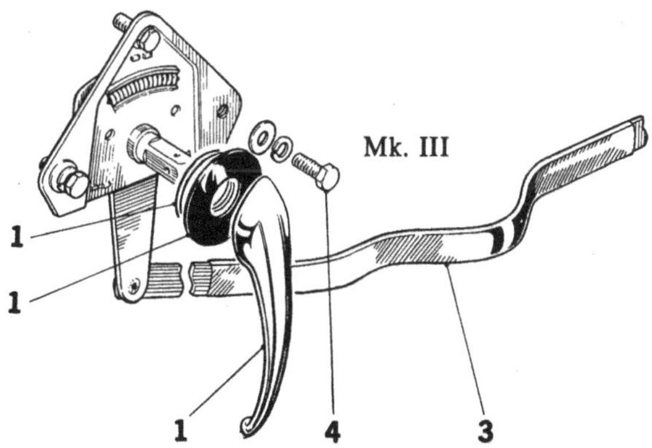

PRIVATE LOCK

– Remove and refit 76.37.39

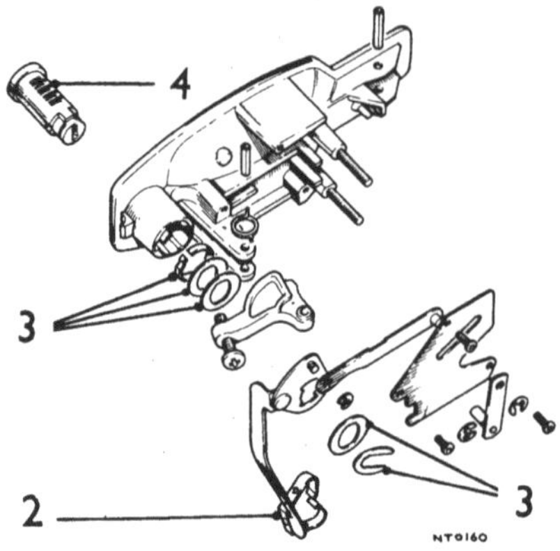

Removing

1. Remove the door handle 76.58.01.
2. Remove the spring clip.
3. Remove the plain washers and waved washer securing the pivot bracket.
4. Push out the lock.

Refitting

5. Reverse instructions 1 to 4.

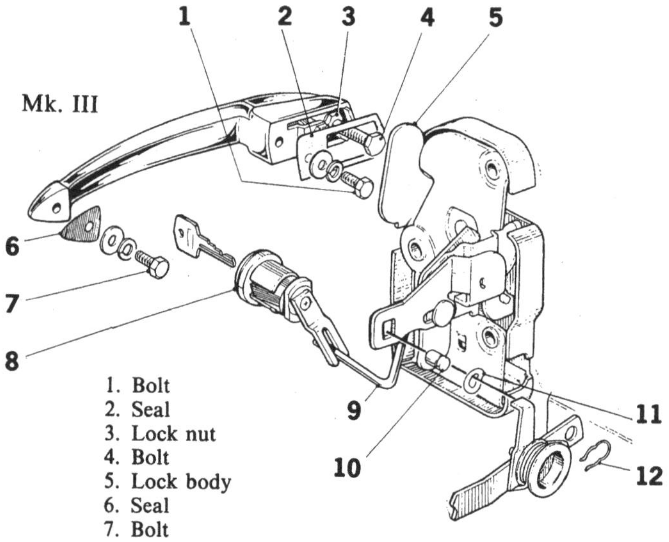

1. Bolt
2. Seal
3. Lock nut
4. Bolt
5. Lock body
6. Seal
7. Bolt
8. Lock cylinder
9. Operating link
10. Pin
11. Spring washer
12. Clip

DOOR CHECK STRAP

– Remove and refit 76.40.27

Removing

1. Isolate the battery.
2. Drill out the rivet.
3. Pull back the dash side carpet and withdraw the check strap.

Refitting

4. Reverse instructions 1 to 3.

FASCIA – DRIVERS SIDE [except Mk. III]

– Remove and refit 76.46.01

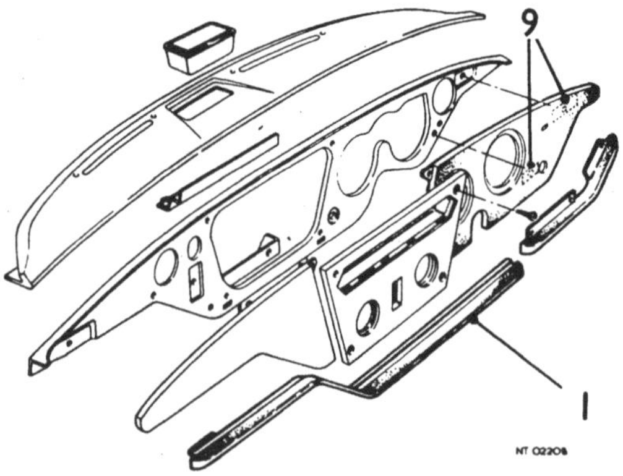

Removing

1. Remove the longer of the two lower padding assemblies – 4 screws and plain washers.
2. Remove the centre fascia 76.46.02.
3. Remove the tachometer 88.30.21.
4. Remove the speedometer 88.30.01.
5. Remove the wiper switch 86.65.38.
6. Remove the mixture control inner cable and bezel 19.20.13, instructions 1 to 3.
7. Remove the hazard switch (if fitted) 86.65.50.
8. Pull out the turn signal bulb holder.
9. Remove the two nuts, spring washers and plain washers from the rear of the fascia.

Refitting

10. Reverse instructions 1 to 9.

76.37.39
76.46.01

Triumph Spitfire Mk IV Manual. Part No. 545254. Issue 1.
Part No. 511243. page 5.517. Issue 1.

FASCIA – CENTRE [except Mk. III]

– Remove and refit 76.46.02

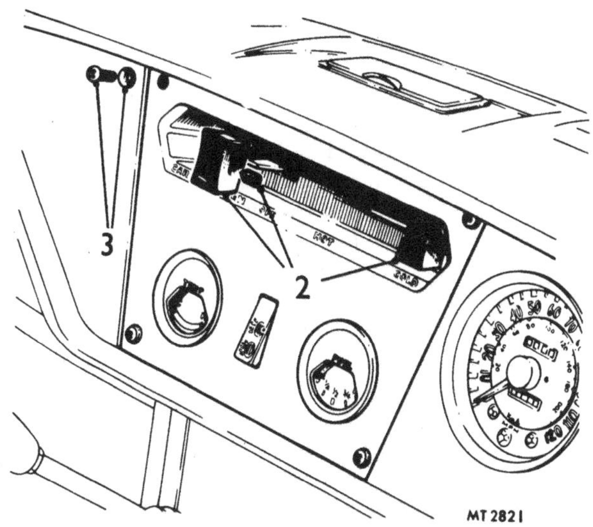

Removing

1. Isolate the battery.
2. Slacken the two grub screws and pull off the heater control knobs.
3. Remove the four screws and cup washers.
4. Lower the fascia centre to the service position.
5. Pull out the two panel light bulb holders.
6. Disconnect the seven Lucar connectors.
7. Remove the fascia centre from the vehicle.
8. Pull out the hazard warning bulb holder if fitted.
9. Pull out the brake warning bulb holder if fitted.

Refitting

10. Reverse instructions 1 to 9.

FASCIA – PASSENGERS SIDE [except Mk. III]

– Remove and refit 76.46.03

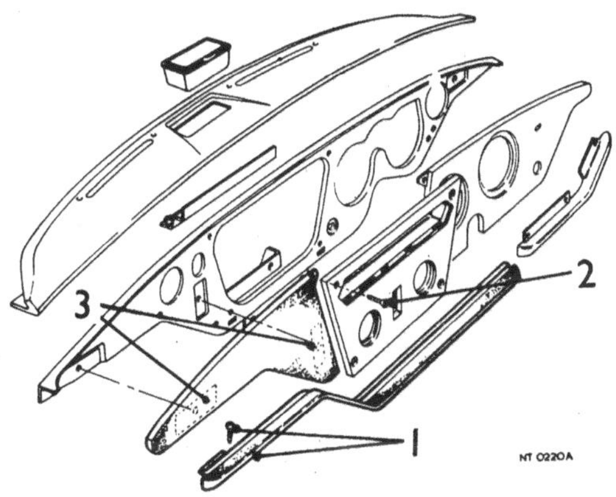

Removing

1. Remove the fascia padding – four screws and plain washers.
2. Remove one screw.
3. Remove the two nuts, spring washers and plain washers from the rear of the fascia.
4. Lift off the fascia.

Refitting

5. Reverse instructions 1 to 4.

FASCIA CRASH PAD [except Mk. III]

– Remove and refit 76.46.04

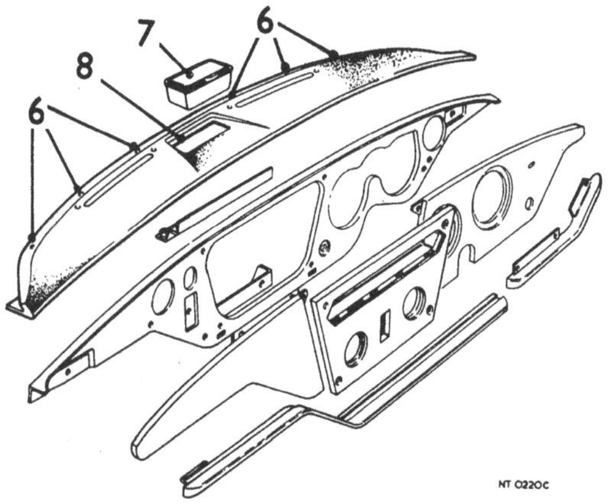

Removing

1. Remove the centre fascia 76.46.02.
2. Remove the driver's side fascia 76.46.01.
3. Remove the passenger's side fascia 76.46.03.
4. Remove the two bolts, spring washers, plain washers and spacers, securing the air flow control to the fascia bracket.
5. Remove the four nuts, plain washers and spring washers securing the demister vents to the body. Remove the demister vents.
6. Remove the six nuts, plain washers and spring washers securing the crash pad to the body.
7. Pull out the ash tray bowl.
8. Straighten the tags securing the ash tray retainer to the body. Lift out the retainer.
9. Carefully raise the crash pad to free the studs from the body. Lift out the crash pad.

Refitting

10. Reverse instructions 1 to 9.

FASCIA SUPPORT BRACKET

– Remove and refit 76.46.09

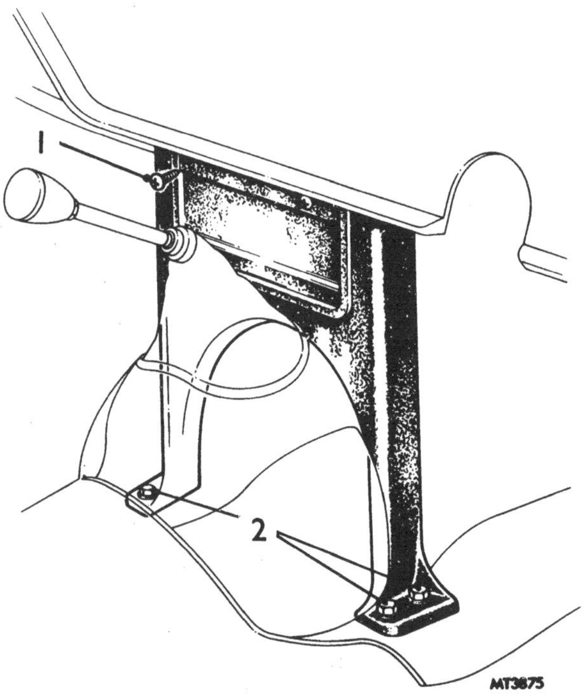

Removing

1. Remove two screws holding support bracket to fascia support rail.
2. Remove four bolts holding fascia support bracket to floor.
3. Remove fascia support bracket.

Refitting

4. Reposition fascia support bracket.
5. Refit four bolts holding fascia support bracket to floor.
6. Refit two screws holding support bracket to fascia support rail.

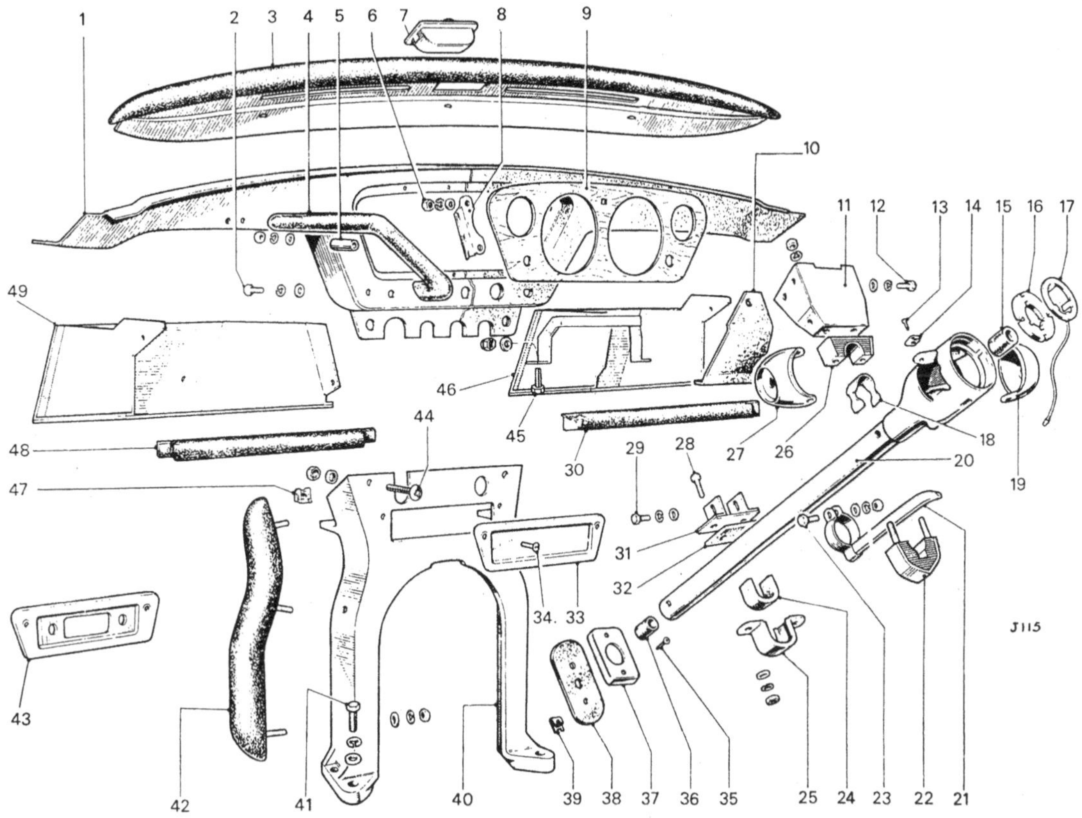

MK. III Fascia

1 Panel assembly
2 Bolt
3 Crash pad
4 Grab handle
5 Distance piece
6 Nut — veneer attachment
7 Ash tray
8 Veneer clamp bracket
9 Veneer instrument panel
10 Filler panel
11 Support bracket
12 Bolt
13 Screw
14 Fix nut
15 Steering column bush
16 Slip ring
17 Cable assembly
18 Clasp ring
19 Escutcheon
20 Steering column cowl
21 Harness cover
22 Upper column clamp
23 Bolt
24 Felt packing
25 Lower clamp
26 Clamp
27 Escutcheon
28 Bolt
29 Bolt
30 Trim roll
31 Support bracket
32 Felt packing
33 Cover plate
34 Screw — cover attachment
35 Screw
36 Bush — lower column
37 Retainer
38 Sealing rubber
39 Fix nut
40 Gearbox support bracket
41 Bolt
42 Trim roll
43 Cover plate
44 Screw
45 Bolt
46 Parcel tray
47 Fix nut
48 Trim roll
49 Parcel tray

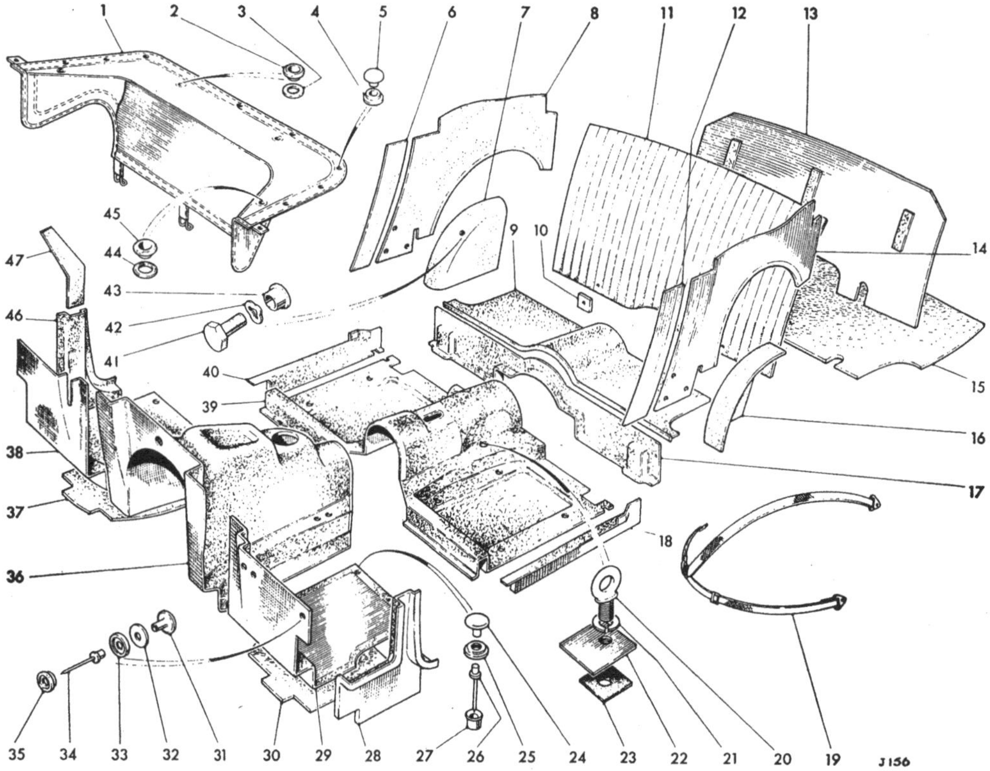

Carpet and interior trim (Mk. III)

1 Hood stowage cover
2 Sail eyelet
3 'B' ring
4 Socket
5 Button
6 Trim finisher 'B' post
7 Wheelarch cover
8 Rear quarter — trim board
9 Rear seat pan carpet
10 Foam pad — squab board
11 Rear compartment — squab board
12 Trim finisher 'B' post
13 Fuel tank — casing board
14 Rear quarter — trim board
15 Luggage floor mat
16 Wheelarch — carpet
17 Heelboard — carpet
18 Sill carpet — L.H.
19 Safety harness
20 Safety harness — eyebolt
21 Lock washer
22 Reinforcement plate — safety harness
23 Mounting pad — safety harness
24 Button
25 Socket
26 Imex rivet
27 Stud
28 Dash side carpet
29 Front floor carpet
30 Front floor felt
31 Button
32 Plain washer
33 Socket
34 Imex rivet
35 Stud
36 Gearbox cover carpet
37 Front floor felt
38 Front floor carpet
39 Rear floor carpet
40 Sill carpet
41 Harness — bolt
42 Waved washer
43 Spacer
44 'B' ring
45 Sail eyelet
46 Dash side carpet
47 'A' post finisher trim

Part No. 511243. page 5.534. Issue 1.

GEARBOX COVER CARPET

– Remove and refit **76.49.01**

Removing

1. Remove gearlever knob.
2. Remove two screws holding trim pads to gearbox cover.
3. Remove four bolts holding fascia support bracket to floor.
4. Remove two screws holding fascia support bracket to fascia rail.
5. Remove fascia support bracket.
6. Remove carpet.

Refitting

7. Reposition carpet over gearbox cover.
8. Reposition fascia support bracket.
9. Refit two screws holding fascia support bracket to fascia.
10. Refit four bolts holding fascia support bracket to floor.
11. Refit trim pads to gearbox cover.
12. Refit gearlever knob.

FLOOR CARPETS-FRONT

– Remove and refit **76.49.02**

Removing

1. Pull R.H. carpet clear of floor pedals.
2. Remove R.H. carpet.
3. Remove L.H. carpet.

Refitting

4. Reposition and fit L.H. carpet.
5. Reposition and fit R.H. carpet.

CARPET – REAR

– Remove and refit **76.49.03**

Removing

1. Remove the seats. 76.70.04/76.70.05.
2. Remove the lower seat belt fixings 76.73.02, operations 1 and 2.
3. Disconnect the four fasteners at the front and lift the carpet clear of the handbrake lever.

Refitting

4. Reverse instructions 1 to 3.

FRONT GRILLE

– Remove and refit **76.55.03**

[**NOTE:** To remove the Mk. III grille, you do not need to remove the underriders. Working inside the engine compartment, remove the eight screws which hold the grille to the bonnet. Collect the four spacers from the four upper screws. Close the bonnet and take the grille out upward from behind the front bumper. Installation is the reverse of removal.]

– Removing

1. Remove one underrider (either side) 76.22.01.
2. Remove one screw and plain washer.
3. Carefully disengage the grille from the second underrider.

Refitting

4. Reverse instructions 1 to 3.

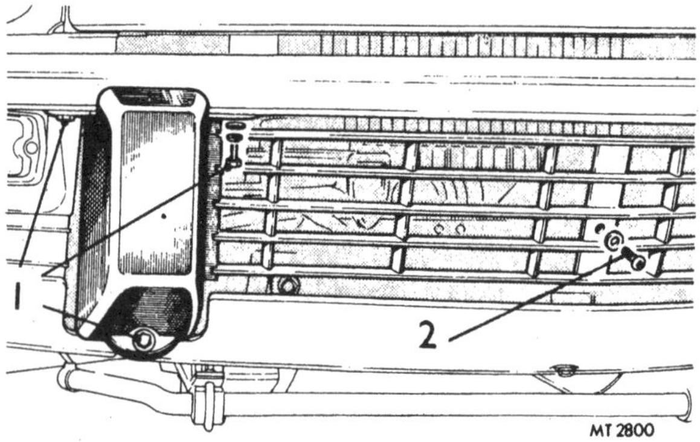

DOOR HANDLE

– Remove and refit **76.58.01**

[**NOTE:** See also the illustrations given in 76.37.39.]

Removing

1. Fully raise the glass.
2. Remove the trim pad. 76.34.01.
3. Remove the spring and the foam pad.
4. Remove the two nuts and plain washers and withdraw the clamping bracket.
5. Disconnect the door lock control rod from the spring clip and carefully withdraw the handle and the seating gasket.

Refitting

6. Reverse instructions 1 to 5.

DOOR PULL

– Remove and refit **76.58.25**

Removing

1. Remove the trim pad 76.34.01.
2. Remove the escutcheon – two screws and spring washers.
3. Remove the backing plate – one screw and spring washer.

Refitting

4. Reverse instructions 1 to 3.

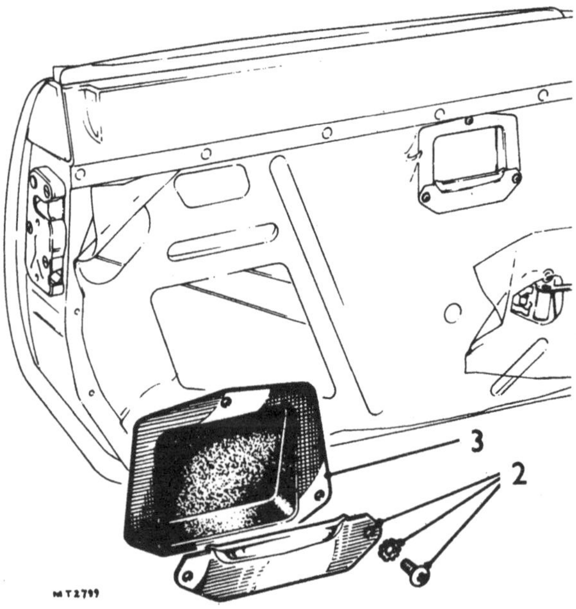

HARDTOP

Remove and refit **76.61.01**

Removing

1. Disconnect the three fasteners securing the hood stowage cover to the rear deck.
2. Remove the two domed bolts, spring washers and plain washers.
3. Remove the two domed bolts, spring washers and plain washers.
4. Remove the two bolts and plain washers.
5. With the aid of an assistant, carefully lift the hardtop from the car.

Refitting

6. Reverse instructions 1 to 5. Ensure that the distance tubes are located in the windscreen header rail.

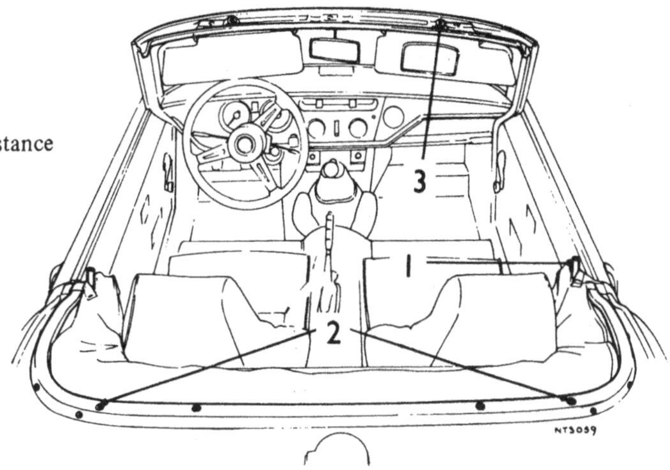

HOOD

– Remove and refit **76.61.08**

Removing

1. Disconnect the eight fasteners.
2. Remove the four bolts, spring washers and plain washers.
3. Remove the two caps, bolts, spring washers and plain
4. Remove the four bolts, spring washers and plain washers.
5. Lift off the hood.

Refitting

6. Reverse instructions 1 to 5.

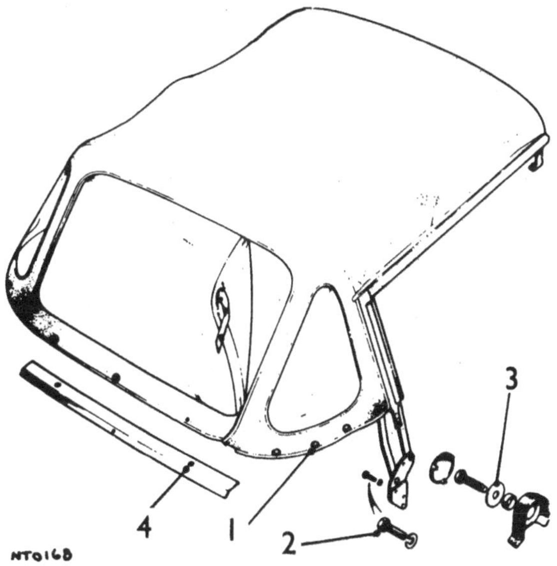

PARCEL SHELF

– Remove and refit – Drivers side **76.67.04**
–Passengers side **76.67.05**

Removing

1. Remove the two screws. [Four screws on Mk. III models.]
2. Remove the two nuts, bolts and plain washers. [Three bolts on the driver's side of Mk. III models. Also, the Mk. III speedometer trip cable must be disconnected.]
3. Lift out the parcel shelf.

Refitting

4. Reverse instructions 1 to 3.

ASHTRAY-FRONT

– Remove and refit **76.67.13**

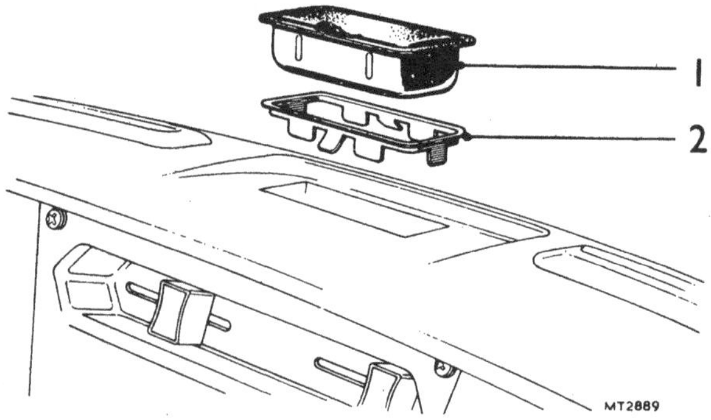

Removing

1. Lever ashtray out of mounting bracket.
2. Straighten legs on mounting bracket and pull bracket clear.

Refitting

3. Place bracket in ashtray aperture.
4. Bend bracket legs to secure mounting bracket in fascia.
5. Pushing securing spring to ashtray, push into mounting bracket.

SEAT CUSHION COVER

–Remove and refit **76.70.02**

Removing

1. Remove the seat(s) 76.70.04/76.70.05.
2. Remove the cushion cover – 32 hog rings.

Refitting

3. Reverse instructions 1 and 2.

SEAT SQUAB COVER

–Remove and refit **76.70.03**

Removing

1. Remove the seat(s) 76.70.04/76.70.05.
2. Remove the seat squab catch 76.70.27.
3. Remove the squab cover, 11 clips.

Refitting

4. Reverse instructions 1 to 3.

SEATS

– Remove and refit

– Drivers seat **76.70.04**

– Passengers seat **76.70.05**

Removing

1. Move the seat fully forward.
2. Remove the two bolts and spring washers.
3. Move the seat fully rearward.
4. Remove the two bolts and spring washers.
5. Lift out the seat complete with runners.

Refitting

6. Ensure that the packing washers are correctly positioned.
7. Reverse instructions 1 to 5.

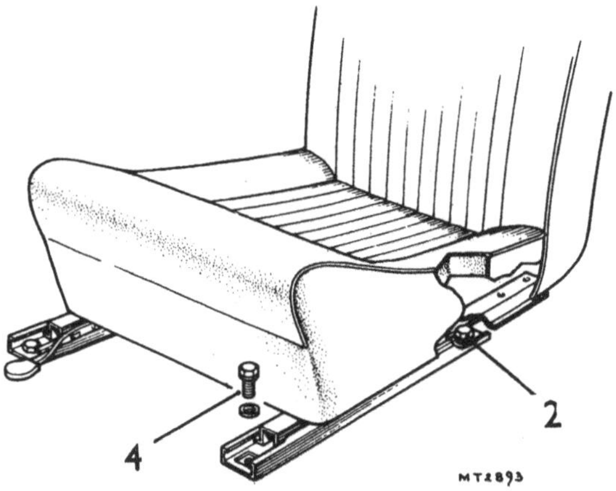

SEAT RUNNERS

Remove and refit **76.70.21**

Removing

1. Remove the seat 76.70.04/76.70.05.
2. Detach the runners – two nuts, bolts and four plain washers.

Refitting

3. Reverse instructions 1 and 2.

SEAT SQUAB CATCH

– Remove and refit 76.70.27

The seat squab catch is secured to the seat frame by two screws, spring washers and plan washers.

SEAT BELTS

– Remove and refit 76.73.02

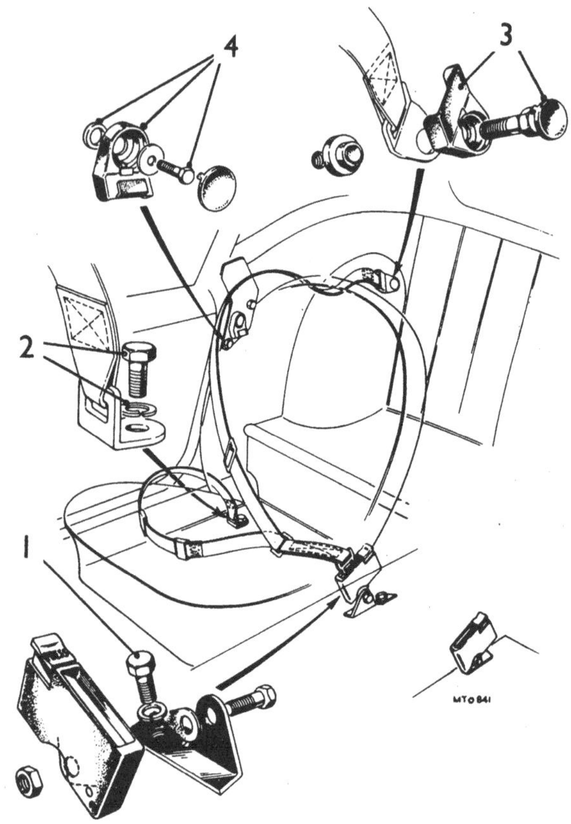

Removing

1. Remove one bolt and spring washer and lift off the tunnel buckle.
2. Remove the bolt and washer securing the anchor bracket to the floor.
3. Remove the cap, bolt, plain washer, cover and spacer securing the swivel bracket to the wheel arch.
4. Remove the cap, bolt, plain washer, stowage hook and spacer.

Refitting

5. Reverse instructions 1 to 4, ensuring that the swivel bracket is left free to swivel.

VALANCE – ENGINE

– Remove and refit – L.H.	76.79.07
– R.H. 2 and 3	76.79.08

Removing

1. Remove one nut, spring washer and plain washer securing the cable harness to the L.H. valance.
2. Remove the two nuts, bolts, spring washers and plain washers.
3. Remove the two nuts, bolts, spring washers and plain washers.

Refitting

4. Reverse instructions 1 to 3.

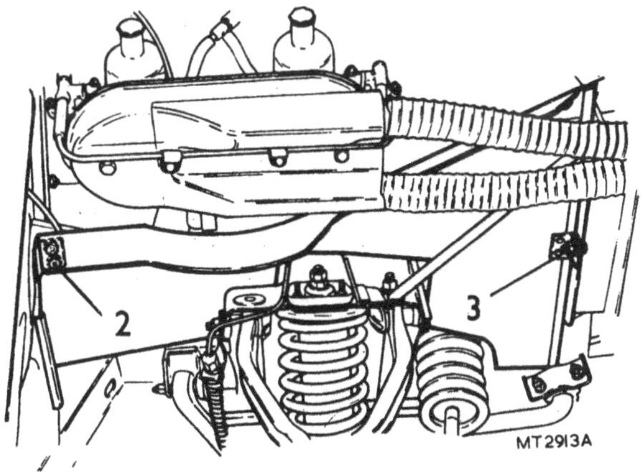

WINDSCREEN

– Remove and refit **76.81.01**

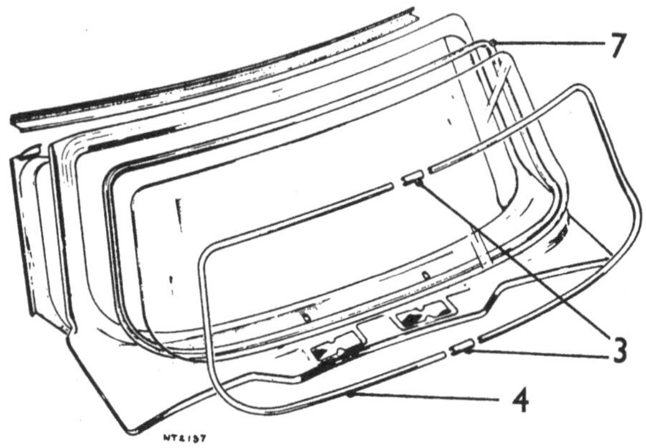

Removing

1. Remove the wiper arms 84.15.01.
2. Remove the interior mirror – two screws.
3. Push the finisher covers to one side to expose the joints, using a suitable piece of wood.
4. Carefully ease the finishers from the weatherstrip.
5. Break the seal, using a suitable blunt tool.
6. Push the glass outwards.

CAUTION: Take care to avoid scratching the glass.

7. Remove the weatherstrip.

Refitting

8. Clean all traces of old sealing compound from the screen aperture flange, using petrol or white spirit.
9. Apply Seelastik SR51 to the glass channel and fit the weatherstrip to the screen.
10. Insert a strong cord into the weatherstrip inner channel allowing the ends to protrude from the lower edge.
11. Have an assistant position the glass centrally in the aperture and maintain a steady pressure whilst the cord ends are pulled to locate the weatherstrip on the body flange.
12. Seal the outer channel of the weatherstrip to the body using Seelastik SR51.
13. Sparingly apply rubber grease or soap/water solution to the weatherstrip finisher recess.
14. Assemble the two finisher covers to the left hand finisher.
15. Fit the finishers using the special tool as follows:
    a. Locate the tool in the recess with open end uppermost and in a vertical position.
    b. Keeping the hook firmly engaged in the recess, draw the tool along to engage the inner flange of the weatherstrip with the finisher.
16. Push the cover finishers centrally over the joins.
17. Fit the interior mirror and wiper arms.

FINISHER STRIP INSERTION TOOL

To fabricate finisher strip insertion tool.

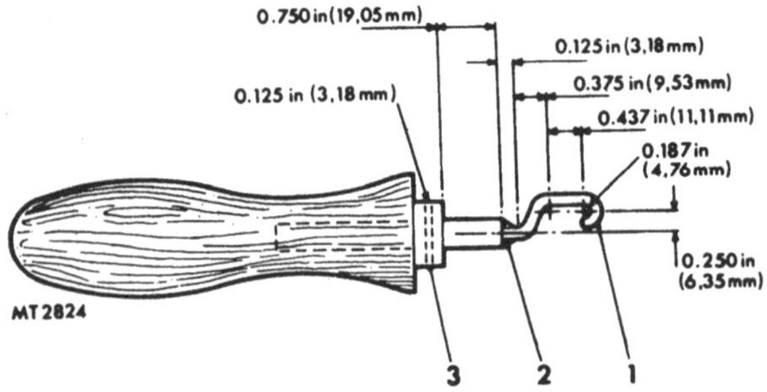

1. Shape a piece of mild steel rod as shown, paying particular attention to the form of the tip. Ensure that the tip is fully rounded and free from burrs.
2. Braze the hook into a metal stock.
3. Fit stock into a file handle and secure with a rivet

HEATING AND VENTILATION OPERATIONS

AIR FLOW CONTROL CABLE

– Remove and refit **80.10.06**

Removing

1. Remove the air flow control 80.10.09.
2. Slacken the trunnion bolt and clamp bolt securing the cable to the heater unit. Remove the cable.

Refitting

3. Reverse instructions 1 and 2.

WATER VALVE CONTROL CABLE

– Remove and refit **80.10.07**

Removing

1. Remove the water valve control 80.10.10.
2. Slacken the trunnion bolt and clamp bolt securing the cable to the water valve.

Refitting

3. Reverse instructions 1 and 2.

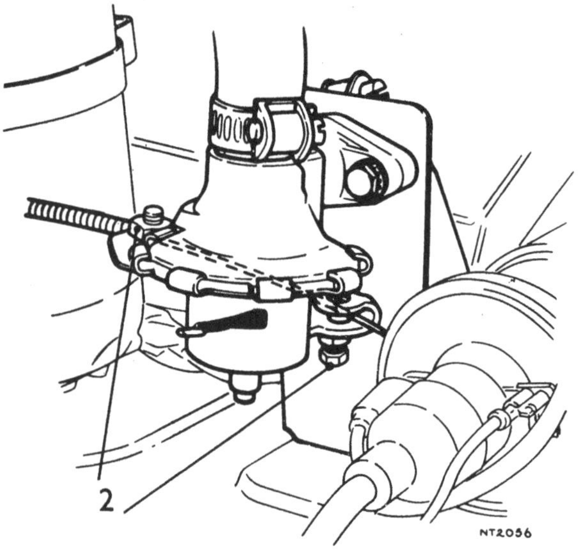

AIR FLOW CONTROL

– Remove and refit **80.10.09**

Removing

1. Lower the centre fascia to the service position 76.46.02. Operations 1 to 4.
2. Remove the two bolts, spring washers, plain washers and spacers securing the control to the fascia bracket.
3. Pull out the snap connector and disconnect the Lucar connector from the fan switch.
4. Slacken the trunnion screw and the cable clamp bolt. Detach the cable from the control.

Refitting

5. Ensure that the control cable is positioned to allow full operation of the flap valve.
6. Reverse instructions 1 to 4.

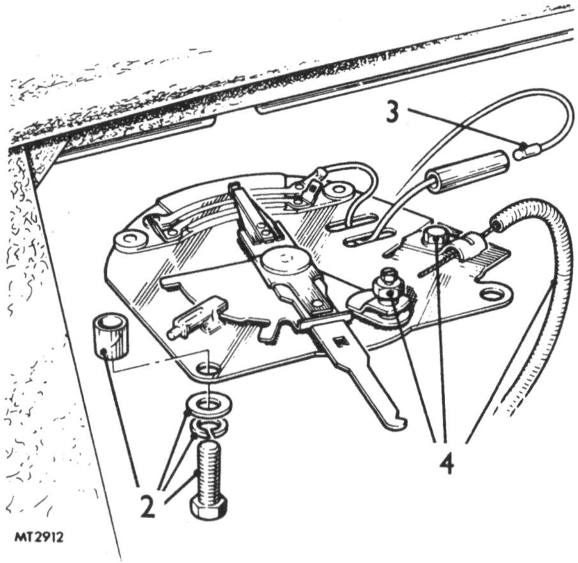

Triumph Spitfire Mk IV Manual. Part No. 545254. Issue 1.

WATER VALVE CONTROL

– Remove and refit **80.10.10**

Removing

1. Lower the centre fascia to the service position 76.46.02. Operations 1 to 4.
2. Remove the two bolts, spring washers, plain washers, and the thick and thin spacers securing the control to the fascia bracket.
3. Slacken the trunnion screw and the cable clamp bolt. Detach the cable from the control.

Refitting

4. Ensure that the control cable is positioned to allow full operation of the water valve.
5. Reverse instructions 1 to 3.

WATER VALVE

– Remove and refit **80.10.16**

Removing

1. Drain the cooling system.
2. Slacken the clips and disconnect the two hoses.
3. Slacken the trunnion screw and cable clamp bolt. Detach the control cable.
4. Remove the two nuts, spring washers and plain washers securing the water valve to the bracket.

Refitting

5. Reverse instructions 1 to 4 ensuring that the control cable is positioned to allow full operation of the water valve.

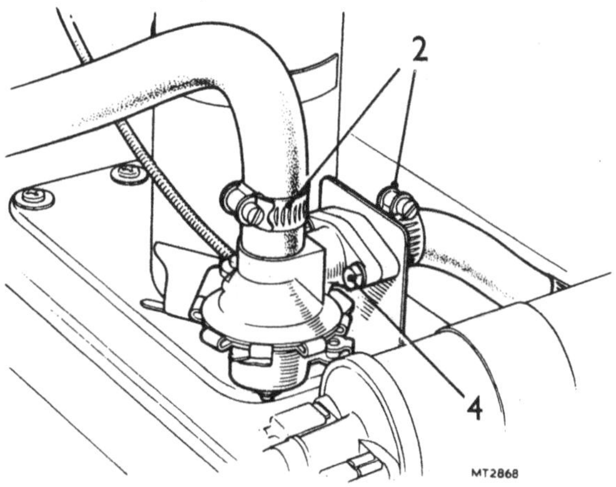

HEATER UNIT

– Remove and refit **80.20.01**

Removing

1. Isolate the battery.
2. Drain the cooling system.
3. Slacken the two clips and disconnect the water hoses. Place one hose in a container and blow down the other to clear the coolant from the heater matrix. Remove the hoses.
4. Remove the passengers side parcel shelf 76.67.05.

[NOTE: On Mk. III models, remove the clamping bracket for the heater and choke cables and disconnect both demister hoses. Then disconnect the tachometer and speedometer drive cables from the backs of the instruments and pull the cables out of the heater box and into the engine compartment. Be careful not to damage the rubber grommets for the cables.]

5. Lower the centre fascia to the service position. 76.46.02. operations 1 to 4.
6. Remove the air flow control. 80.10.09.
7. Remove the air flow control cable. 80.10.06.
8. Remove the passengers side demister hose.
9. Slacken the clip and disconnect the drivers side demister hose from the heater unit.
10. Push the screen washer tubing aside.
11. Remove the four setscrews, lock washers and 'D' washers.
12. Pull the heater unit clear and remove it from the vehicle at the passengers side.

Refitting

13. Apply Seelastik S.R.51 to all mating surfaces of the heater pipe and fan aperture seals. Ensure that seals are correctly positioned.
14. Install the heater unit and replace the four set screws, lock washers and 'D' washers, with the aid of an assistant.
15. Reverse instructions 1 to 10.

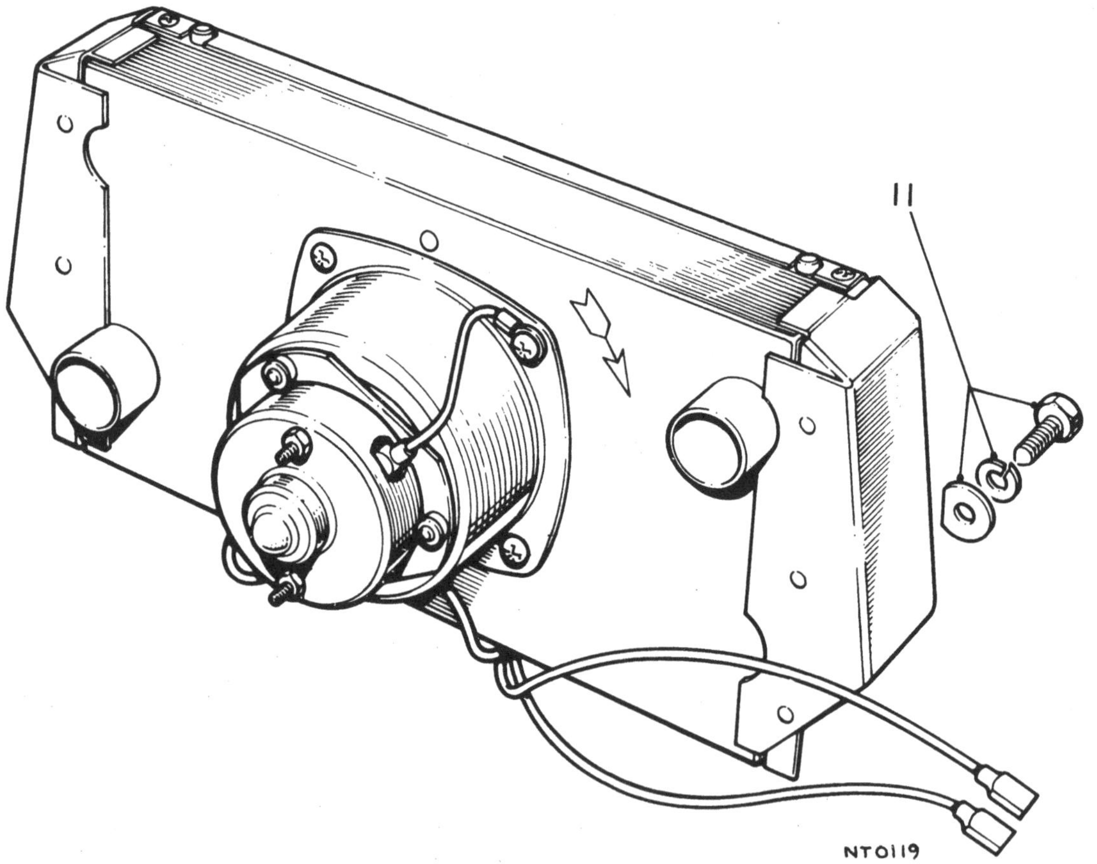

FAN MOTOR

– Remove and refit **80.20.15**

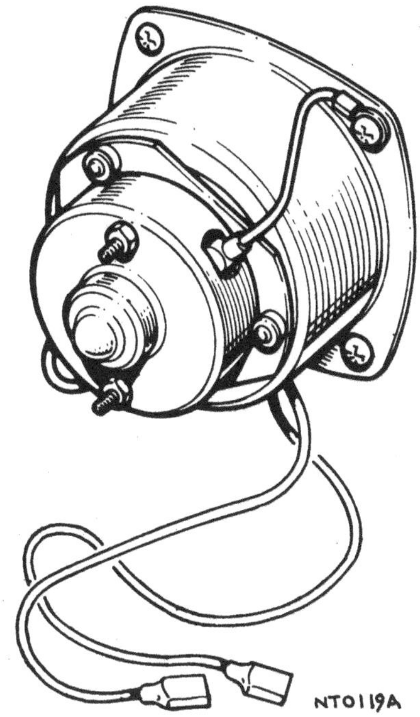
NT0119A

Removing

1. Lower the centre fascia to the service position 76.46.02. operations 1 to 4.

[NOTE: On Mk. III models, you must remove the entire heater unit in order to remove the motor. See 80.20.01. Remove the brass nut on the motor shaft and take off the impeller before you dismount the motor from the heater unit.]

2. Disconnect the Lucar connector and the snap connector.
3. Remove the four screws.
4. Manoeuvre the fan motor clear of the heater unit and remove it from the vehicle on the drivers side.
5. Detach the impellor from the shaft, using a screwdriver to release the retaining clip.

Refitting

6. Refit the impellor, ensuring that ¼ in (6·5 mm) clearance exists between the base of the impellor and the motor.
7. Reverse instructions 1 to 4.

WATER HOSES

– Remove and refit

Hose – cylinder head to water valve (except U.S.A. and Sweden) **80.25.08**

– manifold to water valve (U.S.A. and Sweden only) **80.25.08**

– heater to water pipe **80.25.07**

– water valve to heater **80.25.09**

The heater water hoses are each secured by two clips. Drain the cooling system before removal.

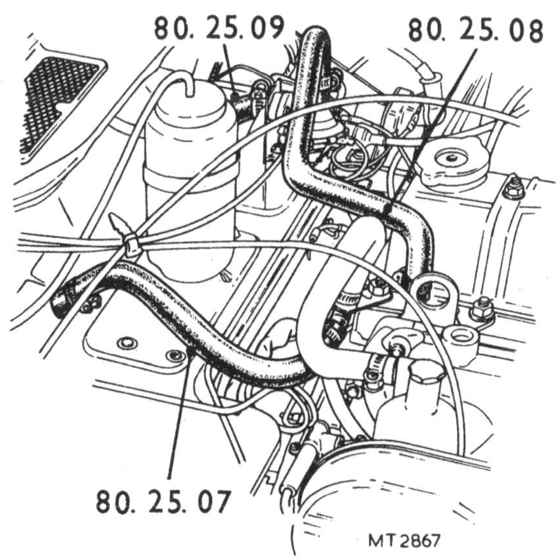

MT2867

WINDSCREEN WIPERS AND WASHERS OPERATIONS

WINDSCREEN WASHER RESERVOIR

– Remove and refit **84.10.01**

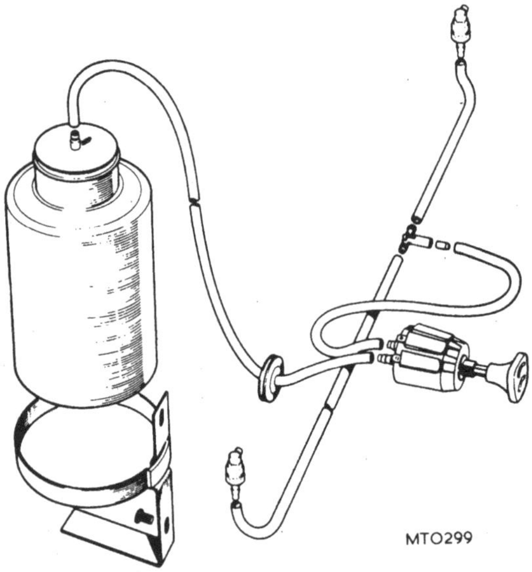

Removing

1. Unscrew the top and pull off the pipe.
2. Manoeuvre the bottle upwards from the carrier.

Refitting

3. Reverse instructions 1 and 2.

WINDSCREEN WASHER JET

– Remove and refit **84.10.09**

Removing

1. Pull the tubing from the jet.
2. Remove the nut and washer.
3. Withdraw the jet.

Refitting

4. Reverse instructions 1 to 3.

WINDSCREEN WASHER PUMP

– Remove and refit **84.10.21**

The manual washer pump and electrical wiper switch is a single integral unit. For removing and refitting refer to 86.65.38.

WINDSCREEEN WIPER SYSTEM [Mk. IV and 1500]

– Data and description **84.15.00**

Motor:

Manufacturer	Lucas
Type	14W
Lucas part No: motor minus gear assembly	75664
gear assembly – normal	54702581
– U.S.A.	54702582
Stanpart No: motor minus gear assembly	517621
gear assembly	520101
Running current – after 60 seconds from cold with connecting rod removed:	
Normal speed	1.5 amp
High speed	2·0 amp
Running speed – final gear after 60 seconds from cold with connecting rod removed:	
Normal speed	46 to 52 rev/min
High speed	60 to 70 rev/min
Armature end float	0·002 to 0·008 in (0·05 to 0·20 mm)
Brush length – new	0·250 in (6 mm)
renew if less than	0·125 in (3 mm) For high speed brush this is when narrow section is worn into full width section.
Brush spring pressure – when compressed so brush bottom is aligned with brushbox slot end	5 to 7 Oz (140 to 200 g)
Maximum permissible force to move cable rack in tubing – arms and blades removed	6 lb (3 kg)

The unit consists of a two speed permanent-magnet motor and a gearbox unit which drives a cable rack mechanism. Rotation of the motor armature is converted to a reciprocating motion of the cable rack by a single stage worm and gear, a connecting rod and a cross-head contained in a guide channel.

Two speed operation is provided by a third brush. When high speed is selected the positive supply is transferred from the normal speed brush to the high speed brush.

A switching feature stops the blades in the park position irrespective of their position when the fascia switch is selected off. This is effected by a two stage limit switch unit attached to the gearbox. The contacts are actuated by a cam on the final gear.

When the fascia switch is selected off, the motor will continue to run until the limit switch first stage contacts open. A momentary period follows during which no contact is made. The second stage contacts then close causing regenerative braking of the armature which maintains constant parking of the blades.

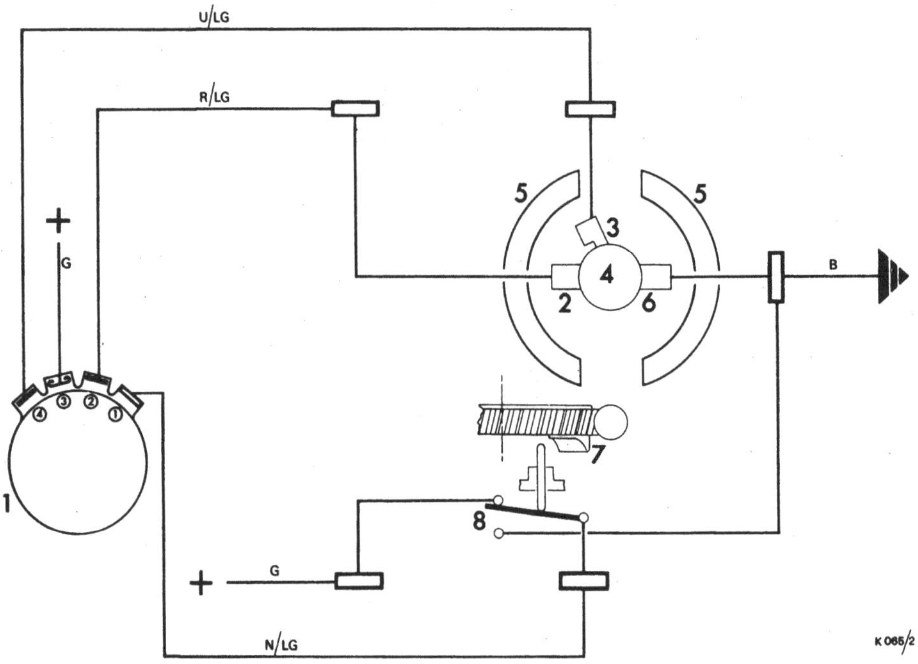

WINDSCREEEN WIPER SYSTEM [Mk. IV and 1500]

Supply

1. Facia switch

OFF	1 to 2
NORMAL	3 to 2
HIGH	3 to 4

2. Normal speed brush
3. High speed brush
4. Commutator
5. Permanent magnet
6. Earth brush
7. Final gear cam
8. Limit switch unit

WINDSCREEN WIPER SYSTEM (Mk. III)

—Data **[84.15.00]**

MOTOR:

Lucas Model DR.3A	Shunt wound single speed.
Light running speed	44 to 48 cycles per minute of wiper blades.
Stall current	13—15 amps.
Light running currents	2·7—3·4 amps. (Measured less cable and rack).
Resistance of field winding at 20°C. (68°F.) ..	8·0—9·5 ohms.
Resistance of armature winding at 20°C. (68°F.) ..	0·29—0·352 ohms. (Measured between adjacent commutation segments).
Brush tension	125—140 grammes.
Maximum permissible force to move rack in protective tubing with wiper motor disconnected and wiper arms removed	6 lbs. (2·7 kgs.).

WINDSCREEN WIPER ARM

– Remove and refit **84.15.01**

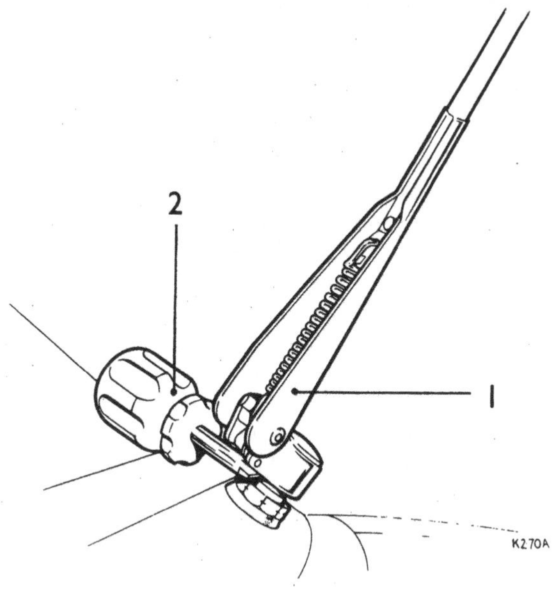

Removing

1. Lift the wiper arm and blade from the screen so that it falls into its service position.
2. Position a screwdriver as shown and impart a twisting action to lift the clip from the spindle groove.
3. The assembly may now be removed by hand.

Refitting

4. Ensure that the spindles are in the 'park' position.
5. Hinge the wiper arm against the spring to adopt its service position.
6. Locate the splines for a suitable 'park' position. Push on to engage the clip to the spindle groove.
7. Lower the wiper arm to the screen.

WINDSCREEN WIPER BLADE

– Remove and refit **84.15.05**

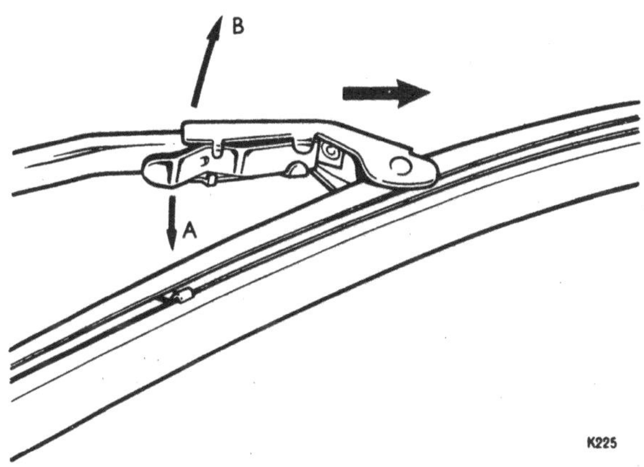

Removing

1. Lift the wiper arm and blade from the screen so that it falls into its service position.
2. Simultaneously lift the clip 'A', tilt cage 'B' and gently pull the wiper blade from the arm.

Refitting

3. Locate the cage and clip assembly to the wiper arm. Push on to engage 'pip'.
4. Lower the wiper arm to the screen.

WIPER MOTOR [except Mk. III]

– Remove and refit 84.15.12

[NOTE: Information on Mk. III models is included on p. 383.]

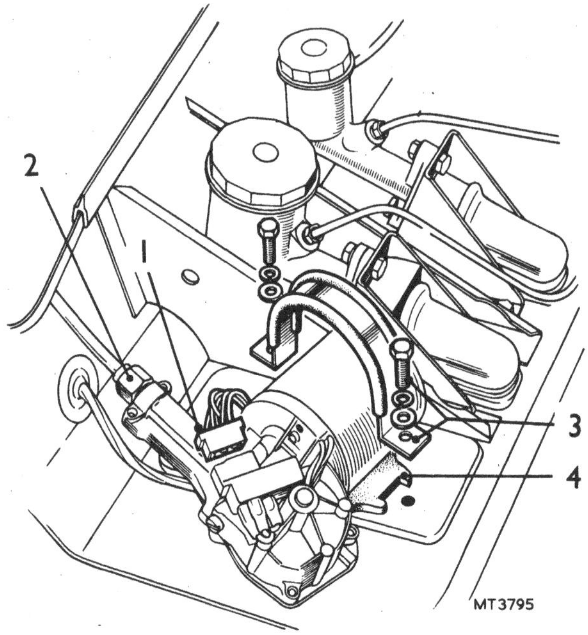

Removing

1. Remove the harness plug.
2. Unscrew the nut securing the tubing to the gear assembly.
3. Remove the strap and sleeve and detach the earth lead. – two bolts, spring washers and plain washers.
4. Lift off the rubber mounting pad.
5. Remove the gearbox cover – four screws.
6. Remove the crank-pin spring clip by withdrawing sideways. Remove the washer.
7. Carefully withdraw the connecting rod and remove the washer.
8. Detach the rack assembly and remove the motor and gear assembly.

Refitting

9. Lubricate the crankpin with Shell Turbo 41 oil.
10. Lubricate the cross-head end of the connecting rod, including pin with Ragosine Listate grease.
11. Reverse instructions 1 to 8.

WINDSCREEN WIPER SYSTEM [except Mk. III]

– Motor overhaul **84.15.18**

[NOTE: Information on Mk. III models is included on p. 383.]

General

If the motor is not operating correctly, first check electrical supply of 12 volts on terminal 5 with normal speed selected and terminal 3 with high speed selected.

If electrical supply is satisfactory, perform the following operations to determine if the fault is in the motor or in the rack, tube and wheelbox assembly, resulting in the motor being required to drive an excessive load.

Running current

1. Remove four screws. Lift off gearbox cover.
2. Remove the crankpin spring clip by withdrawing sideways. Remove washer.
3. Carefully withdraw connecting rod. Remove washer.
4. Connect ammeter suitable for running current (see Data) in supply circuit.
5. Allow motor to run for 60 seconds. Ammeter reading should then be as given in Data for normal speed and high speed respectively.

If the reading is not as stated, a fault in the motor is indicated.

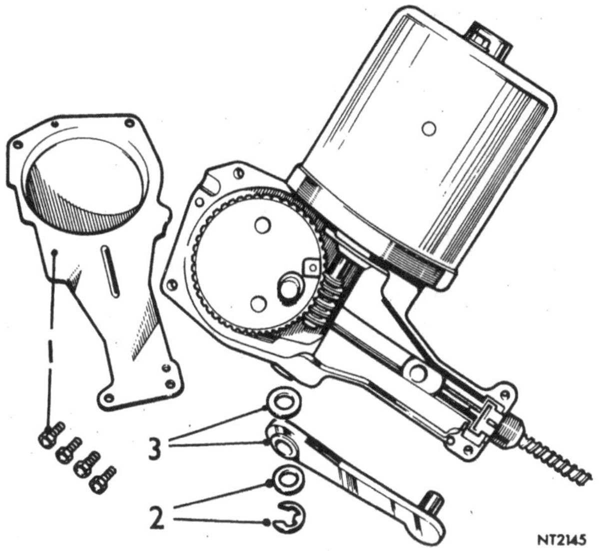

Running speed

1. Remove four screws. Lift off the gearbox cover.
2. Remove crankpin spring clip by withdrawing sideways. Remove washer.
3. Carefully withdraw connecting rod. Remove washer.
4. Allow motor to run for 60 seconds. Speed of final gear should then be as given in Data for normal speed and high speed respectively.

If the speed is not as stated, a fault in the motor is indicated.

Force to move rack in tube and wheelbox assembly.

1. Remove four screws. Lift off gearbox cover.
2. Remove crankpin spring clip by withdrawing sideways. Remove washer.
3. Carefully withdraw connecting rod. Remove washer.
4. Remove wiper arms 84.15.02 and 84.15.03.
5. Attach a suitable spring scale to hole in cross-head. Maximum permissible force to move rack is given in Data.

If the required force is greater than stated, a fault in the rack, tube and wheelbox assembly is indicated.

WINDSCREEN WIPER SYSTEM [except Mk. III]

–Motor-overhaul 84.15.18

Dismantling

1. Remove four screws. Lift off gearbox cover.
2. Remove crankpin spring clip by withdrawing sideways. Remove washer.
3. Carefully withdraw connecting rod. Remove washer.
4. Remove final gearshaft spring clip by withdrawing sideways. Remove washer.
5. Ensure shaft is burr-free and withdraw. Remove dished washer.
6. Remove thrust screw or thrust screw and lock nut as fitted.
7. Remove through bolts.
8. Carefully withdraw cover and armature about 0·2 in (5 mm). Continue withdrawal, allowing bushes to drop clear of commutator. Ensure three brushes are not contaminated with grease.
9. Pull armature from cover against action of permanent magnet.
10. Remove three screws to release brush assembly. Lift assembly from recess.
11. Lift and slide limit switch out sideways to release spring clip. Remove limit switch and brush assembly joined together by wires.

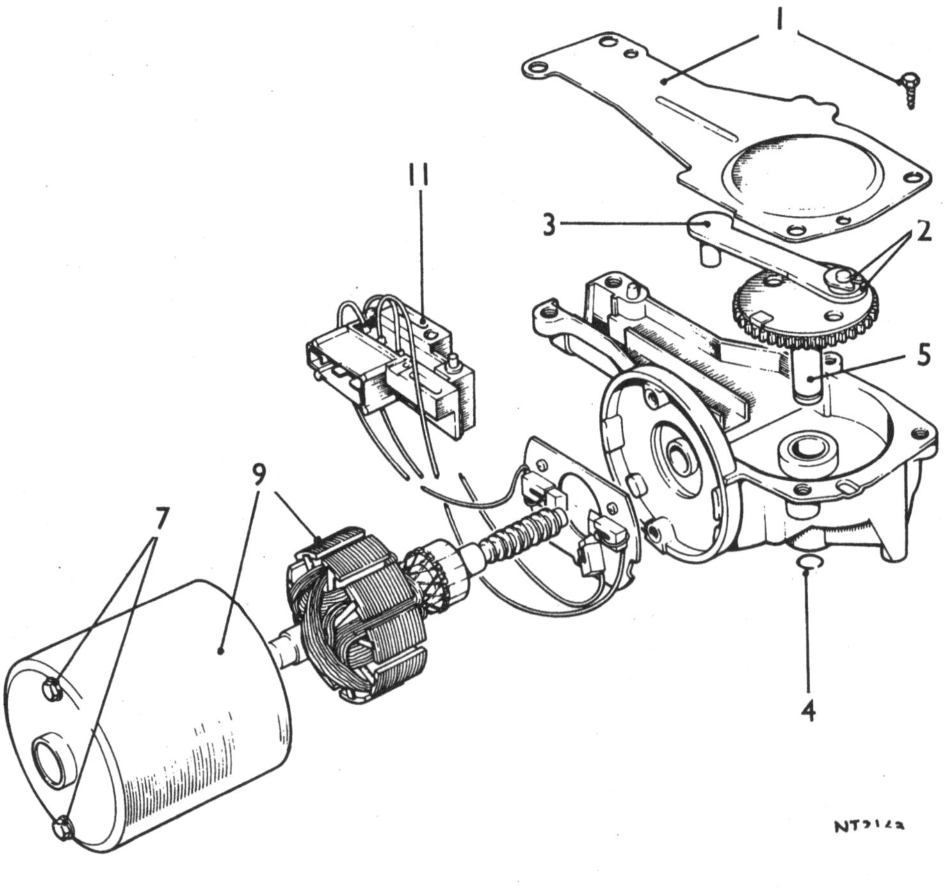

NT2142

Reassembling

NOTE: The following lubricants are required during assembly: Shell Turbo 41 oil and Ragosine Listate grease.

12. Slide limit switch in sideways to secure with spring clip.
13. Position brush assembly. Secure with three screws.
14. Lubricate cover bearing and saturate cover bearing felt washer with Shell Turbo 41 oil. Position armature to cover against action of permanent magnet.
15. Lubricate self aligning bearing with Shell Turbo 41 oil. Carefully insert armature shaft through bearing. Ensure that brushes are not contaminated with lubricant. Push three brushes back to clear commutator.
16. Seat cover against gearbox. Turn cover to align marks shown. Fit through bolts.
17. Fit thrust screw or thrust screw and lock nut as fitted.
18. If a non adjustable thrust screw is fitted check armature end float as follows:
 Position a feeler gauge between armature shaft and thrust screw. Push armature towards cover. End float should be 0·002 – 0·008 in. In the unlikely event of adjustment being required end float may be increased by fitting shim washer/washers under thrust screw head or reduced by mounting thrust screw in lathe and removing metal from underside of head.
19. If an adjustable thrustscrew and lock nut is fitted adjust armature end float as follows: Slacken lock nut. Screw thrust screw in until resistance is felt. Screw thrust screw out quarter of a turn – maintain in this position and tighten lock nut.
20. Lubricate final gear bushes with Shell Turbo 41 oil. Lubricate final gear cam with Ragosine Listate grease. Fit dished washer with concave surface facing final gear. Insert shaft.
21. Fit washer. Fit spring clip by inserting sideways.
22. Pack Ragosine Listate grease around worm gear, final gear and into cross-head guide channel.
23. Fit washer. Lubricate final gear crankpin with Shell Turbo 41 oil. Lubricate cross-head end of connecting rod, including pin, with Ragosine Listate grease. Carefully insert connecting rod.
24. Fit washer. Fit spring clip by inserting sideways.
25. Position gearbox cover. Secure with four screws.

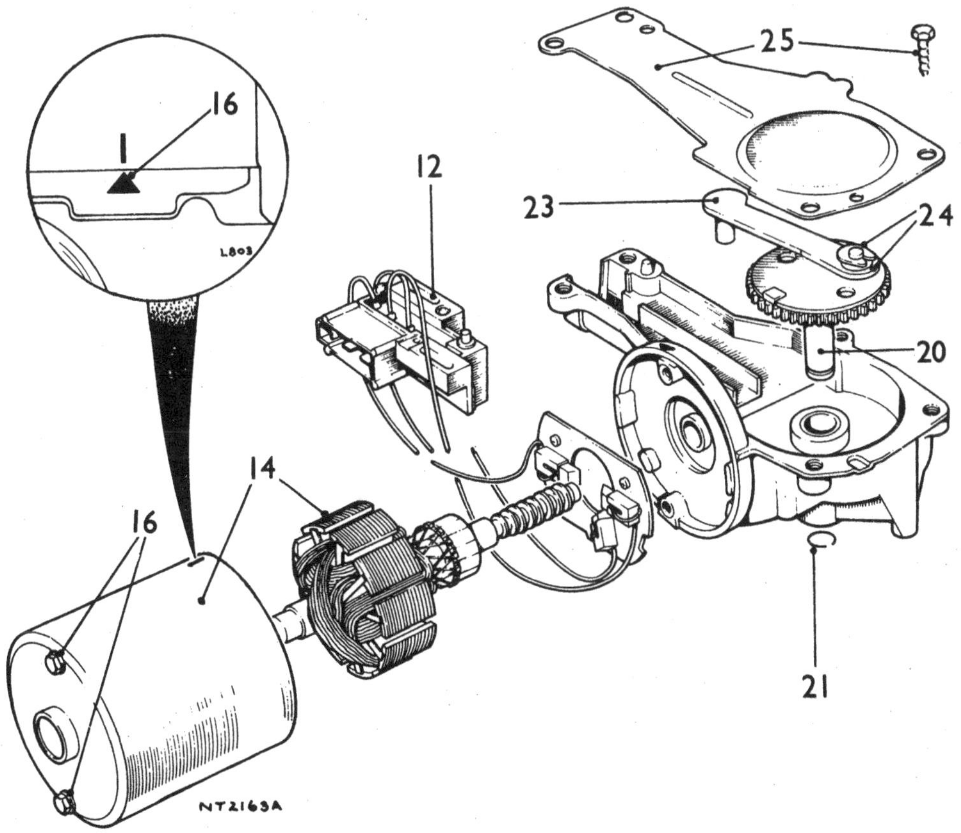

WINDSCREEN WIPER WHEELBOX – DRIVERS (R.H. STEER) [except Mk. III]

–Remove and refit 84.15.28

[NOTE: Information on Mk. III models is included on p. 383.]

Removing

1. Isolate the battery.
2. Remove the wiper motor 84.15.12.
3. Remove the rack 84.15.24.
4. Remove the drivers side parcel shelf 76.67.04.
5. Remove the wiper switch 86.65.38.
6. Remove the tachometer 88.30.21.
7. Remove the flasher warning light bulb holder.
8. Remove the drivers side demister vent and capping – two nuts, bolts, spring washers and plain washers.
9. Lower the centre fascia to the service position 76.46.02, instructions 2 to 4.
10. Remove the two bolts, spring washers, plain washers, and the thick and thin spacers securing the water valve control to the fascia bracket.
11. Remove the two nuts securing the wheelbox rear plate.
12. Place the tube ends aside.
13. Remove the spindle nut and withdraw the wheelbox.

Refitting

14. Reverse instructions 1 to 13.

WINDSCREEN WIPER WHEELBOX – PASSENGERS [except Mk. III]

–Remove and refit 84.15.29.

Removing

1. Remove the wiper motor 84.15.12.
2. Remove the rack 84.15.24.
3. Remove the two nuts securing the wheelbox rear plate.
4. Place the tube ends aside.
5. Remove the spindle nut and withdraw the wheelbox.

Refitting

6. Reverse instructions 1 to 5.

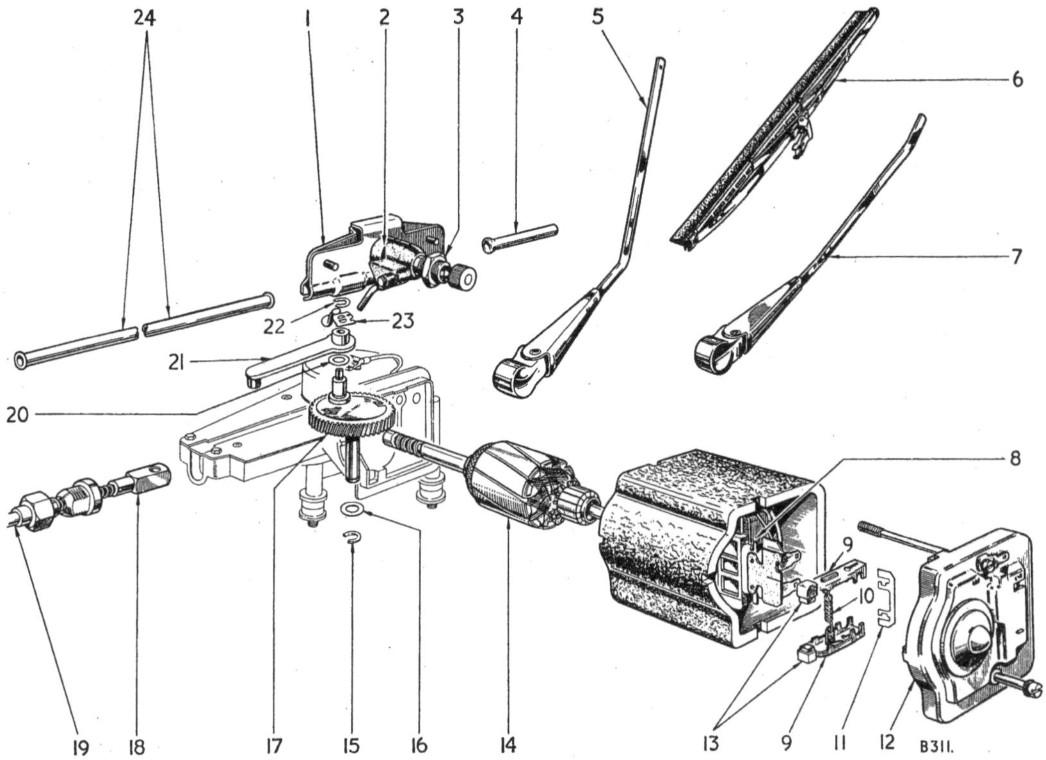

1 Wheel box
2 Jet and bush assembly
3 Nut
4 Rigid tubing—right-hand side
5 Wiper arm
6 Blade
7 Wiper arm
8 Field coil assembly
9 Brushgear
10 Tension spring and retainers
11 Brushgear retainer
12 End cover
13 Brushes
14 Armature
15 Circlip
16 Washer
17 Final drive wheel
18 Cable rack
19 Rigid tubing—left-hand side
20 Spacer
21 Connecting rod
22 Circlip
23 Parking switch contact
24 Rigid tubing—centre section

WIPER MOTOR (Mk. III)

—Remove and refit **[84.15.12]**

Removing

1. Remove the wiper arms and blades. 84.15.01
2. Unscrew the large nut (19) which secures the outer tubing to the gearbox.
3. Remove the three bolts that secure the motor mounting bracket to the dash panel. Then withdraw the motor complete with the inner cable rack.

Refitting

4. Lubricate the crankpin with Shell Turbo 41 oil.
5. Lubricate the cross-head end of the cable and the pin on the connecting rod with Ragosine Listate grease.
6. Reverse instructions 1 to 3.

WINDSHIELD WIPER SYSTEM (Mk. III)

—Motor overhaul **[84.15.18]**

Dismantling

1. Mark the domed limit switch cover in relation to the gearbox lid. Then take out the four screws and remove the lid.
2. Pry off the circlip (22) and then lift off the parking switch contact (23).
3. Lift off the connecting rod (21) and cable rack (18). Note the spacer (20) between the connecting rod (21) and the final drive wheel (17).
4. Remove the two bolts and then lift off the end cover (12).
5. Lift out the brushgear retainer (11) and remove the brushgear (9).
6. Remove the field coil assembly and housing as a unit; the red earth cable is long enough to permit the body to be lifted clear of the armature.
7. Remove the armature.

continued

8. If further dismantling is required, remove the circlip (15) and washer (16). Use a fine file to remove any burrs from around the circlip groove. Then remove the final drive wheel (17).
9. Mark the yoke and field coil relative to each other. Remove the two screws and then withdraw the field coil pole piece and the field coil.

Reassembling

10. Check the brush tension. It should be between 125 and 140 g.
11. Reverse instructions 1 to 9. Lightly lubricate the bearings, final drive wheel, and armature worm with motor oil; the brushgear and commutator must be kept free of oil or grease. If the gearbox has been washed clean, refill it with 25 to 35 cc of Ragosine Listate grease.
12. Set the adjusting screw in the side of the gearbox to permit 0.008 to 0.012 in (0.203 to 0.305 mm) end play of the armature. Then firmly set the locknut.
13. Before reconnecting the cable rack (18) to the gearbox, push the rack fully into the tubing. Then, using a spring scale attached to the rack, withdraw the rack from the tubing. The force required should not exceed 6 lbf (2.7 kg).

WINDSHIELD WIPER WHEELBOXES (Mk. III)

—Remove and refit **[84.15.28]** **[84.15.29]**

Removing

1. Disconnect the battery cables; disconnect the wires from the wiper motor. Note the wire colors relative to the motor terminals.
2. Take off the wiper arms complete with blades 84.15.01.
3. Remove the wiper motor 84.15.12.
4. Remove the nut and the rubber bush that secure each wheelbox to the lower windshield rail. Then push the boxes into the car.
5. Working inside the car, withdraw the boxes sufficiently to permit removal of the screws that secure the tubing to each wheelbox. Then withdraw the box.

Refitting

6. Reverse instructions 1 to 5.

ELECTRICAL OPERATIONS

***All bracketed material ([]) in this section has been added by Robert Bentley, Inc.**

BULB CHART — SPITFIRE MK. III

Lamp	Watts	Lucas Part No.	Stanpart No.	
Headlamps—L.H. Dip ..	60/45	54521872	512231	*
R.H. Dip—Normal	45/40	410	510218	
France	45/40	411	510219	
U.S.A.	50/40	54522231	——	*
Front parking lamps	6	207	57591	
Front flasher lamps	21	382	502379	
Tail/stop lamps	6/21	380	502287	
Rear flasher lamps	21	382	502379	
Reverse lamps	21	382	502379	
Plate illumination lamp—Normal	6	939	59467	
U.S.A.	4	222	501436	
Instrument illumination ..	2·2	987	59492	
Warning lights	2·2	987	59492	

*Sealed beam light unit

BULB CHART — SPITFIRE MK. IV and 1500

	Watts	Unipart No.	Stanpart No.	
Headlamps:				
L.H. dip–Normal	60/45	GLU 101	512231	†
R.H. dip–Normal	60/50		215735	†
France	45/40		510219	
U.S.A.	50/40		–	†
Front parking lamps	5	GLB 207	57591	
Front flasher lamps	21	GLB 382	502379	
Front marker lamps	4		501436	
Rear marker lamps	4		501436	
Tail/stop lamps	5/21	GLB 380	502287	
Rear flasher lamps	21	GLB 382	502379	
Reverse lamp	21	GLB 382	502379	
Plate illumination lamp	5	GLB 207	57591	
Instrument illumination	2·2	GLB 987	59492	
Courtesy light	2·2	GLB 987	59492	
Warning lights	2·2	GLB 987	59492	

†–Sealed beam light unit

WIRING DIAGRAM — SPITFIRE MK. IV

RIGHT HAND STEER THROUGH MODEL YEAR 1972 — DIAGRAM 1

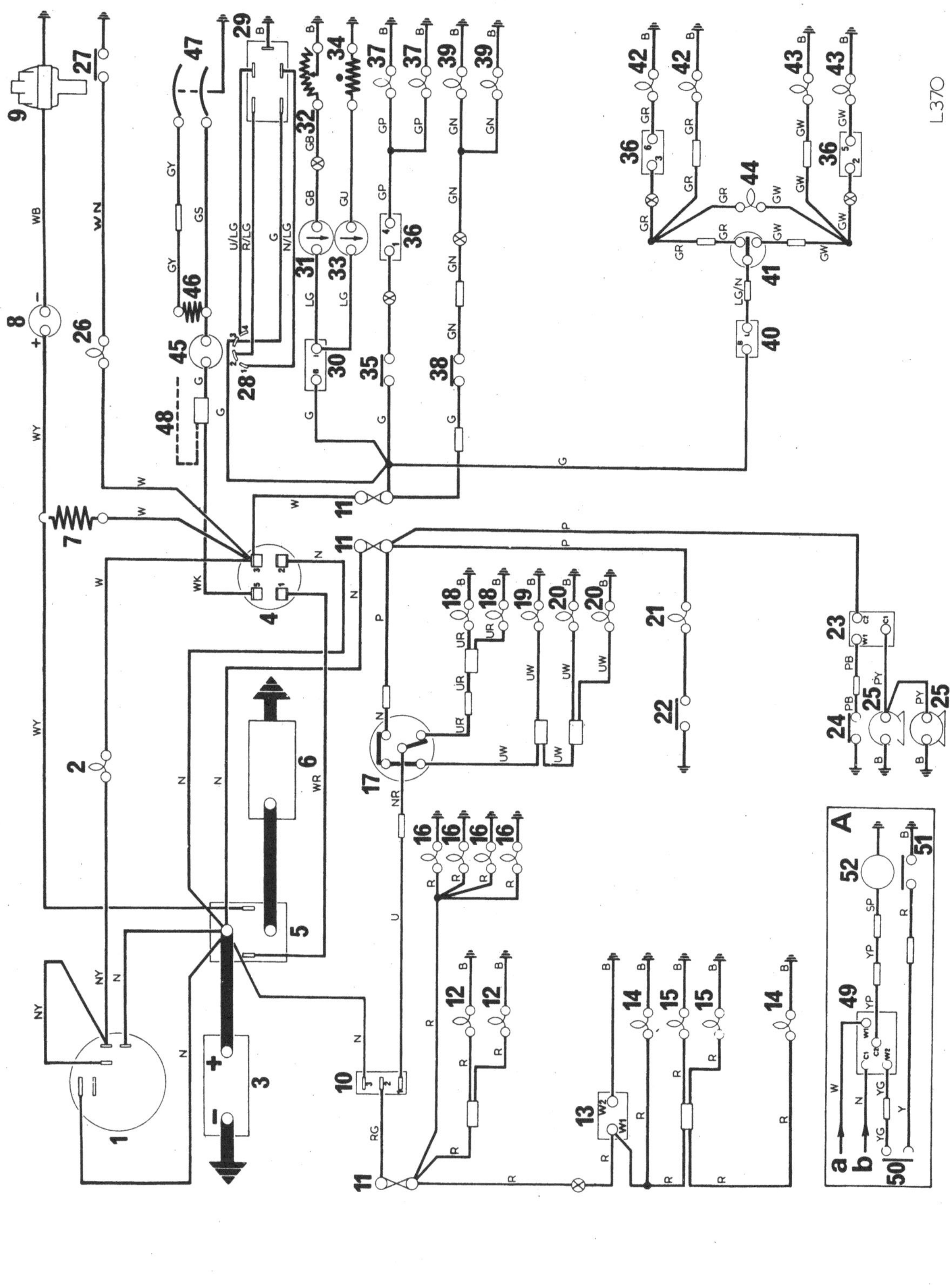

KEY TO WIRING DIAGRAM — SPITFIRE MK. IV

RIGHT HAND STEER THROUGH MODEL YEAR 1972 — DIAGRAM 1

1. Alternator
2. Ignition warning light
3. Battery
4. Ignition/starter switch
5. Starter solenoid
6. Starter motor
7. Ballast resistor
8. Ignition coil – 6 volt
9. Ignition distributor
10. Master light switch
11. Fuse
12. Front parking lamp
13. Night dimming relay winding
14. Tail lamp
15. Plate illumination lamp
16. Instrument illumination
17. Column light switch
18. Dip beam
19. Main beam warning light
20. Main beam
21. Courtesy light
22. Door switch
23. Horn relay
24. Horn push
25. Horn
26. Oil pressure warning light
27. Oil pressure switch
28. Windscreen wiper switch
29. Windscreen wiper motor
30. Voltage stabilizer
31. Fuel indicator
32. Fuel tank unit
33. Temperature indicator
34. Temperature transmitter
35. Stop lamp switch
36. Night dimming relay
37. Stop lamp.
38. Reverse lamp switch
39. Reverse lamp
40. Turn signal flasher unit
41. Turn signal switch
42. L.H. flasher lamp
43. R.H. flasher lamp
44. Turn signal warning light
45. Heater motor
46. Heater rheostat
47. Heater switch
48. Radio facility

A. Overdrive (optional extra)

49. Overdrive relay
50. Overdrive gearbox switch
51. Overdrive gear lever switch
52. Overdrive solenoid

a. From ignition/starter switch – terminal 3.
b. From ignition/starter switch – terminal 2.

WIRING DIAGRAM — SPITFIRE MK. IV

RIGHT HAND STEER FROM MODEL YEAR 1973 — DIAGRAM 2

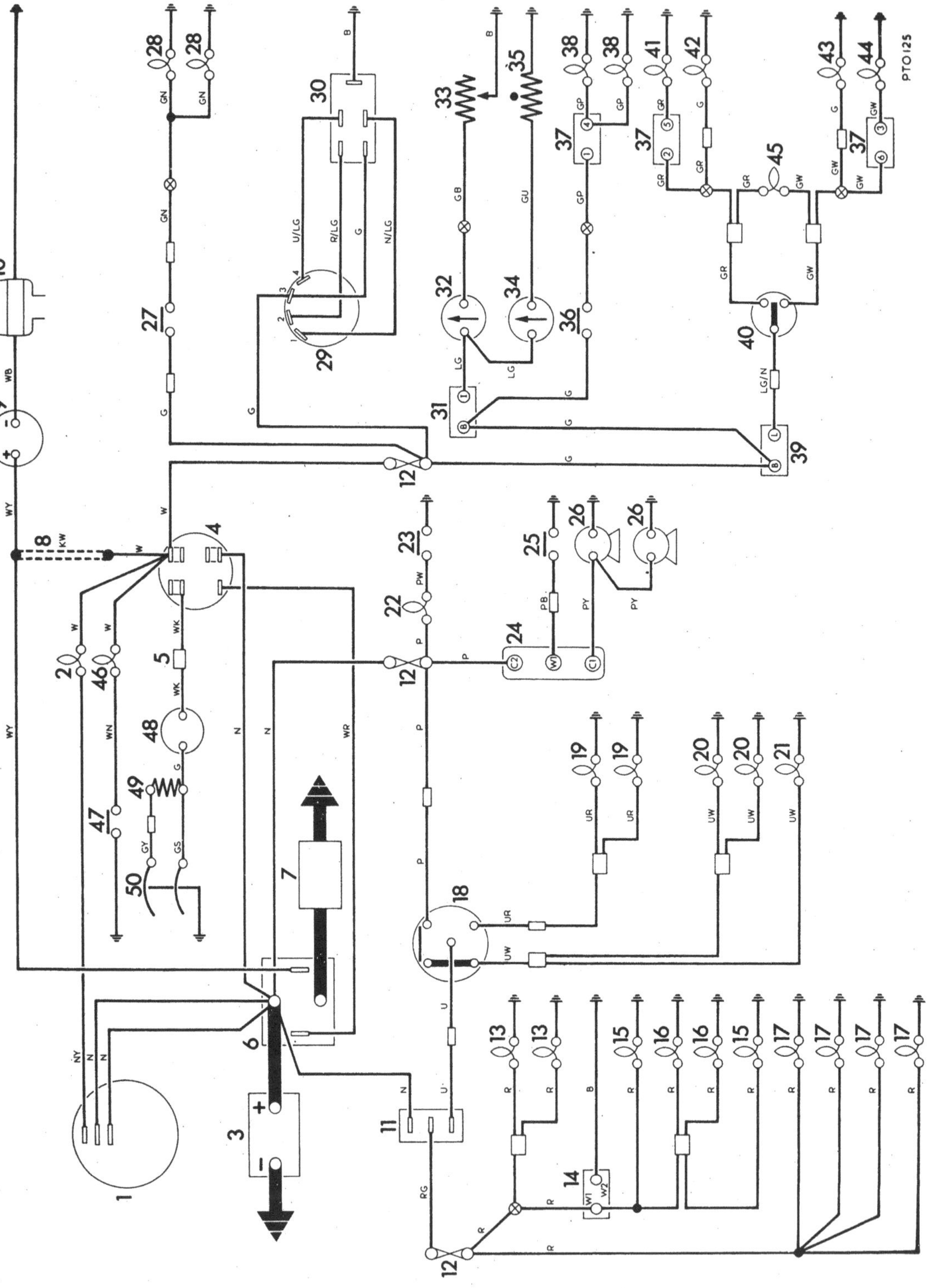

KEY TO WIRING DIAGRAM — SPITFIRE MK. IV

RIGHT HAND STEER FROM MODEL YEAR 1973 — DIAGRAM 2

1. Alternator
2. Ignition warning light
3. Battery
4. Ignition/starter switch
5. Radio supply
6. Starter solenoid
7. Starter motor
8. Ballast resistor wire
9. Ignition coil – 6 volt
10. Ignition distributor
11. Master light switch
12. Fuse
13. Front parking lamp
14. Night dimming relay winding
15. Tail lamp
16. Plate illumination lamp
17. Instrument illumination
18. Column light switch
19. Dip beam
20. Main beam
21. Main beam warning light
22. Key light
23. Door switch
24. Horn relay
25. Horn push
26. Horn
27. Reverse lamp switch
28. Reverse lamp
29. Windscreen wiper switch
30. Windscreen wiper motor
31. Voltage stabilizer
32. Fuel indicator
33. Fuel tank unit
34. Temperature indicator
35. Temperature transmitter
36. Stop lamp switch
37. Night dimming relay contacts
38. Stop lamp
39. Turn signal flasher unit
40. Turn signal switch
41. L.H. rear flasher lamp
42. L.H. front flasher lamp
43. R.H. front flasher lamp
44. R.H. rear flasher lamp
45. Turn signal warning light
46. Oil pressure warning light
47. Oil pressure switch
48. Heater motor
49. Heater resistor
50. Heater switch

WIRING DIAGRAM — SPITFIRE MK. III
LEFT HAND STEER FROM MARCH 1967 — DIAGRAM 3

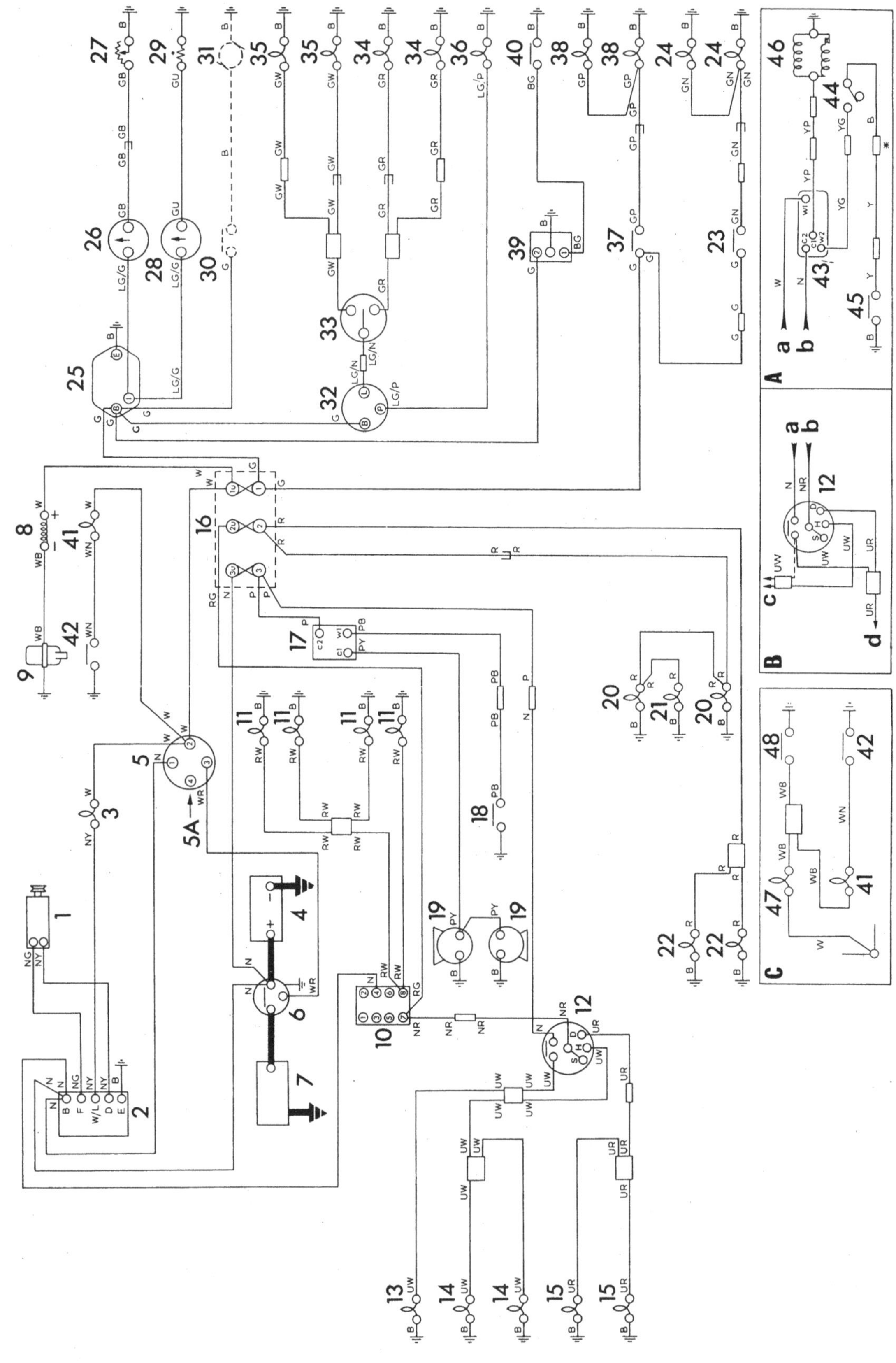

Part No. 511243. page 6.308. Issue 1.

KEY TO WIRING DIAGRAM — SPITFIRE MK. III

LEFT HAND STEER FROM MARCH 1967 — DIAGRAM 3

1 Generator
2 Control box
3 Ignition warning light
4 Battery
5 Ignition/starter switch
5A Ignition/starter switch radio supply connector
6 Starter solenoid
7 Starter motor
8 Ignition coil
9 Ignition distributor
10 Master light switch
11 Instrument illumination
12 Column light switch
13 Main beam warning light
14 Main beam
15 Dip beam
16 Fuse assembly
17 Horn relay
18 Horn push
19 Horn
20 Tail lamp
21 Plate illumination lamp
22 Front parking lamp
23 Reverse lamp switch
24 Reverse lamp
25 Voltage stabilizer
26 Fuel indicator
27 Fuel tank unit
28 Temperature indicator
29 Temperature transmitter
30 Heater switch (optional extra)
31 Heater motor (optional extra)
32 Flasher unit
33 Flasher switch
34 L.H. Flasher lamp
35 R.H. Flasher lamp
36 Flasher warning light
37 Stop lamp switch
38 Stop lamp
39 Windscreen wiper motor
40 Windscreen wiper switch
41 Oil pressure warning light
42 Oil pressure switch

A Overdrive (optional extra)
43 Overdrive relay
44 Overdrive column switch
45 Overdrive gearbox switch
46 Overdrive solenoid
a From ignition/starter switch – connector 2
b From ignition/starter switch – connector 1

B Dip beam flasher (Italy only)
12 Column light switch
a From fuse assembly
b From master light switch
c To main beam circuit
d To dip beam circuit

C Modifications to comply with U.S. Federal standards
41 Oil pressure warning light
42 Oil pressure switch
47 Brake line failure warning light
48 Brake line failure switch

COLOUR CODE

N.	Brown	**LG**	Light Green
U.	Blue	**W.**	White
R.	Red	**Y.**	Yellow
P.	Purple	**S.**	Slate
G.	Green	**B.**	Black

WIRING DIAGRAM — SPITFIRE MK. IV

LEFT HAND STEER THROUGH MODEL YEAR 1971 (U.S.A.) — DIAGRAM 4

1972 (elsewhere)

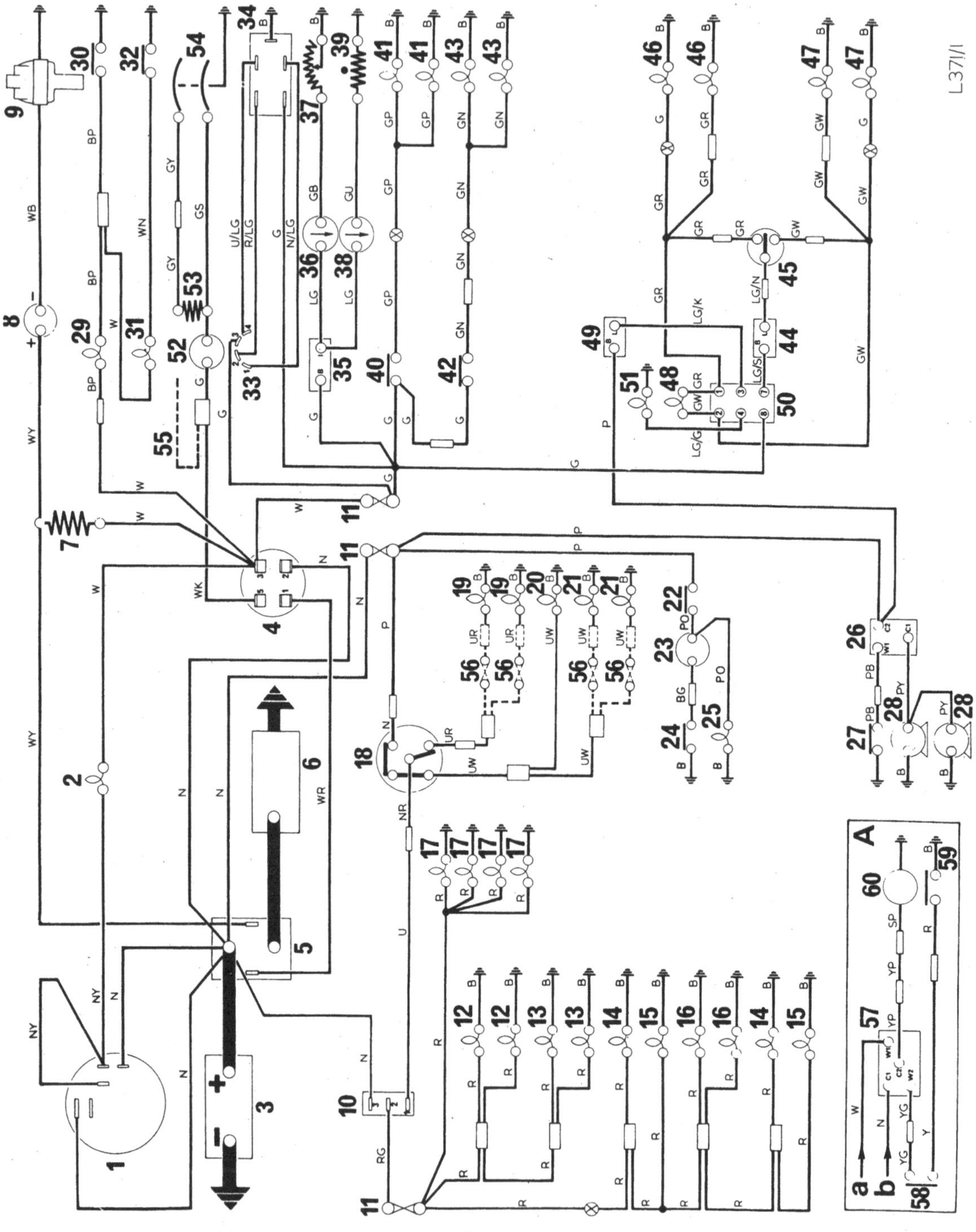

KEY TO WIRING DIAGRAM — SPITFIRE MK. IV

LEFT HAND STEER THROUGH MODEL YEAR 1971 (U.S.A.) — DIAGRAM 4

1972 (elsewhere)

1. Alternator
2. Ignition warning light
3. Battery
4. Ignition/starter switch
5. Starter solenoid
6. Starter motor
7. Ballast resistor
8. Ignition coil – 6 volt
9. Ignition distributor
10. Master light switch
11. Fuse
12. Front parking lamp
13. Front marker lamp
14. Rear marker lamp
15. Tail lamp
16. Plate illumination lamp
17. Instrument illumination
18. Column light switch
19. Dip beam
20. Main beam warning light
21. Main beam
22. Door switch
23. Buzzer
24. Key switch
25. Key light
26. Horn relay
27. Horn push
28. Horn
29. Brake warning light
30. Brake line failure switch
31. Oil pressure warning light
32. Oil pressure switch
33. Windscreen wiper switch
34. Windscreen wiper motor
35. Voltage stabilizer
36. Fuel indicator
37. Fuel tank unit
38. Temperature indicator
39. Temperature transmitter
40. Stop lamp switch
41. Stop lamp
42. Reverse lamp switch
43. Reverse lamp
44. Turn signal flasher unit
45. Turn signal switch
46. L.H. flasher lamp
47. R.H. flasher lamp
48. Turn signal warning light
49. Hazard flasher unit
50. Hazard switch
51. Hazard warning light
52. Heater motor
53. Heater rheostat
54. Heater switch
55. Radio facility
56. Line fuse – 25 amp (Italy only)

A. Overdrive (optional extra)

57. Overdrive relay
58. Overdrive gearbox switch
59. Overdrive gear lever switch
60. Overdrive solenoid

a. From ignition/starter switch – terminal 3
b. From ignition/starter switch – terminal 2

WIRING DIAGRAM — SPITFIRE MK. IV

LEFT HAND STEER FROM MODEL YEAR 1972 (U.S.A. only) — DIAGRAM 5

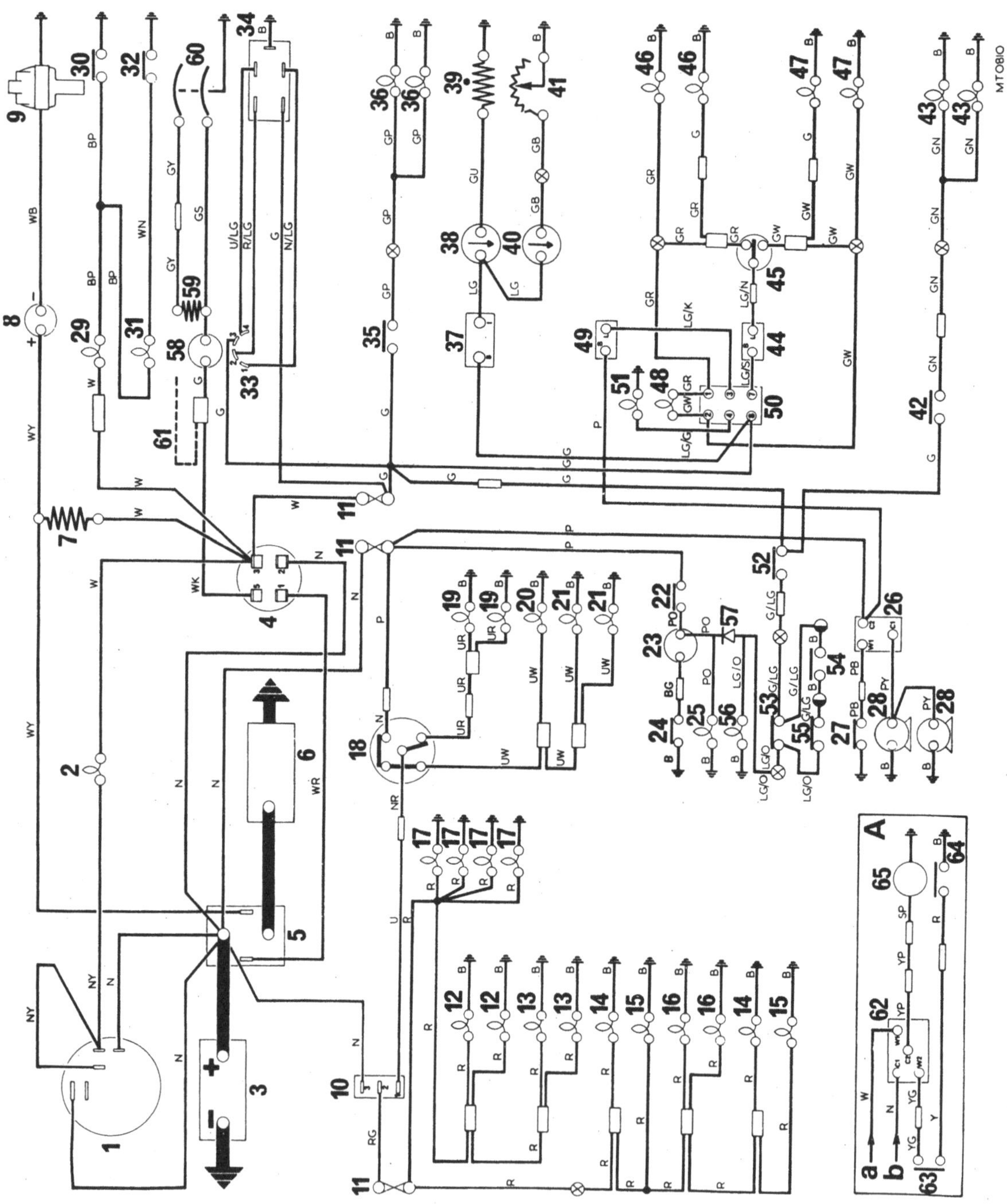

KEY TO WIRING DIAGRAM — SPITFIRE MK. IV

LEFT HAND STEER FROM MODEL YEAR 1972 (U.S.A. only) — DIAGRAM 5

1. Alternator
2. Ignition warning light
3. Battery
4. Ignition/starter switch
5. Starter solenoid
6. Starter motor
7. Ballast resistor
8. Ignition coil – 6 volt
9. Ignition distributor
10. Master light switch
11. Fuse
12. Front parking lamp
13. Front marker lamp
14. Rear marker lamp
15. Tail lamp
16. Plate illumination lamp
17. Instrument illumination
18. Column light switch
19. Dip beam
20. Main beam warning light
21. Main beam
22. Door switch
23. Buzzer
24. Key switch
25. Key light
26. Horn relay
27. Horn push
28. Horn
29. Brake warning light
30. Brake line failure switch
31. Oil pressure warning light
32. Oil pressure switch
33. Windscreen wiper switch
34. Windscreen wiper motor
35. Stop lamp switch
36. Stop lamp
37. Voltage stabilizer
38. Temperature indicator
39. Temperature transmitter
40. Fuel indicator
41. Fuel tank unit
42. Reverse lamp switch
43. Reverse lamp
44. Turn signal flasher unit
45. Turn signal switch
46. L.H. flasher lamp
47. R.H. flasher lamp
48. Turn signal warning light
49. Hazard flasher unit
50. Hazard switch
51. Hazard warning light
52. Belt warning gearbox switch
53. Drivers belt switch
54. Passengers seat switch
55. Passengers belt switch
56. Belt warning light
57. Diode
58. Heater motor
59. Heater rheostat
60. Heater switch
61. Radio facility

A. Overdrive (optional extra)

62. Overdrive relay
63. Overdrive gearbox switch
64. Overdrive gear lever switch
65. Overdrive solenoid

WIRING DIAGRAM — SPITFIRE MK. IV

LEFT HAND STEER FROM MODEL YEAR 1973 (except U.S.A.) — DIAGRAM 6

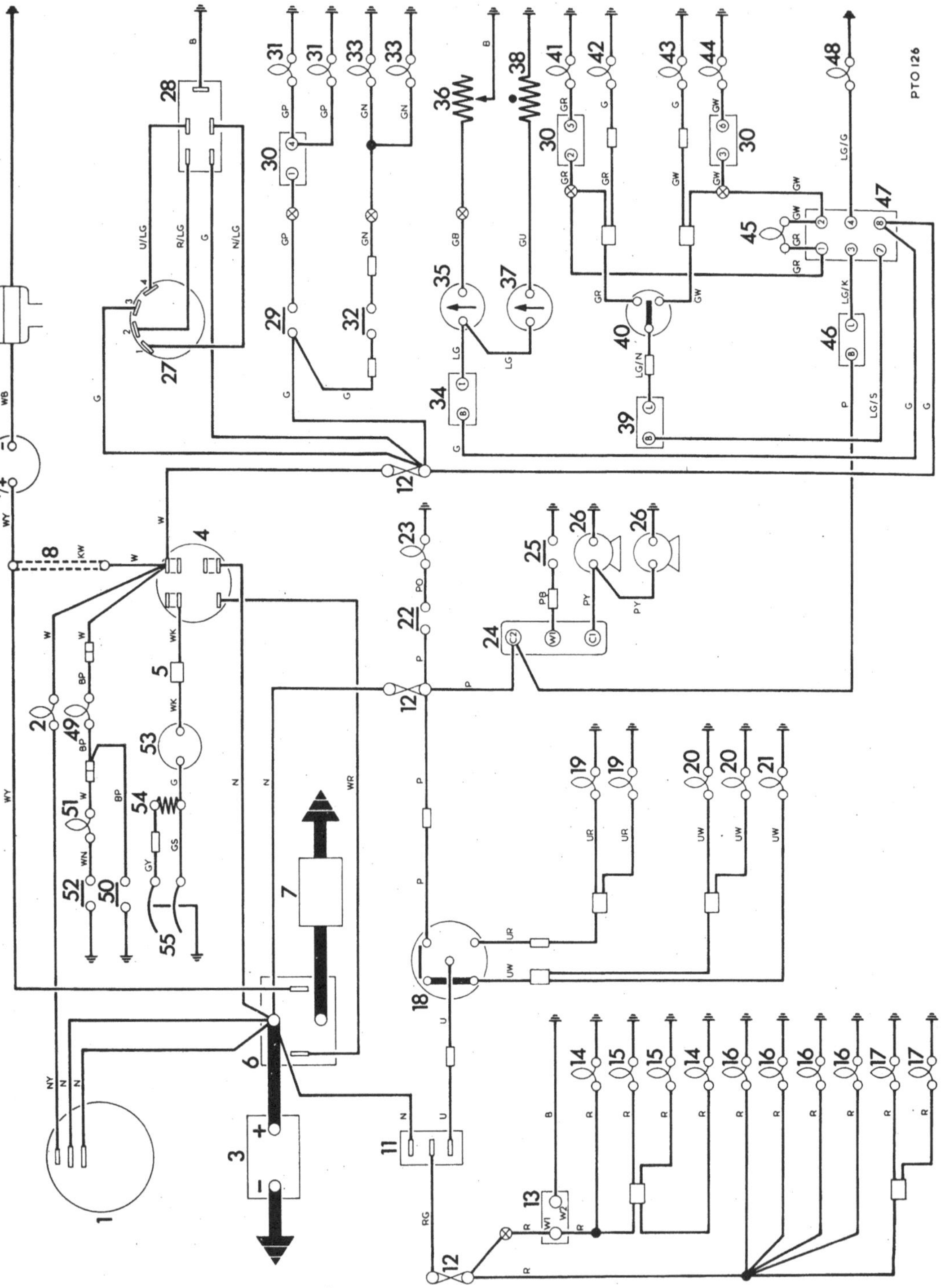

KEY TO WIRING DIAGRAM — SPITFIRE MK. IV

LEFT HAND STEER FROM MODEL YEAR 1973 (except U.S.A.) — DIAGRAM 6

1. Alternator
2. Ignition warning light
3. Battery
4. Ignition/starter switch
5. Radio supply
6. Starter solenoid
7. Starter motor
8. Ballast resistor wire
9. Ignition coil – 6 volt
10. Ignition distributor
11. Master light switch
12. Fuse
13. Night dimming relay winding
14. Tail lamp
15. Plate illumination lamp
16. Instrument illumination
17. Front parking lamp
18. Column light switch
19. Dip beam
20. Main beam
21. Main beam warning light
22. Door switch
23. Key light
24. Horn relay
25. Horn push
26. Horn
27. Windscreen wiper switch
28. Windscreen wiper motor
29. Stop lamp switch
30. Night dimming relay contacts
31. Stop lamp
32. Reverse lamp switch
33. Reverse lamp
34. Voltage stabilizer
35. Fuel indicator
36. Fuel tank unit
37. Temperature indicator
38. Temperature transmitter
39. Turn signal flasher unit
40. Turn signal switch
41. L.H. rear flasher lamp
42. L.H. front flasher lamp
43. R.H. front flasher lamp
44. R.H. rear flasher lamp
45. Turn signal warning light
46. Hazard flasher unit
47. Hazard switch
48. Hazard warning light
49. Brake line failure warning light
50. Brake line failure switch
51. Oil pressure warning light
52. Oil pressure switch
53. Heater motor
54. Heater resistor
55. Heater switch

WIRING DIAGRAM — SPITFIRE 1500
LEFT HAND STEER FROM MODEL YEAR 1973 — DIAGRAM 7

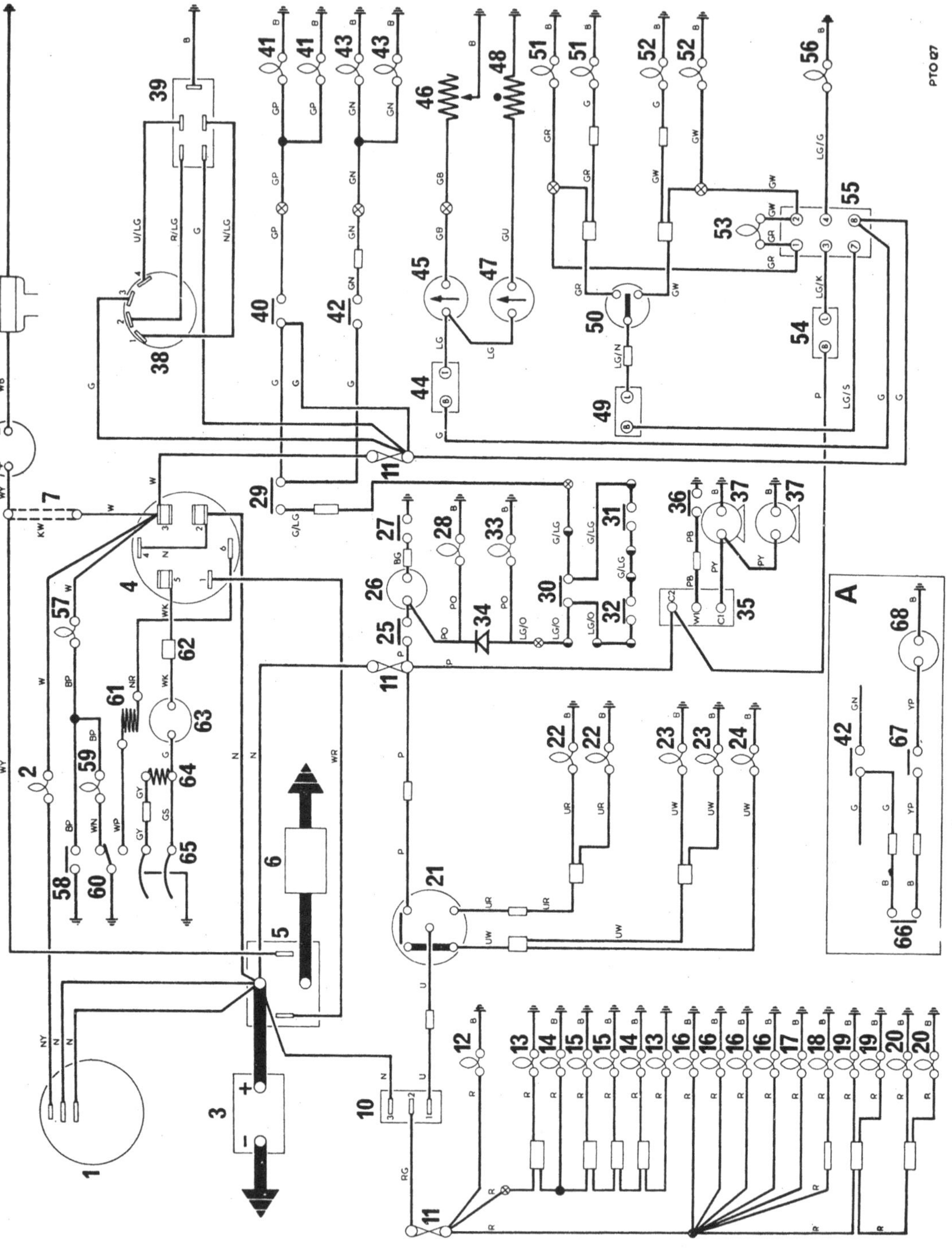

KEY TO WIRING DIAGRAM — SPITFIRE 1500

LEFT HAND STEER FROM MODEL YEAR 1973 — DIAGRAM 7

1. Alternator
2. Ignition warning light
3. Battery
4. Ignition/starter switch
5. Starter solenoid
6. Starter motor
7. Ballast resistor wire
8. Ignition coil – 6 volt
9. Ignition distributor
10. Master light switch
11. Fuse
12. Wipe/wash switch identification light
13. Rear marker lamp
14. Tail lamp
15. Plate illumination lamp
16. Instrument illumination
17. Hazard switch identification light
18. Heater control identification light
19. Front parking lamp
20. Front marker lamp
21. Column light switch
22. Dip beam
23. Main beam
24. Main beam warning light
25. L.H. door switch
26. Buzzer
27. Key switch
28. Key light
29. Belt warning gearbox switch
30. Drivers belt switch
31. Passengers seat switch
32. Passengers belt switch
33. Fasten belts warning light
34. Diode
35. Horn relay
36. Horn push
37. Horn
38. Windscreen wiper switch
39. Windscreen wiper motor
40. Stop lamp switch
41. Stop lamp
42. Reverse lamp switch
43. Reverse lamp
44. Voltage stabilizer
45. Fuel indicator
46. Fuel tank unit
47. Temperature indicator
48. Temperature transmitter
49. Turn signal flasher unit
50. Turn signal switch
51. L.H. flasher lamp
52. R.H. flasher lamp
53. Turn signal warning light
54. Hazard flasher unit
55. Hazard switch
56. Hazard warning light
57. Brake warning light
58. Brake line failure switch
59. Oil pressure warning light
60. Oil pressure switch
61. Anti run on valve
62. Radio facility
63. Heater motor
64. Heater rheostat
65. Heater switch

A. Overdrive (optional extra)

66. Overdrive gear lever switch
67. Overdrive gearbox switch
68. Overdrive solenoid

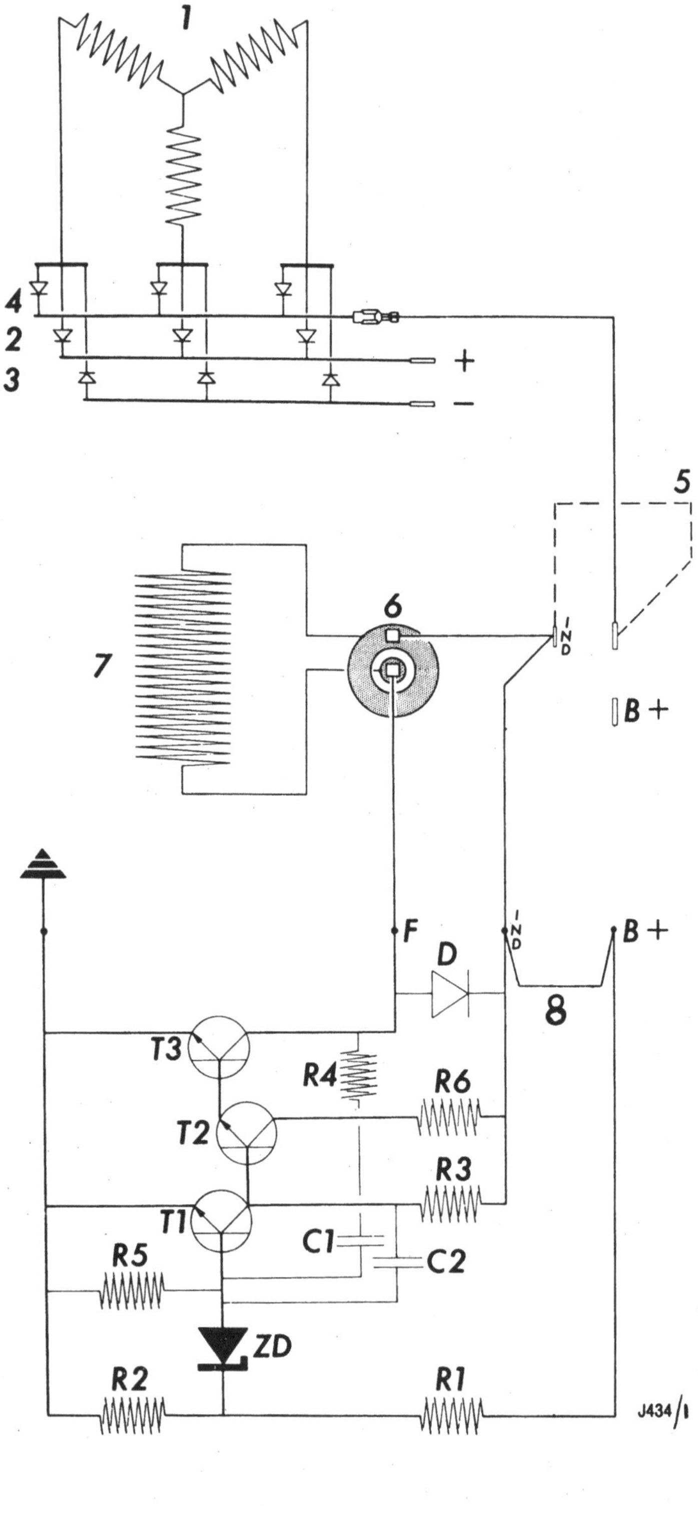
1
4
2
3
+
−
5
6
7
IND
B +
F
D
8
T3
R4
R6
T2
R3
T1
C1
C2
R5
ZD
R2
R1
J434/1

KEY TO ALTERNATOR WIRING DIAGRAM

1.	Stator windings	
2.	Live side output diodes	
3.	Earth side output diodes	
4.	Field winding supply diodes	
5.	Harness loop	Circuit is made when multi-socket connector is fitted and broken when connector is removed.
6.	Brushes and slip rings	
7.	Field winding	
8.	Internal B+ connection	
R3.	Resistor	Restricts T2 base current supplied from 'field winding supply' diodes.
T2.	Intermediate transistor	Controls T3 base current direct.
R6.	Resistor	Restricts T3 base current supplied from 'field winding supply' diodes.
T3.	Output transistor	Controls field winding earth return circuit.
R1 and R2.	Potential divider	Senses battery reference voltage.
ZD.	Zener diode	Voltage sensitive component. Opposes passage of current until breakdown voltage – approximately 8 volts – is reached. Controls T1 base current direct.
T1.	Input transistor	Controls T2 base current by diverting current passing through R3 to earth when ZD is conducting.
C1 and R4.	Capacitor and Resistor	Prevents transistor overheating by providing a positive feed back circuit to ensure quick switching of transistors from 'fully on' to 'fully off'.
R5.	Resistor	Path for small leakage current which may pass through ZD at high temperatures.
D.	Surge quench diode	Connected across field winding. Protects T3 from field winding high induced voltage surge and smoothes field winding current.
C2.	Condenser	Radio interference suppression.

ALTERNATOR

– Functional check 86.10.01

This operation must be performed in two parts. The first to prove the alternator's capacity to produce current, while the second is to prove the performance of the integral control unit.

Check capacity to produce current

NOTE: The stated output may be exceeded slightly when the alternator is cold. To avoid misleading results, the check should be performed with the unit as near normal operating temperature as possible.

1. Check drive belt adjustment 86.10.05
2. Disconnect the multi-socket connectors.
3. Remove the moulded cover.
4. Provide a test circuit as shown.

CAUTION: The alternator contains polarity-sensitive components that may be irreparably damaged if subjected to incorrect polarity. Observe polarity of alternator and battery terminals.

5. Do not connect the variable resistor across the battery for longer than is necessary to perform the check.
6. Run the engine.
7. Gradually increase the speed. At 1500 alternator rev/min (620 engine rev/min) the light should be extinguished.
8. Hold the speed at approximately 6,000 alternator rev/min (2,480 engine rev/min) Adjust the variable resistor so that the voltmeter reads 14 volts. The ammeter reading should now be approximately 28 amps.
9. If the ammeter reading is not approximately 28 amps, the indication is that the alternator requires overhaul or replacement.

continued

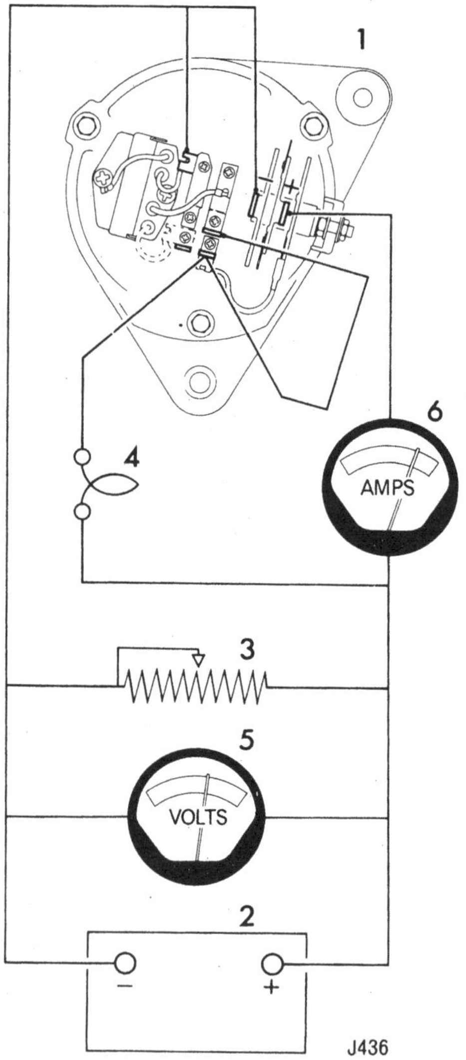

1. Alternator
2. Battery . . . . 12 volt
3. Variable resistor . . . 0–15 ohm – 35 amp
4. Light . . . . . . 12 volt – 2·2 watt
5. Voltmeter . . . . 0–20 volt
6. Ammeter . . . . 0–40 amp

Triumph Spitfire Mk IV Manual. Part No. 545254. Issue 1.

Check control unit

NOTE: The stated output may be exceeded slightly when the alternator is cold. To avoid misleading results, the check should be performed with the unit as near to its normal operating temperature as possible.

10. Check drive belt adjustment 86.10.05.
11. Disconnect multi-socket connectors.
12. Provide test circuit as shown.

CAUTION: The alternator contains polarity-sensitive components that may be irreparably damaged if subjected to incorrect polarity. Observe polarity of alternator and battery terminals.

13. Run the engine.
14. Gradually increase the speed. At 1,500 alternator rev/min (620 engine rev/min) the light should be extinguished.
15. Hold the speed at approximately 6,000 alternator rev/min (2,480 engine rev/min). The voltmeter reading should now be steady at 14·0 to 14·4 volts.
16. If the voltmeter reading is not steady at the above figure – and a satisfactory check of capacity to produce current has been performed – the indication is that the control unit should be replaced.

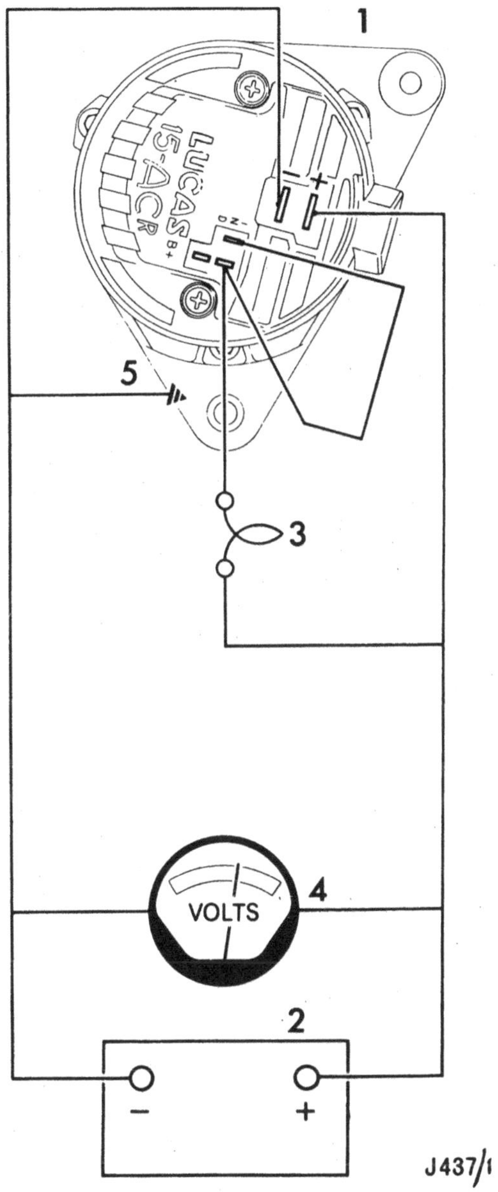

1. Alternator
2. Battery . . . . 12 volt
3. Light . . . . . . 12 volt – 2·2 watt
4. Voltmeter . . . . 0–20 volt
5. Earth connection to alternator body.

ALTERNATOR

– Remove and refit 86.10.02

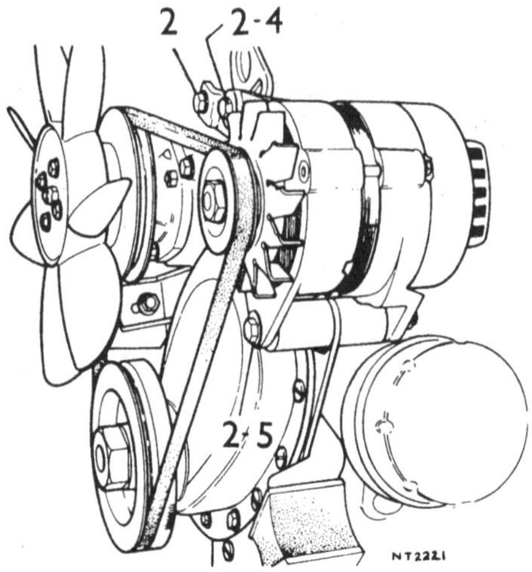

Removing

1. Disconnect the multi-socket connectors.
2. Slacken the main mounting bolt and the two adjustment bracket bolts.
3. Push the alternator towards the engine and remove the drive belt from the pulley.
4. Remove the outer adjustment bracket bolt.
5. Support the weight of the alternator and remove the main mounting bolt and spacer.

Refitting

6. Position the alternator. Fit the main mounting bolt and spacer.
7. Fit the outer adjustment bracket bolt.
8. Push the alternator towards the engine and fit the drive belt to the pulley.
9. Adjust the drive belt 26.20.01
10. Connect the multi-socket connectors.

ALTERNATOR

– Overhaul 86.10.08

[CAUTION: Please read the instructions thoroughly before you attempt alternator repairs. If you lack the necessary skills or tools we suggest you leave such work to an Authorized Dealer or other qualified shop. Before taking a removed alternator in for repair, wipe clean its exterior but do not undertake any disassembly.]

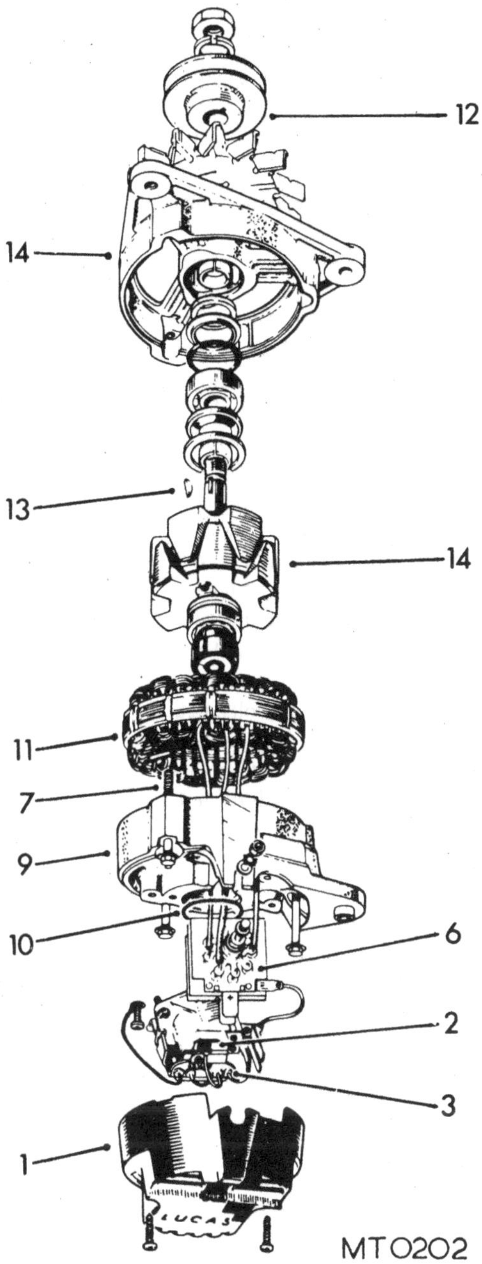

Dismantling

1. Remove the moulded cover.
2. Remove the brush box and control unit assembly by disconnecting the Lucar type connector from the rectifier pack and unscrewing three screws.
3. If required, the control unit may be detached from the assembly. Note the position of the three wire eyelets. Withdraw the screw to release the control unit and three screws to release the wire eyelets.
4. Note the position of the three stator wires on the rectifier pack.
5. Unsolder the three stator wire connections. Do not overheat the diodes or bend the diode pins. Solder quickly and provide a heat sink by gripping the diode pin with pliers.
6. Slacken the nut and withdraw the rectifier pack.
7. Remove the through bolts.
8. Provide an extractor tool as shown.
9. To remove the slip ring end bracket, position the extractor tool to engage with the outer journal of the slip ring end bearing. Employ a second operator to support the slip ring end bracket by hand. Carefully tap the extractor tool to drive the bearing from the housing.

NOTE: It may be necessary to carefully file away surplus solder from the two field winding connections on the slip ring moulding if the extractor tool will not pass over the moulding.

10. The rubber 'O' ring fitted in the slip ring end bracket bearing housing may remain in situ unless replacement is contemplated.
11. Remove the stator windings from the drive end bracket.
12. Prevent the rotor turning by wrapping a scrap fan belt round the pulley and retaining by hand or vice. Remove the nut, spring washer, pulley and fan. If necessary, use a suitable extractor.
13. Remove the key.
14. Using a suitable press, remove the rotor from the drive end bracket.

CAUTION: Do not attempt to remove the rotor by applying hammer blows to the shaft end. Such action may burr over and damage the thread.

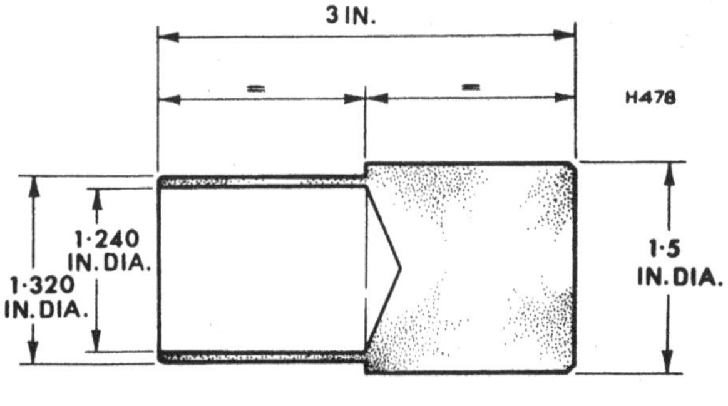

Reassembling

15. Using the spacer (arrowed) and a suitable tube, fit the rotor to the drive end bracket by applying pressure to the bearing inner journal.

CAUTION: Do not use the drive end bracket as a support while fitting the rotor. If the spacer is not employed, the felt ring may be damaged.

16. Fit the key.
17. Fit the fan, pulley, spring washer and nut. Prevent the rotor turning by wrapping a scrap fan belt round the pulley and retaining by hand or vice. Torque load the nut to 25 to 30 lbf ft. (3·46 to 4·15 kgf. m).
18. Observe the relationship of the stator windings to the drive end bracket determined by the stator wire connections, the rectifier pack position on the slip ring end bracket, the alignment of the mounting lugs on the end brackets and the through bolt clearances on the stator windings.
19. Position the stator windings to the drive end bracket.
20. Ensure that the rubber 'O' ring is fitted correctly in the slip ring end bracket bearing housing.
21. Fit the slip ring end bracket by carefully pushing the bearing into the housing.
22. Fit the through bolts, tightening evenly.
23. Ensure that the rubber locating piece is correctly fitted to the rectifier pack. Position the rectifier pack and secure it with the nut.
24. Position the three stator wires on rectifier pack as noted operation 4.
25. Solder the three stator wire connections. Note the precautions stated in operation 5 and use 'M' grade 45-55 tin-lead solder.
26. If required, attach the control unit to the brush box. Position the three wire eyelets on the brush box as noted in operation 3. Insert the screw to secure the control unit and three screws to secure the wire eyelets.
27. Ensure that the brushes are entered correctly in the brush box. Fit the brush box and control unit assembly by inserting three screws and connecting the Lucar type connector to the rectifier pack.
28. Fit the moulded cover.

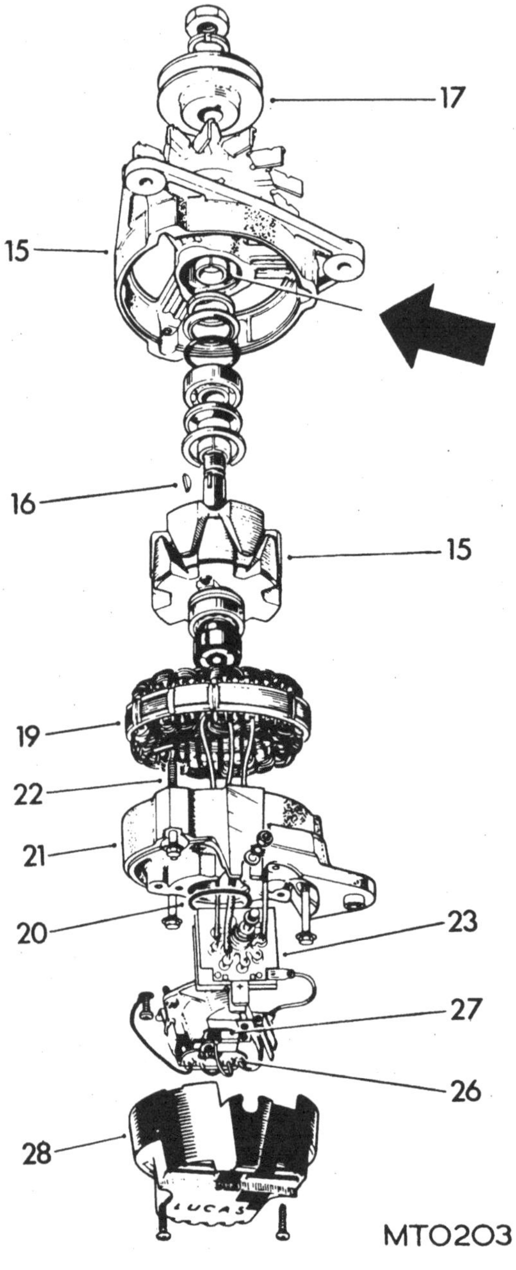

GENERATOR

—Charging circuit check **[86.11.01]**

Checking general condition

[NOTE: Before disturbing any electrical or mechanical settings, carry out the following checks to ensure that the fault does not lie outside the control box.

1. In the event of reported undercharging, ascertain that the condition is not due to short-trip driving.
2. Check the battery either by substituting a good battery or by using a hydrometer and a heavy discharge tester.
3. Inspect the generator drive belt. It should be just taut enough to drive without slipping.
4. Inspect the charging circuit wiring. Carry out continuity tests between the generator and the control box.
5. Check for good earth connections — particularly at the control box.
6. If none of these faults is found, proceed with electrical and mechanical adjustments, striving to obtain the nominal settings.

CAUTION: Please read the instructions thoroughly before you attempt charging system tests and adjustments. If you lack the necessary skills or tools we suggest you leave such work to an Authorized Dealer or other qualified shop.

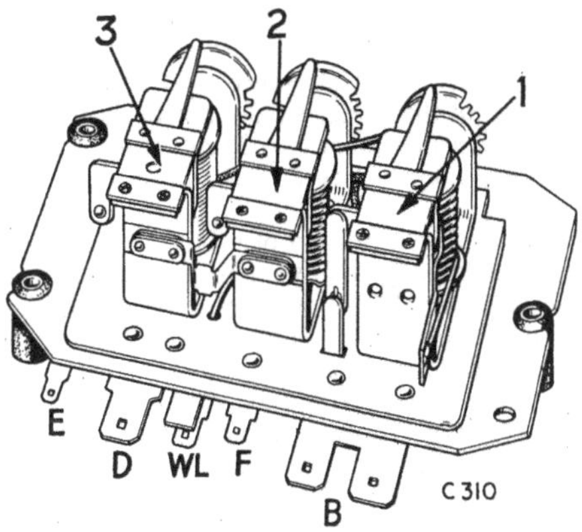

1 Cut out
2 Current regulator
3 Voltage regulator

1 Swamp resistors
2 Cut-out relay coil
3 Cut-out relay current coil
4 Cut-out relay contacts
5 Current/control relay contacts
6 Current control relay coil
7 Contacts resistor
8 Voltage control relay contacts
9 Voltage control relay coil
10 Battery
11 Generator field coils
12 Generator armature

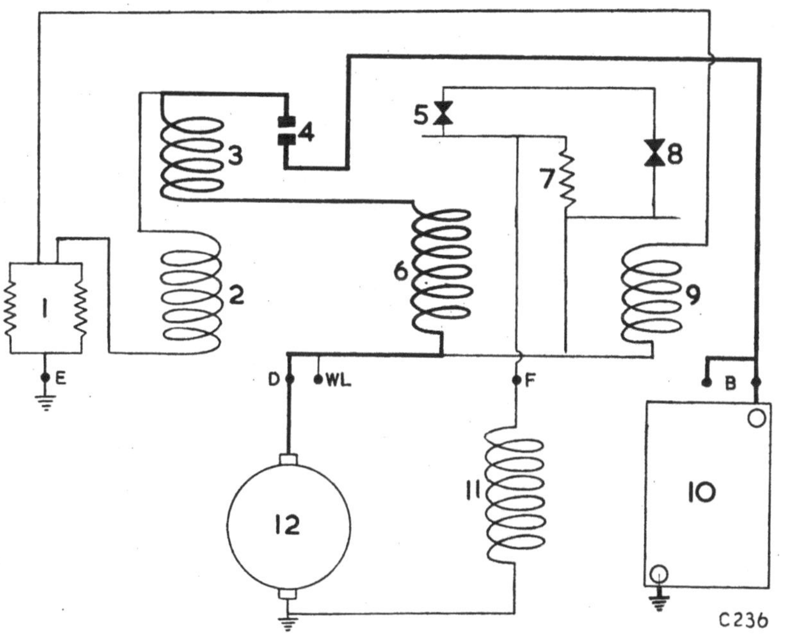

Generator changing circuit diagram

Checking and adjusting open circuit setting

NOTE: Complete this voltage regulator check and make adjustments as rapidly as possible to avoid errors arising from overheating of the operating coil.

1. Remove the control box cover. Insert a piece of cardboard between the armature and the core of the cut-out to prevent the contacts from closing.
2. Connect a first-grade 0 to 20 volt moving-coil voltmeter between control box terminal D and a good earthing point.

NOTE: A convenient method of making this connection is to withdraw the ignition warning light cable from terminal WL and clip the voltmeter positive wire to the exposed small terminal blade. This terminal is electrically common with terminal D.

3. Start the engine and run the generator at 3,000 rev/min.
4. Observe the voltmeter reading. This should be between the limits given in Table A, according to the temperature.

Table A

Ambient Temperature		Open Circuit Voltage SETTING
10°C. (50°F.)	..	14·9 — 15·5
20°C. (68°F.)	..	14·7 — 15·3
30°C. (86°F.)	..	14·5 — 15·1
40°C. (104°F.)	..	14·3 — 14·9

NOTE: An unsteady reading (i.e., one fluctuating more than ± 0.3 volt) may be due to unclean contacts. Clean dirty contacts with a fine carborundum stone or silicon carbide paper followed by denatured alcohol. If the reading is steady but occurs outside the appropriate limits, adjust as described in the following instructions.

5. Using a special tool (2 in the illustration), turn the voltage regulator adjustment until the correct reading is obtained. Turn the tool clockwise to increase voltage, counterclockwise to decrease voltage.
6. Restore the original connection and remove the cardboard.

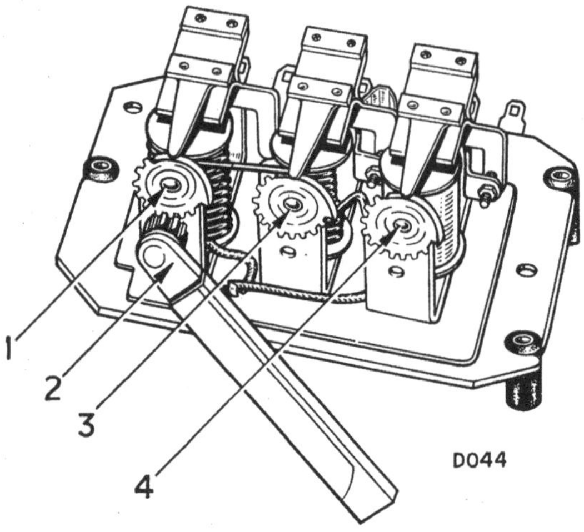

1 Cut-out
2 Special tool
3 Current regulator
4 Voltage regulator

Adjusting cut-out

Checking and adjusting cut-in setting

NOTE: Complete this cut-out check and make adjustments as rapidly as possible to avoid errors arising from overheating of the operating coil.

7. Connect a first-grade moving-coil voltmeter between control box terminal WL and a good earthing point.
8. Switch on an electrical load, such as the headlights. Start the engine and, slowly increasing the speed, observe the voltmeter reading. The voltage should rise steadily and then drop slightly at the instant of contact closure. The cut-in voltage is that which is indicated immediately before the pointer drops back. The drop should occur between the limits given in Table A.

NOTE: If the cut-in occurs outside the appropriate limits, reduce the engine speed to below the cut-in value and adjust as described in the following instructions. If it is necessary to clean the contacts, do so with fine glass paper — never with a carborundum stone or emery cloth.

9. Using the special tool, turn the cut-out relay adjustment cam clockwise to increase the cut-in voltage, counterclockwise to decrease the cut-in voltage.
10. Switch off the engine. Then, if no further test or adjustment is required, restore the original connections and refit the control box cover.

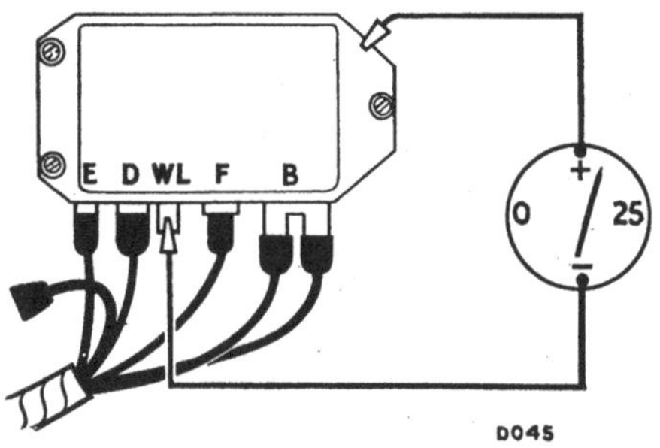

Checking cut-in voltage

Part No. 511243. page 6.120. Issue 1.
Part No. 511243. page 6.122. Issue 1.

Checking and adjusting drop-off voltage

11. Disconnect the wires from terminal B-B and connect the S.W. terminal on the coil to the battery. Connect a first-grade moving-coil voltmeter between control box terminal B-B and earth.
12. Start the engine and run it up to approximately 3,000 rev/min. While slowly decelerating the engine, observe the voltmeter reading.

NOTE: Opening of the contacts is indicated when the voltmeter pointer drops to zero. This should occur between the limits given in Table A. If the drop-off occurs outside the appropriate limits, adjust as described in the following instructions.

13. Stop the engine and, if necessary, remove the control box cover.
14. Adjust the drop-off voltage by carefully bending the fixed contact bracket. Reducing the contact gap will raise the drop-off voltage; increasing the gap will lower the drop-off voltage.

NOTE: Bending the bracket should result in a contact "follow through" or blade deflection of 0.010 to 0.020 in (0.25 to 0.51 mm).

Checking and adjusting current setting

NOTE: The generator must be developing its maximum rated output at the time of setting.

15. If necessary, remove the control box cover.
16. Insert a piece of cardboard between the armature and the core face of the voltage regulator to prevent the contacts of the regulator from opening.
17. Withdraw the wires from the control box terminal blades B-B and connect the wires from terminals B-B to the load side of the first-grade 0 to 40 ampere moving-coil ammeter. Ensure that terminal B carries only this one connection.
18. Switch on all lights and equipment. Run the engine at approximately 3,000 rev/min and observe the ammeter reading, which should be steady and indicate the maximum rated output of the generator (22 amperes).

NOTE: An unsteady reading (i.e., one fluctuating more than ± 1 ampere) may be due to unclean contacts. Clean the contacts with a fine carborundum stone or silicon carbide paper, followed by denatured alcohol. If the reading is too high or too low, adjust as described in the following instructions.

19. Using the special tool, turn the current adjustment cam clockwise to raise the ampere reading or counterclockwise to lower the reading.
20. Switch off the engine and restore the original connections.
21. Remove the cardboard. If no further checks or adjustments are required, refit the control box cover.

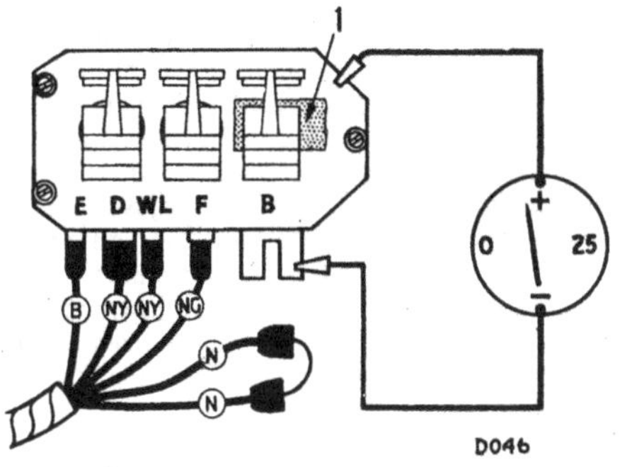

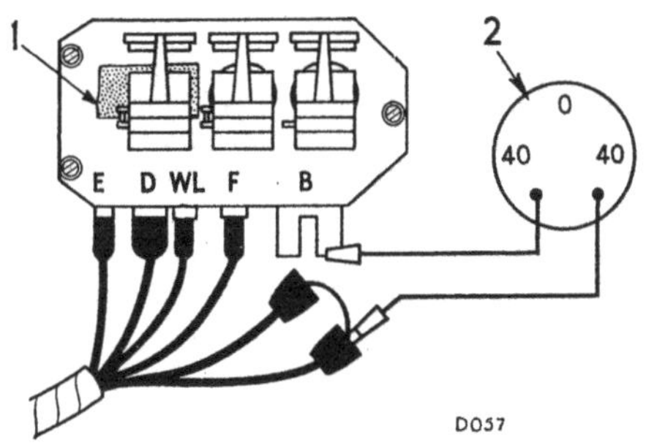

1 Cardboard under cut-out armature
2 0 - 40 ammeter

Checking current setting

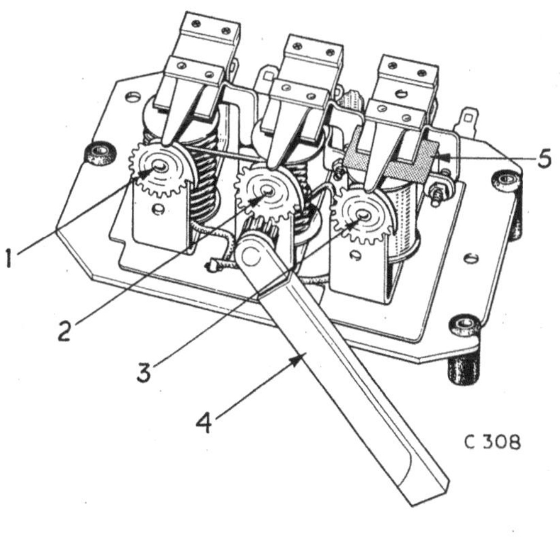

1 Cut-out cam
2 Current control cam
3 Voltage regulator cam
4 Special adjusting current control
5 Cardboard under voltage regulator

Adjust current control cam

Checking and adjusting air gap settings

NOTE: After long service, it may be necessary to adjust the air gap settings on the control box as described in the following instructions. Such adjustments should be applied to the armature-to-bobbin core gaps of the voltage regulator and current regulator only. The air gaps should be checked if previous adjustments cannot be made accurately. Always carry out all previously described adjustments after resetting the air gaps.

22. Using the special tool, turn the adjustment cams on the voltage regulator and the current regulator counter-clockwise for minimum lift of the armature tensioning springs.
23. Slacken the adjustable contact locking nut and screw back the adjustable contact. Insert a flat steel feeler gauge of 0.045 in (0.04 mm) thickness between the armature and the copper separator on the core face, taking care not to turn up or damage the copper. The gauge should be inserted as far back as the two rivet heads on the underside of the armature.
24. Retaining the feeler gauge in position, press squarely down on the armature and screw in the adjustable contact until it just touches the armature contact.
25. Readjust the electrical settings as previously described.

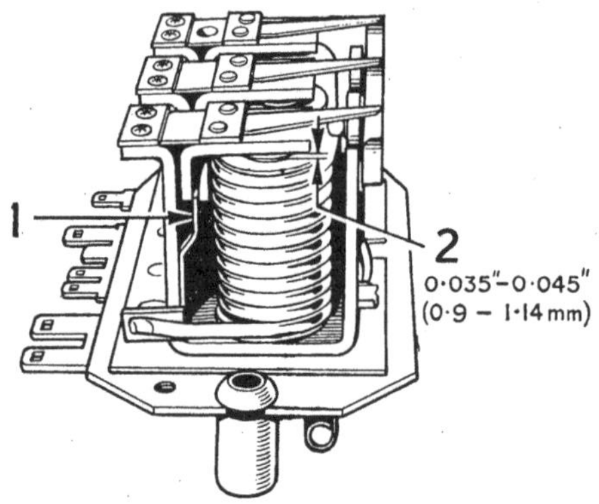

1 Voltage regulator contacts
2 Airgap setting

Voltage regulator airgap setting

Checking and adjusting cut-out relay air gap and contact "follow-through"

26. Press the armature squarely down against the copper separator on the core face.
27. Adjust the fixed contact bracket to give 0.010 to 0.020 in (0.25 to 0.51 mm) "follow-through" or blade deflection of the moving contact.
28. Release the armature and adjust the armature back stop to give a core gap of 0.035 to 0.045 in (0.90 to 1.04 mm).
29. Readjust the cut-in and drop-off voltage settings.

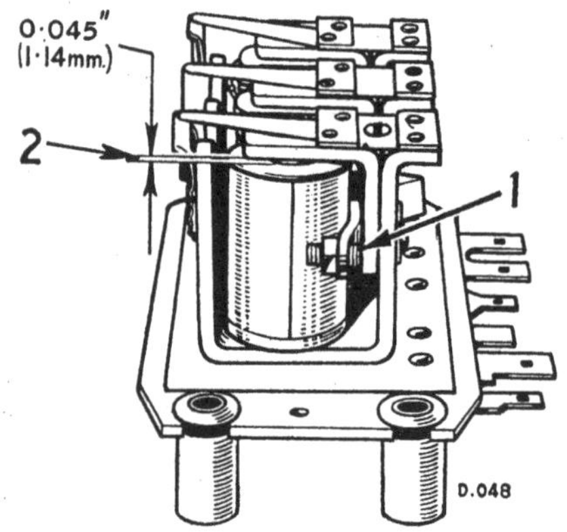

1 Backstop adjustment
2 Airgap settings

Cut-out airgap settings

Checking control box stability

30. Connect a voltmeter as described in "Checking and adjusting open circuit setting". Connect an ammeter as described in "Checking and adjusting current setting".
31. Run the generator at 4,500 rev/min.
32. Switch on and off a lamp load equivalent to 75 per cent of the maximum output of the generator.

NOTE: Assuming the generator and external circuits to be in good order, instability (i.e., violent fluctuations of the voltage and current reactions to the conditions imposed) could be due to (1) narrow air gap settings, (2) foreign matter in the air gaps, (3) faulty internal connections causing an intermittent open circuit.

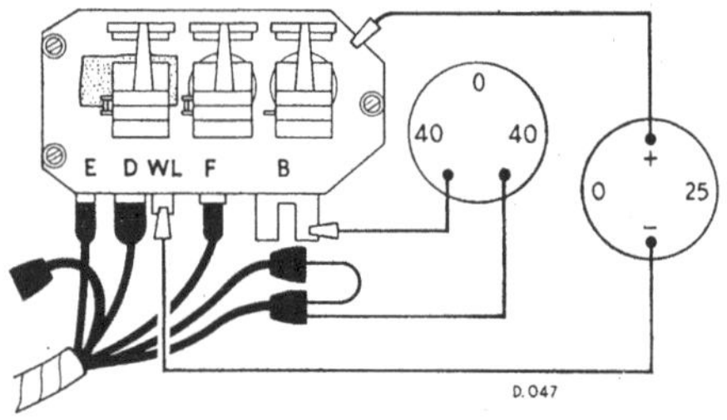

Control box stability test

Part No. 511243. page 6.124. Issue 1.

GENERATOR

—Remove and refit [86.11.02]

Removing

1. Isolate the battery. Then disconnect the two wires from the generator.
2. Slacken the two lower mounting bolts and the two adjustment bracket bolts.
3. Push the generator towards the engine and remove the drive belt from the pulley.
4. Remove the outer adjustment bracket bolt.
5. Support the weight of the generator and remove the two lower mounting bolts.

Refitting

6. Position the generator on its lower mounting bracket. Fit the two lower mounting bolts.
7. Fit the outer adjustment bracket bolt.
8. Push the generator towards the engine and fit the drive belt to the pulley.
9. Adjust the drive belt 26.20.01.
10. Reconnect the generator wires and the battery cables.

GENERATOR

—Overhaul [86.11.03]

CAUTION: Please read the instructions thoroughly before you attempt generator repairs. If you lack the necessary skills or tools we suggest you leave such work to an Authorized Dealer or other qualified shop. Before taking a removed generator in for repair, wipe clean its exterior but do not undertake any disassembly.

Dismantling

1. Remove the generator [86.11.02].
2. Remove the nut and the spring washer, then extract the pulley and take out the woodruff key.
3. Remove the two bolts. Then withdraw the commutator end bracket from the yoke. Note the fiber thrust washer adjacent to the commutator.
4. Withdraw the armature and the drive end bracket complete with bearing.
5. Support the bearing retainer plate and press the shaft from the drive end bracket.

1 Bolts
2 Brush
3 Felt ring and aluminium sealing disc
4 Brush spring
5 Bearing bush
6 Commutator end bracket
7 Field coils
8 Rivet
9 Bearing retainer plate
10 Corrugated washer
11 Felt washer
12 Driving end bracket
13 Pulley retainer nut
14 Bearing
15 Woodruff key
16 Armature

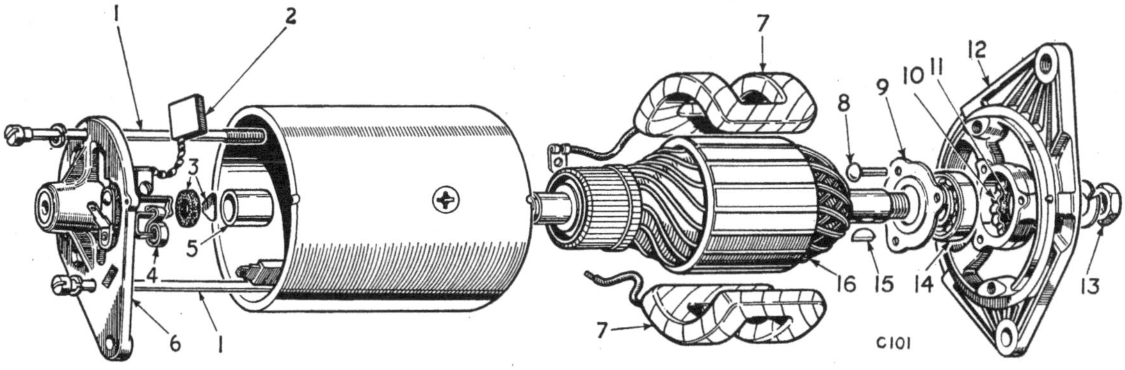

[86.11.02 Sheet 1]
[86.11.03 Sheet 1]

6. Inspect the commutator. It should be smooth and free of pits. Slight burning can be removed with fine glass-paper; badly worn commutators should be turned on a lathe. After turning, undercut the insulators between the commutator segments to a depth of 1/32 in (0.80 mm).
7. Inspect the brushes. Replace those which are less than 11/32 in (8.70 mm) in length. Refit used brushes in their original positions only. The brushes must move freely in their holders.
8. Using a spring scale, test the brush spring tension. Fit new springs if the tension is below 15 oz (425 g).
9. Inspect the bearing bush in the commutator end bracket. If it is worn or otherwise damaged, screw a 5/8 in (or 16 mm) tap squarely into the bush. Then, by pulling on the tap, withdraw the bush. Insert the felt ring and the aluminum sealing disc. Then, using a suitable bearing driver, press in a new bush until it is flush with the inner face of the end bracket.
10. Inspect the ball bearing at the driving end. If it is noisy, stiff, or feels rough when turned, replace it as described in instructions 11 through 15.
11. To remove the ball bearing, first drill out the rivets and remove the bearing retainer plate.
12. Press the ball bearing from the end bracket and remove the corrugated washer, the felt washer, and the oil retaining washer.
13. Clean the replacement bearing and then pack it with a high melting point grease, such as Energrease RBB.3 or equivalent.
14. Place the oil retaining washer, felt washer, and corrugated washer in the bearing housing. Then press in the new bearing.
15. Fit and rivet the retaining plate to the end bracket.
16. Inspect the field coils. If the insulation is damaged the coils should be replaced as described in instructions 17 through 24. Also replace the coils if they have lost electrical continuity or if the field resistance is significantly less or greater than 6 ohms.
17. To replace the field coils, first drill out the rivet securing the field terminal assembly to the yoke. Then unsolder the field coil connections.
18. Remove the insulation piece which prevents the junctions of the field coils from contacting the yoke.
19. Mark the yoke and the pole shoes so that they can be refitted to their original positions.
20. Unscrew the pole shoe retaining screws, remove the pole shoes and lift off the coils.
21. Fit the new field coils over the pole shoes and re-position them inside the yoke.
22. Locate the pole shoes and field coils by lightly tightening the retaining screws; fully tighten them by using a wheel-operated screwdriver. Lock the screws by caulking.
23. Replace the insulation piece between the field coil connections and the yoke.
24. Re-solder the field coil connections to the field coil terminal tags and rivet the assembly to the yoke.

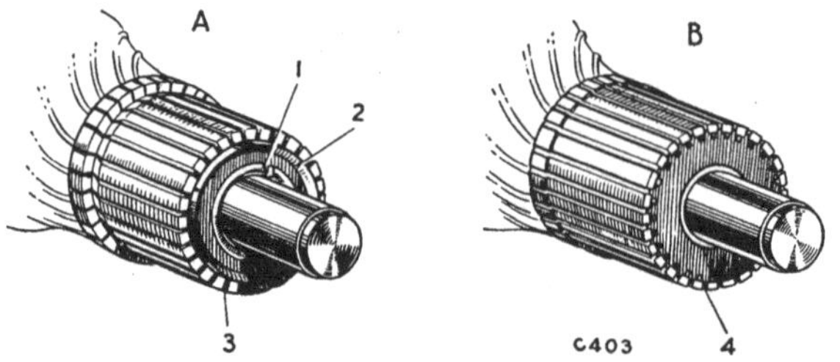

A. Fabricated commutator. B. Moulded commutator.
1 Metal roll-over
2 Insulating cone
3 Slot depth—0·032″ (0·81 mm.) maximum
4 Slot depth — 0·02″ to 0·035″ (0·508 to 0·89 mm.).

Commutator details

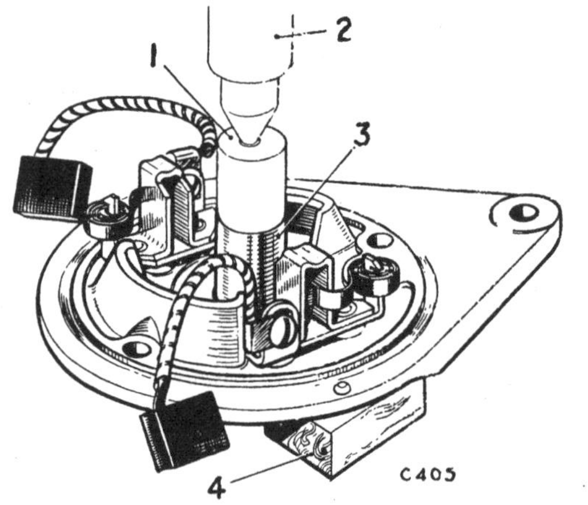

1 Mandrel
2 Press
3 Bush
4 Wood blocks

Fitting a new bearing to the commutator end bracket

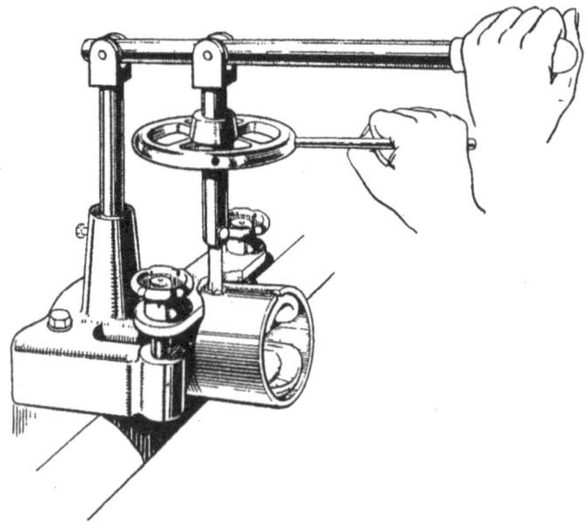

Removing the pole pieces from the yoke

Reassembling

25. Supporting the inner journal of the ball bearing to prevent damage, press the armature through the bearing — the bearing having been previously installed in the drive end bracket.
26. Assemble the armature and end bracket to the yoke.
27. Hold the brushes up by positioning each brush spring at the side of its brush.
28. Fit the commutator end bracket on the armature shaft until the brush boxes are partly over the commutator. Press each brush down on the commutator and move its spring to the operating position.
29. Fit the commutator end bracket to the yoke and refit the bolts.

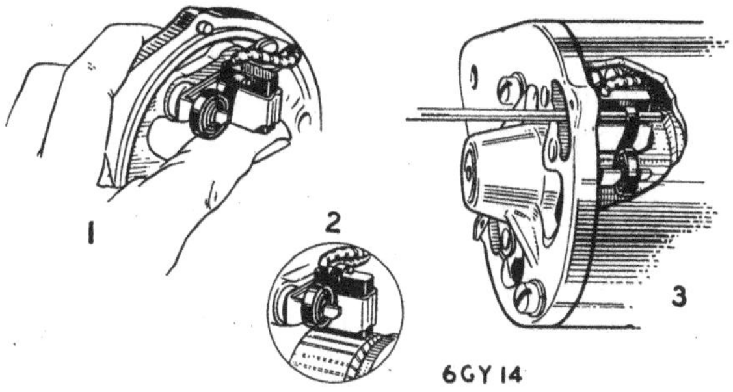

1 Method of trapping brush in raised position with spring
2 Normal working position
3 Method of releasing brush on to commutator

Fitting commutator end bracket to "windowless" yoke generator

BATTERY

– Remove and refit 86.15.01

[CAUTION: Keep sparks and open flame away from the top of the battery. Hydrogen gas from the battery could explode violently. When you work with battery electrolyte, wear goggles and do not allow the liquid to contact your skin or clothing. Electrolyte is corrosive and can cause severe burns. If it should spill onto your skin, flush the area of contact immediately with large quantities of water. Spilled electrolyte can be neutralized with a strong baking soda solution.]

Removing

1. Remove the battery leads.
2. Slacken the nuts and swing down the battery retaining assembly.
3. Lift the battery from the vehicle.

Refitting

4. Lift the battery into the tray.
5. Swing up the battery retaining assembly. Tighten the nuts.
6. Fit the battery leads. Do not hammer the terminals on to the terminal posts.
7. Coat the terminals with petroleum jelly (Vaseline) to prevent corrosion.

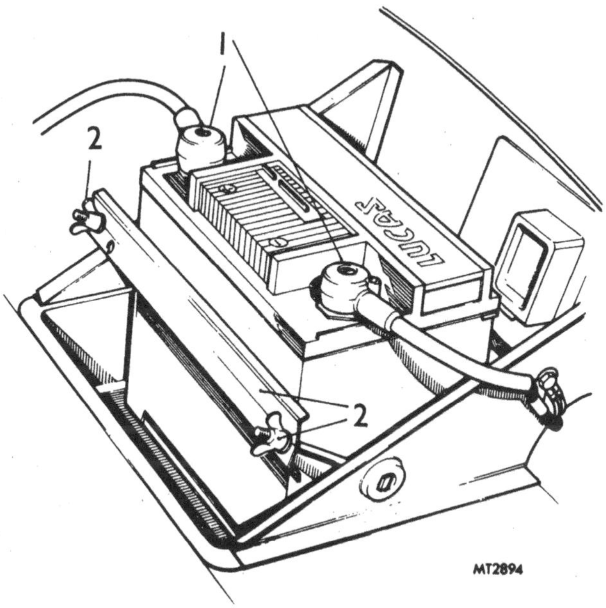

The Spitfire Mark 4 model range is equipped with Delco Remy distributor. These vehicles are built with variations in power unit according to current market specification requirements. Distributor units are employed with different centrifugal advance, vacuum advance and retard unit characteristics to match each power unit.

An 'Ignition distributor applicability, ignition timing and features chart is provided below. Data information for each unit is given on the following pages.

IGNITION DISTRIBUTOR APPLICABILITY, IGNITION TIMING AND FEATURES CHART

	NORMAL		U.S.A.		SWEDEN	
	Fitted up to Engine No. FH 25000 HE or LE	Fitted from Engine No. FH25001 HE or LE	Fitted up to Engine No. FK25000 HE or LE	Fitted from Engine No. FK25001 HE	Fitted up to Engine No. FL 2 HE	Fitted from Engine No. FH 25001 HE
Ignition distributor Stanpart No.	214088	218099	311761	312222	311761	218099
Ignition timing – static	6 degree B.T.D.C.	8 degree B.T.D.C.	6 degree A.T.D.C.	8 degree B.T.D.C.	6 degree A.T.D.C.	8 degree B.T.D.C.
			Use the static adjustment only to start the engine if the ignition timing setting has been lost. A dynamic adjustment is necessary to produce the accuracy of ignition timing required to meet the legal requirements in respect of air pollution throughout the speed range. The above statement is especially true of these distributors. This is because to meet the legal requirements in respect of air pollution the timing is exceptionally retarded at idle. At cruising speeds normal timing is required. To meet these two demands the distributor features an exceptionally steep advance curve below 1000 rev/min.			
Ignition timing – at idle	–	–	2 degree A.T.D.C.	2 degree A.T.D.C.	2 degree A.T.D.C.	–
Idle speed	–	–	800 to 850 rev/min	800 to 850 rev/min	800 to 850 rev/min	–
Centrifugal advance	X	X	X	X	X	X
Vacuum advance	X	X				X
Retard unit				X		
Micrometer adjustment nut	X					

IGNITION DISTRIBUTOR

Data

Manufacturer	Delco Remy
Series	D 204
Delco Remy part No.	7953460
Stanpart No.	214088
Contact gap	0·016 ± 0·001 in (0·38 to 0·43 mm)
Rotation – viewed on rotor	Anticlockwise
Firing angles	90 ± 1 degree
Dwell angle	39 ± 1 degree
Open angle	51 ± 1 degree
Moving contact spring tension	19 to 24 ozf (540 to 680 g)
Condenser capacity	0·18 to 0·23 microfarad
Engine firing order	1–3–4–2

Centrifugal advance

Distributor r.p.m.	Degs. distributor advance		Crankshaft r.p.m.	Degs. crankshaft advance	
	Minimum	Maximum		Minimum	Maximum
400	0	1·90	800	0	3·8
725	6·00	8·00	1450	12·0	16·0
1100	7·50	9·50	2200	14·0	19·0
1500	9·00	11·00	3000	18·0	22·0
2250	11·25	13·25	4500	22·5	26·5
2500	13·25	13·25	5000	26·5	26·5

Vacuum advance

Inches of mercury vacuum	Degs. distributor advance		Degs. crankshaft advance	
	Minimum	Maximum	Minimum	Maximum
4.0	No advance to occur			
6·2	0	3·1	0	6·2
10·7	5·5	7·5	11·0	15·0
20·0	7·5	7·5		

IGNITION DISTRIBUTOR

Data

Manufacturer	Delco Remy
Series	D 204
Delco Remy part No	7992128
Stanpart No.	218099
Contact gap	0·016 ± 0·001 in (0·38 to 0·43 mm)
Rotation – viewed on rotor	Anticlockwise
Firing angles	90 ± 1 degree
Dwell angle	39 ± 1 degree
Open angle	51 ± 1 degree
Moving contact spring tension	19 to 24 ozf (540 to 680 g)
Condenser capacity	0·18 to 0·23 microfarad
Engine firing order	1–3–4–2

Centrifugal advance

Distributor r.p.m.	Degs. Distributor advance		Crankshaft r.p.m.	Degs. crankshaft advance	
	Minimum	Maximum		Minimum	Maximum
400	No advance to occur		800	No advance to occur	
700	3·00	5·00	1400	6·00	10·00
750	4·00	6·00	1500	8·00	12·00
1250	5·40	7·49	2500	10·80	14·98
2000	7·71	9·71	4000	15·42	19·42
2500	10·00	10·00	5000	20·00	20·00

Vacuum advance

Ins. of mercury vacuum	Degs. distributor advance		Degs. crankshaft advance	
	Minimum	Maximum	Minimum	Maximum
3	No advance to occur			
5	0	0·84	0	1·68
8	1·52	3·52	3·04	7·04
10	3·20	5·20	6·40	10·40
15	4·00	6·00	8·00	12·00
20	6·00	6·00	12·00	12·00

IGNITION DISTRIBUTOR

Data

Manufacturer	Delco Remy
Series	D 204
Delco Remy part No.	7953944
Stanpart No.	311761
Contact gap	0·016 ± 0·001 in (0·38 to 0·43 mm)
Rotation – viewed on rotor	Anticlockwise
Firing angles	90 ± 1 degree
Dwell angle	39 ± 1 degree
Open angle	51 ± 1 degree
Moving contact spring tension	19 to 24 ozf (540 to 680 g)
Condenser capacity	0·18 to 0·23 microfarad
Engine firing order	1–3–4–2

Centrifugal advance

Distributor r.p.m.	Degs. distributor advance		Crankshaft r.p.m.	Degs. crankshaft advance	
	Minimum	Maximum		Minimum	Maximum
300	0	1·00	600	0	2·00
600	6·20	8·20	1200	12·40	16·40
750	9·80	11·80	1500	19·60	23·60
1000	10·58	12·58	2000	21·16	25·16
1600	12·44	14·44	3200	24·88	28·88
2100	14·00	16·00	4200	28·00	32·00
2500	16·00	16·00	5000	32·00	32·00

IGNITION DISTRIBUTOR

Data

Manufacturer	Delco Remy
Series	D 204
Delco Remy part No.	7992115
Stanpart No.	312222
Contact gap	0·016 ± 0·001 in (0·38 to 0·43 mm)
Rotation – viewed on rotor	Anticlockwise
Firing angles	90 ± 1 degree
Dwell angle	39 ± 1 degree
Open angle	51 ± 1 degree
Moving contact spring tension	19 to 24 ozf (540 to 680 g)
Condenser capacity	0·18 to 0·23 microfarad
Engine firing order	1–3–4–2

Centrifugal advance

Distributor r.p.m.	Degs. distributor advance		Crankshaft r.p.m.	Degs. crankshaft advance	
	Minimum	Maximum		Minimum	Maximum
400	No advance to occur		800	No advance to occur	
700	3·0	5·0	1400	6·0	10·0
750	4·0	6·0	1500	8·0	12·0
1600	6·6	8·6	3200	13·2	17·2
2400	9·0	11·0	4800	18·0	22·0
2800	11·0	11·0	5600	22·0	22·0

Retard unit

Ins of mercury vacuum	Degs distributor retard		Degs. crankshaft retard	
	Minimum	Maximum	Minimum	Maximum
4	No advance to occur			
6	0	2	0	4
10	4	6	8	12
15	4	6	8	12
20	6	6	12	12

IGNITION DISTRIBUTOR — SPITFIRE MARK 3

Normal component

Data – normal compression ratio engine 9·0 : 1

Manufacturer	Delco Remy
Series	D200
Delco Remy part No.	7953460
Stanpart No.	214088
Contact gap	0·015 in.
Rotation – viewed on rotor	Anticlockwise
Firing angles	90 degs.
Dwell angle	40 – 42 degs.
Open angle	48 – 50 degs.
Moving contact spring tension	22 – 26 ozs.
Condenser capacity	0·18 – 0·23 mfd.
Engine firing order	1 – 3 – 4 – 2

Centrifugal advance

Distributor r.p.m.	Degs. distributor advance		Crankshaft r.p.m.	Degs. crankshaft advance	
	Minimum	Maximum		Minimum	Maximum
300	No advance to occur		600	No advance to occur	
400	0	1·90	800	0	3·8
725	6·00	8·00	1450	12·0	16·0
1100	7·50	9·50	2200	15·0	19·0
1500	9·00	11·00	3000	18·0	22·0
2500	—	13·25	5000	—	26·5

Vacuum advance

Ins. of mercury vacuum	Advance
4·0 – 6·2	Distributor must start to advance
7·7 – 10·7	5·5 degs. distributor advance – 11 degs. crankshaft advance – must occur
20	7·5 degs. distributor advance – 15 degs. crankshaft advance – must not be exceeded

IGNITION DISTRIBUTOR — SPITFIRE MARK 3

Emission control component

Data

Manufacturer	Delco Remy
Series	D200
Delco Remy part No.	7953557
Stanpart No.	214799
Contact gap	0·015 in.
Rotation – viewed on rotor	Anticlockwise
Firing angles	90 degs.
Dwell angle	40 – 42 degs.
Open angle	48 – 50 degs.
Moving contact spring tension	17 – 21 ozs.
Condenser capacity	0·18 – 0·23 mfd.
Engine firing order	1 – 3 – 4 – 2

Centrifugal advance

Distributor r.p.m.	Degs. distributor advance		Crankshaft r.p.m.	Degs. crankshaft advance	
	Minimum	Maximum		Minimum	Maximum
258	No advance to occur		516	No advance to occur	
300	0	1·00	600	0	2·00
750	9·80	11·80	1500	19·60	23·60
1000	10·58	12·58	2000	21·16	25·16
1600	12·44	14·44	3200	24·88	28·88
2000	14·00	16·00	4000	28·00	32·00
2500	—	16·00	5000	—	32·00

Vacuum advance

Run distributor at 500 RPM maximum in the appropriate direction of rotation.

Ins. of mercury vacuum	Advance
0 – 4·0	No advance to occur
4·0 – 6·2	Distributor must start to advance
7·7 – 10·7	5·5 degs. distributor advance – 11 degs. crankshaft advance – must occur
20	7·5 degs. distributor advance – 15 degs. crankshaft advance – must not be exceeded

IGNITION COIL AND BALLAST RESISTOR

Ignition coil	Mk. III	Mk. IV	1500
Manufacturer	Lucas	Lucas	Lucas
Type	HA12	16C6	15C6
Lucas part No.	5195	45232	45243
Stanpart No.	—	154272	158830
Primary winding resistance . .	3·1 to 3·5 ohms	1·43 to 1·58 ohms	1·30 to 1·45 ohms

Ballast resistor unit — [Mk. IV]

Manufacturer	Lucas
Type	3BR
Lucas part No.	47170
Stanpart No.	134176
Resistance	1·3 to 1·4 ohms

Ballast resistor wire — [1500]

Resistance	1·3 to 1·5 ohms
Wire length	63 in (160 cm)

Mk. IV vehicles only —

This system is designed to assist engine starting under adverse conditions. A ballast resistor is positioned in series in the normal supply to the ignition coil. This unit causes a voltage drop in the circuit so that the 12-volt supply from the ignition switch may be employed to power the nominally rated 6-volt ignition coil.

1500 vehicles only —

This system is designed to assist engine starting under adverse conditions. A ballast resistor wire built into the harness between two crimped joints is electrically positioned in series in the normal supply to the ignition coil. This wire causes a voltage drop in the circuit so that the 12 volt supply from the ignition switch may be employed to power the nominally rated 6 volt ignition coil.

All vehicles [except Mk. III] —

During engine start the resistor is by-passed and the battery voltage (reduced from 12-volt by the starter motor load) is applied to the coil direct from the starter solenoid. This slight voltage overload provides an increased high tension voltage at the spark plugs.

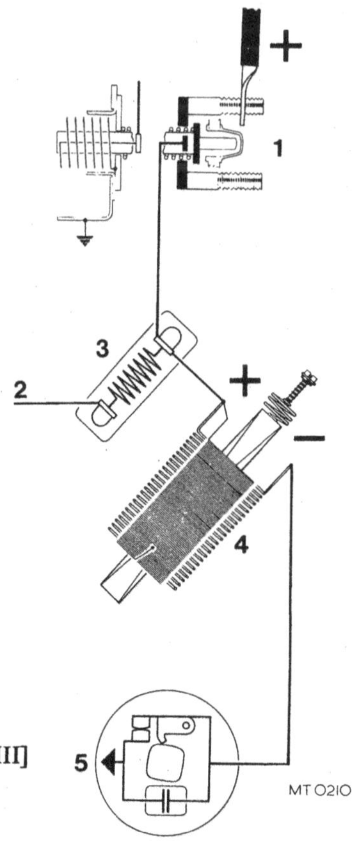

1. Starter solenoid
2. Normal ignition supply
3. Ballast resistor unit – shown
   or
   Ballast resistor wire – not shown
4. Ignition coil — 6 volt [12 volt on Mk. III]
5. Ignition distributor

IGNITION DISTRIBUTOR

Contact assembly

– Remove and refit 86.35.13

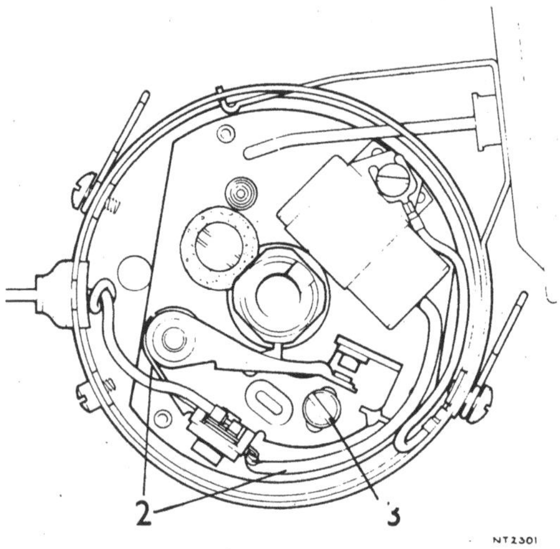

Removing

1. Remove the distributor cover and rotor.
2. Push the moving contact spring away from the terminal post and detach the two leads.
3. Remove the screw securing the fixed contact and lift out the contact assembly.

Refitting

4. Wipe preservative from new contact faces.
5. Reverse instructions 1 to 3.
6. Adjust contact gap 86.35.14

IGNITION DISTRIBUTOR

– Contact gap 86.35.14

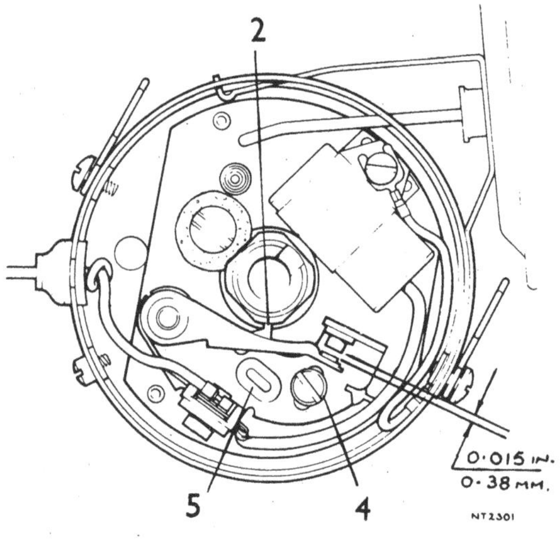

Adjust

1. Remove the distributor cover and rotor.
2. Rotate the crankshaft to position the contact heel on a cam peak.
3. If the contact gap is correct, an 0·015 in (0·38 mm) feeler gauge will just slide between the contacts.
4. If correction is required, slacken the screw securing the fixed contact.
5. Move the fixed contact to obtain the correct gap.
6. Retighten the fixed contact securing screw.
7. Refit the rotor and distributor cover.

IGNITION DISTRIBUTOR

– Ignition timing – adjust 86.35.15

For normal market vehicles a static adjustment is satisfactory.

For all U.S.A. market vehicles, also Swedish market vehicles up to Engine No. FL2 HE use the static adjustment only to start the engine if the ignition timing setting has been lost. A dynamic adjustment is necessary to produce the accuracy of ignition timing required to meet the legal requirements in respect of air pollution throughout the speed range.

Static

1. Adjust the contact gap 86.35.14.
2. Disconnect the distributor l.t. lead from the coil.
3. Provide a test lamp circuit as shown.
4. Rotate the crankshaft in the engine run direction to bring the pulley notch nearly up to the edge of the scale. The test lamp should now be illuminated.
5. Carefully rotate the crankshaft further until the lamp just goes out.
6. If the timing is correct the pulley notch will be aligned with the scale at the static figure given in the chart 86.35.00.
7. If correct operations 8 to 10 may be ignored.
8. Distributor fitted with micrometer adjustment nut only – if a small correction is required rotate the micrometer adjustment nut to advance or retard the timing and repeat operation 4 onwards. If a large correction is required centre the micrometer adjustment nut and perform operation 9 onwards.
9. Slacken the clamp bolt. Align the pulley notch with the scale at the static figure given in the chart 86.35.00. Rotate the distributor body anticlockwise past the test lamp illumination position. Carefully rotate clockwise until the lamp just goes out. Tighten the clamp bolt with the unit in this position.
10. Repeat operation 4 onwards.

continued

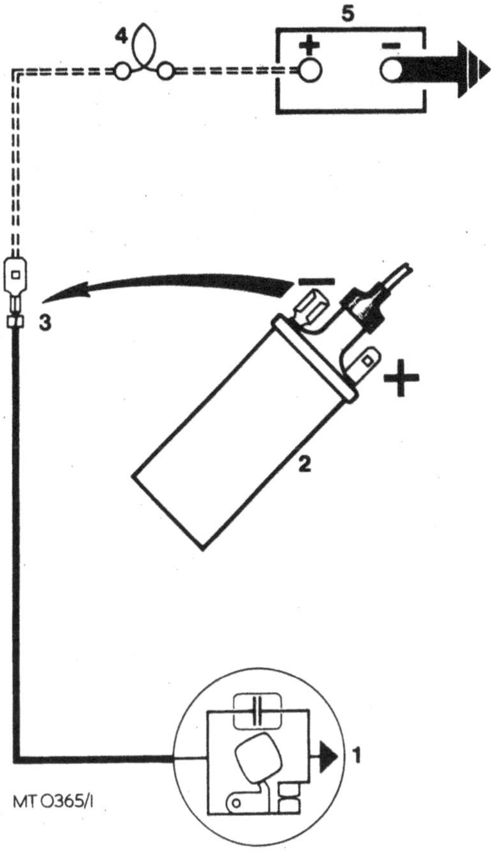

1. Distributor – diagrammatic layout
2. Ignition coil
3. Distributor lead removed from coil
4. Test lamp – 12 volt
5. Vehicle battery

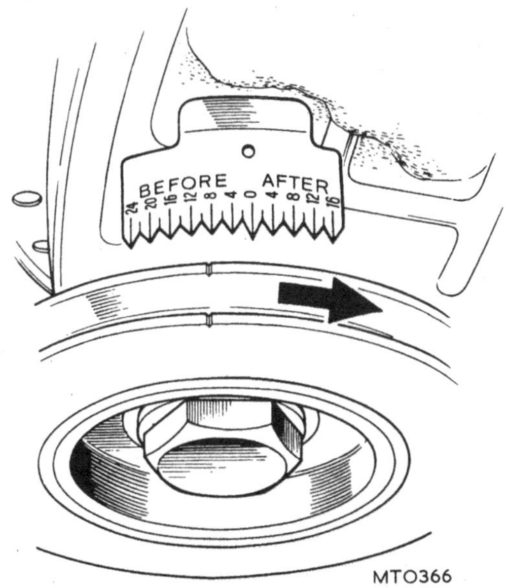

Dynamic

WARNING: **Ensure that personnel and equipment keep clear of the rotating cooling fan while using the timing light.**

11. Adjust the contact gap 86.35.14.
12. Pull off the retard unit pipe if fitted.
13. Connect timing light as instructed by the manufacturer. The engine is timed on No. 1 cylinder which is located at the front of the engine.
14. Run the engine at idle speed 800 to 850 revs/min.
15. Position timing light to illuminate crankshaft pulley and scale.
16. If the timing is correct the equipment will show the 'at idle' figure given in the chart 86.35.00.
17. If correct operation 18 may be ignored.
18. With the engine still running at 800 to 850 revs/min use a second operator to slacken the clamp bolt and carefully rotate the distributor body as required until the equipment shows the 'at idle' figure given in the chart 86.35.00. Tighten the clamp bolt with the unit in this position.
19. Replace the retard unit pipe if fitted.

IGNITION DISTRIBUTOR

– Lubrication 86.35.18

1. Remove the distributor cover and rotor.
2. Apply a few drops of thin oil to the felt plug, oil holes and contact mounting post. Carefully wipe off any excess oil.
3. Lightly grease the cam (or lubricator pad, if fitted) with Shell Alvania No. 2 grease or equivalent.

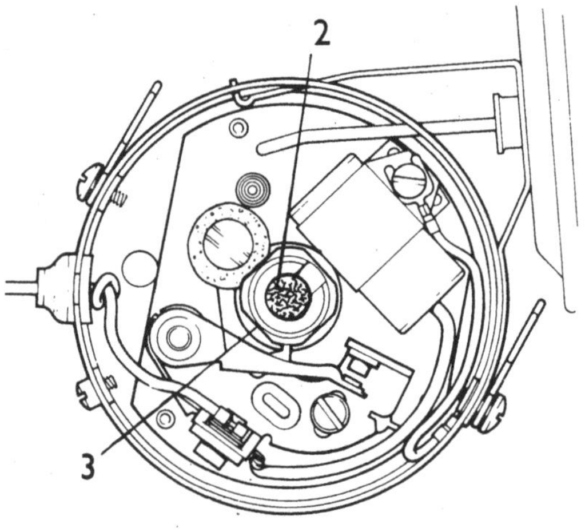

IGNITION DISTRIBUTOR

– Remove and refit 86.35.20

Removing

1. Remove the distributor cover.
2. Pull off the distributor l.t. connector from the coil.
3. Pull off the vacuum control pipe (if fitted).
4. Unscrew the knurled connector securing the tachometer drive.
5. Remove one bolt and spring washer securing clamp bracket to pedestal.
6. Withdraw the distributor.

NOTE: To facilitate re-timing do not slacken the clamp bolt.

Refitting

7. Rotate the crankshaft until No. 1 piston is at T.D.C. and the offset distributor drive gear slot is in the position shown.
8. Reverse instructions 1 to 6.
9. Adjust ignition timing if necessary 86.35.16.

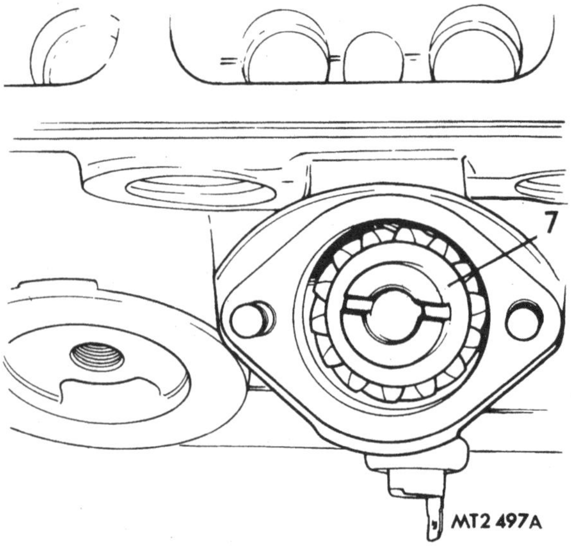

IGNITION DISTRIBUTOR

– Overhaul 86.35.26

Dismantling

1. Remove the contact assembly 86.35.13.
2. Remove the capacitor – one screw and spring washer.
3. Pull off the cam lubricator pad if fitted.
4. Remove the screws and detach the earth lead and the cap securing clips if fitted.
5. Withdraw the contact breaker base plate and detach the vacuum control if fitted.
6. Pierce the staked plug and prise it out.
   CAUTION: Do not perform this operation unless a new plug is available.
7. Withdraw the tachometer drive gear and thrust washer.
8. File and tap out the coupling pin. Remove the coupling and thrust washer.
   CAUTION: Do not perform this operation unless a new coupling pin is available.
9. Withdraw the shaft assembly and thrust washer.
10. Remove the control springs.

Reassembling

11. Fit new control springs.
    CAUTION: Do not over extend the springs.
12. Lubricate and install the shaft assembly and thrust washer.
    NOTE: If fitting a new undrilled shaft, install and drill so that the end float is 0·002 to 0·005 in (0·02 mm to 0·13 mm) and relationship of the coupling offset key to the rotor locating slot is as shown viewed from the coupling end.
13. Lubricate the tachometer drive gear with Shell Alvania No. 2 grease or equivalent. Install the gear and thrust washer.
14. Fit a new plug and stake in six equidistant positions.
15. Reverse instructions 1 to 5.

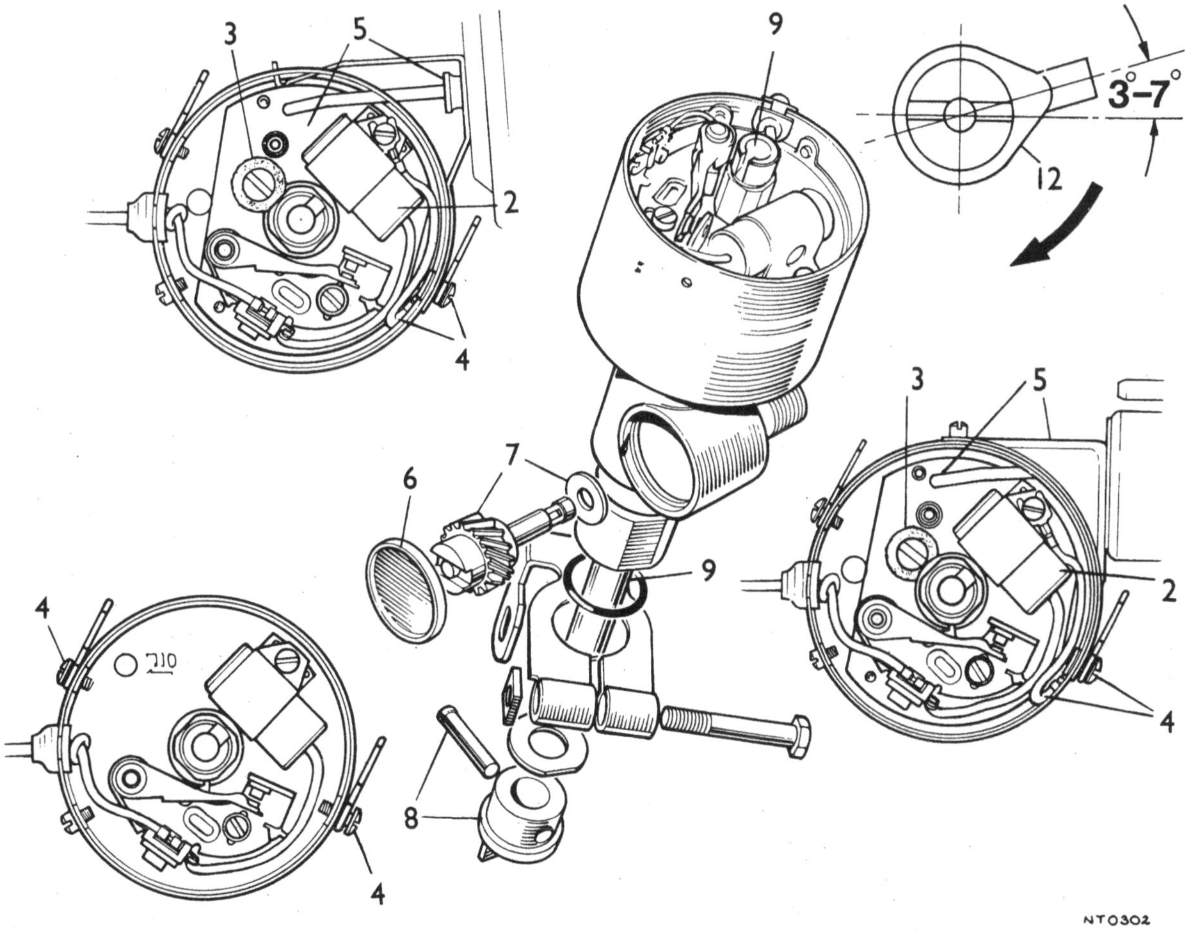

IGNITION COIL AND BALLAST RESISTOR

– Ignition coil – remove and refit **86.35.32**

Removing

1. Disconnect the two low tension Lucar connectors.
2. Pull off the high tension lead.
3. Remove the two screws, spring washers and washers, and lift out the coil.

Refitting

4. Position the coil and secure it with two washers, spring washers and screws. Include in the left hand screw assembly the ballast resistor lug.
5. Push on the high tension lead.
6. Connect the two low tension Lucar connectors as follows:
   White/yellow wire to positive terminal. [White wire on Mk. III.]
   White/black wire to negative terminal.

IGNITION COIL AND BALLAST RESISTOR [Mk. IV onl]

Ballast resistor – remove and refit **86.35.33**

Removing

1. Locate the ballast resistor on the left hand engine bay valance adjacent to the ignition coil.
2. Disconnect the Lucar connectors.
3. Remove the screw, spring washer and washer and lift out the ballast resistor.

Refitting

4. Position the ballast resistor and secure it with the washer, spring washer and screw.
5. Connect the Lucar connectors as follows:
   White wire to rear terminal
   White/yellow wire to front terminal.

LAMPS

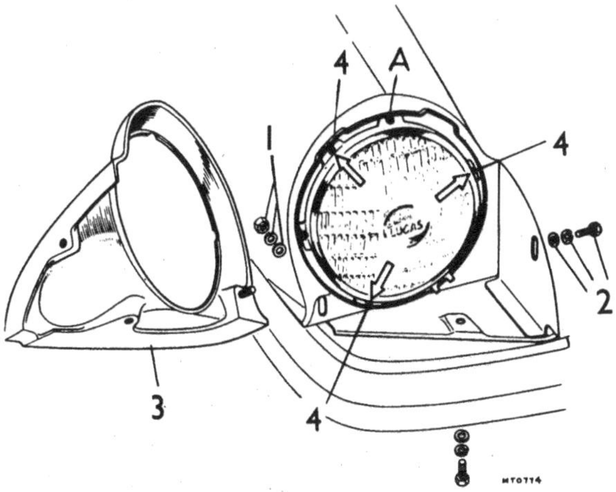

Headlamp – Remove and refit 86.40.02

Removing

1. Remove the nut, spring washer and plain washer.

[NOTE: On Mk. III cars, carefully prise off the snap-on rim surrounding the headlamp. Take out the rubber seal, then proceed with instruction number 4.]

2. Remove the two bolts, spring washers and plain washers.
3. Lift off the headlamp surround.
4. Remove three screws and lift off the light unit retaining rim.

NOTE: Do not disturb the two screws 'A' unless beam alignment is intended.

5. Pull the connector from the light unit.

Refitting

6. Reverse instructions 1 to 5.

LAMPS

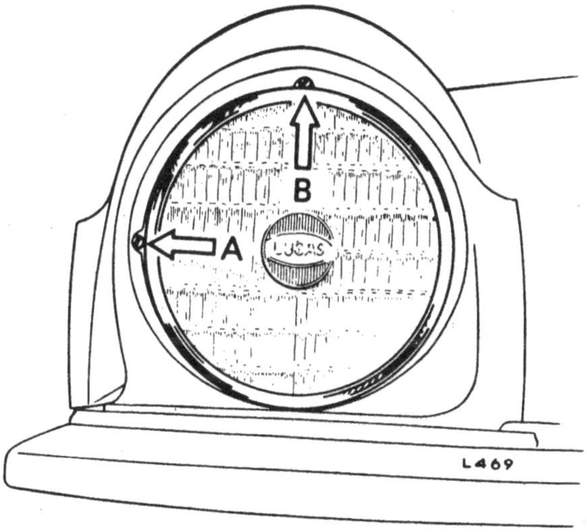

– Headlamp beam alignment 86.40.17

Screw 'A' positions the beam in the horizontal plane. Screw 'B' controls beam height.

Beam aiming can best be accomplished using equipment such as Lucas 'Beam-setter' or 'Lev-L-Lite'. This service is available at Triumph distributors or dealers and will ensure maximum road illumination with minimum discomfort to other road users.

Triumph Spitfire Mk IV Manual. Part No. 545254. Issue 1.

LAMPS

– Front parking and flasher lamp – remove and refit86.40.26

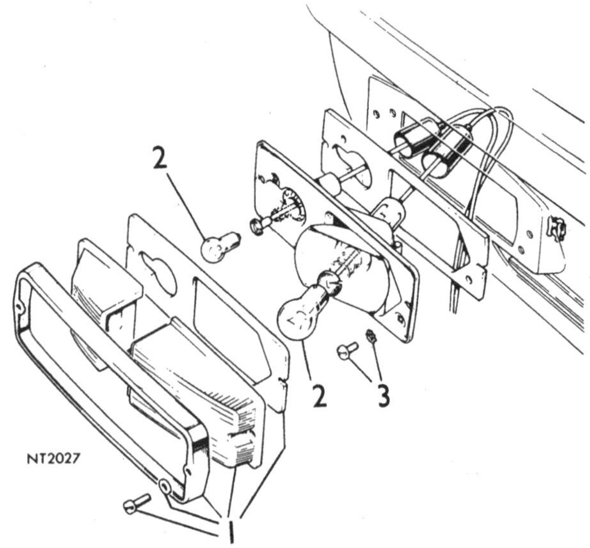

Removing

1. Remove the two screws and withdraw the rim, two lenses and gasket.
2. Remove the two bulbs from the bayonet fittings.
3. Remove the two screws and withdraw the lamp base and gasket.
4. Disconnect the earth wire from the lamp base.
5. Disconnect the two wires from the snap connectors.

Refitting

7. Reverse instructions 1 to 5.

LAMPS

– Front parking and flasher lamp (U.S.A. only) – remove and refit 86.40.26

Removing

1. Remove the two screws and withdraw the rim and lens.
2. Remove the bulb from the bayonet fitting.
3. Remove the two screws and withdraw the lamp base and gasket.
4. Disconnect the earth wire from the lamp base.
5. Disconnect the two wires from the snap connecters.

Refitting

6. Reverse instructions 1 to 5.

LAMPS

Front marker lamp (U.S.A. only)

– Remove and refit 86.40.59

Removing

1. Using a screwdriver, carefully prise the rim out of the lamp base.
2. Carefully pull out the lamp lens.
3. Remove the bulb from the bayonet fitting.
4. Carefully pull out the reflector lens.
5. Raise the bonnet.
6. Disconnect the red wire from the snap connector centrally located at the front of the bonnet.
7. Disconnect the black wire from the lamp.
8. Remove the three nuts, plain and rubber washers.
9. Withdraw the lamp base together with the bulb holder, wire and reflector base.
10. Remove the reflector base.
11. Remove the bulb holder and wire.

Refitting

12. Reverse instructions 1 to 11.

LAMPS

Rear marker lamp (U.S.A. only)

– Remove and refit 86.40.64

Removing

1. Using a screwdriver, carefully prise the rim out of the lamp base.
2. Carefully pull out the lamp lens.
3. Remove the bulb from the bayonet fitting.
4. Carefully pull out the reflector lens.
5. Open the luggage compartment lid.
6. Disconnect the red wire from the snap connector.
7. Disconnect the black wire from the lamp.
8. Remove the three nuts, plain washers and rubber washers.
9. Withdraw the lamp base together with the bulb holder, wire and reflector base.
10. Remove the reflector base.
11. Remove the bulb holder and wire.

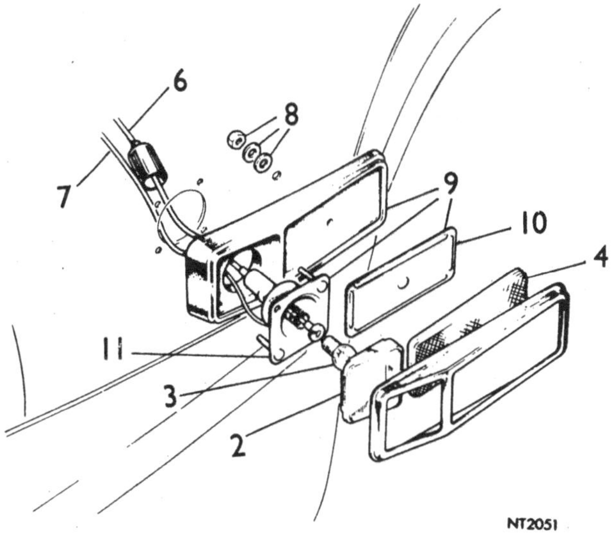

NT2051

Refitting

12. Reverse instructions 1 to 11.

LAMPS

– Rear tail/stop, flasher and reverse lamp
– remove and refit **86.40.70**

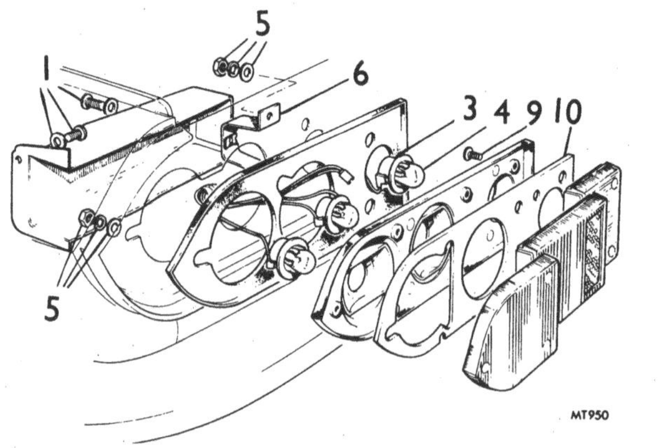

[NOTE: On Mk. III cars, the tail light lens is secured by screws in addition to being held in place by rubber. Remove the screws. Then, using a screwdriver, turn back the rubber and remove the rim. This permits the glass lens to be similarly removed for access to the bulb and the lamp mounting components.]

Removing

1. Remove the two screws and plain washers securing the lamp cover.
2. Disconnect the Lucar connector.
3. Pull out the three bulbholders.
4. Remove the three bulbs from the bayonet fittings.
5. Remove the five nuts, spring washers and plain washers securing the lamp to the body.
6. Detach the cover mounting bracket and withdraw the lamp.
7. Remove the flasher lamp lens – three screws.
8. Remove the tail/stop lamp lens – two screws.
9. Remove the reverse lamp lens – two screws.
10. Remove the lens seating gasket.

Refitting

11. Reverse instructions 1 to 10.

[Removing (Later U.S.A. cars only)

1. **Remove two screws and washers and withdraw the lens assembly.**
2. **If required, carefully remove the festoon bulb.**
3. **Disconnect two leads behind the lamp in the luggage compartment.**
4. **Remove two nuts and plain washers and withdraw the lamp from the car.**

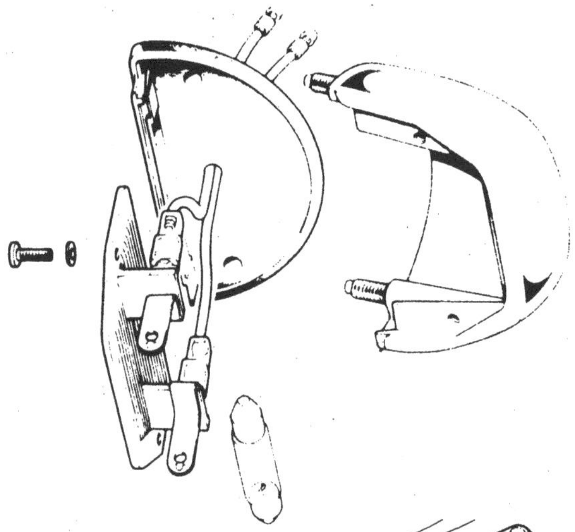

Refitting (Later U.S.A. cars only)

5. **Reverse instructions 1 to 4 ensuring a good weather-resistant seal.]**

REAR NUMBER PLATE LAMP

– Remove and refit **86.40.86**

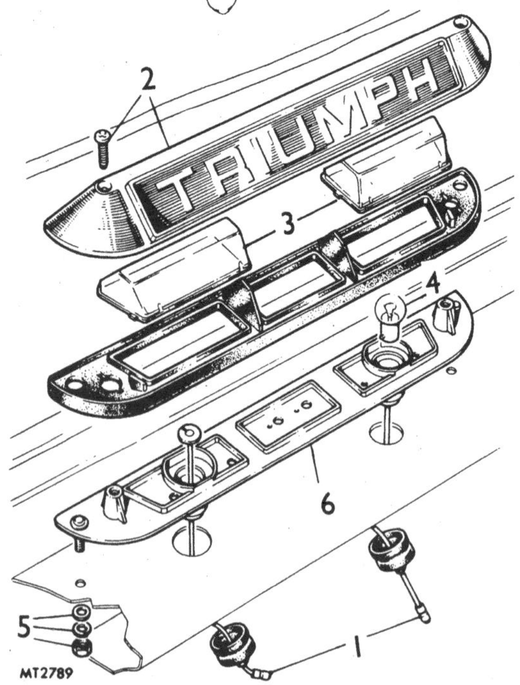

Removing

1. Disconnect the two leads from inside the luggage compartment and pull them through the panel grommet.
2. Remove two screws and lift off the cover.
3. Remove the lenses.
4. Remove the bulbs.
5. Remove the two nuts, spring washers and plain washers.
6. Remove the lamp base from the bumpers.

Refitting

7. Reverse instructions 1 to 6.

86.40.70
86.40.86

RELAYS

– Night dimming relay [except Mk. III]

Data and description **86.55.00**

Manufacturer	Lucas
Type	11RA
Lucas part No.	33329
Stanpart No.	217177
Body colour	Yellow
Winding resistance	36 ohm
Pull in voltage	4 to 10 volt
Release voltage	0 to 1 volt
R1 resistance	3 ohm
R2 resistance	3 ohm
R3 resistance	30 ohm
R4 resistance	1.5 ohm

This relay dims stop lamps and rear flasher lamps when the parking lamps are illuminated. At night minimum discomfort to other road users is ensured. During daylight the safety of full intensity is provided.

The relay winding is controlled by the parking lamp circuit. Three sets of normally closed contacts with associated resistors permanently connected in parallel are featured. The pair of stop lamps, the left hand rear flasher lamp and the right hand rear flasher lamp are each associated with one contact/resistor set.

When the relay is not energized the contacts are closed and the lamps operate at full intensity. When the relay is energized the contacts are open. The resistors are positioned in series with the lamps which then operate at reduced voltage.

To maintain the correct frequency of operation of the turn signal flasher unit the current must be kept constant. To provide current compensation for either flasher circuit a fourth resistor is used. This resistor is brought into circuit by the double contact set between terminals 2 and 5. The switching is such that when either flasher circuit is selected, a parallel path exists through R3, the resistor of the unselected flasher circuit and across the filament of the unselected rear flasher lamp to earth. While providing compensation, this current is not large enough to illuminate the unselected lamp.

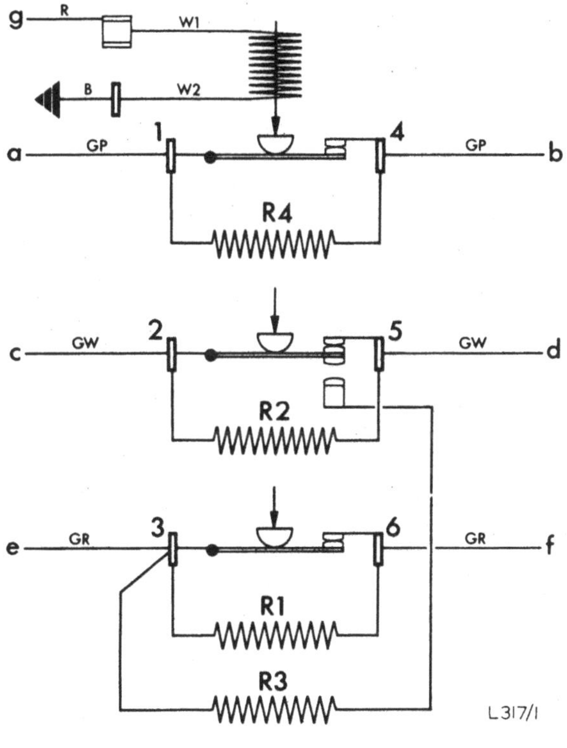

a. From stop lamp switch
b. To pair of stop lamps
c. From turn signal switch
d. To L.H. rear flasher lamp
e. From turn signal switch
f. To R.H. rear flasher lamp
g. From master light switch

RELAYS

– Night dimming relay [except Mk. III]

– Remove and refit 86.55.03

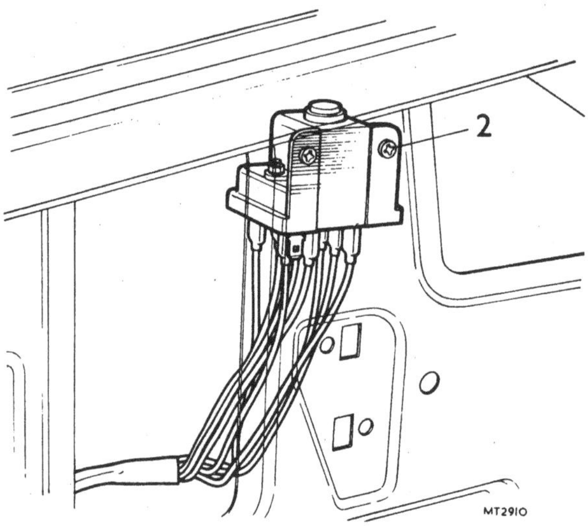

Removing

1. Open the luggage compartment lid and locate the relay behind the left hand wheelarch outer panel.
2. Remove the two screws and withdraw the relay.
3. Disconnect the eight Lucar connectors.

Refitting

4. To ensure the correct operation of the compensation circuit the following test is necessary to establish which green/red wire is from the turn signal switch and which is to the left hand rear flasher lamp. The pair of green/white wires must similarly be identified.
5. In order to perform the test a positive supply may be obtained in the area by connecting a short slave wire to the red wire and switching on the side lamp circuit.
6. Using the live slave wire touch each green/red wire in turn. Illumination of the left hand rear flasher lamp will identify the wire to be connected to terminal 6. Connect the remaining green/red wire to terminal 3.
7. Using the live slave wire touch each green/white wire in turn. Illumination of the right hand rear flasher lamp will identify the wire to be connected to terminal 5. Connect the remaining green/white wire to terminal 2.
8. Connect the green/purple wires to terminals 1 and 4. They may be fitted either way round.
9. Remove the slave wire and connect the red wire to terminal W1.
10. Connect the black wire to terminal W2.
11. Reverse instructions 1 and 2.

RELAYS

– Overdrive relay

– Remove and refit 86.55.04

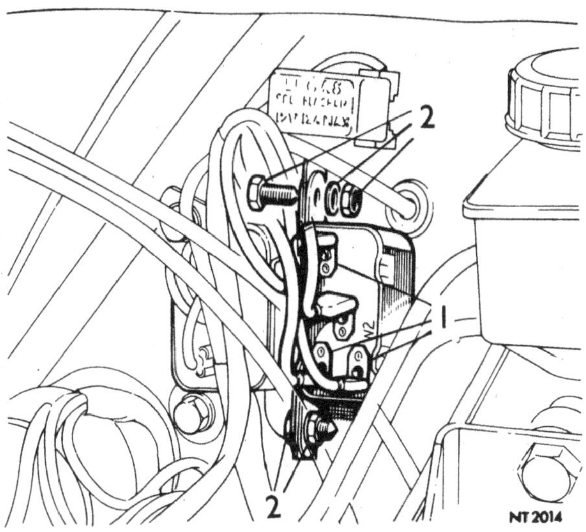

Removing

1. Disconnect the four Lucar connectors.
2. Remove the two nuts, bolts and plain washers securing the horn and overdrive relays to the bulkhead.

Refitting

3. Reverse instructions 1 and 2. Connect the Lucar connectors as follows:
   Brown wire to terminal C1.
   Yellow/purple wire to terminal C2.
   White wire to terminal W1.
   Yellow/green wire to terminal W2.

RELAYS

– Starter solenoid – remove and refit 86.55.05

Removing

1. Isolate the battery.
2. Note the four wire colour codes and positions.
3. Disconnect the four Lucar connectors.
4. Note the position of the battery lead and starter motor lead. This is important to ensure correct operation of the ballast resistor starter system.
5. Pull back the rubber protector.
6. Remove the two nuts and spring washers. Disconnect the battery lead and starter motor lead.
7. Note the position of the solenoid.
8. Remove the two screws, washers and spring washers, and lift the solenoid from the vehicle.

Refitting

9. Reverse instructions 1 to 8. Ensure that the battery lead, starter motor lead and four Lucar connectors are positioned as shown. Ensure good electrical contact between the solenoid and the vehicle body.

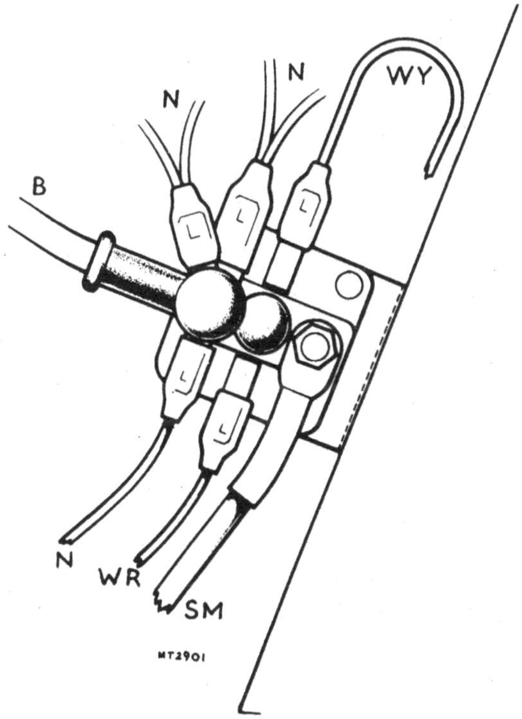

B Battery lead
N Brown wire
WY White/Yellow wire
N Brown wire
WR White/Red wire
SM Starter motor lead

RELAYS

– Horn relay – remove and refit 86.55.09

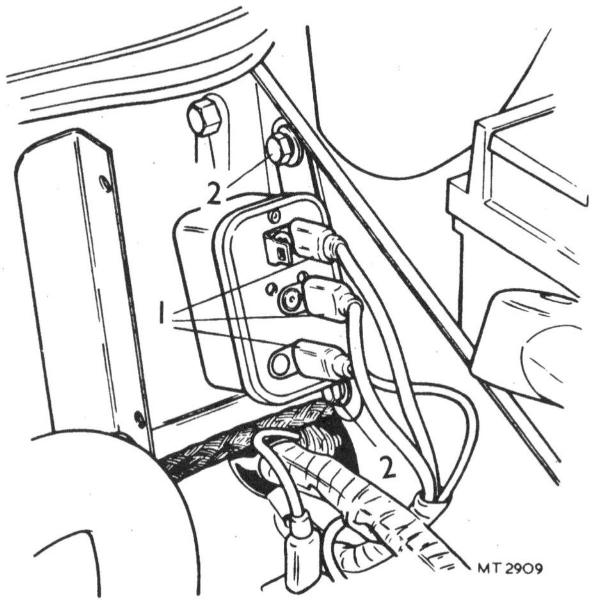

Removing

1. Disconnect the three Lucar connectors.
2. Remove the two bolts, spring washers and plain washer securing the relay to the bulkhead.

Refitting

3. Reverse instructions 1 and 2. Connect the Lucar connectors as follows:

   Purple/yellow wire to terminal C1
   Purple wires to terminal C2
   Purple/black wires to terminal W1

FLASHER UNIT

– Turn signal flasher unit – remove and refit 86.55.11

Removing

1. Locate the flasher unit mounted in clip on the forward face of the fascia panel beneath the speedometer on right hand steer cars.

[NOTE: On Mk. III cars, the flasher unit is located next to the rear side of the fuse box.]

2. Pull the flasher unit from the clip.
3. Disconnect the two Lucar connectors.

Refitting

4. Connect the Lucar connectors: Green wire to terminal 'B'. Light green/brown wire to terminal 'L'.
5. Fit the flasher unit to the clip.

FLASHER UNITS

Hazard flasher unit – remove and refit 86.55.12

The hazard flasher unit (if fitted) is located on the left hand side of the bulkhead.

Triumph Spitfire Mk IV Manual. Part No. 545254. Issue 1.

BUZZERS

– Key warning buzzer – remove and refit 86.55.13

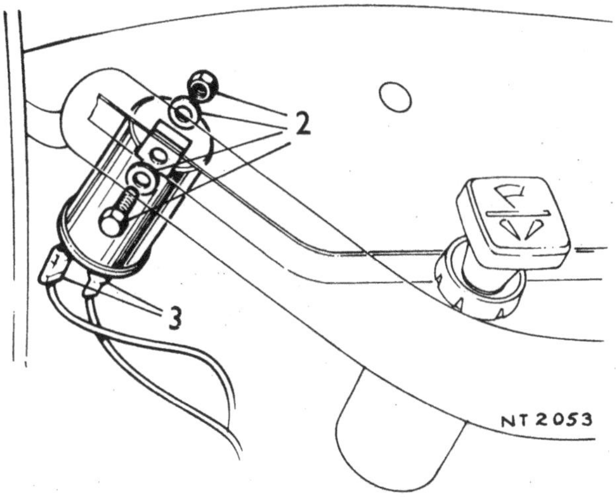

Removing

1. Locate the buzzer on the extreme left hand side of the fascia panel.
2. Remove the nut, bolt and plain washers.
3. Disconnect the two Lucar connectors.

Refitting

4. Reverse instructions 1 to 3. Connect the Lucar connectors as follows:
   Purple/Orange wire to terminal W1
   Black/Green wire to terminal C1

NIGHT DIMMING RELAY [except Mk. III]

– Contacts adjust 86.55.14

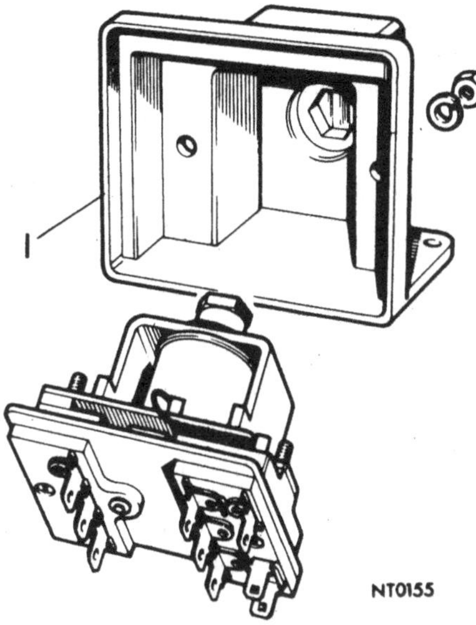

The contacts are set during production and normally require no attention in service. If the relay cover has been removed and the hexagon rotated, re-set as follows:

1. Remove the cover.
2. Provide a test circuit as shown.
3. With the winding not energised rotate the hexagon by hand fully clockwise.
4. Check that the three contact sets are correctly closed.
5. Energise the winding by applying 12 volts between W1 and W2.
6. Rotate the hexagon anticlockwise until the test lamp just illuminates.
7. Rotate the hexagon a further 90 degrees anticlockwise.
8. Observe the recess for the hexagon in the cover. If necessary rotate the hexagon a little further anticlockwise to achieve alignment.
9. Fit the cover.

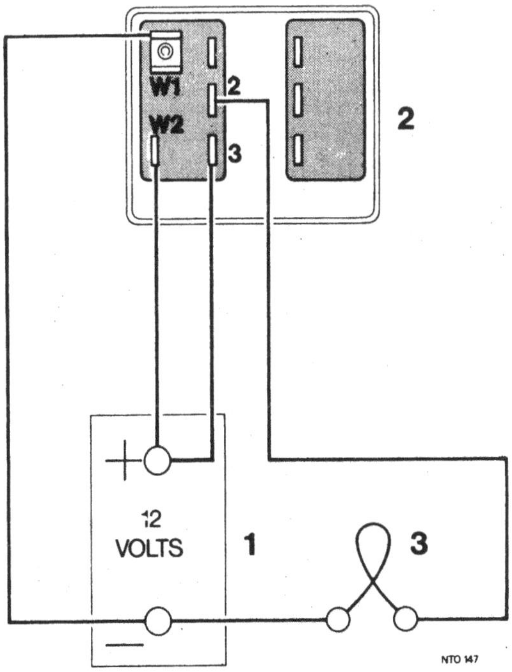

STARTER MOTOR

– Data and description 86.60.00

Manufacturer	Lucas
Type	M35J **[M35G on Mk. III]**
Lucas Part No.	25149
Stanpart No.	200535
Yoke diameter	3·5 in (88·90 mm)
Light running – current	65 amp
– speed	8,000 to 10,000 rev/min
Running torque – force	4·4 lbf. ft. (0·60 kgf. m)
– current	260 to 275 amp
– speed	1,000 rev/min
Lock torque – force	7·0 lbf. ft. (0·95 kgf. m)
– current	350 to 375 amp

M35G Performance Data

ARMATURE SPEED	TORQUE		CURRENT CONSUMPTION	
	lbs. ft.	Kgms.	Amperes	Volts
Locked	10	1·38	420—440	7·9—7·3
1,000 r.p.m.	5·4	0·75	250—270	9·3—8·9
7,400—8,500 r.p.m. ..	No load		45	12

Commutator minimum skimming thickness	0·080 in (2·03 mm)
[Brush length (Type M35G)	0·315 in (8·00 mm)]

	[Mk. III]		[Mk. IV and 1500]	
Brush length – new	0·625 in	15.88 mm	0·500 in	12·70 mm
– renew if less than	0·375 in	9.53 mm	0·375 in	9·53 mm

Brush spring pressure 28 oz (800 g)

[Description (M35J)]

A series-wound, four-pole, four-brush motor with an extended shaft which carries a conventional inertia drive.

The armature shaft rotates in two porous bronze bushes. A squared extension of the shaft protrudes to enable the shaft to be rotated to clear any jamming between the inertia drive and engine flywheel ring gear. The armature features a face-type moulded commutator.

A plastic brush box is riveted to the commutator end bracket. It holds four wedge-shaped brushes and captive coil springs. The brushes are keyed to ensure correct fitting.

The field winding is a continuously wound strip with no joints. One end is attached to two brush flexibles, while the other is attached to a single flexible which is earthed to the yoke.

[On late models], the yoke is windowless and has no through-bolts. The commutator end bracket is secured by four screws which align with tappings in the yoke. The drive end bracket is attached by two slot-headed bolts which screw into tappings provided in the end faces of two of the pole-shoes.

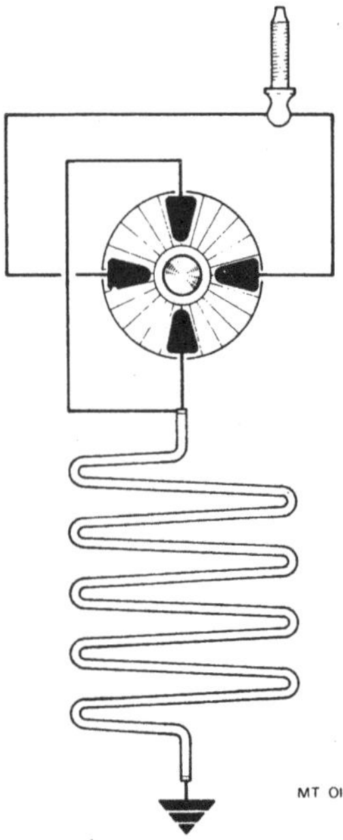

Triumph Spitfire Mk IV Manual. Part No. 545254. Issue 1.
Part No. 511243. page 6.102. Issue 1.

STARTER MOTOR

– Remove and refit 86.60.01

[CAUTION: Be extremely careful in removing the terminal nut from early-type starters which are identifiable by the presence of two through bolts. On such motors, the terminal is attached to the field coils. If the terminal stud is allowed to turn as the nut is removed, the field coils may be ruined.]

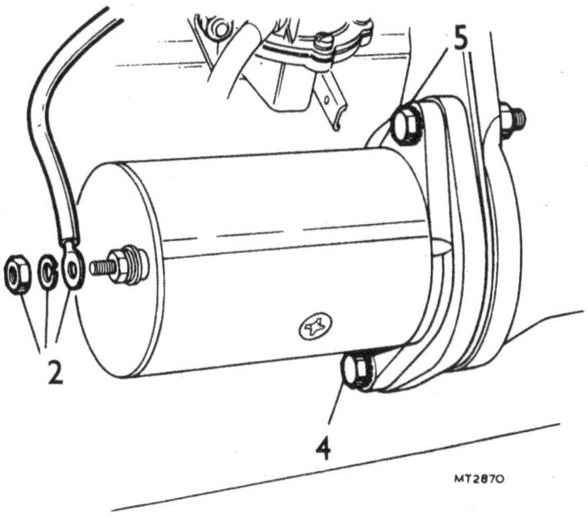

Removing

1. Isolate the battery.
2. Remove the nut and spring washer. Disconnect the lead from the terminal post.
3. Note the relationship between the starter motor, shims if fitted, packing and clutch housing.
4. Working from below the engine, remove the lower mounting bolt.
5. Working from above the engine, remove the upper mounting bolt.
6. Working from below the engine, withdraw the starter motor downwards from the vehicle complete with the packing and shims if fitted.

Refitting

7. Reverse instructions 1 to 6.

STARTER MOTOR

– Inertia drive – remove and refit 86.60.06

Removing

1. Dismantle the starter motor 86.60.13 operations 1 to 5.
2. Using a suitable press, compress the main spring and ease the jump ring from the shaft. Churchill hand press S4221A and adaptor S4221A-14 may be employed to perform this operation.
3. Withdraw the inertia drive components from the shaft.

Refitting

4. Lightly lubricate the drive end bracket bush with thin engine oil. Position the drive end bracket.
5. Fit the inertia drive components to the shaft as shown. The screwed sleeve may be fitted either way round.
6. Using a suitable press, compress the main spring and fit the jump ring to the shaft.
7. Reassemble the starter motor. 86.60.13 operations 22 to 26.

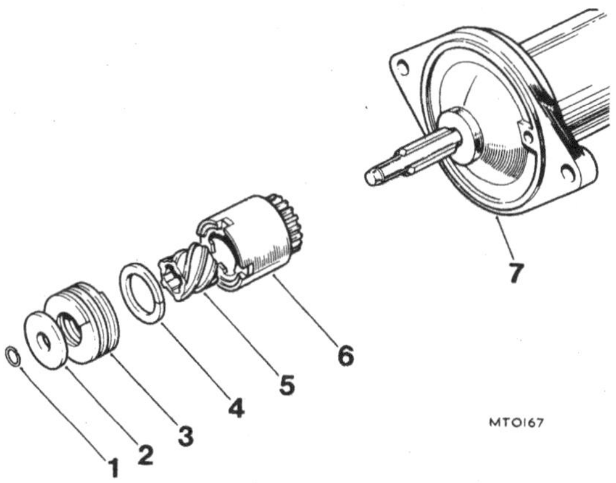

1. Jump ring
2. Shaft collar
3. Main spring
4. Buffer washer
5. Screwed sleeve
6. Pinion and barrel
7. Drive end bracket

Exploded View of early Mk. III Starter

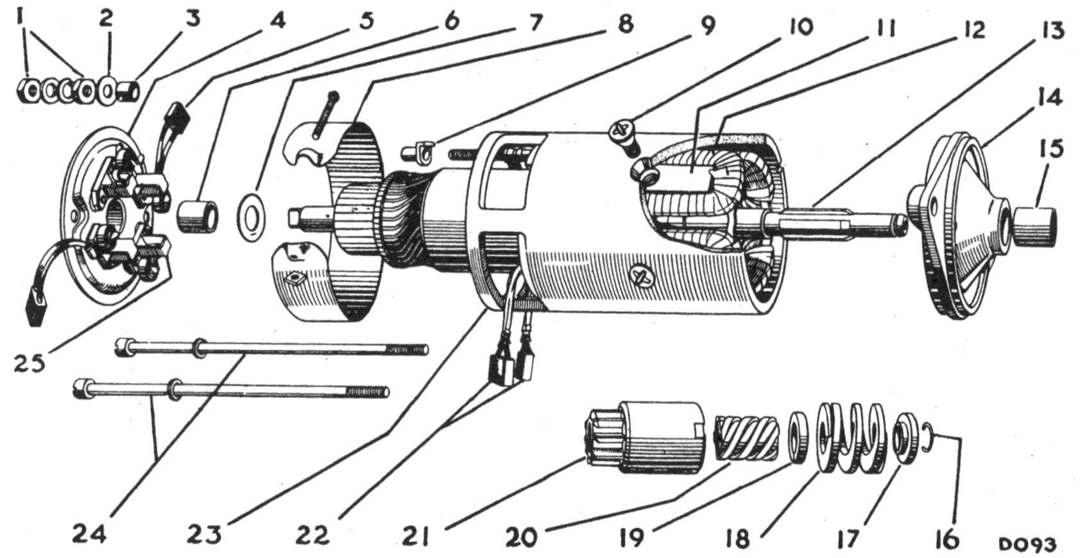

1 Terminal nuts and washers
2 Insulating washer
3 Insulating bush
4 End plate
5 Brush
6 Bush
7 Thrust washer
8 Cover band
9 Insulating bush
10 Pole securing screw
11 Pole piece
12 Field coil
13 Shaft
14 End bracket
15 Bush
16 Jump ring
17 Retainer
18 Main spring
19 Thrust washer
20 Sleeve
21 Pinion and barrel assembly
22 Brushes
23 Yoke
24 Through bolts
25 Brush box

Exploded View of Mk. IV and 1500 Starter

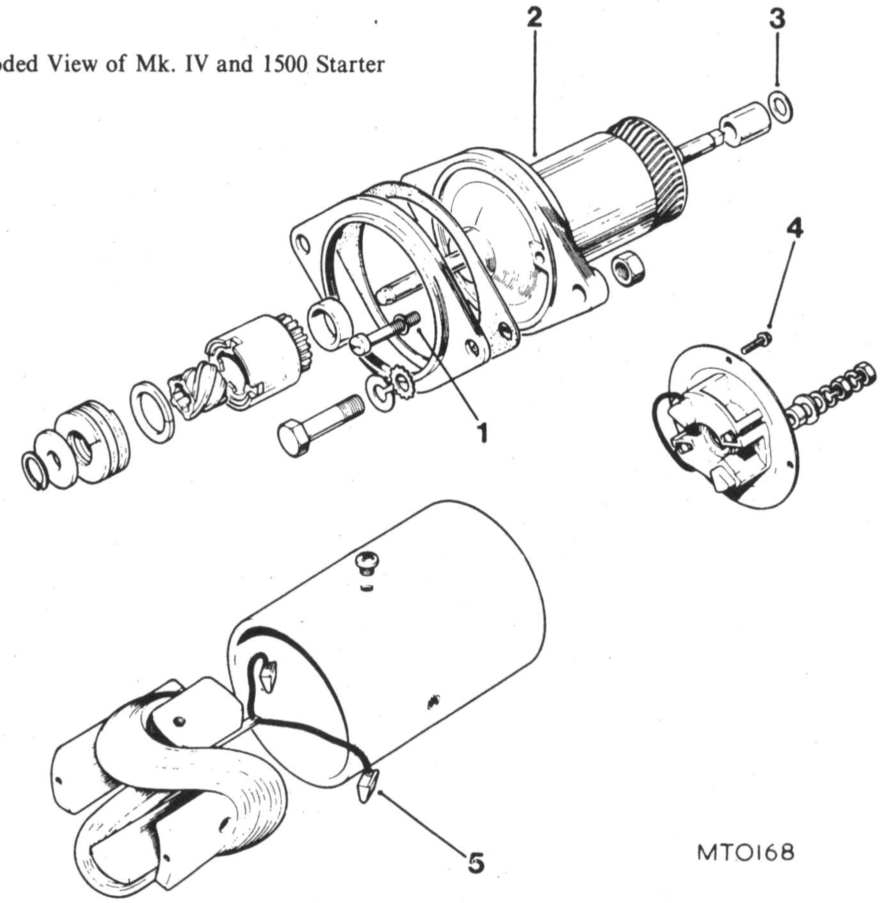

Triumph Spitfire Mk IV Manual. Part No. 545254. Issue 1.
Part No. 511423. page 6.126. Issue 2.

STARTER MOTOR

– Overhaul **86.60.13**

[CAUTION: Please read the instructions thoroughly before you attempt starter repairs. If you lack the necessary skills or tools we suggest you leave such work to an Authorized Dealer or other qualified shop. Before taking a removed starter in for repair, wipe clean its exterior but do not undertake any disassembly.]

Dismantling

1. [On Mk. III models, remove the terminal nuts and washers — ensuring that the terminal stud does not turn in the end plate. Then remove the two through bolts. On Mk. IV and 1500 models], remove the two drive end bracket bolts and spring washers
2. Withdraw the drive end bracket, armature and inertia drive assembly complete from the yoke.
3. Remove the thrust washer.
4. [On Mk. IV and 1500 models] remove the four 4B.S. bolts and lift aside the commutator end bracket.
5. Lift out the two field winding brushes from the brush box to separate the commutator end bracket from the yoke.

Armature

6. To separate the armature from the drive end bracket and inertia drive, perform the following: Remove the inertia drive, 86.60.06 and slide the drive end bracket from the shaft.
7. Inspect the laminations for score marks. These may indicate a bent shaft, worn bearings or a loose pole-shoe.
8. Clean the commutator with a petrol-moistened cloth. If the commutator is in good condition it will be smooth and free from pits or burned spots.
9. If necessary, polish the commutator with fine glass-paper.
10. If necessary, skim the commutator. Mount the armature in a lathe and rotate at high speed. Using a very sharp tool, take a light cut. Polish with fine glass-paper. Do not cut below the minimum skimming thickness given in Data. Do not undercut insulators between segments.

Bearings

11. Inspect the porous bronze bearing bushes for wear.
12. If necessary, renew the commutator end bracket bush. Drill out the two rivets and discard the plate and felt seal. Screw a ½ in tap squarely into the bush and withdraw. Prepare the porous bronze bush by immersing it in thin engine oil for 24 hours. Using a highly polished, shouldered mandrel suitably dimensioned and a suitable press, fit the bush. Do not ream the bush after fitting or its porosity may be inpaired. Assemble the brush box, commutator end bracket, felt seal and plate. Secure with two rivets.
13. If necessary, renew the drive end bracket bush. Remove the inertia drive, 86.60.06, and slide the drive end bracket from the shaft. Support the bracket and press out the bush. Prepare the porous bronze bush by immersing it in thin engine oil for 24 hours. Using a highly polished, shouldered mandrel suitably dimensioned and a suitable press, fit the bush. Do not ream the bush after fitting or its porosity may be impaired.

continued

Brushes

14. Clean the brushes and brush box with a petrol-moistened cloth.
15. Check that the brushes move freely in the brush box.
16. **[On Mk. IV and 1500 models] check the brush spring pressure as shown. Position a new brush so that the top protrudes 0·060 in. (1·50 mm) above the brush box. Brush spring pressure should be as given in Data. Repeat for the remaining three springs. If the pressure is low, renew the commutator end bracket assembly.**
17. Check the brush length. Renew the brushes if less than the length given in Data.
18. If necessary, renew the commutator end bracket brushes. Brushes are supplied attached to a new terminal post. Withdraw the two brushes from the brush box. Remove the terminal post outer attachments. Withdraw the terminal post and remove the insulation piece. Reverse to assemble. Retain the longer flexible under the clip.
19. If necessary, renew the field winding brushes. Brushes are supplied attached to a common flexible. Cut the old flexibles 0·250 in (6 mm) from the joint. Solder the new flexible to the ends of the old flexible. Do not attempt to solder direct to the field winding strip as the strip may be produced from aluminium.

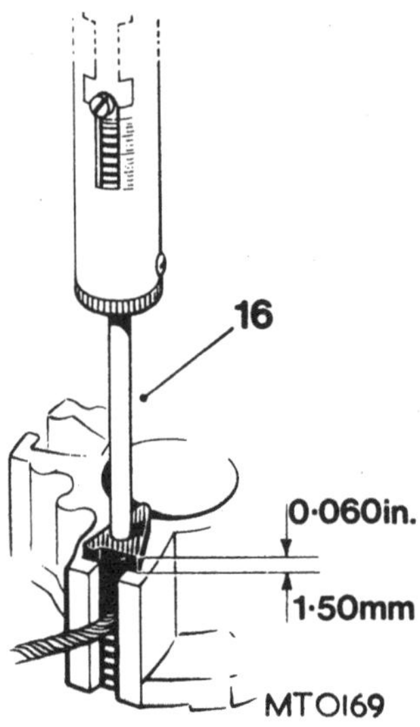

Field winding

[NOTE: On Mk. III motors, it is unnecessary to drill out a rivet as described in instruction 20. Continuity can be tested by applying one test probe to the terminal stud and the other to one or the other of the brushes.]

20. If necessary, check the fields winding insulation from the yoke as follows: Drill out the rivet at the earth connection. Apply the normal 110 volt a.c. test lamp circuit to the field winding and yoke.
    Do not attempt to disconnect the flexible from the field winding strip as the strip may be produced from aluminium.
21. If necessary, renew the field winding. Drill out the rivet at the earth connection. Using a pole-shoe screwdriver, slacken the four pole-shoe screws. Remove the two diametrically opposite screws and pole-shoes. Slacken the remaining two screws sufficient to allow the field winding to be withdrawn from the yoke. Reverse to assemble.

Reassembling

[NOTE: In assembling the Mk. III motor, the brush springs must not be positioned against the brushes until the armature has been installed. See the instructions given in GENERATOR — Overhaul 86.11.03. To reassemble this early starter motor, reverse instructions 1 to 5.]

22. Insert the two field winding brushes into the brush box with the flexibles positioned as shown.
23. Position the commutator end bracket and secure it with four 4 B.A. bolts.
24. Fit the thrust washer.
25. Insert the drive end bracket, armature and inertia drive assembly complete into the yoke.
26. Fit the two drive end bracket bolts and spring washers.

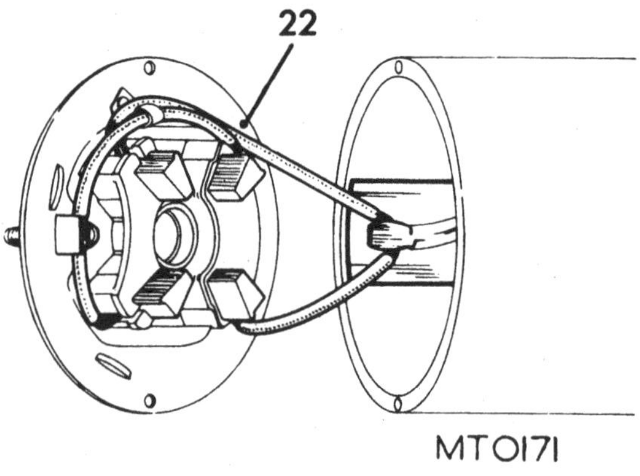

SWITCHES [except Mk. III]

Data 86.65.00

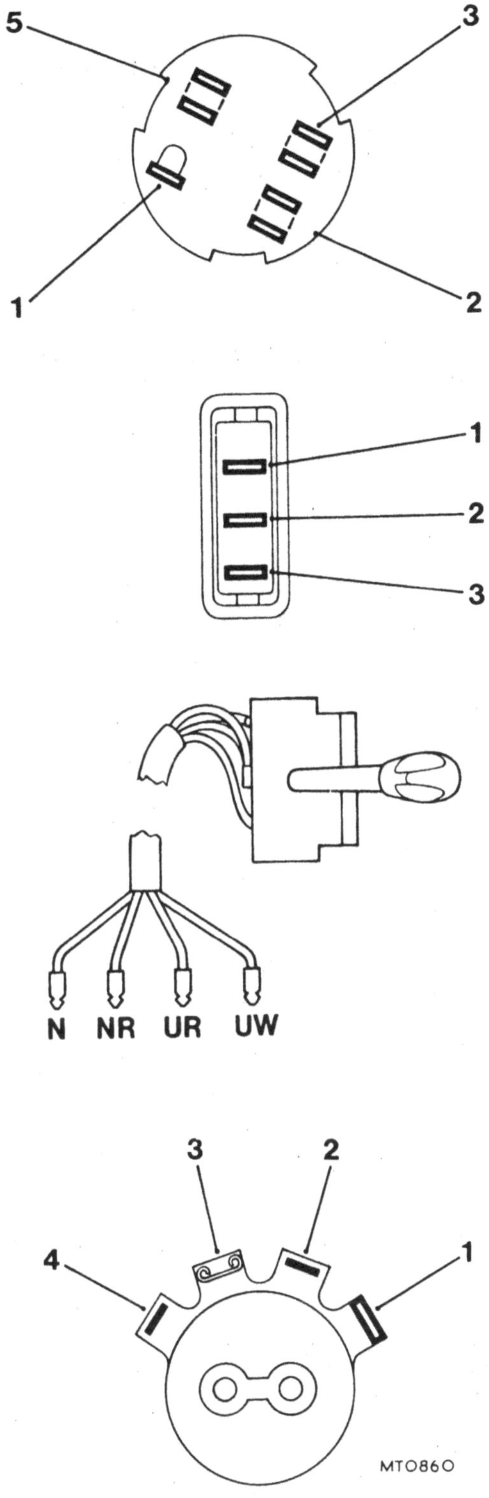

Ignition/starter switch

Position 0	Off	No connections
Position 1	Auxiliary	2 to 5
Position 2	Ignition	2 to 5 to 3
Position 3	Start	2 to 3 to 1

Master light switch

Position	Off	No connections
Position	Side	3 to 2
Position	Head	3 to 2 to 1

Column light switch

Position	Head main	U to UW
Position	Head dip	U to UR
Position	Head flash	P to UW

Windscreen wiper switch

Used when manual windscreen washer system is fitted to vehicle.

Position	Park	1 to 2
Position	Normal speed	3 to 2
Position	High speed	3 to 4

continued

Heater switch

Position	Off	No connections
Position	Low speed	GY to earth
Position	High speed	GS to earth

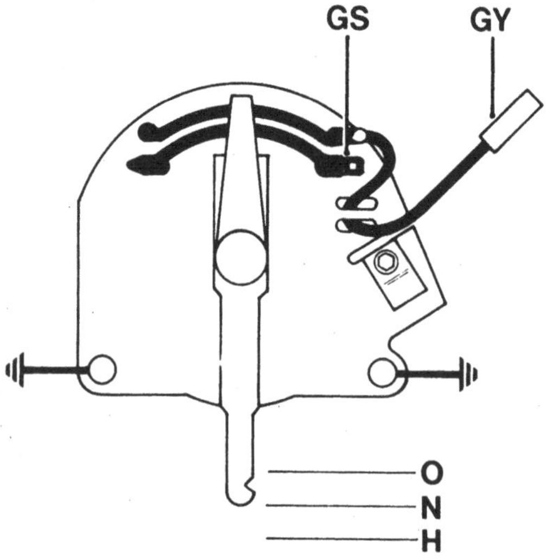

Turn signal switch

Position	Central	No connections
Position	L.H. turn signal	LG/N to GR
Position	R.H. turn signal	LG/N to GW

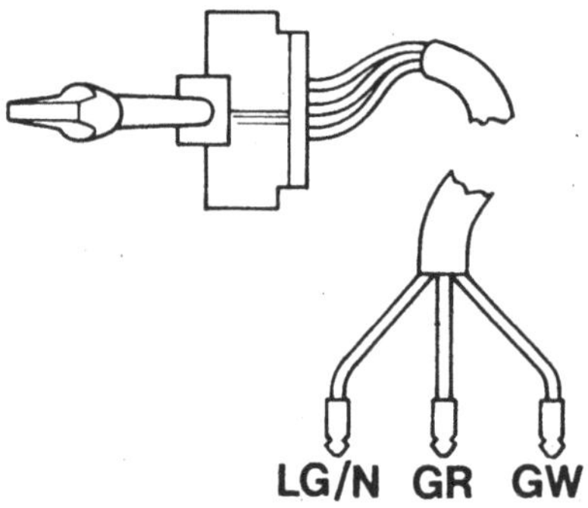

Hazard switch

Used when hazard warning system is fitted to vehicle.

Position	Off	8 to 7
Position	Hazard	3 to 1 to 2 to 4

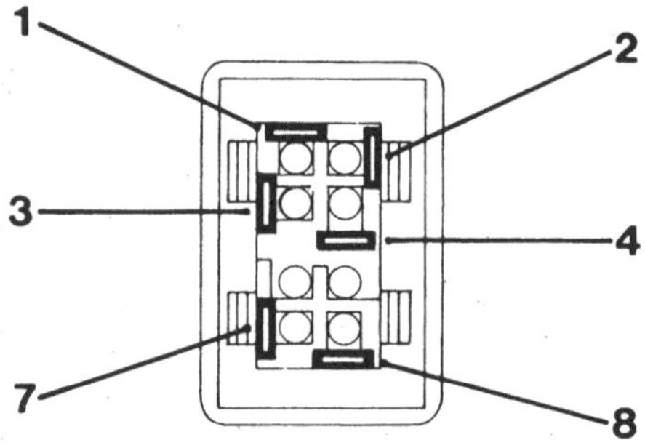

Door switch

Used when key warning system is not fitted to vehicle.

Position	Door closed	No connections
Position	Door open	Terminal to earth

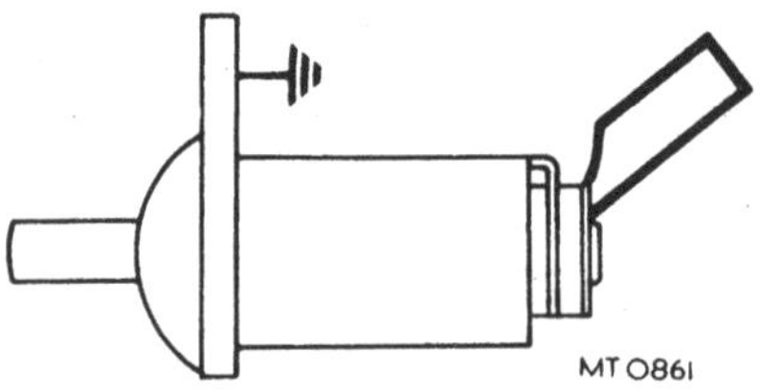

continued

Door switch

Used when key warning system is fitted to vehicle.

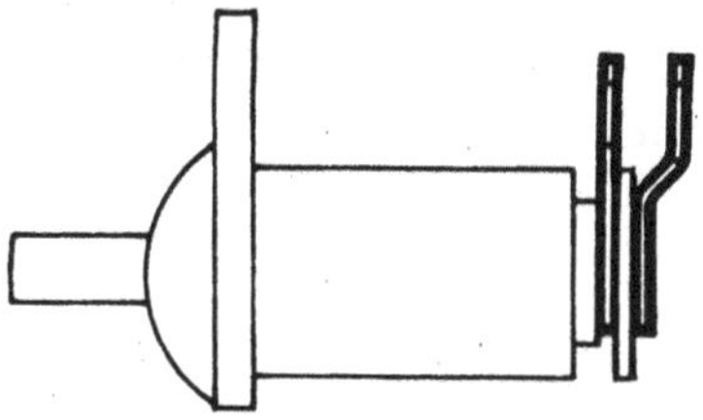

Position	Door closed	No connections
Position	Door open	Two Lucar blades connected

Windscreen washer/wiper switch

Used when electric washer pump is fitted to vehicle.

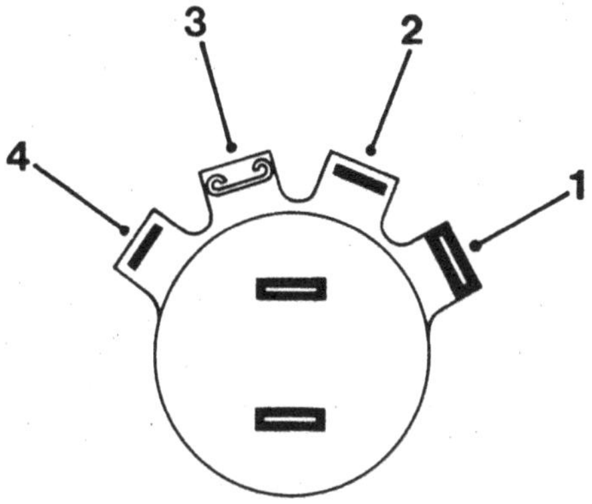

MT 0897

Position	Park	1 to 2
Position	Normal speed	3 to 2
Position	High speed	3 to 4
Position	Wash	Two centre terminals connected

MK. III FASCIA CONNECTIONS

—Remove and refit **[86.65.01]**

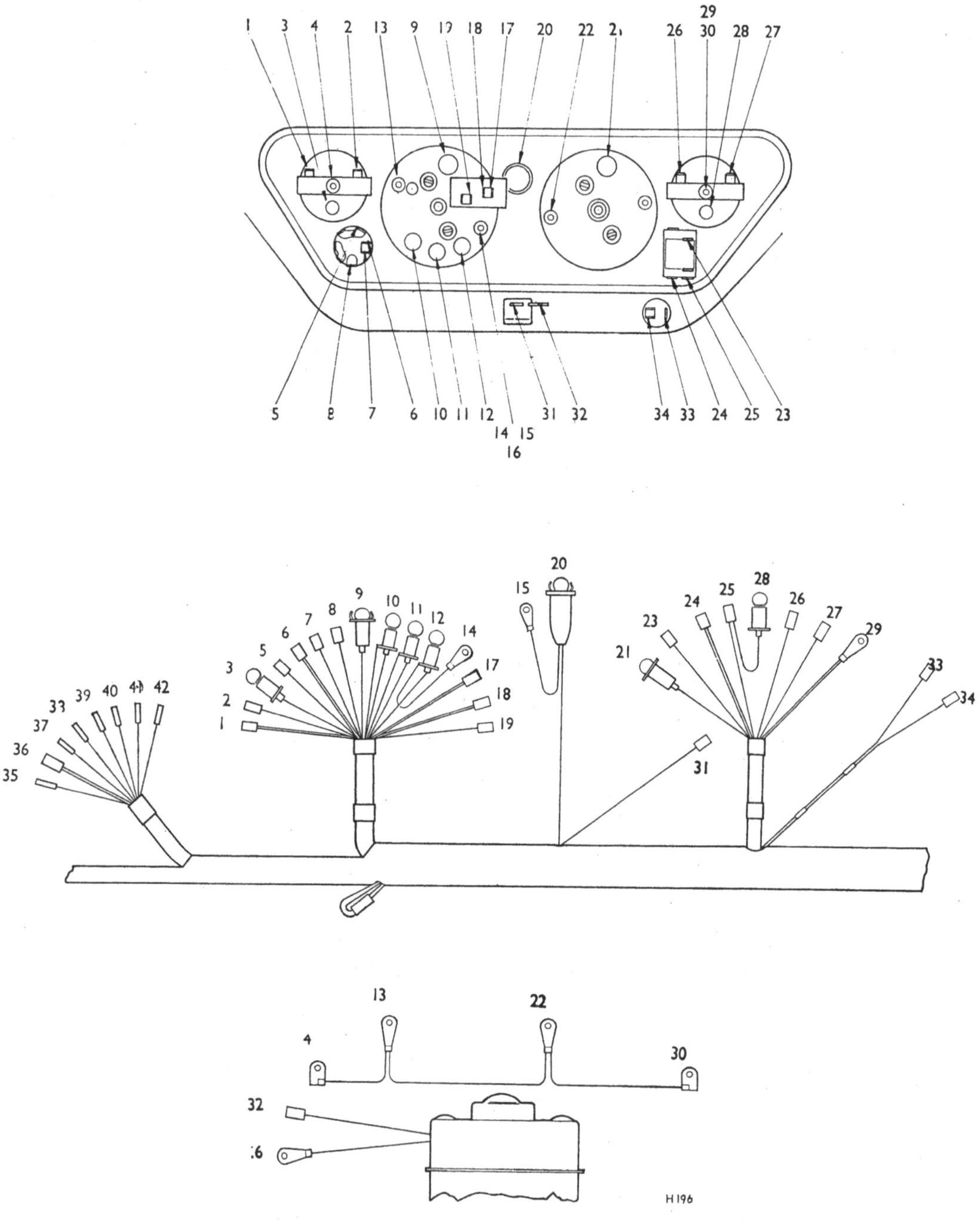

Fascia Connections—Mk. III

Part No. 511243. page 6.314. Issue 1.

Key to Fascia Connections—Mk. III

NO.	COLOUR	CONNECTION	COMPONENT
1	LG/G	Lucar – 2 wire	Temperature indicator
2	GU	Lucar	Temperature indicator
3	RW	Bulb holder	Temperature indicator
4	B	Flag eyelet	Temperature indicator
5	N	Lucar	Ignition/starter switch
6	W	Lucar – 2 wire	Ignition/starter switch
7	W	Lucar	Ignition/starter switch
8	WR	Lucar	Ignition/starter switch
9	RW	Bulb holder	Speedometer
10	W and NY	Bulb holder	Speedometer – ignition warning light
11	W and WN	Bulb holder	Speedometer – oil pressure warning light
12	UW and B	Bulb holder	Speedometer – main beam warning light
13	B	Eyelet – 2 wire	Speedometer
14	B	Eyelet	Speedometer
15	B	Eyelet	Speedometer
16	B	Eyelet	Speedometer
17	G	Lucar – 2 wire	Voltage stabilizer
18	G	Lucar – 2 wire	Voltage stabilizer
19	LG/G	Lucar	Voltage stabilizer
20	LG/P and B	Bulb holder	Flasher warning light
21	RW	Bulb holder	Tachometer
22	B	Eyelet – 2 wire	Tachometer
23	N	Lucar	Master light switch
24	NR and RG	Lucar – 2 wire	Master light switch
25	RW	Lucar – 2 wire	Master light switch
26	GB	Lucar	Fuel indicator
27	LG/G	Lucar	Fuel indicator
28	RW	Bulb holder	Fuel indicator
29	B	Eyelet – 2 wire	Fuel indicator
30	B	Flag eyelet	Fuel indicator
31	G	Lucar	Heater switch
32	B	Lucar	Heater switch
33	BG	Lucar	Windscreen wiper switch
34	B	Lucar	Windscreen wiper switch
35	NR	Snap connector	Column light switch
36	UW	Double snap connector – 2 wire	Column light switch
37	UR	Snap connector	Column light switch
38	P	Snap connector	Column light switch
39	LG/N	Snap connector	Flasher switch
40	GR	Snap connector	Flasher switch
41	GW	Snap connector	Flasher switch
42	PB	Snap connector	Horn push

SWITCHES [except Mk. III]

– Ignition/Starter switch – remove and refit 86.65.02

Removing

1. Isolate the battery.
2. Pull back the switch cover.
3. Withdraw the switch.
4. Disconnect the five Lucar connectors.

Refitting

5. Reverse instructions 1 to 3. When inserting the switch into the steering column lock assembly, note the keyway and ensure that the lock shaft and switch are aligned for correct engagement.
6. Connect the wires as follows:
   White/red to terminal 1
   Brown to terminal 2
   White to terminal 3
   White/pink to terminal 5

SWITCHES – [except Mk. III]
Master light switch

– Remove and refit 86.65.09

Removing

1. Lower the centre fascia to the service position 76.46.02 operations 1 to 4.
2. Disconnect the three Lucar connectors.
3. Compress the springs and push out the switch.

Refitting

4. Reverse instructions 1 to 3.

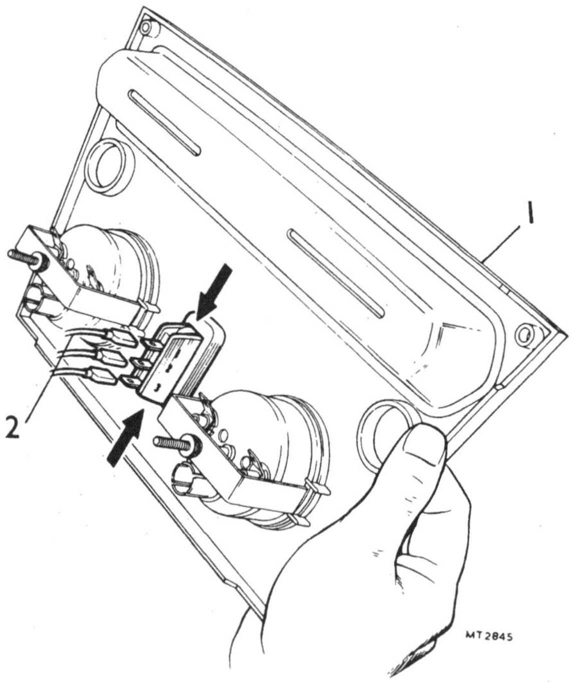

SWITCHES

– Door switch – remove and refit 86.65.14

Removing

1. Isolate the battery.
2. Remove the screw.
3. Remove the switch.
4. Disconnect the terminal(s).

Refitting

5. Reverse instructions 1 to 4.

SWITCHES–
Column light switch

– Remove and refit 86.65.17

Removing

1. Isolate the battery.
2. Remove the switch fairings – two screws.
3. Remove the two Allen screws and spring washers securing the steering column clamp. Remove the clamp and the harness cover.
4. Disconnect the four snap connectors.
5. Remove the two screws and spring washers.
6. Carefully withdraw the switch and leads.

Refitting

7. Reverse instructions 1 to 6. Ensure that the wires are connected as follows:-
   Blue/Red to Blue/Red
   Blue/White to Blue/White
   Brown/Red to Blue
   Brown to Purple
   The inconsistencies enable a common switch to be used on different circuits.

SWITCHES–
Horn push

– Remove and refit 86.65.18

Removing

1. Pull off the steering wheel centre pad.
2. Prise out the horn push.
3. Withdraw the horn brush.

Refitting

4. Reverse instructions 1 to 3, ensuring that the horn push contact strip is positioned over the horn brush.

SWITCHES

– Reverse lamp switch – remove and refit 86.65.20

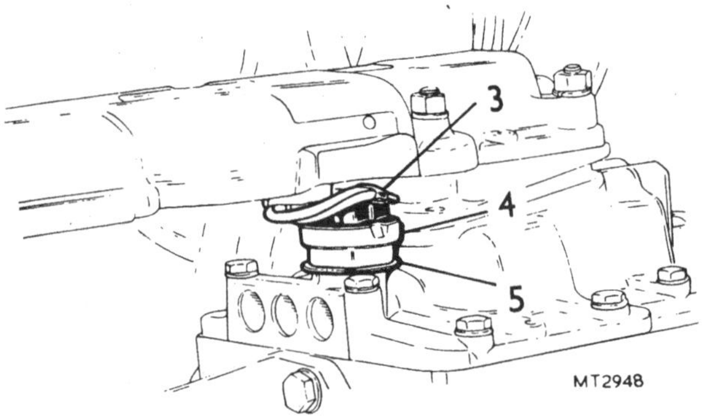

Removing

1. Remove the gearbox tunnel cover. 76.25.07.
2. Remove the top cover extension. 37.12.10.
3. Disconnect the two Lucar connectors.
4. Unscrew the reverse lamp switch.
5. Remove the packing washers.

Refitting

6. Reverse instructions 1 to 5.

SWITCHES

– Oil pressure switch – remove and refit 86.65.30

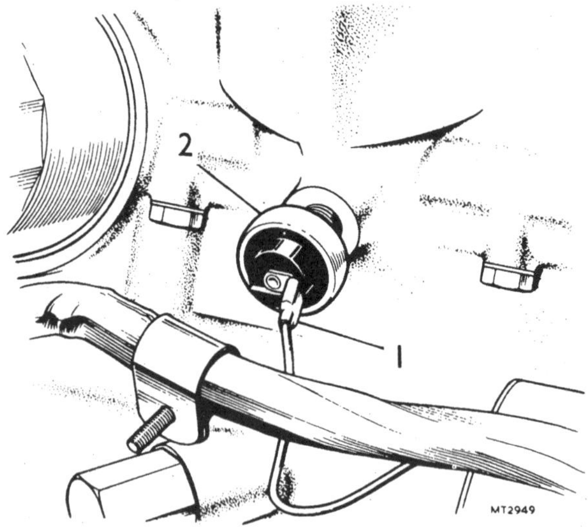

Removing

1. Disconnect Lucar connector.
2. Using spanner unscrew switch from block.

Refitting

3. Screw switch into block and tighten to 11 to 14 lbf. ft (1·5 to 2·0 kg. m). Thread is tapered. Do not attempt to seat switch shoulder.
4. Connect Lucar connector.

SWITCHES

—Seat belt — driver's belt switch — remove and refit [86.65.31]
—Seat belt — passenger's belt switch — remove and refit [86.65.32]

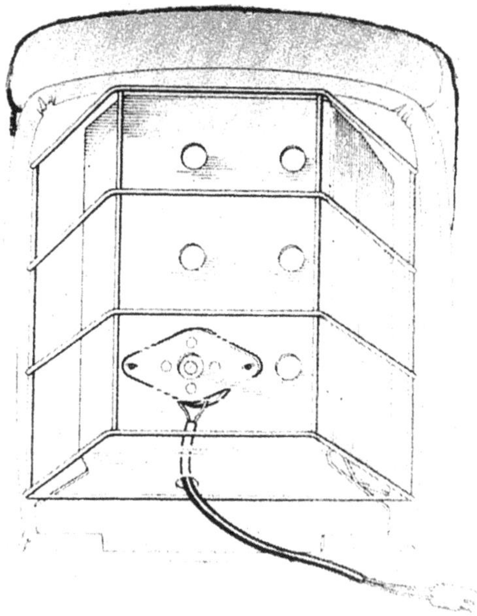

Removing

1. Lift the carpet at the rear of the seatwell and disconnect the electrical harness plug.
2. Remove single bolt and spring washer. Lift out the buckle and switch unit.

Refitting

3. Reverse instructions 1 to 2. If the sealing has been disturbed, seal the large plain washer to the floor panel with an approved sealer to ensure a waterproof joint.

SWITCHES

– Overdrive gearbox switch – remove and refit 86.65.33

Removing

1. Remove the gearbox tunnel cover. 76.25.07.
2. Disconnect the two Lucar connectors.
3. Unscrew the switch.

Refitting

4. Reverse instructions 1 to 3.

SWITCHES

– Overdrive gear lever switch – remove and refit 86.65.34

Removing

1. Prise the switch from the gear lever knob (a slot at the reat of the knob enables a blade to be inserted).
2. Disconnect the two wires from the switch.

Refitting

3, Connect the two wires.
4. Clip the switch onto the gear lever knob.

SWITCHES– Windscreen wiper switch

– Remove and refit 86.65.38

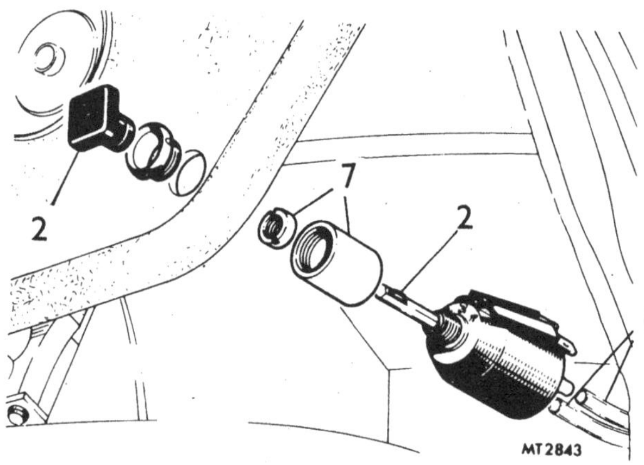

Removing

1. Isolate the battery.
2. Depress the retainer by inserting a suitable probe into the hole in the underside of the knob and pull the knob from the shaft.
3. Slacken the bezel using a suitable tool. Support the switch behind the fascia and unscrew the bezel.
4. Withdraw the switch and lower it into a visible position.
5. Disconnect the four Lucar connectors noting colour codes and positions for reconnection.
6. Pull off the two pipes, noting positions for reconnection.

Refitting

7. Ensure that the switch is securely attached to the spacer. If not tighten the slotted ring.
8. Reverse instructions 1 to 6.

SWITCHES–

Brake line failure switch – remove and refit **86.65.47**

Removing

1. Pull back the retaining clips and withdraw the socket from the switch.
2. Unscrew the switch.

Refitting

3. Reverse instructions 1 and 2. Ensure that the earth connection spring is correctly located with the outer end pointing in an anti-clockwise direction as shown.

SWITCHES

– Hazard switch – remove and refit **86.65.50**

Left hand steer vehicles only.

Removing

1. Isolate the battery.
2. To gain access remove either the speedometer 88.30.01 or the tachometer 88.30.21. The choice depends entirely on which operation is preferred by the individual fitter.
3. Push inwards two plastic clips on the switch and withdraw the switch from the panel.
4. Disconnect the Lucar connectors.

Refitting

5. Reverse instructions 1 to 4.

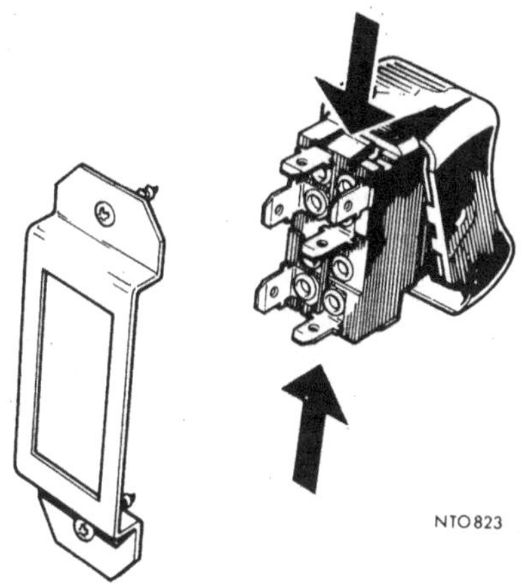

NTO 823

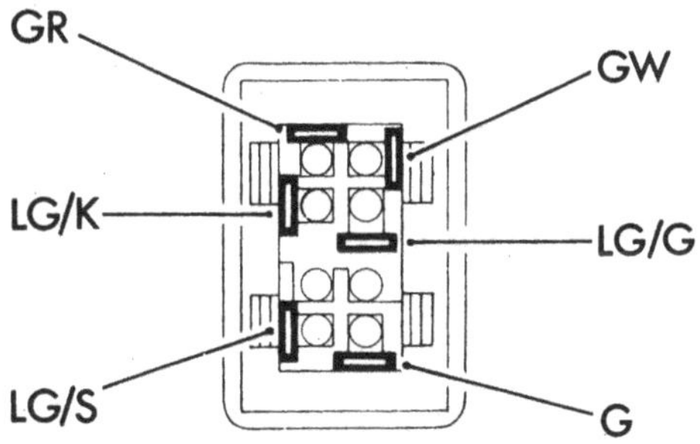

PTO 392

SWITCHES

Stop lamp switch – remove and refit **86.65.51**

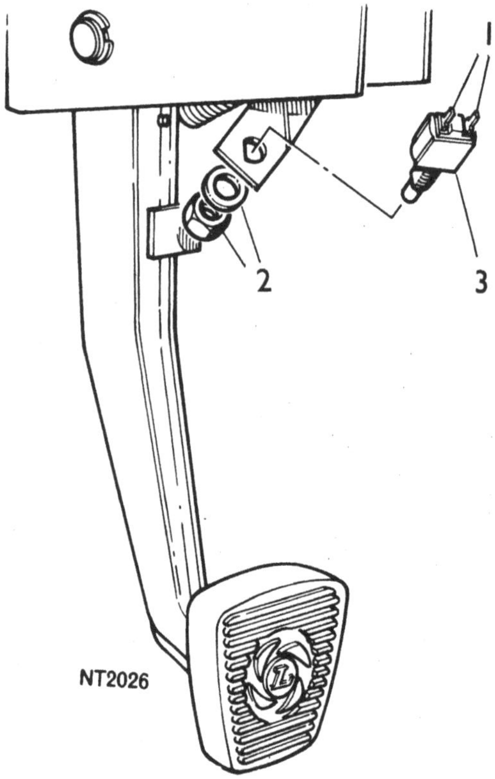

NT2026

Removing

1. Disconnect the two Lucar connectors.
2. Remove the nut and plain washer.
3. Withdraw the switch from the mounting bracket.

Refitting

4. Reverse instructions 1 to 3.

SWITCHES

Turn signal switch – remove and refit **86.65.54**

Removing

1. Remove the switch fairings – two screws.
2. Remove the two Allen screws and spring washers securing the steering column clamp. Remove the clamp and the harness cover.
3. Disconnect the three snap connectors.
4. Remove the two screws and spring washers.
5. Carefully withdraw the switch and leads.

Refitting

6. Reverse instructions 1 to 5.

FUSE SYSTEM

– Fuse – remove and refit **86.70.02**

Removing

1. Lift the bonnet and locate the fusebox on the left hand side of the bulkhead.
2. Pull off the plastic cover.
3. Identify the defective fuse.
4. Carefully lever the fuse from the contacts.

Refitting

5, Reverse instructions 1 to 4.

INSTRUMENTS OPERATIONS

INSTRUMENTS, SWITCHES AND CONTROLS (MK. III ONLY)

—Remove and refit **[88.20.25]**

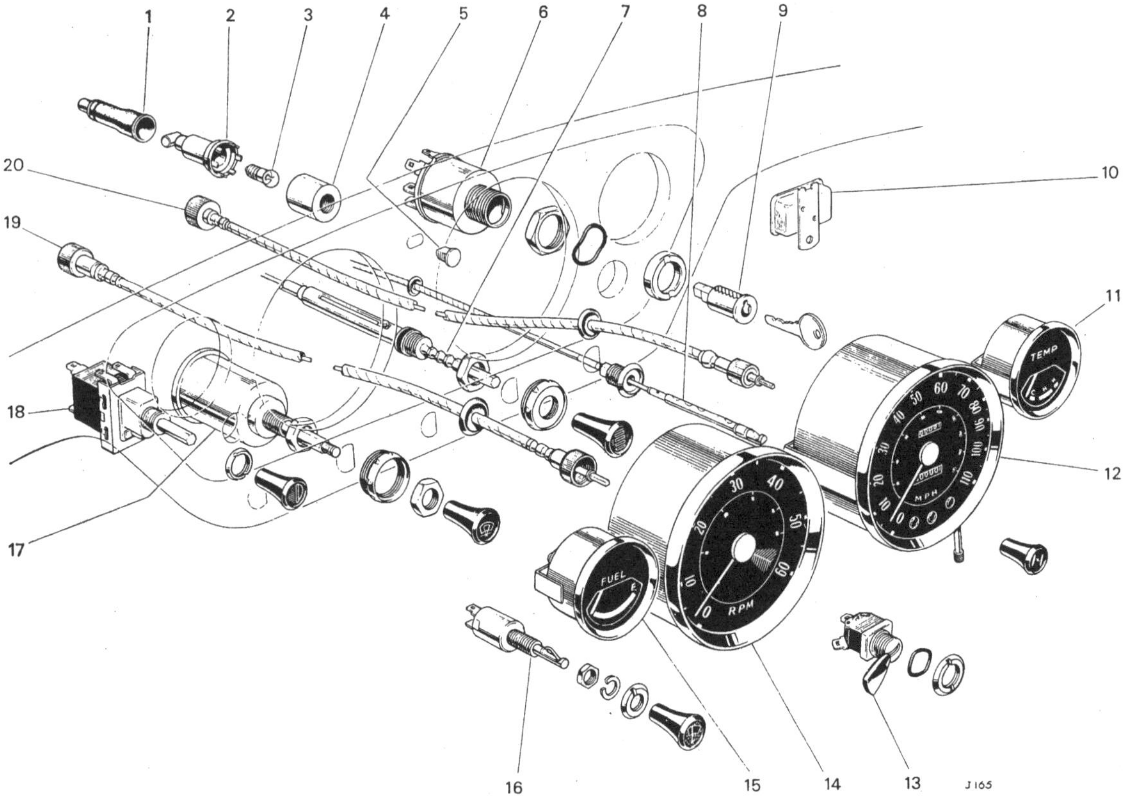

1	Sleeve	11	Temperature gauge
2	Bulb holder	12	Speedometer
3	Bulb	13	Blower switch
4	Lamp body	14	Tachometer
5	Lens	15	Fuel gauge
6	Starter/ignition switch	16	Wiper switch
7	Heat control cable	17	Windscreen washer pump
8	Choke cable	18	Lighting switch
9	Ignition barrel	19	Tachometer drive cable
10	Speedo voltage stabilizer	20	Speedometer drive cable

VOLTAGE STABILISER

– Remove and refit **88.20.26**

Removing

1. Remove the speedometer 88.30.01.
2. Remove one screw and lift off the voltage stabiliser.

Refitting

3. Reverse instructions 1 and 2. Connect the Lucar connectors as follows:
   Green wire to terminal B. Light green wire to terminal I.

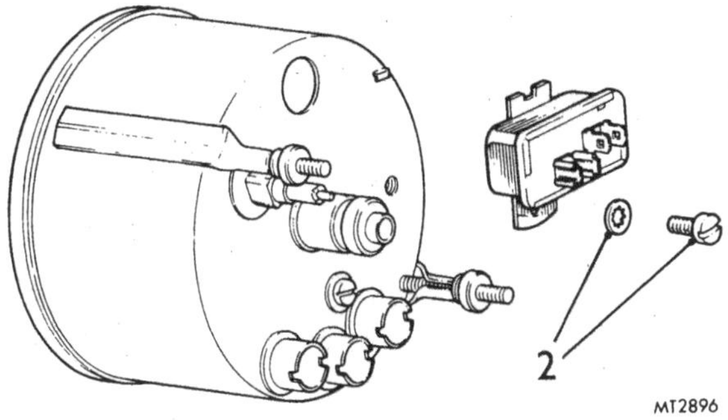

TEMPERATURE INDICATOR

– Remove and refit **88.25.14**

Removing

1. Lower the fascia centre to the service position 76.46.02, operations 1 to 4.
2. Pull out the panel light bulb holder.
3. Disconnect the two Lucar connectors.
4. Remove the knurled nut and spring washer.
5. Withdraw the indicator from the fascia centre.

Refitting

6. Reverse instructions 1 to 5.

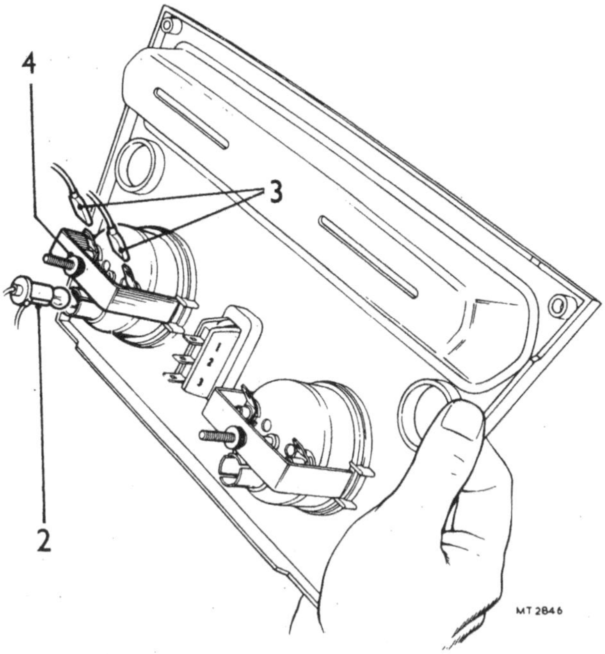

TEMPERATURE TRANSMITTER

– Remove and refit **88.25.20**

Removing

1. Drain part of the coolant 26.10.01.
2. Disconnect the Lucar connector.
3. Unscrew the transmitter.

Refitting

4. Reverse instructions 1 to 3.

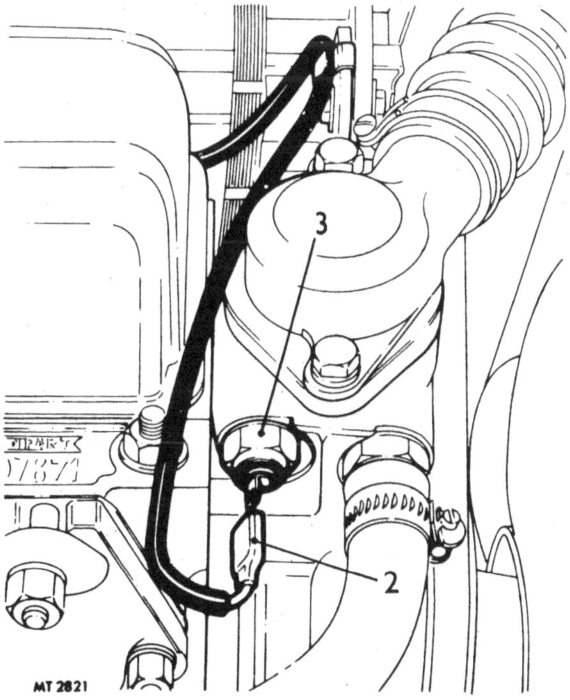

FUEL INDICATOR

– Remove and refit 88.25.26

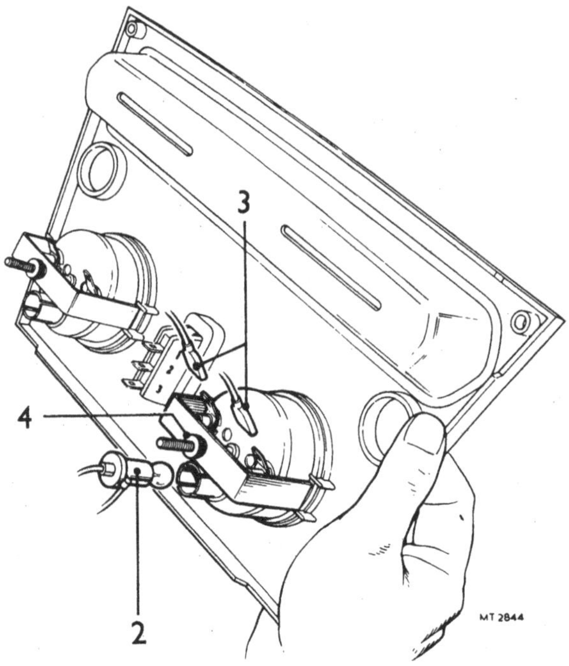

Removing

1. Lower the fascia centre to the service position 76.46.02, operations 1 to 4.
2. Pull out the panel light bulb holder.
3. Disconnect the two Lucar connectors.
4. Remove the knurled nut and spring washer.
5. Withdraw the indicator from the fascia centre.

Refitting

6. Reverse instructions 1 to 5.

FUEL TANK UNIT

– Remove and refit 88.25.32

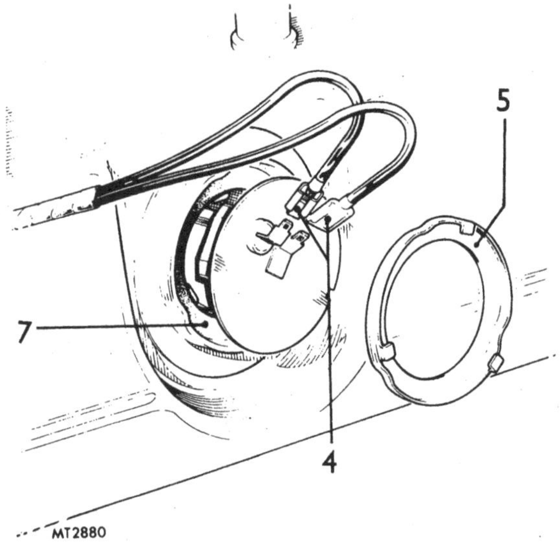

Removing

1. Isolate the battery and extinguish all naked lights.
2. Siphon the fuel out of the tank.
3. Remove the petrol tank rear casing board – seven screws, spring washers and plain washers.
4. Disconnect the two Lucar connectors noting positions for reconnection.
5. Release the locking ring by tapping anti-clockwise. Remove the locking ring.
6. Carefully withdraw the tank unit.
7. Remove the sealing washer.

Refitting

8. Reverse instructions 1 to 7.

SPEEDOMETER

– Remove and refit **88.30.01**

Removing

1. Remove the centre fascia for access to the rear of the speedometer 76.46.02.
2. Remove the tachometer for access to the rear of the speedometer 88.30.21.
3. Depress the lever to release the catch from the annular groove in the boss. Detach the speedometer cable from the instrument.
4. Unscrew the trip reset knurled nut.
5. Remove the two knurled nuts, spring washers and clamp legs. Detach the earth wire.
6. Disconnect the two Lucar connectors from the voltage stabiliser
7. Pull out the three warning light bulb holders.
8. Pull out the panel light bulb holder.

Refitting

9. Reverse instructions 4 to 8.
10. Using long nosed pliers, withdraw the inner cable about 1 in (25 mm).
11. Engage the inner cable to the instrument.
12. Push the outer cable over the boss, ensuring that the catch engages in the annular groove.
13. Refit the tachometer 88.30.21.
14. Refit the centre fascia.

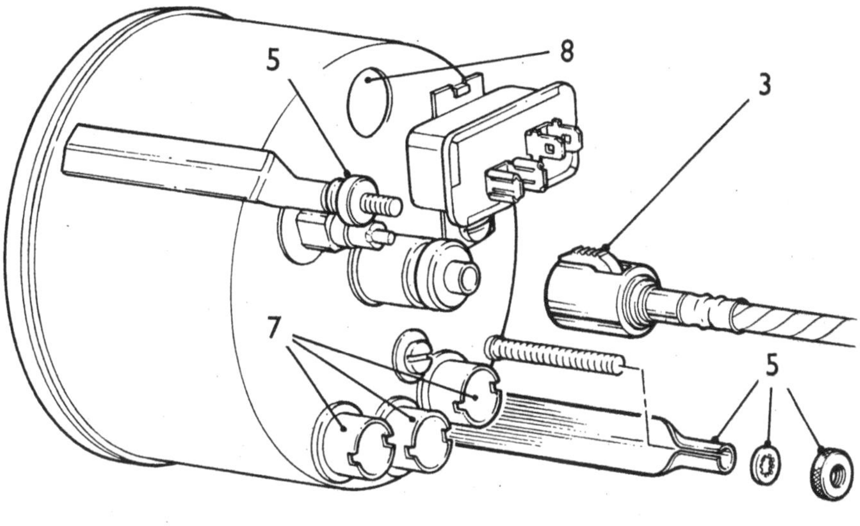

SPEEDOMETER CABLE – COMPLETE

– Remove and refit (non-overdrive models only) 88.30.06

Removing

1. Place the vehicle on a ramp.
2. Remove the centre fascia for access to the rear of the speedometer 76.46.02.
3. Remove the tachometer for access to the rear of the speedometer 88.30.21.
4. Depress the lever to release the catch from the annular groove in the boss. Pull the speedometer cable from the instrument.
5. Working from below the vehicle, unscrew the speedometer cable knurled nut from the gearbox extension.
6. To assist refitting, carefully note the cable position relative to other components.
7. Manoeuvre the cable downwards through the grommet aperture and detach it from the vehicle.

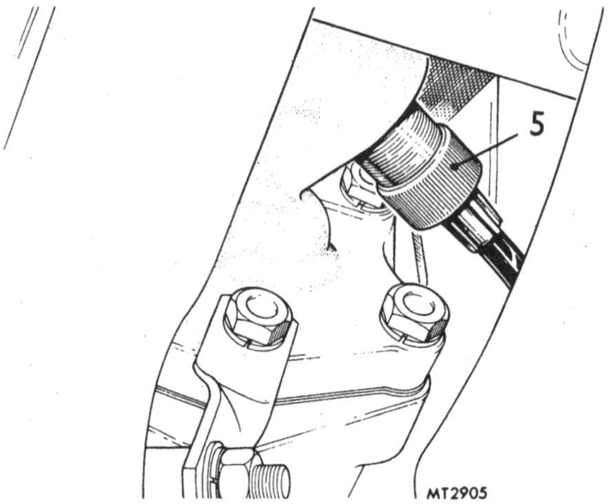

Refitting

8. Reverse instructions 5 to 7. Seal the grommet to the bulkhead with seelastik SR51.
9. Withdraw the inner cable about 1 in (25 mm).
10. Engage the inner cable to the instrument.
11. Push the outer cable over the boss, ensuring that the catch engages in the annular groove.
12. Refit the tachometer 88.30.21.
13. Refit the centre fascia 76.46.02.

SPEEDOMETER CABLE – COMPLETE

– Remove and refit (overdrive models only) 88.30.06

Removing

1. Remove the centre fascia for access to the rear of the speedometer 76.46.02.
2. Remove the tachometer for access to the rear of the speedometer 88.30.21.
3. Depress the lever to release the catch from the annular groove in the boss. Pull the speedometer cable from the instrument.
4. Remove the gearbox tunnel cover 76.25.07.
5. Unscrew the speedometer cable knurled nut from the gearbox extension.
6. To assist refitting, carefully note the cable position relative to the other components.
7. Manoeuvre the cable downwards through the grommet aperture and detach it from the vehicle.

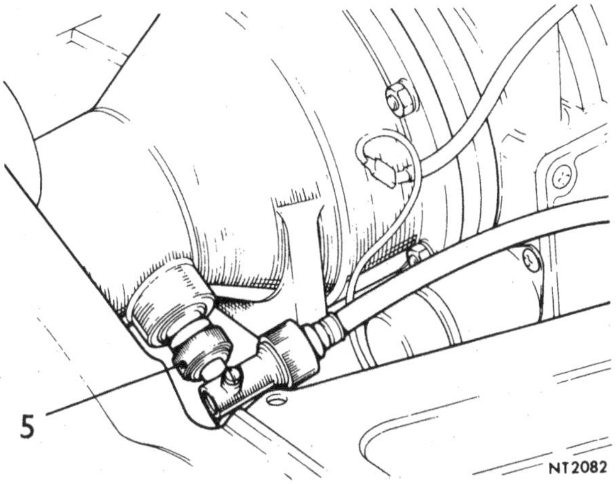

Refitting

8. Reverse instructions 5 to 7. Seal the grommet to the bulkhead with Seelastik SR51.
9. Withdraw the inner cable about 1 in (25 mm).
10. Engage the inner cable to the instrument.
11. Push the outer cable over the boss, ensuring that the catch engages in the annular groove.
12. Refit the tachometer 88.30.21.
13. Refit the centre fascia 76.46.02.

SPEEDOMETER CABLE – INNER

– Remove and refit **88.30.07**

Removing

1. Remove the centre fascia for access to the rear of the speedometer 76.46.02.
2. Remove the tachometer for access to the rear of the speedometer 88.30.21.
3. Depress the lever to release the catch from the annular groove in the boss. Pull the speedometer cable from the instrument.
4. Using long nosed pliers, withdraw the inner cable. Take care not to contaminate the upholstery or fittings with grease.

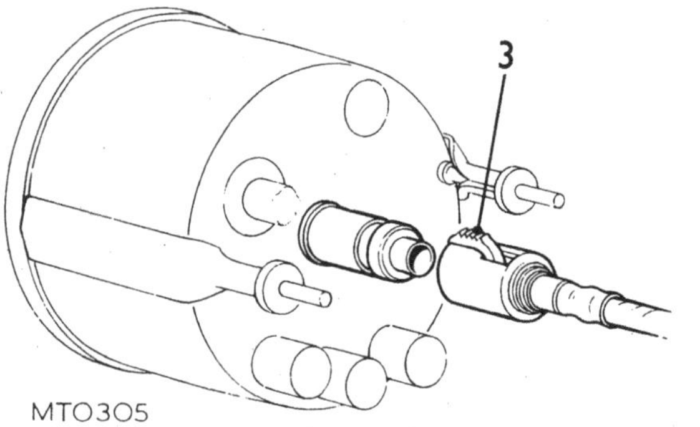

MTO305

Refitting

5. Sparingly grease the inner cable. Do not use oil.
6. Feed the inner cable into the outer cable, rotating it slightly to assist operation.
7. Withdraw the inner cable about 8 in (200 mm) and wipe off surplus grease. Re-insert the inner cable, rotating it slightly to assist engagement of the squared end to the drive gear.
8. Withdraw the inner cable about 1 in (25 mm).
9. Engage the inner cable to the instrument.
10. Push the outer cable over the boss, ensuring that the catch engages in the annular groove.
11. Refit the tachometer 88.30.21.
12. Refit the centre fascia, 76.46.02.

Triumph Dolomite Manual. Part No. 545209. Issue 1.

TACHOMETER

– Remove and refit **88.30.21**

Removing

1. Remove the two knurled nuts, spring washers and clamp legs.
2. Pull out the bulb holder.
3. Unscrew the knurled retainer securing the cable to the tachometer.
4. Withdraw the tachometer from the fascia.

Refitting

5. Reverse instructions 1 to 4.

TACHOMETER CABLE – COMPLETE

– Remove and refit **88.30.23**

Removing

1. Unscrew the knurled retainer securing the cable to the tachometer.
2. Unscrew the knurled retainer securing the cable to the distributor.
3. To assist refitting, carefully note the cable position relative to other components.
4. Manoeuvre the cable downwards through the grommet aperture and detach it from the vehicle.

Refitting

5. Reverse instructions 1 to 4. Seal the grommet to the bulkhead with Seelastik SR51.

TACHOMETER CABLE – INNER

– Remove and refit **88.30.24**

Removing

1. Unscrew the knurled retainer securing the cable to the tachometer.
2. Using long nosed pliers, withdraw the inner cable. Take care not to contaminate the upholstery or fittings with grease.

Refitting

3. Sparingly grease the inner cable. Do not use oil.
4. Feed the inner cable into the outer cable, rotating it slightly to assist operation.
5. Withdraw the inner cable about 8 in (200 mm) and wipe off surplus grease. Re-insert the inner cable, rotating it slightly to assist engagement of the squared end to the drive gear.
6. Engage the inner cable to the tachometer.
7. Refit the knurled retainer.

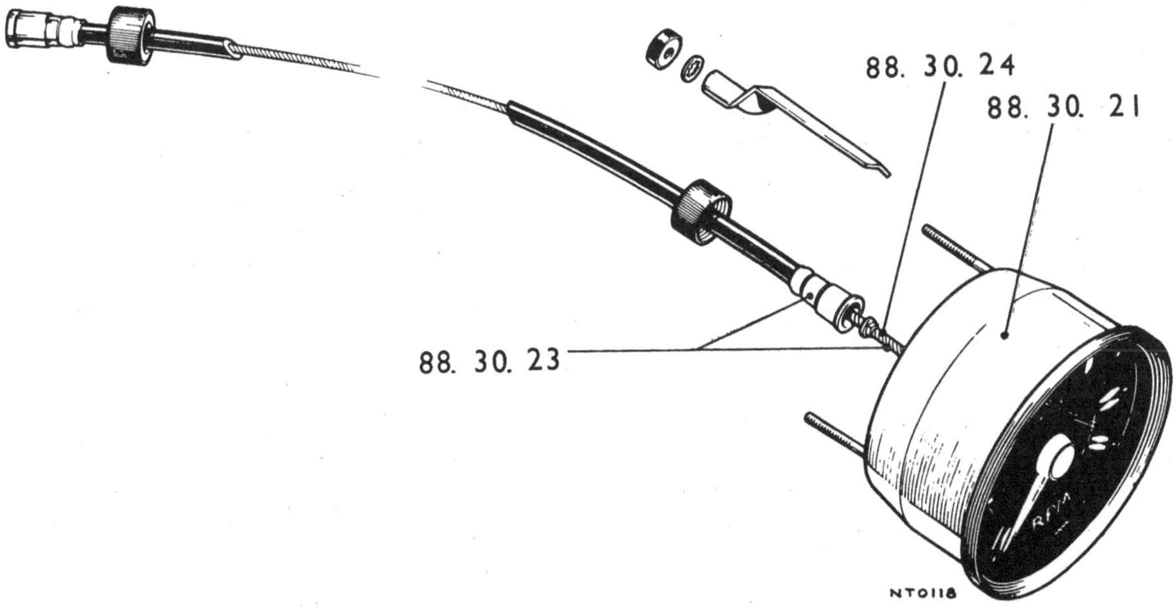

SERVICE TOOLS

Tool No.	Description
18G.47	Hand Press
18G.47BD	Differential Bearing Carrier Remover
18G.106	Valve Spring Compressor
18G.131C	Differential Case Spreader
18G.131E	Differential Case Spreader Adaptor Set
18G.134DH	Differential Bearing Replacer
18G.191	Pinion Height Gauge
18G.191M	Dummy Pinion
60A	Valve Guide Remover/Replacer
S.60A-2A	Valve Guide Remover/Replacer Adaptor
S.101	Differential Case Spreader
S.109C	Rear Hub Remover
S.144	Gearbox Mainshaft Circlip Remover
S.145	Gearbox Mainshaft Circlip Replacer
S.160	Ball Joint Separator
S.300A	Rear Hub Needle Bearing Remover/Replacer
S.304	Rear Hub Bearing Replacer
S.306	Brake Adjusting Tool
S.314	Mainshaft Ball Bearing Replacer
335	Con Rod Aligning Jig
S.336-4	Con Rod Arbor Adaptor
S.337	Flange Holder
S.341	Rack Mounting Compressor
S.353	Carburetter Adjusting Tool
RG.421	Adjustable Flange Holding Wrench
550	Driver Handle
3072A	Slide Hammer

Tool No.	Description
S.4221A	Hand Press
S.4221A-5	I.F.S. Coil Spring Remover/Replacer Adaptor
S.4221A-7B	Inner Axle Shaft Bearing Remover/Replacer Adaptor
S.4221A-14	Rear Hub Bearing Remover Adaptor
S.4221A-17	Pinion Head Bearing Remover/Replacer Adaptor
S.4221A-19	Gearbox Mainshaft Ball Race Remover/ Replacer Adaptor
4235A	Slide Hammer
S.4235A-2	Constant Pinion Remover Adaptor
6118B	Valve Spring Compressor
S.6118B-1	Valve Spring Compressor Adaptor

OVERDRIVE TOOLS

L.178	Freewheel Assembly Ring
L.183A	Pump Barrel Remover
L.183A-2A	Adaptor
L.188	Hydraulic Test Equipment
L.201	Dummy Drive Shaft
L.202A	Tailshaft Ballrace Remover
L.206A	Oil Pump Body Replacer
L.213	Oil Pump Body Key

All Service Tools mentioned in this Manual must be obtained direct from the manufacturers:

Messrs. V.L. Churchill & Co. Ltd.
P.O. Box No. 3
London Road,
Daventry, Northants

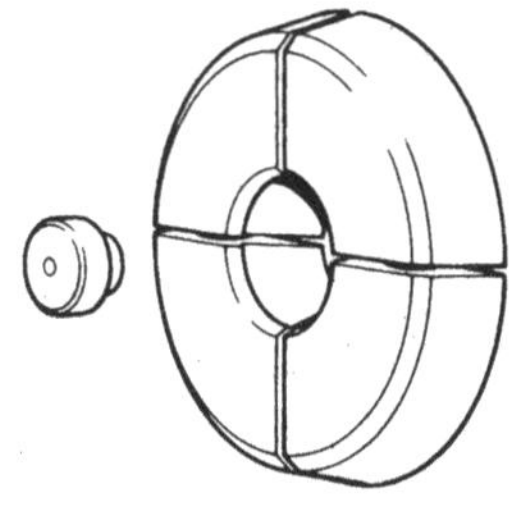

18G.47BD

18G.106

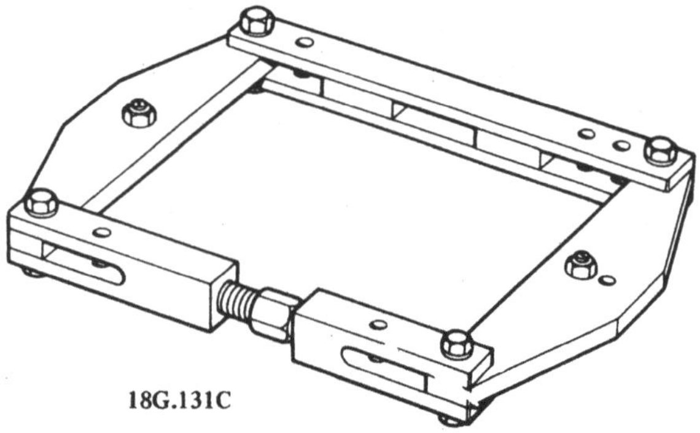

18G.131C

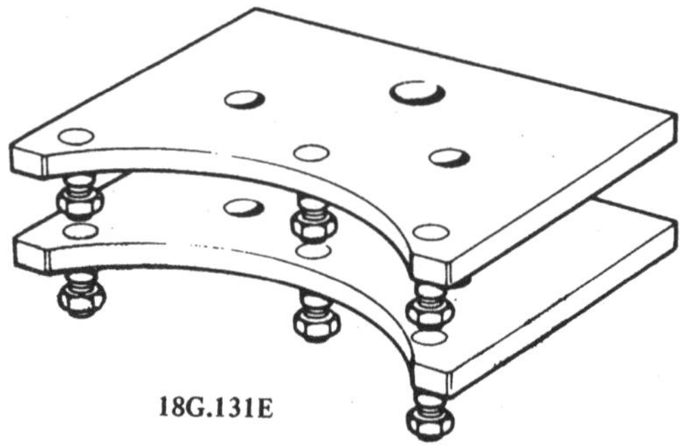

18G.131E

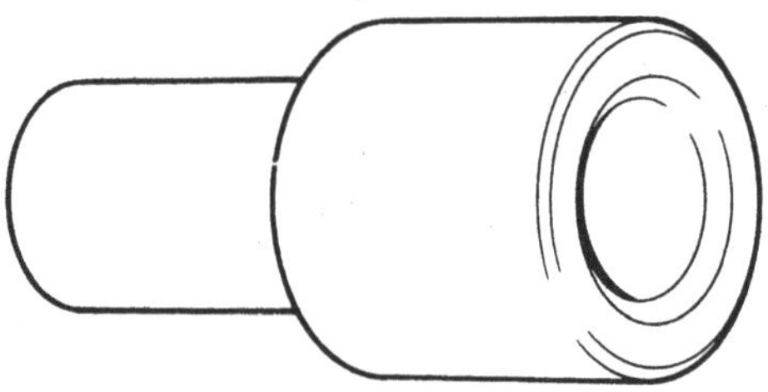

18G.134DH

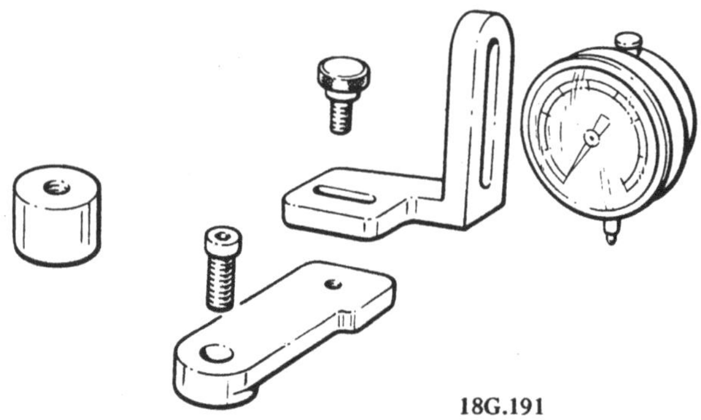

18G.191

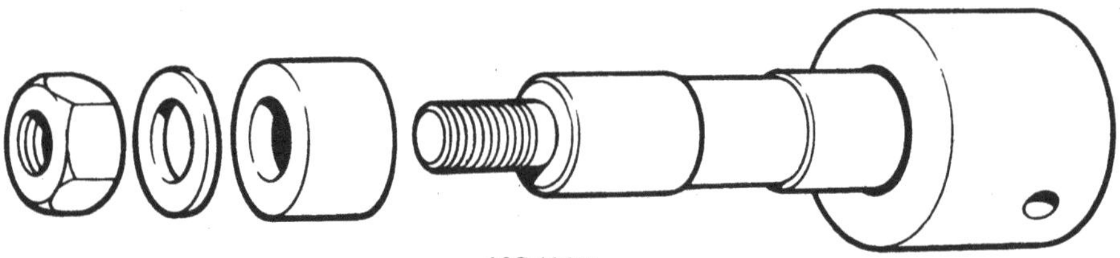

18G.191M

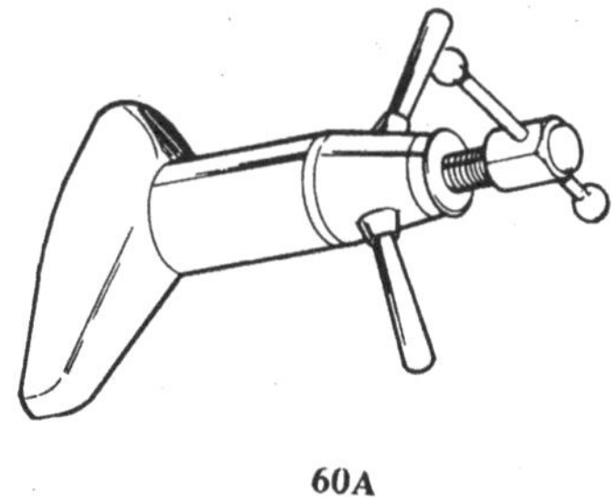

60A

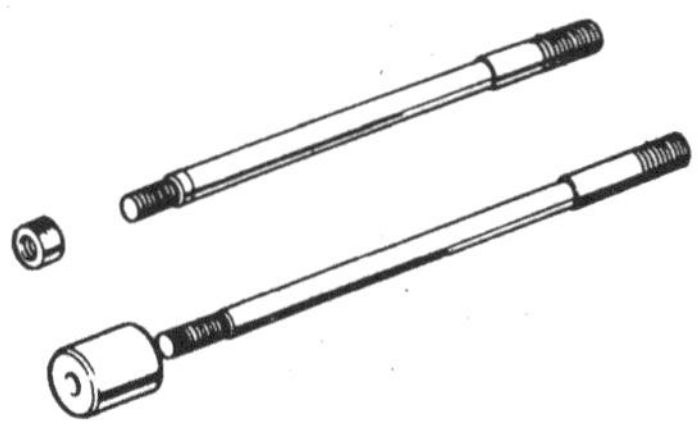

S60A-2

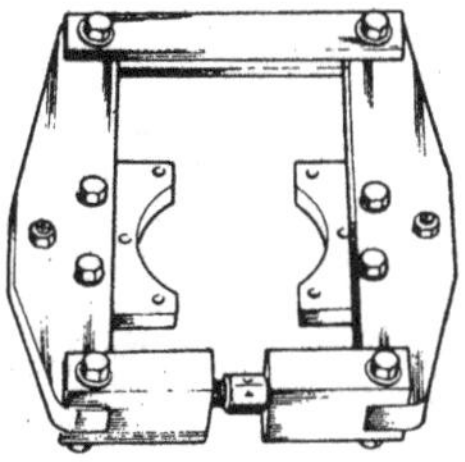
S.101

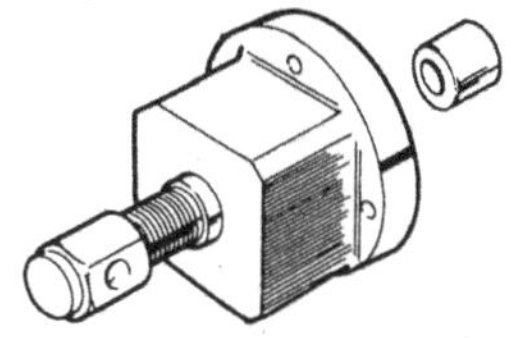
S.109C

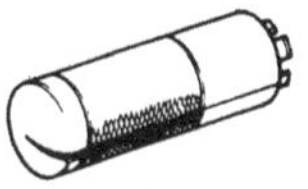
S.144

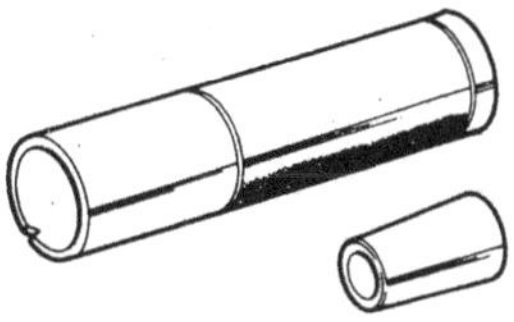
S.145

S.160

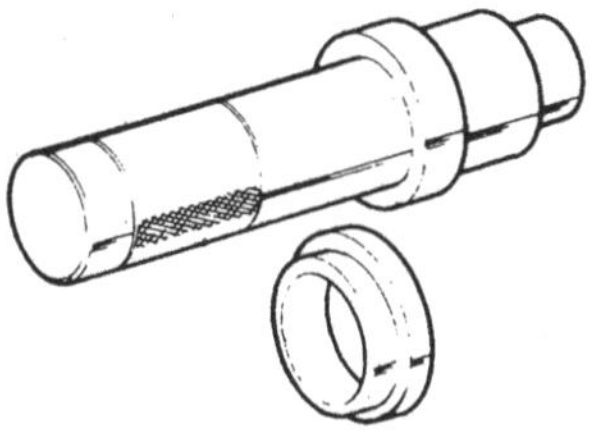
S.300A

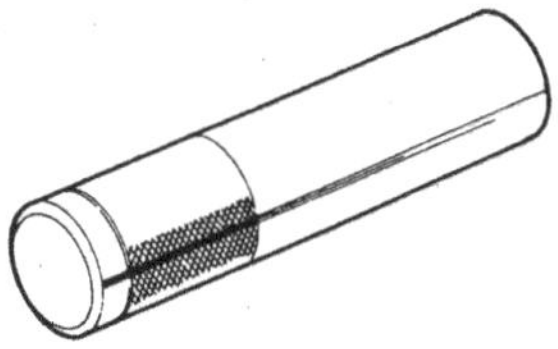
S.304

S.306

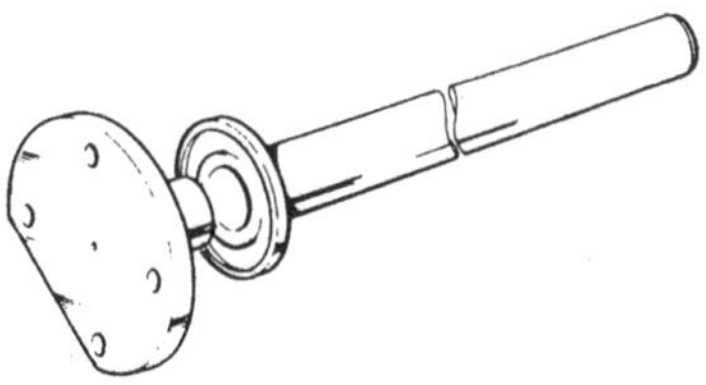
S.314

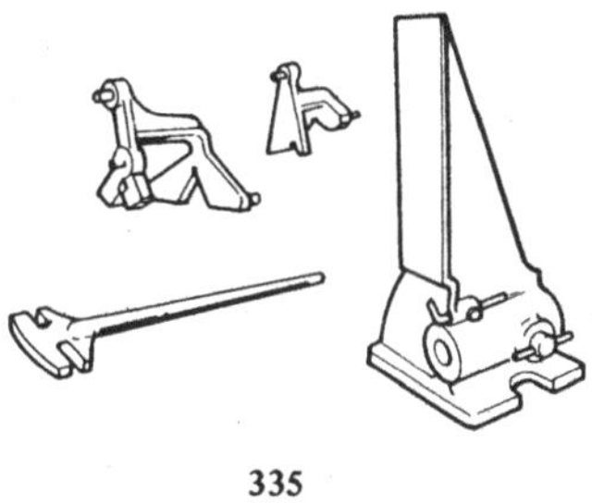
335

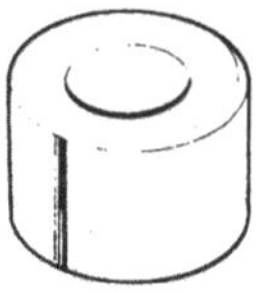
S.336-4

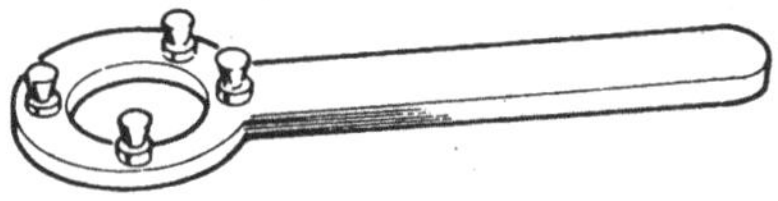

S.337

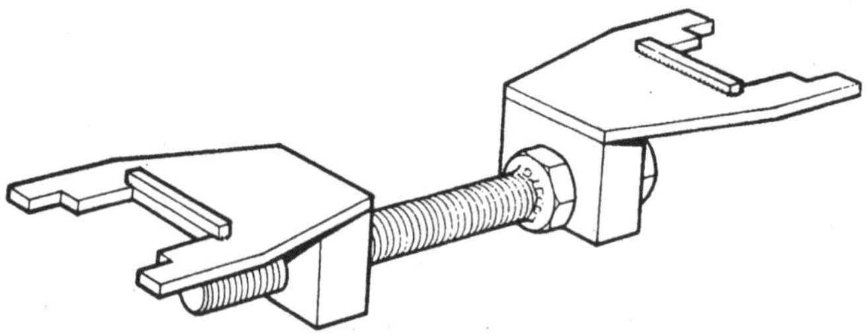

S.341

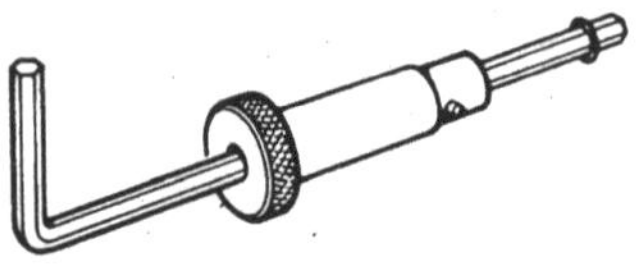

S.353

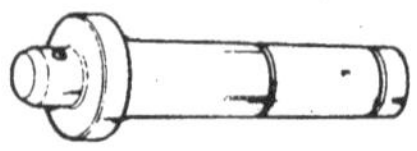

550

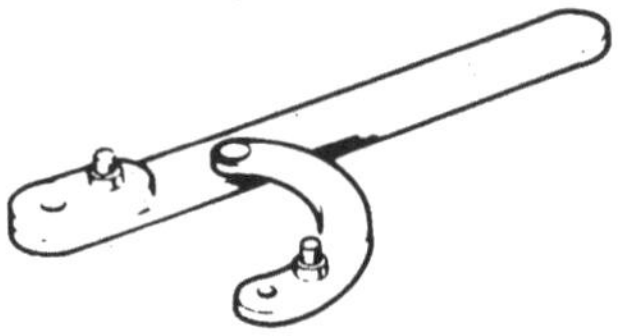

RG.421

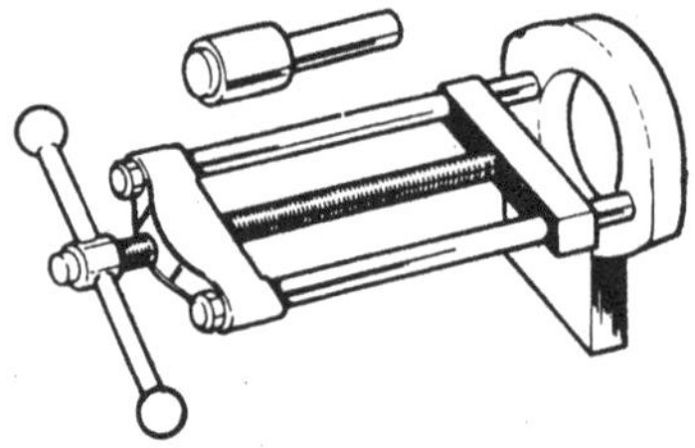

S.4221A

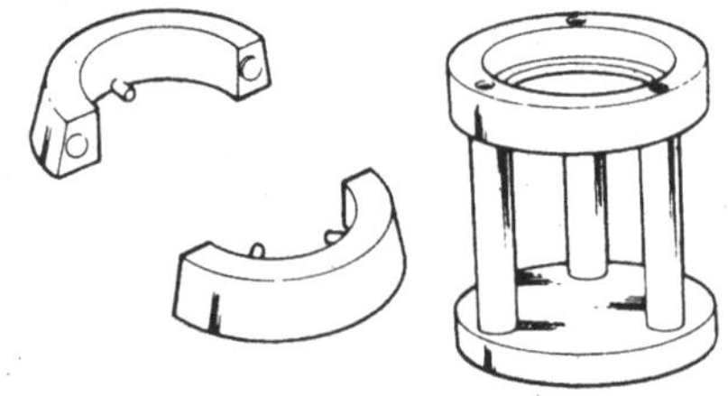

S.4221A-5

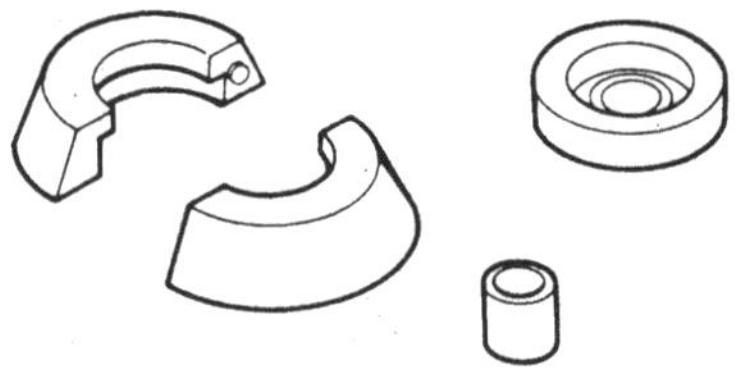

S.4221A-7B

S.4221A-14

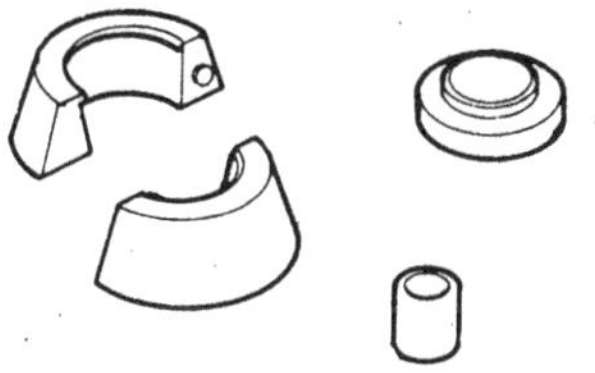

S.4221A-17

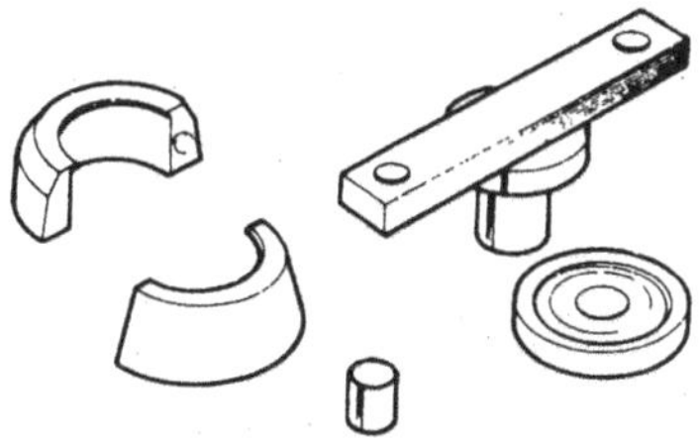

S.4221A-19

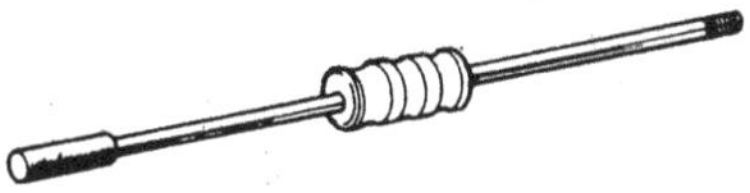

4235A

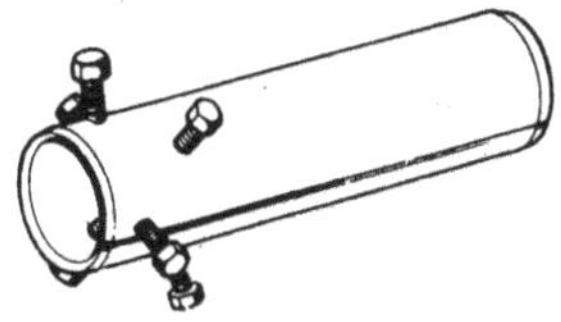

S.4235A-2

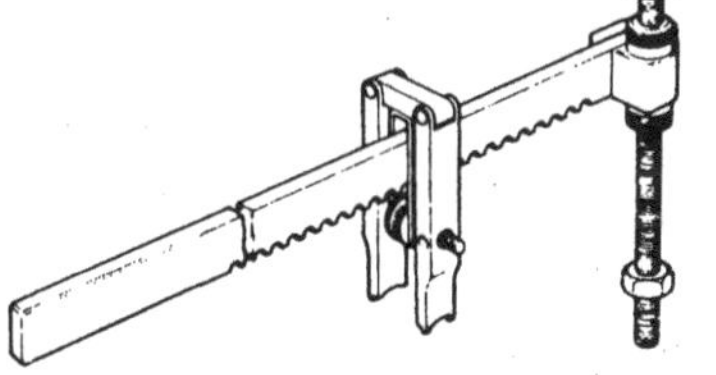

6118B

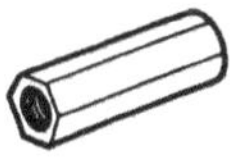

S.6118-1

SERVICE TOOLS

L.178

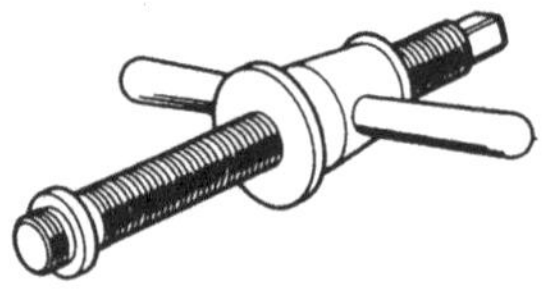

L.183A

L.183A-2A

L.188

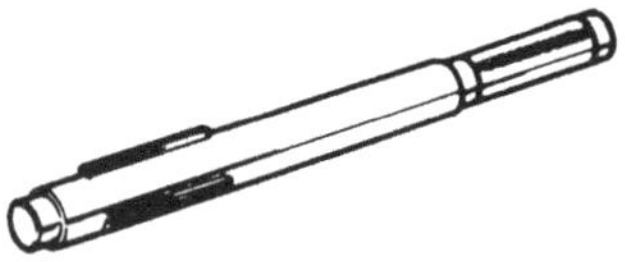

L.201

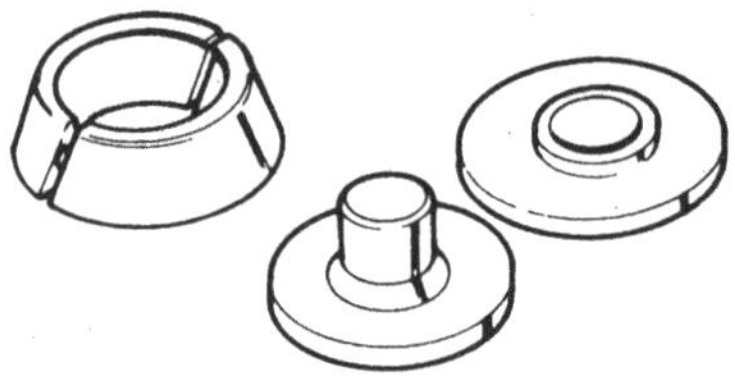

L.202A

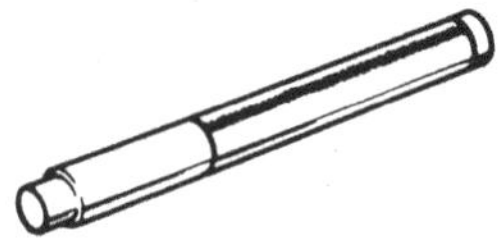

L.206A

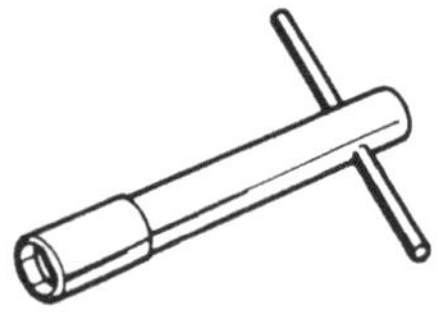

L.213